AF352911

Advances in Food and Non-Food Biomass Production, Processing and Use in Sub-Saharan Africa

Advances in Food and Non-Food Biomass Production, Processing and Use in Sub-Saharan Africa

Towards a Basis for a Regional Bioeconomy

Special Issue Editors

Daniel Callo-Concha
Hannah Jaenicke
Christine B. Schmitt
Manfred Denich

MDPI • Basel • Beijing • Wuhan • Barcelona • Belgrade • Manchester • Tokyo • Cluj • Tianjin

Special Issue Editors

Daniel Callo-Concha
Center for Development
Research (ZEF),
University of Bonn
Germany

Hannah Jaenicke
Center for Development
Research (ZEF),
University of Bonn
Germany

Christine B. Schmitt
Center for Development
Research (ZEF),
University of Bonn
Germany

Manfred Denich
Center for Development
Research (ZEF),
University of Bonn
Germany

Editorial Office
MDPI
St. Alban-Anlage 66
4052 Basel, Switzerland

This is a reprint of articles from the Special Issue published online in the open access journal *Sustainability* (ISSN 2071-1050) (available at: https://www.mdpi.com/journal/sustainability/special_issues/Food_Biomass_Production).

For citation purposes, cite each article independently as indicated on the article page online and as indicated below:

LastName, A.A.; LastName, B.B.; LastName, C.C. Article Title. *Journal Name* **Year**, *Article Number*, Page Range.

ISBN 978-3-03928-668-3 (Pbk)
ISBN 978-3-03928-669-0 (PDF)

Cover image courtesy of Manfred Denich.

© 2020 by the authors. Articles in this book are Open Access and distributed under the Creative Commons Attribution (CC BY) license, which allows users to download, copy and build upon published articles, as long as the author and publisher are properly credited, which ensures maximum dissemination and a wider impact of our publications.

The book as a whole is distributed by MDPI under the terms and conditions of the Creative Commons license CC BY-NC-ND.

Contents

About the Special Issue Editors

Daniel Callo-Concha is scientist at the Institute for Environmental Sciences (iES) University of Koblenz-Landau, and associated researcher at the Center for Development Research (ZEF) University of Bonn. Has worked for 20 years in Latin America and Africa. His current research focuses on social-ecological systems resilience, climate change adaptation, food and nutrition security, urbanization, and sustainable mining.

Hannah Jaenicke is a plant physiologist with wide experience in agricultural research for development. She has lived and worked in several African and Asian countries, focusing on underutilized fruit and vegetable species. She is currently the coordinator of the Horticulture Competence Center at Bonn University and an associate scientist at the Center for Development Research (ZEF).

Christine B. Schmitt is a landscape ecologist studying plant diversity and human uses in forest landscapes worldwide. Her research underpins the conservation and sustainable use of biodiversity and biological resources. She has worked in transdisciplinary contexts and in international policy deliberation and currently is a senior researcher with the Center for Development Research (ZEF), University of Bonn, Germany.

Manfred Denich is a senior scientist in the Ecology and Natural Resource Management department of the Center for Development Research (ZEF), University of Bonn, Germany. He holds degrees in biology and agriculture and has worked for more than 30 years in the tropics and subtropics worldwide. His working fields are land use systems (agroforestry), conservation and use of biodiversity, biomass production, food security, renewable energy, as well as transdisciplinary development research.

 *sustainability* 

Editorial

Food and Non-Food Biomass Production, Processing and Use in sub-Saharan Africa: Towards a Regional Bioeconomy

Daniel Callo-Concha *, Hannah Jaenicke, Christine B. Schmitt and Manfred Denich

Center for Development Research (ZEF), University of Bonn, 53113 Bonn, Germany;
h.jaenicke@uni-bonn.de (H.J.); cschmitt@uni-bonn.de (C.B.S.); m.denich@uni-bonn.de (M.D.)
* Correspondence: d.callo-concha@uni-bonn.de; Tel.: +49-228-731795

Received: 2 March 2020; Accepted: 4 March 2020; Published: 6 March 2020

Abstract: The bioeconomy concept has the aim of adding sustainability to the production, transformation and trade of biological goods. Though taken up throughout the world, the development of national bioeconomies is uneven, especially in the global South, where major challenges exist in Sub-Saharan Africa with respect to implementation. The BiomassWeb project aims to underpin the bioeconomy concept by applying the 'value web' approach, which seeks to uncover complex interlinked value webs instead of linear value chains. The project also aimed to develop intervention options to strengthen and optimize the synergies and trade-offs among different value chains. The special issue "Advances in Food and Non-Food Biomass Production, Processing and Use in Sub-Saharan Africa: Towards a Basis for a Regional Bioeconomy" compiles 22 articles produced in this framework. The articles are grouped in four sections: the value web approach; the production side; processing, transformation and trade; and global views. The synthesis presented in this paper introduces the challenges of the African bioeconomy and the value web approach, and outlines the contributing articles.

Keywords: Biomass-based value web; biological goods; bio-based economy; food and non-food; circular economy

---

1. Introduction

1.1. The Sub-Saharan African Biomass Sector

The rising global demand for biomass as food, feed, industrial raw material, and a source of energy is putting increasing pressure on the agricultural sector. This is particularly true for Sub-Saharan Africa (SSA), where many countries are confronted with growing regional and global demands for biomass-derived products while not yet having solved their national demands for food and non-food biomass [1–5].

Though food and nutrition security has improved globally in the last few decades, around 30% of the population in SSA is still faced with various forms of food insecurity. The number of undernourished in SSA has risen from 177 million in 2005 to 237 million in 2017 [5]. Associated indicators, such as the rates of anemia in women of reproductive age and stunting and wasting in children under the age of five, have increased [6,7]. With regard to energy supply, the major source of domestic fuel in SSA is fuelwood, which is primarily collected from forests, woodlands, and parklands. Due to rapid urbanization and a lack of alternatives, fuelwood is in demand not only in rural but also in urban areas, where up to 90% of households depend on it. With an average consumption of 1 kg fuelwood per person and day and a population of one billion people with an annual growth rate between 3% and 4%, the 'fuelwood gap' remains an ecological and socioeconomic challenge [8–11]. In

SSA, modern biomass processing is still in an early stage, and the production of food is not harmonized with the production of biomass-based raw materials. This was evidenced during the recent boom of the biofuel industry that attracted governments to promote the large-scale cultivation of oil palm, jatropha, cassava, and sugarcane despite warnings about the risks of competition for land [12–16].

Many of these matters are rooted in the agricultural sector, which is the focus of this special issue. Farming in SSA falls roughly into two major categories: (i) subsistence or semi-subsistence farming by smallholders to cover their own demands and to market surpluses, and (ii) commercial farming managed by estates, large enterprises, emerging medium-size farms, or organized small farmers under government programs (contract farming), many of which are devoted to producing export-oriented crops, such as cotton, coffee, cocoa, flowers or vegetables [17,18]. Both categories face challenges that hamper and limit their development. While subsistence farming is severely limited by rural poverty, institutional and technical weaknesses, ecological fragility (aggravated by climate change), and political instability [19–21], commercial farms, by their economic focus, have been accused of undermining environmental and social standards, and they are challenged by the volatility of international prices due to their dependence on external markets [22].

1.2. Challenges for an African Bioeconomy

Bioeconomy, or bio-based economy, is defined as the " … knowledge-based production and use of biological resources to provide products, processes and services in all economic sectors within the frame of a sustainable economic system." It is based on the expectation of expanding biomass production and processing sectors going beyond the production of food, feed, fiber, fuel, and other basic products towards the production of value-added goods and services that are demanded by other economic sectors such as the industry, energy, pharmacy, and chemical sectors [16].

A biomass-based economy is increasingly envisaged by many countries as a path to follow. While most countries of the global North are investigating and developing new technologies, establishing large and sophisticated biorefineries, focusing on maximizing benefits, minimizing waste, and even reorganizing their institutions accordingly, progress in the global South is limited ([23–25] in this special issue).

In most regions of SSA, biophysical features such as the wide availability of productive land and a constant solar radiation are the major comparative advantages for biomass production, and they represent a great natural potential to increase the amount of biomass that is used for food and industrial raw material (non-food, including energy). Nonetheless, currently, only 15% of the net primary production (NPP) of the continent is used (human appropriation of net primary production—HANPP), and the growth rates of this use are much lower than population growth [26,27]. In Europe, for instance, 35% of the NPP is appropriated by humans. Accordingly, there are compelling opportunities for SSA's further development based on the more intensive production and use of biomass in the context of a comprehensive African bioeconomy. On the other hand, major challenges for a regional bioeconomy are the weak economic, technical, and institutional conditions that restrain the production, post-harvest and processing sectors. The extent and diversity of these challenges and the pressure on SAA countries to catch up with global trends require diversified strategies and coordinated actions to simultaneously handle these challenges [16,28–30].

A broad consensus is that an African bioeconomy agenda should prioritize (i) the encouragement and enhancement of the productive sector under the premises of ecological sustainability, social equity, and fair economic return to farmers; (ii) the further development of the processing sector by generating, promoting, and adopting innovations, technologies, and techniques to convert biomass into goods of higher value; and (iii) linking producers with processors and with local, national, and international markets to guarantee reliable incomes [16].

1.3. Biomass-based Value Web Approach

In the context of an emerging African bioeconomy, the project "Improving food security in Africa through increased system productivity of biomass-based value webs (BiomassWeb)" aimed at understanding and enhancing food and non-food biomass production, processing, and trade in Ghana, Nigeria, and Ethiopia. These countries were chosen because of their regional importance and potential as case studies with relevance to other SSA countries. The key crops considered were maize, cassava, plantain/banana/enset, and bamboo, which were selected according to their regional relevance as sources of food and non-food biomass.

The overarching concept was that of the 'biomass-based value webs', i.e., complex systems of interlinked value chains in which biomass products and by-products are produced, processed, traded, and consumed (Figure 1). Though introduced two decades ago [31,32], the value web approach is still innovative, as it more realistically describes the value that is added in the biomass sector in comparison to the linear supply and value chain concepts. Value webs not only depict material and cash flows, they also connect supply and value chains with their actors, e.g., through information flows, the effects of policy decisions, or innovative developments in the production and processing of biomass, as well as via the effects of national and international market events. The resulting dynamic, hyper-connected, and collaborative relationships are country- and situation-specific and can only be disclosed in cooperation with local stakeholders and experts. We argue that the value web is a useful scientific approach for investigating SSA biomass-related activities in view of its current and forthcoming challenges.

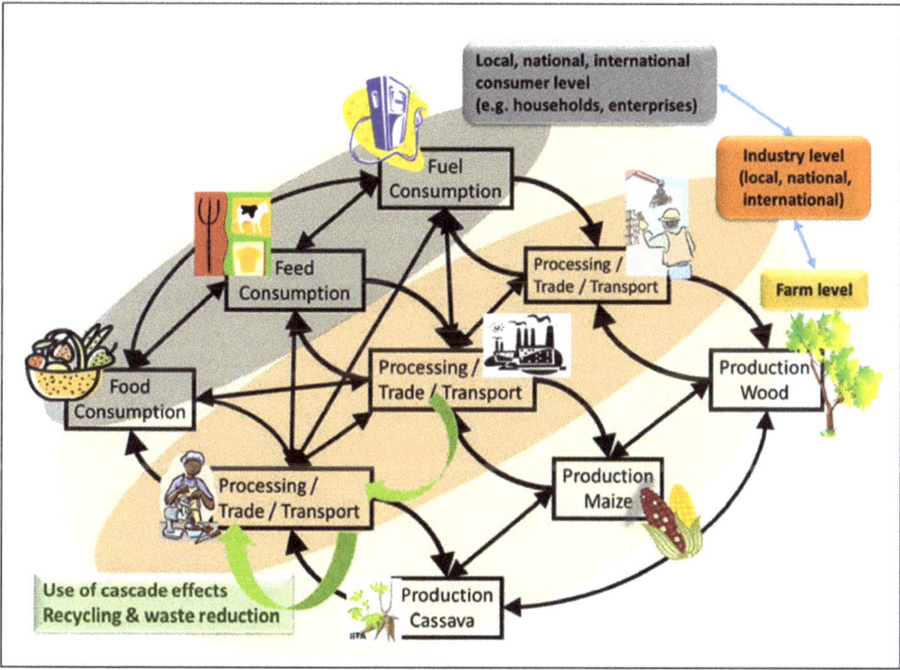

Figure 1. Interactions, trade-offs, and synergies in a biomass-based value web (schematic) [33].

The BiomassWeb project objectives were: (i) to investigate the potential interventions to strengthen biomass value web production, processing and trade, and (ii) to identify the synergies and trade-offs among them. Along this process, BiomassWeb had a strong foresight character in identifying and facilitating the current and future synergies and trade-offs among biomass uses.

BiomassWeb was co-led by the Center for Development Research (ZEF) of the University of Bonn and the Forum for Agricultural Research in Africa (FARA). The consortium included a network of universities, national research institutions, international agricultural research organizations, and partners from the private sector in Ghana, Nigeria, Ethiopia, and Germany.

2. Summary of Articles Included in this Special Issue

This special issue summarizes some of the results of the BiomassWeb project, together with other selected research results regarding exploring, developing, and testing innovative approaches to produce, process, and trade food and non-food biomass in SSA.

The 22 articles in this volume cover stand-alone and aggregated studies, disciplinary, inter- and transdisciplinary approaches, and ex-post and foresight-oriented investigations. They are compiled into four major sections: (i) overarching studies contributing to the value web approach; (ii) the production side; (iii) processing, transformation and trade; and (iv) the global view.

2.1. The Value Web Approach

A few articles are featured that look at the value web approach itself from different perspectives: (i) complex systems analysis, (ii) as the basis for analyzing a supply chain, and (iii) describing a demand-driven research and development concept to identify potential interventions to strengthen the effectiveness and efficiency of biomass-based value webs.

Concerning food security, Anderson et al. model and analyze biomass-based value webs of selected crops in Ghana, Nigeria, and Ethiopia by applying the systems analysis software iMODELER in participatory stakeholder workshops. In all three countries, the transdisciplinary mapping of the different crop-value chains clearly reveals widely ramified systems with a complex web character having food security as the overall target. In contrast to the initial hypothesis, the value chains of the different crops considered do not show relevant direct links between each other in their matter and capital flows. However, they are connected through nonspecific factors (corresponds to nodes or variables in other systems approaches) such as communication, governmental interventions, extension services, agricultural innovations, global food prices, and others. Results from a generic model allow for a critical reflection on the relation between value web dynamics and food security policy in SSA. Current policy-making trends targeting the market integration, mechanization, and reduction of post-harvest losses are supported by the model results.

In a case study, Lin et al. focus on the current market challenges and opportunities for the future development of the northern Ethiopian bamboo producing and processing sectors. The results show that bamboo producers are constrained by the lack of local demand and markets for higher-value bamboo products. This also leads to less product diversification on the local markets and reduces the innovative capacity of the manufacturing sector. It is recommended that local and regional governments support specific training programs on bamboo production and processing. Additionally, the market access of small bamboo producers may be improved through the establishment of cooperatives and the development of contractual arrangements that protect local producers, processors, and traders.

A demand-driven research and development (DDRD) program was one of the innovative aspects of the BiomassWeb project. Funds were provided by the donor agency for implementing research and development activities that emerged from alliances with stakeholders of the biomass producing, processing and trading sectors during the project lifetime. Jatta et al. discuss the concept and application of DDRD in Ghana, Nigeria, and Ethiopia based on six projects that were selected and supervised by the Forum for Agricultural Research in Africa (FARA): (i) using cassava peels for mushroom production, (ii) the development of plantain biomass into composite flour for traditional foods and bakery products, (iii) bamboo leaves as fodder for livestock production, (iv) the production of bio-plastics and bio-gels from agricultural waste, (v) the mass and energy balance analysis of pneumatic dryers for cassava, and (vi) exploring potentials of the bamboo sector for employment and food security.

2.2. The Production Side

The contributions in this section cover a broad array of biomass-based production systems and highlight the need to tackle food insecurity by using a variety of approaches.

Several studies focus on smallholder crop production systems. Scheiterle and Birner show that in Ghana, maize production systems with above-average yields of 1.5 Mt/ha are profitable at the household level, while production systems below this threshold report negative social profits. The use of fertilizers that are sponsored through national subsidies, however, does not increase the likelihood to produce above-average yields, while the use of improved seeds and herbicides does. For Ethiopia, Srivastava et al. employ a modelling approach to test the effect of different intensification scenarios on maize yields. They report that a combination of higher mineral fertilizer rates and the incorporation of crop residues are the most successful, while the rotation of maize with groundnut could help to increase economic profits. Legesse et al. use a modelling framework to evaluate the effects of a higher efficiency of fertilizer application in a variety of cropping systems in Ethiopia. Higher fertilizer application increases annual yields at the average farm and is profitable despite a price reduction on the markets, which has a positive effect on the welfare gains for both rural farming and non-farming households. Finally, Poku et al. show that in the case of cassava outgrower schemes in Ghana, contract farming could benefit both farmers and agribusiness firms if contracts sustained long-term supportive business agreements.

Next to field cropping systems, agroforestry systems can make a strong contribution to food security through food and non-food biomass production. For Ethiopia, Jemal et al. demonstrate that smallholder farmers can benefit from 120 plant species that grow in home gardens, in multistory coffee systems, and in farmland systems with multipurpose trees. Similarly, Kelboro and Stellmacher underline that family farms in Ethiopia rely on ad-hoc agricultural production systems to achieve food security at the household level, and this needs to be taken into consideration when developing agricultural intensification schemes. In Ghana, Akoto et al. assess the local acceptance of bamboo use in agroforestry systems in a dry forest zone. They show that farmers who have traditional knowledge of multipurpose trees and bamboo are more inclined to adopt these systems for combined charcoal, fodder leaf, and crop production. By using a transdisciplinary approach, Mbeche et al. analyze the application of the push–pull technology in Ethiopia to control stemborer pests and *Striga* weed in maize and demonstrate that transdisciplinary approaches can be efficient in tackling real-life problems.

Moving from the rural to the urban setting, Nero et al. demonstrate the potential of food trees in Accra, Ghana, to contribute to food security in African cities. They report that the diversity of tree species with food uses is higher in poorer neighborhoods than in wealthier neighborhoods, but their abundance is lower in the former than in the latter. They conclude that policies to promote food trees can support several goals, such as achieving food security and raising the quality of living.

2.3. Processing, Transformation and Trade

Processing, transformation, and trade are important elements of the biomass-based value web. Several articles explore opportunities and also challenges of the conversion of by-products or waste into valuable products.

Loos et al. apply participatory methods, expert interviews and group discussions to evaluate the potential of plantain residues as a resource for industrial raw materials (fiber) in Ghana. They report that key stakeholders and structures exist that could boost the establishment of a sustainable plantain-based value web. However, pilot activities and technology transfer of suitable innovations from other countries would be required.

In their article, Chala et al. explore the potential of by-products from coffee processing (husk, pulp and mucilage) for biogas production in Ethiopia. The authors estimate that the anaerobic digestion of these products could generate as much as 68×10^6 m^3 methane per year, which could be converted into 238,000 MWh of electricity and 273,000 MWh of thermal energy. Both electricity and thermal energy are used by coffee processing facilities and, accordingly, biogas production would lead to energy cost savings.

Intani et al. take a critical look into the use of corncob biochar, which is a sought-after resource for soil amendment. As phytotoxicity has been observed, the experiments carried out by the authors demonstrate that the phytotoxicity of fresh corncob biochar can be effectively mitigated by washing and heat treatments.

Several articles address production and use of cassava and its processing by-products. Ayetigbo et al. provide a review in which they compare properties of cassava root, flour, and starch from white-flesh and biofortified yellow-flesh variants. The companion papers by Adeyemo and Okoruwa on the effects of value addition on the productivity of cassava farming households and by Adeyemo et al. on determinants of the intensity of cassava utilization—both in Nigeria—demonstrate that the prospect of adding value through processing determines production decisions. Better extension services, training, and enterprise regulation, as well as asset acquisition, improving land quality, and the encouragement of social capital development among smallholders, are important drivers.

Looking at waste management options amongst cassava processors in Nigeria, Omilani et al. report that public expenditure on training for processors in waste management options would empower them to use solid-waste conversion technologies to generate value-added products. Besides generating additional income, this would lead to social benefits including a lower exposure to environmental toxins from the air, streams, and groundwater.

2.4. Global Views

This section offers a global view on the trends, challenges, and opportunities that countries are confronted with when developing their own bioeconomies.

Given the countries' differences in potential, priority setting and strategies to develop their own bioeconomies, Biber-Freudenberger et al. propose a classification and then sort them into primary, advanced, high-tech, and mixed categories that they later use to gauge their performance in terms of sustainability. The authors find that countries with more sophisticated bioeconomies are more diversified in terms of innovations and policies that promote them. In contrast, countries with incipient bioeconomies are based on less varied alternatives and concentrate on bioenergy, but they tend to explore and expand towards high-tech strategies. Interestingly, the efforts of the former are not necessarily accompanied by more sustainable performance.

One step ahead, Beuchelt and Nassl, under the premise of the UN Sustainable Development Goals that suggest the satisfaction of multiple demands instead of the optimization of a single or a few of them, model the trends of several bioeconomy operation plans and their weighing of economic sectors. They report a worrying imbalance that tends to prioritize the uses of biomass for the generation of energy at the expense of other material uses and even the satisfaction of basic needs like food production.

Finally, Dietz et al. examine the governance strategies of the 41 countries that lead the progression into a biomass-based economy. Contextualizing their analysis into the UN Sustainable Development Goals, they identify four potential pathways and foresee sets of governance measures to enable or constrain their development. Based on the unevenness of outcomes, the authors conclude that advocating for the establishment of political structures (when nonexistent) to put national bioeconomies into operation and for the creation of global frameworks to coordinate and guarantee the sustainability of these structures are key issues.

3. Concluding Remarks and Outlook

To date, only a few national or regional strategies for innovative uses of biomass in Africa exist. The papers in this special issue underpin the fact that there is great potential for food and non-food biomass production, use, and trade in Africa. These may encompass production systems, such as bamboo intercropping, underutilized plant species in agroforestry systems, and fruit trees in urban settings, as well as processing techniques, e.g., biochar production, starch uses, and agricultural residues as industrial raw materials or energy sources. It clearly appears that further research is needed on implementing the findings in practice, while the results and thoughts that are presented by the authors of this volume already make a contribution to this process.

Additional efforts, however, are needed to disseminate the results to practitioners and thus to contribute to rural development. In general, targeted biomass-related policies and governance measures at the local and national levels, also considering the UN Sustainable Development Goals (SDGs), are recommended to support efforts to help achieve food security and improved quality of living. In particular, rural policies should focus on extension services as well as capacity building and training for biomass producers, processors, and traders. Furthermore, market opportunities and access to markets, in combination with the establishment of cooperatives, have to be developed. Other recommendations for promoting a biomass-based economy range from enterprise regulations, asset acquisition, and contractual issues to social capital development in rural environments.

The examples show that the establishment of biomass-based economies faces a multitude of challenges. The dilemma is that the respective research-based recommendations—as can be seen above—are of a general nature. The implementation of research outputs in practice, however, requires more detailed instructions for action. These can only be developed through systematic implementation research that is transdisciplinary in nature and, accordingly, based on stakeholder involvement. Implementation research aims to identify and overcome the barriers to the implementation of research outputs. Its activities can of course be considerably reduced if newly developed concepts and technologies to be implemented are based on demand-driven research.

Implementation research is the basis for the piloting and up-scaling of innovations based on research and development. In the case of biomass-based innovations, their implementation should take place in the context of a circular bioeconomy. The integration of biomass production, processing, and trade into a circular system requires that value chains of food and non-food biomass and the value webs that are derived from them are analyzed in their system context. Particular attention must be paid to cause-and-effect relationships between the system components. Knowledge of these relationships helps to identify intervention options that contribute to optimizing effectiveness and efficiency of value chains and webs in the biomass sector. Additionally, the meaningful implementation research should consider not only economic, socio-cultural and technical aspects but also aspects such as political structures or the personal resources and capacities of the members of the target groups.

Last but not least, when developing a regional bioeconomy, it must be borne in mind that the nation states concerned have their own policy priorities and strategies that face different development trends, challenges and potentials. Accordingly, different information needs must be met, for which an efficient science communication system has to be established. In this context, the interactive online expert network BiomassNet (https://www.biomassnet.org/) provides a forum for information exchange on biomass-related aspects in Africa.

Funding: This research was funded by the German Federal Ministry of Education and Research (BMBF) and Ministry for Economic Cooperation and Development (BMZ), grant numbers FKZ 031A258A-I.

Conflicts of Interest: The authors declare no conflict of interest.

References

1. FAO. Biofuels: Prospects Risks and Opportunities. In *The State of Food and Agriculture 2008*; United Nations Food and Agriculture Organization: Rome, Italy, 2008. [CrossRef]
2. FAO High-Level Expert Forum. Global Agriculture towards 2050. United Nations Food and Agriculture Organization 2009. Available online: http://www.fao.org/fileadmin/templates/wsfs/docs/Issues_papers/HLEF2050_Global_Agriculture.pdf (accessed on 11 July 2019).
3. Alexandratos, N.; Bruinsma, J. World Agriculture towards 2030/2050: The 2012 Revision. United Nations Food and Agriculture Organization. 2012. Available online: http://www.fao.org/3/a-ap106e.pdf (accessed on 11 July 2019).

4. van Ittersum, M.K.; van Bussel, L.G.J.; Wolf, J.; Grassini, P.; van Wart, J.; Guilpart, N.; Claessens, L.; de Groot, H.; Wiebe, K.; Mason-D'Croz, D.; et al. Can Sub-Saharan Africa Feed Itself? *Proc. Natl. Acad. Sci.* **2016**, *113*, 14964–14969. [CrossRef] [PubMed]

5. FAO, IFAD, UNICEF, WFP and WHO. The State of Food Security and Nutrition in the World 2018. Building climate resilience for food security and nutrition. Rome, FAO. Available online: http://www.fao.org/3/i9553en/i9553en.pdf (accessed on 11 July 2019).

6. Leathers, H.D.; Foster, P. *The World Food Problem: Toward Understanding and Ending Undernutrition in the Developing World*; Lynn Rienner Publishers: Boulder, CO, USA, 2017; ISBN-13: 978-1626374515.

7. *FAO Statistical Pocketbook 2015: World Food and Agriculture*; Food and Agriculture Organization of the United Nations: Rome, Italy, 2015. Available online: http://www.fao.org/3/a-i4691e.pdf (accessed on 11 July 2019).

8. Ouedraogo, B. Household Energy Preferences for Cooking in Urban Ouagadougou, Burkina Faso. *Energy Policy* **2006**, *34*, 3787–3795. [CrossRef]

9. Hiemstra-van der Horst, G.; Hovorka, A.J. Fuelwood: The 'Other' Renewable Energy Source for Africa? *Biomass Bioenergy* **2009**, *33*, 1605–1616. [CrossRef]

10. Konare, S.; Ninomya, I.; Kobayashi, O.; Shimamura, T. Rural Community Perception of Fuelwood Usage by Families Living in Wassorola, Mali: Interview with Women as Main Fuelwood Collectors. *J. Agri. Crop Res.* **2013**, *1*, 76–83.

11. Callo-Concha, D.; Harou, L.; Krings, L.; Ngonjock, J.; Ziemacki, J. Ziemacki. Farming Adaptation in the Western Sudanese Savannah: Lessons Learnt and Challenges Ahead. In *Advancing Climate Change Research in West Africa. Trends, Impacts, Vulnerability, Resilience, Adaptation and Sustainability Issues*; Kokoye, S., Yegmebey, R., Awoye, O., Eds.; Nova Publishers: Harpark, NY, USA, 2019; pp. 95–131.

12. Gnansounou, E.; Panichelli, L.; Villegas, J.D. Sustainable Liquid Biofuels for Transport: The Context of the Southern Africa Development Community. 2007. Available online: https://infoscience.epfl.ch/record/121498/files/Sustainable_Liquid_Biofuels_for_Transport-SADC.pdf (accessed on 11 July 2019).

13. Amigun, B.; Musango, J.K.; Stafford, W. Biofuels and Sustainability in Africa. *Renew. Sust. Energ. Rev.* **2011**, *15*, 1360–1372. [CrossRef]

14. van Zyl, W.H.; Chimphango, A.F.A.; den Haan, R.; Görgens, J.F.; Chirwa, P.W.C. Next-Generation Cellulosic Ethanol Technologies and Their Contribution to a Sustainable Africa. *Interface Focus* **2011**, *1*, 196–211. [CrossRef] [PubMed]

15. Popp, J.; Lakner, Z.; Harangi-Rákos, M.; Fári, M. The Effect of Bioenergy Expansion: Food, Energy, and Environment. *Renew. Sust. Energ. Rev.* **2014**, *32*, 559–578. [CrossRef]

16. Bioeconomy Council. *Global Bioeconomy Summit 2018*; Conference Report; Federal Ministry for Education and Research: Bonn, Germany, 2018; Available online: www.gbs2018.com (accessed on 11 July 2019).

17. Porter, G.; Phillips-Howard, K. Comparing Contracts: An Evaluation of Contract Farming Schemes in Africa. *World Dev.* **1997**, *25*, 227–238. [CrossRef]

18. Hall, R.; Scoones, I.; Tsikata, D. Plantations, Outgrowers and Commercial Farming in Africa: Agricultural Commercialisation and Implications for Agrarian Change. *J. Peasant Stud.* **2017**, *44*, 515–537. [CrossRef]

19. Challinor, A.; Wheeler, T.; Garforth, C.; Craufurd, P.; Kassam, A. Assessing the Vulnerability of Food Crop Systems in Africa to Climate Change. *Clim. Chang.* **2007**, *83*, 381–399. [CrossRef]

20. Morris, M.; Binswanger, H.; Byerlee, D.; Savanti, P.; Staatz, J. *Awakening Africa's Sleeping Giant: Prospects for Commercial Agriculture in the Guinea Savannah Zone and Beyond*; The World Bank: Washington, DC, USA, 2009; ISBN-13: 978-0821379417.

21. Callo-Concha, D.; Gaiser, T.; Webber, H.; Tischbein, B.; Müller, M.; Ewert, F. Farming in the West African Sudan Savanna: Insights in the Context of Climate Change. *Afr. J. Agric. Res.* **2013**, *8*, 4693–4705. [CrossRef]

22. von Braun, J.; Tadesse, G. Food Security, Commodity Price Volatility, and the Poor. In *Institutions and Comparative Economic Development*; Aoki, M., Kuran, T., Roland, G., Eds.; International Economic Association Series; Palgrave Macmillan: London, UK, 2012; pp. 298–312. [CrossRef]

23. Aguilar, A.; Wohlgemuth, R.; Twardowski, T. Perspectives on Bioeconomy. *N. Biotechnol.* **2018**, *40*, 181–184. [CrossRef] [PubMed]

24. Dietz, T.; Börner, J.; Förster, J.J.; von Braun, J. Governance of the Bioeconomy: A Global Comparative Study of National Bioeconomy Strategies. *Sustainability* **2018**, *10*, 3190. [CrossRef]

25. Biber-Freudenberger, L.; Kumar, B.A.; Bruckner, M.; Börner, J. Sustainability Performance of National Bio-Economies. *Sustainability* **2018**, *10*, 2705. [CrossRef]

26. Krausmann, F.; Erb, K.-H.; Gingrich, S.; Lauk, C.; Haberl, H. Global patterns of socioeconomic biomass flows in the year 2000: A comprehensive assessment of supply, consumption and constraints. *Ecol. Econ.* **2008**, *65*, 471–487. [CrossRef]

27. Krausmann, F.; Erb, K.-H.; Gingrich, S.; Haberl, H.; Bondeau, A.; Gaube, V.; Lauk, C.; Plutzar, C.; Searchinger, T.D. Global human appropriation of net primary production doubled in the 20th century. *PNAS* **2013**, *110*, 10324–10329. [CrossRef] [PubMed]

28. El-Chichakli, B.; von Braun, J.; Lang, C.; Barben, D.; Philp, J. Policy: Five Cornerstones of a Global Bioeconomy. *Nat. News* **2016**, *535*, 221–223. [CrossRef] [PubMed]

29. Virchow, D.; Beuchelt, T.D.; Kuhn, A.; Denich, M. Biomass-Based Value Webs: A Novel Perspective for Emerging Bioeconomies in Sub-Saharan Africa. In *Technological and Institutional Innovations for Marginalized Smallholders in Agricultural Development*; Gatzweiler, F., von Braun, J., Eds.; Springer: Berlin, Germany, 2016; ISBN 978-3-319-25718-1.

30. Virgin, I.; Morris, E.J. (Eds.) *Creating Sustainable Bioeconomies—The bioscience revolution in Europe and Africa*; Routledge: London, UK; New York, NY, USA, 2017; ISBN 978-1-138-81853-8.

31. Stabell, C.B.; Fjeldstad, O.D. Configuring value for competitive advantage: on chains, shops, and networks. *Strateg. Manag. J.* **1998**, *19*, 413–437. [CrossRef]

32. Smith, R.W.; Broxterman, W.E.; Murad, D.S. *Understanding Value Webs as a New Business Modeling Tool: Capturing & Creating Value in Adhesives*; The Adhesive & Sealant Council: Las Vegas, NV, USA, 2000.

33. Virchow, D.; Beuchelt, T.; Denich, M.; Loos, T.K.; Hoppe, M.; Kuhn, A. The value web approach—so that the South can also benefit from the bioeconomy. *Rural* **2014**, *21*, 3.

© 2020 by the authors. Licensee MDPI, Basel, Switzerland. This article is an open access article distributed under the terms and conditions of the Creative Commons Attribution (CC BY) license (http://creativecommons.org/licenses/by/4.0/).

sustainability

Article

Identifying Biomass-Based Value Webs for Food Security in Sub-Saharan Africa: A Systems Modeling Approach

Carl C. Anderson [1,2,*], Manfred Denich [1], Kai Neumann [3], Kwadwo Amankwah [4] and Charles Tortoe [5]

[1] Center for Development Research (ZEF), D-53115 Bonn, Germany; uzef0008@uni-bonn.de
[2] School of Interdisciplinary Studies, University of Glasgow, Dumfries, Scotland DG1 4ZL, UK
[3] Consideo GmbH, D-23562 Lübeck, Germany; neumann@consideo.com
[4] Faculty of Agriculture, Kwame Nkrumah University of Science and Technology (KNUST), Kumasi, Ghana; kojo116@yahoo.com
[5] Council for Scientific and Industrial Research-Food Research Institute, P. O. Box M20 Accra, Ghana; ctortoe@yahoo.co.uk
* Correspondence: c.anderson.4@research.gla.ac.uk; Tel.: +44-(0)-1387-702033

Received: 8 September 2018; Accepted: 20 May 2019; Published: 21 May 2019

Abstract: Food security in Sub-Saharan Africa (SSA) is dependent on complex networks of interconnected actors and the flows of resources (biomass, capital) and information among them. However, the degree to which actors and value chains of different crops are in fact interconnected and their current systemic influence on food security are unclear. Therefore, the concept of "value webs" to better capture the complexity within the networks emerges. Biomass-based value webs of selected crops in Ghana, Nigeria, and Ethiopia are modeled using the systems analysis software iMODELER and by eliciting factors as well as their interconnections through participatory stakeholder workshops. Furthermore, a generic model was created compiling the country models to identify overarching system dynamics with supporting and hindering factors impacting food security in SSA. Findings from the country models show highly complex value webs, suggesting that the predominant value chain approach may oversimplify actual structures and resource flows in real life settings. However, few interconnections within the value webs link the actors and flows of different crops, contradicting predictions emerging from other research. Results from the generic model allow for a critical reflection on the relation between value web dynamics and food security policy in SSA. Current national and regional policy trends targeting market integration, mechanization, and reduction of post-harvest losses are supported by model results.

Keywords: availability; access; Ghana; Nigeria; Ethiopia; value chain

1. Introduction

Improving food security in Sub-Saharan Africa (SSA) is central to the achievement of objectives behind the Malabo Declaration [1], the Sustainable Development Goals (SDGs) [2], and Africa's Agenda 2063 [3]. There are significant gains to be made before achieving these goals, however, as SSA continues to have the highest proportion of food-insecure people of any region globally at 23% undernourished in 2016, the situation having worsened since 2010 [4]. In addition, according to the latest progress report on achieving the SDGs [5], thirteen countries in SSA experienced high or moderately high domestic food prices relative to historic levels in 2016, reducing people's access to food.

The role of agricultural systems in improving food security in SSA is undeniable and has long been the focus of policy efforts, e.g., [6]. Agriculture is the most important sector for African economies. On average, it accounts for more than 25% of GDP [7], employs around 57% of the continent's total labor force, and constitutes the primary source of income for 90% of the rural population [8]. The potential for development and inclusive growth within this sector thus represents an opportunity for both increasing the availability of food as well as generating incomes through food and non-food biomass-based products [9–11]. Because biomass is a renewable resource, its increasing role is promising for sustainable growth and food security [11,12].

Biomass can be classified into food, feed, sources of energy, and raw materials for industry, the increasing demand of which represents both a unique opportunity and critical challenge for SSA countries [11]. Although there is an abundance of both potential human labor and natural resources that can be leveraged for economic growth in the region [13–15], demand for biomass-based products is also closely tied to food security through the crops from which they are derived. Food security, in turn, is embedded within a network of other overarching factors such as health, welfare, employment, income, environmental degradation, markets, and culture [9,10,16].

Policy interventions for increasing food security are always parts of complex systems and must face inevitable trade-offs and unintended effects. The need to understand these systems is thus crucial, particularly with a predicted increase in demand and economic integration of biomass-based products and the potential for newly emerging interconnections in value chains [11,12]. The flows of goods and capital between actors specializing in agricultural production, collecting, transport, storage, processing, sale, and consumption are expected to merge and form more complex systems with feedback loops and non-linear interactions [11]. Rather than value chains, the complexity described may require adopting the concept of a biomass-based value web, defined by Virchow et al. [11] (p. 229) as, " … the production and processing of both food and non-food biomass from locally adapted crops within flexible, efficient and sustainable production, processing, trading and consumption systems".

Despite the expected utility of the concept of value webs and its emergence and establishment in business-oriented fields [17–20], there is only limited empirical evidence to support its adoption in the context of biomass production and food security, though this is growing. Poku et al. [21] mapped the cassava value web in Ghana and Adeyemo et al. [22] that of Nigeria, with findings from both works suggesting that increased actor coordination and market integration could have positive outcomes for food security. Scheiterle et al. [23] use systems modeling and in-depth interviews to examine the case of sugarcane in Brazil's bioeconomy. The concept and method proved effective at identifying a need for targeted incentives and improvement in terms of linking knowledge institutions with industry. Loos et al. [24] look at the demand and market integration of natural fibers from plantain residue from a value web perspective in Ghana, while Lin et al. [25] perform a similar analysis focusing on the bamboo sector in Ethiopia. Neumann et al. [26] examine the systemic effects of changes in labor and land productivity in relation to money flows and food security.

The research described demonstrates that a more holistic understanding of actors and their interconnections using the value web approach is possible and can allow for evidence-based policy recommendations. However, the use of value chains remains more prevalent [27–32]. Furthermore, the aforementioned studies have focused only on one crop, disregarding the emerging potential of more integrated multicrop value webs leveraging demand for food, feed, energy, or industrial material [11]. Along with limiting the concept of value webs to single crops, research has also been limited only to primary actors and flows of resources.

Our research thus aims to address these gaps by using an explorative systems modeling approach to determine the degree to which complex multicrop biomass-based value web structures have formed and their influence on food security. To the authors' knowledge, there has been no prior research using a participatory causal systems modeling approach to map and analyze multicrop biomass-based value webs in relation to food security.

The research presented here is thus guided by the following key questions: (1) Are biomass-based value webs a more appropriate conceptualization of exchanges among actors than traditional value chains? (2) Are there interactions among factors describing multicrop value webs and to what degree are they integrated? These two research questions are addressed by the creation of country-specific value webs for Ghana, Nigeria, and Ethiopia. A third research question is addressed by merging the individual country models into a generic SSA model and further analyzing it through the lens of food security: (3) What are the current supporting and hindering factors within biomass-based value webs that influence food security across SSA, and how do they relate to current policy efforts?

Based on prior research [23–25], we expect the concept of the value web to perform well in terms of revealing otherwise neglected interactions and the complexity of interconnections among actors and flows. We aim to test the hypothesis that a greater integration of value web factors across crops has occurred. This is a reasonable assumption in light of increasing demand for multipurpose biomass uses and because a greater integration of resource and information flows and actors should improve market efficiency. The relations between specific factors within the value webs in terms of supporting or hindering progress towards food security in relation to current policy is unknown. It is crucial to both create baselines and generate data that can be used for future monitoring and evaluation [33,34], as well as determine the causal effect of factors that support or hinder food security and their relation to current or potential policy options across SSA. By using an explorative modeling approach driven by stakeholder knowledge, we aim to produce such evidence.

2. Methods

2.1. Participatory Systems Analysis

The systems analysis software iMODELER [35] is used to create qualitative systems models of biomass-based value webs in relation to food security. First, participatory stakeholder workshops were conducted in Ghana, Nigeria, and Ethiopia to describe the factors (*factor* is the terminology used for iMODELER—it is synonymous with *node* or *variable* in other systems modeling approaches) and interconnections of their respective national biomass-based value webs. Mapping causal connections between key factors identified in the workshops provides insight regarding the degree to which value chains of selected crops used for food and non-food products have become interconnected into value webs. These country models are aggregated to allow for the creation of a generic model to provide insight into the impacts of policy options for food security across SSA. The influence of overarching supporting and hindering factors on food availability and access are identified through systems modeling. Lastly, the generic model results are compared to expert rankings to better understand the relations between model output and current policy trends.

2.2. Case Study Countries

The biomass-based value webs of Ghana, Nigeria, and Ethiopia were selected for the systems modeling analysis (Figure 1). The selection of these African case study countries is based on a combination of three criteria: (1) high potential for biomass production, (2) prevalence of hunger, and (3) number of agro-food companies and markets. Large swathes of Ghana and Nigeria are within the Sudanian Savanna zone that crosses the African continent in an East–West direction directly south of the Sahelian zone and is characterized by ideal climatic conditions for biomass production. Likewise, the extensive agroforestry systems and bamboo market of Ethiopia also have high potential for an increased role in improving food security in SSA [28,36].

These three Sub-Saharan African countries continue to suffer from food insecurity despite sharing high rates of overall national economic growth (averages of 6.24%, 9.02%, and 6.6% annual percentage GDP growth rate 2000–2017, respectively) [37]. In the Global Food Security Index rankings of 113 countries [38], Ghana ranks 76th overall, while Nigeria and Ethiopia are at 92nd and 99th, respectively, with significantly poorer ranks related to food access rather than food availability in all three countries. However, there is a high potential for benefiting from increasing demand in biomass-based products since the agricultural sectors account for 23%, 24%, and 46% of total GDP in Ghana, Nigeria, and Ethiopia, respectively [29,39], with a growing proportion derived from non-food biomass products [9]. Significant investments in the countries' agricultural sectors along with an increasing number of active agro-food companies increase the need for inclusive policy options with an emphasis on food security [9].

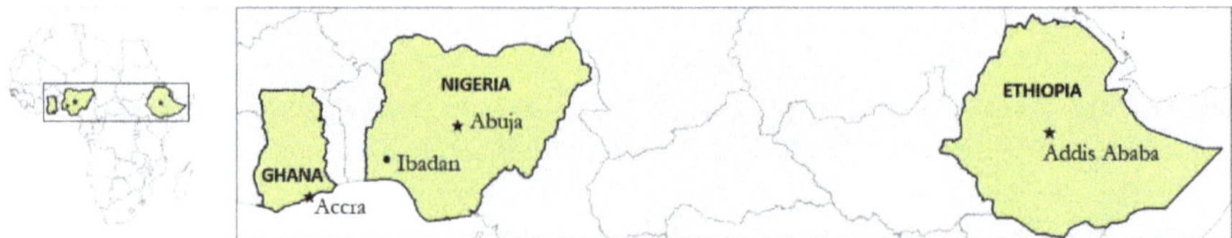

Figure 1. Sub-Saharan African case study countries. Participatory stakeholder workshops were conducted in Accra, Ghana; Ibadan, Nigeria; and Addis Ababa, Ethiopia to create the biomass-based value web models.

The value webs of Ghana and Nigeria included maize, plantain, and cassava, and in Ethiopia, maize, enset, and bamboo. These crops, despite differences such as seasonal cycles, nutritional value, and cultural importance, can be considered the most important crops in terms of their availability for consumption and ability to generate income in the case study countries, considering both food and non-food uses [29,31,32,40].

2.3. Biomass-Based Value Web Modeling

2.3.1. Country Models: Ghana, Nigeria, and Ethiopia

Two 2-day workshops were conducted in Ghana (Accra) and Ethiopia (Addis Ababa), and one in Nigeria (Ibadan), in the period between 2014 and 2017. The three-year research process allowed for iterative methodological adjustment based on stakeholder feedback and data analysis following each workshop, which consisted of between 10 and 20 stakeholders. Stakeholders remained in contact regarding data analysis, and reviews of past findings were presented during follow-up workshops. Participants were invited through local partners and by exploiting existing institutional networks in an effort to capture a diversity of stakeholder perspectives. A wide breadth of knowledge was drawn from, representing farmers' associations, ministries of food and agriculture, food research institutes, food marketing institutions, food policy programs, seed enterprises, NGOs, and university institutes, among others.

Two different 2-stage approaches were applied to adequately develop the models through participation of the invited stakeholders. Most participants were retained for the second stage of model creation, although some changes in stakeholders present did occur. In Ethiopia, the second workshop was used to continue building and elaborating the model generated during the first workshop. In Ghana, the second workshop used a model with a reduced set of factors from the first workshop as a starting point and was not limited to the prior three model crops. There were therefore two separate models describing the value web of Ghana, one crop-specific (Ghana 1) and one generalized (Ghana 2), as well as one model each from Nigeria and Ethiopia (see Appendix A).

It should be noted that the discrepancies in approaches to model creation reflect a research design for model building that was both explorative and pragmatic in terms of answering the research questions. Such a design was necessary given the novelty of the research approach and corresponding lack of best-practice methodological evidence on which it could otherwise be based. Thus, different approaches were used in order to exclude the possibility of biased results from a single constant yet potentially flawed methodology. In Ethiopia, a higher degree of specificity in the model was achieved by further building on the same model in each workshop. It was hypothesized that revealing any interconnections among crops within the value web may be dependent on such specificity. The second model in Ghana (Ghana 2) was created without limiting stakeholder input to specific crops, to likewise test the hypothesis that the value web may yield different findings.

The iMODELER software allows for the directional connection of factors to represent causality by using the basic question of "What (may) lead(s) to more or less of the factor now or in the future?" [41]. By starting with the target factors of *food security*, *availability*, and *access*, the workshop participants were asked by a moderator to consider what leads to more and what leads to less of these targets, while considering actors and flows of biomass, capital, and information.

Accordingly, the modeling starts with general factors and connections that are further detailed in the following modeling steps (Figure 2).

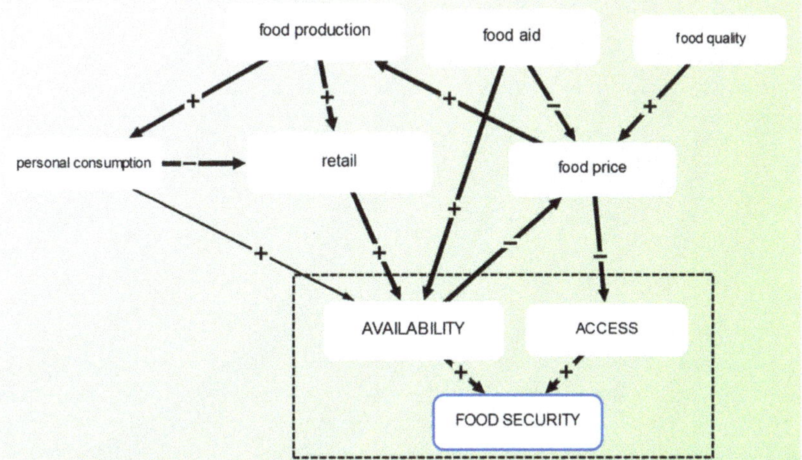

Figure 2. Example of factors and factor connections leading to *food security*. The dashed box includes pre-existing targets into which relevant causal factors could be connected. Such causal factors were added to the model in a step-wise process based on inputs from workshop participants.

Here, for example, an increase in *food production* leads to more *retail*, which leads to more (food) *availability*. Each connection is evaluated only on its immediate connections, therefore rather than attempting to logically create flows, they emerge through the sum of individual connections [29]. For example, an increase in *food aid* leads to less (lower) *food price*, and a higher *food price* leads to less *access*.

Throughout the workshops, the models were projected onto a screen and new factors and connections were added by a moderator. Stakeholder discussion and agreement allowed for continuous expansion until input for possible further causal connections was exhausted. Through the process of consensus and exhaustive "tracing backward" from *food security*, the factors and connections were completed. The models created in the stakeholder workshops were finalized through a desk-top validation to check for erroneous connections and factor names. By visually examining and tracing the connections among factors representing actors and flows of the crops included in the modeling, the degree to which the individual value chains are in fact integrated within a value web system is determined.

2.3.2. Generic Model: Sub-Saharan Africa

The original country models were aggregated into a generic model representing overarching factors and connections of SSA biomass-based value webs. The generic model describes and allows for an analysis of the current (generalized) system dynamics regarding supporting and hindering factors for food security. Rather than focusing on specific crops, it is based on the identification of cross-cutting factors for extraction from the country models. The participatory and explorative nature of the modeling led to a number of factors and connections that would not duly represent an aggregated SSA biomass-based value web. Therefore, a process of reducing the number of factors in the models was carried out using the three criteria of (1) relevance, (2) level of detail, and (3) redundancy.

Factors were extracted, grouped, and again subjected to possible removal based on relevance to food security and biomass production, resulting in 82 final factors. These were then used as the input for a matrix in order to map out all pre-existing causal connections from the country models and determine potential further connections. Because some of the factors from the country models are unique, there was no basis for any prior causal connection for these factors from the workshops, and therefore the authors' judgment was relied upon. The factors and connections in the matrix were converted into iMODELER to create the generic model with 82 factors and 213 connections (Table 1).

Table 1. Process steps for creating the generic model of Sub-Saharan Africa, based on factors from the four country models (with one from each of the two workshops in Ghana).

	Original Factors	Factors after Reduction and Standardization	Extracted Factors	Total Factors	Factors Further Reduced Based on Relevance	Final Model Factors
Ghana 1	153	140	9			
Ghana 2	111	82	85			
Nigeria	113	77	44	189	125	82
Ethiopia	211	203	51			

The generic model is analyzed using the qualitative systems analysis features of iMODELER to determine the influence of each factor within the web on the targets of *availability* and *access* leading to *food security*. iMODELER allows for the relative weighting of one factor on another based on the degree of causal influence between them. This means that the model does not assume that all factors and factor connections are equal in terms of causal importance, but rather, that they can be defined by stakeholder input.

Weights, when aggregated along a chain towards the target factors, provide a total value representative of that factor's influence on the target. Total weight absolute value sums of 100 for factors that connect to or cause the same subsequent factor are assumed in this analysis. For example, both of the two incoming factors *retail* and *food production* are given an equal weight (50) in terms of (positive) causal influence regarding increasing *availability*. At only one hierarchical level, factor weights are equal to their influence on the target (Figure 3).

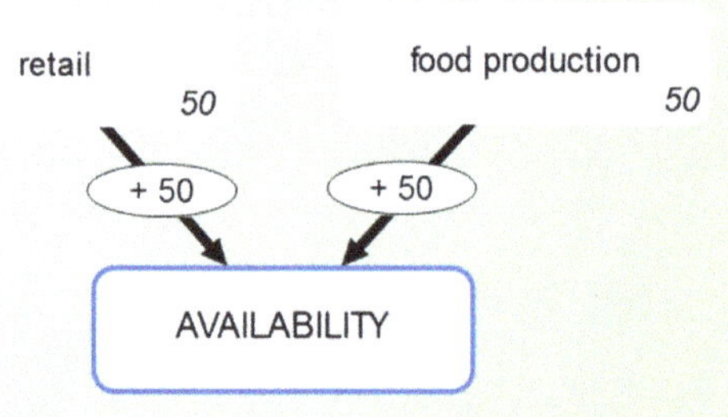

Figure 3. Weights of *retail* and *food production* on the target *availability* are shown in the ovals embedded in the arrows and are both 50. Because there are no intermediary factors, the influence (shown within the rectangular factor boxes) of *retail* and *food production* on the target remain unchanged.

At subsequent hierarchical levels, the influence of any given factor (*INF*) on the target is multiplied by the weight(s) (divided by 100) of factors along the path towards the target. Because there can be many paths in a system, this is calculated as the aggregated sum of each individual path (*X*):

$$INF = \sum_{i=1}^{n} X_i$$

$$X = \left(SW \times \frac{W_1}{100} \times \frac{W_2}{100} \times \frac{W_3}{100} \times \frac{W_n}{100}\right)$$

The starting factor weight (*SW*) is multiplied by the weights of subsequent factors along the path (*W*). This means that scores are both a function of (1) a factor's distance in number of intermediary connections to the target factors, (2) weights of the intermediary connections, (3) sign (+/−) of these connections, and (4) number of paths to be summed.

Thus, at the next hierarchical level, the factors *personal consumption, restricted access to markets, arable land,* and *challenge of seasonality* are filtered through *retail* and *food production* before reaching *availability*. In this example, with only one path from each factor to the target, *arable land* was identified by stakeholders to be four times more important in terms of "causing" *food production* when compared to *challenge of seasonality* (Figure 4).

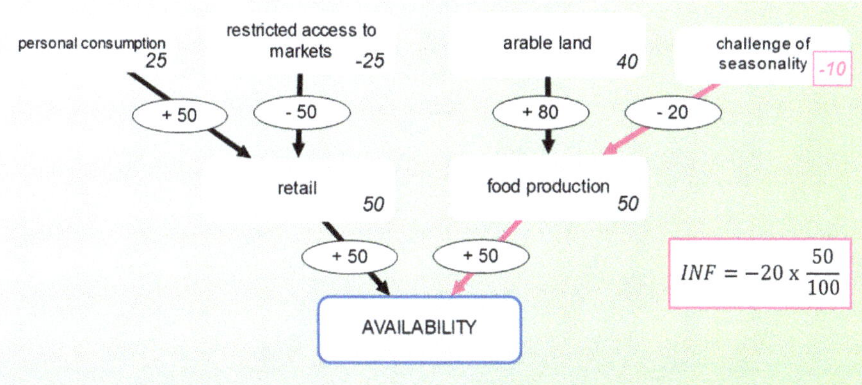

Figure 4. Weights of all factors along the second hierarchical level are (left to right in ovals) 50, −50, 80, and −20, and their influence on *availability* is reduced by the weights of the intermediary factors *retail* (50) and *food production* (50) along the paths towards the target *availability*. The calculation of influence (*INF*) for the factor *challenge of seasonality* is provided in the rectangle on the right.

Weighting is conducted in the generic model by standardizing to sums of 100 for factors with multiple incoming connections (2 factors = 50 each, 3 factors = 33.3 each, etc.). For exogenous factors (factors with no incoming connections on the outer edges of the model), weights are reduced from 100 to either 75 or 50 depending on the weights originally identified by stakeholders during the creation of the country models and by determining the potential for missing factors.

To analyze the relative influence of the system's factors in relation to food security, first all of the unique factors in the web that can be the focus of targeted policy interventions to improve food security are identified. These are labeled as either "supporting" or "hindering", based on their known relation to food security. All other factors that are not readily manageable through policy intervention or largely represented by other considered factors are deemed "neutral". For example, *increase access to market for smallholders* was identified as a supporting factor, *post-harvest loss* as a hindering factor, and *transportation* as a neutral factor. Although transportation could potentially be promoted through policy, it is labeled as neutral because it is both broad and already largely represented by the supporting factor *road infrastructure*. The five most influential and least influential supporting factors and the five most influential and least influential hindering factors with relation to the targets of *availability* and *access* (20 total factors) within the generic systems model were extracted and analyzed. This provides a relative description of their current influence within a generalized value web in relation to food security.

Results from the generic model were tested using a ranking exercise conducted by stakeholders of the most and least influential supporting and hindering factors for both *availability* and *access*. This allows for a reflection on relevant policy in SSA based on the generic model results. By asking stakeholders to rank the factors in terms of what is most important for supporting or hindering food security, their perception of the best options for policy strategies was elicited. The stakeholders are experts representing government and research entities dedicated to the agricultural sector of Ghana and familiar with the wider SSA context. Although the generic model is not limited to input from the workshops held in Ghana, the output strongly reflects the Ghanaian context of food security, allowing for a regional extrapolation of results.

3. Results and Interpretation

3.1. Multicrop Value Web Models

Extensive system models of the biomass-based value webs in the three case study countries were created. The participatory format and explorative approach yielded large models with a high degree of complexity of interconnections (Table 2). Differences among country models are a function of the national biomass-based value webs, the crops modeled, and the expertise of the stakeholders. Moreover, the Ethiopian model was expanded within an additional workshop, accounting for its larger size.

Table 2. Number of factors and connections from the four models created for the three case study countries—Ghana, Nigeria, and Ethiopia.

	Factors	Connections
Ghana 1	153	285
Ghana 2	111	186
Nigeria	113	161
Ethiopia	211	490

The numerous factors with non-linear casual connections within the value web models suggest that the linear value chain concept may be an oversimplification. Thus, adopting a value web approach has merit in terms of describing the actual interconnections among actors and flows of matter, capital, and information as a basis for understanding complex problems like food security. Connections among crop-specific (e.g., maize) actors such as smallholders, collectors, processors, and consumers are interconnected and form feedback loops with food and non-food products. In the Ethiopian model alone, there are 18,564 loops, some incorporating as many as 15 different factors.

Despite the complexity of the models, findings show very few direct interconnections within the value webs describing different crops in the case study countries. The factors representing actors and flows of products for each crop are generally ramified within the models, with e.g., cassava smallholders, collectors, transporters, and processors in Nigeria practicing their livelihoods and limiting exchanges with other crop-specific actors. By assigning colors to factors directly relevant to the different crops and arranging the factors by clusters of connections, the cassava factors (a; white) stand alone and are mostly disconnected from those of other crops, as is the case for the internal interconnections of maize (b; yellow) and plantain (c; green) factors (Figure 5).

The factors relevant to each crop are connected through more general supporting factors (blue) that would benefit webs of multiple crops. For example, *increased use of capacity of machines* can lead to increased production of all of the crops, while *fewer seasonal layoffs of processors* would influence jobs related to maize and plantain but is not connected to the cassava web because cassava allows for more flexible planting and harvesting times [31]. However, the majority of actors and flows pertain to crop-specific factors, exhibiting only sparse integration among crops.

The importance of such general overarching factors that do in fact connect across crops in the value webs should not be discounted. For example, the factor *transportation infrastructure* from the Ethiopian model is highly embedded in the system as it provides for smallholders, collectors, government agencies, enables export trade, and allows for increased market access for rural actors [42] (Figure 6). The importance of transportation and connectivity for SSA value webs is well documented, e.g., [21,43], and similar factors can also be found in the Ghanaian and Nigerian models.

The country models provide a resource for further research or targeted interventions with specific components of the value webs (e.g., *transportation infrastructure*) while also representing a baseline for better understanding the systemic role of biomass in shifting national economies.

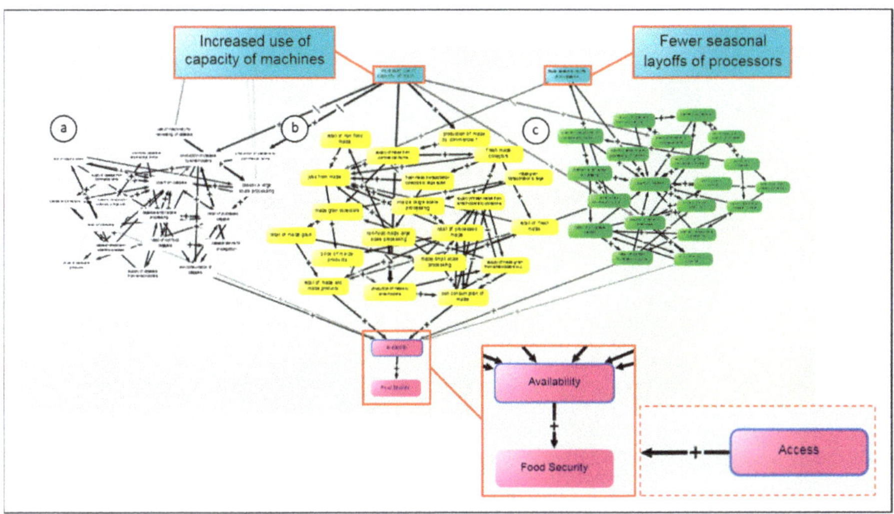

a

Figure 5. *Cont.*

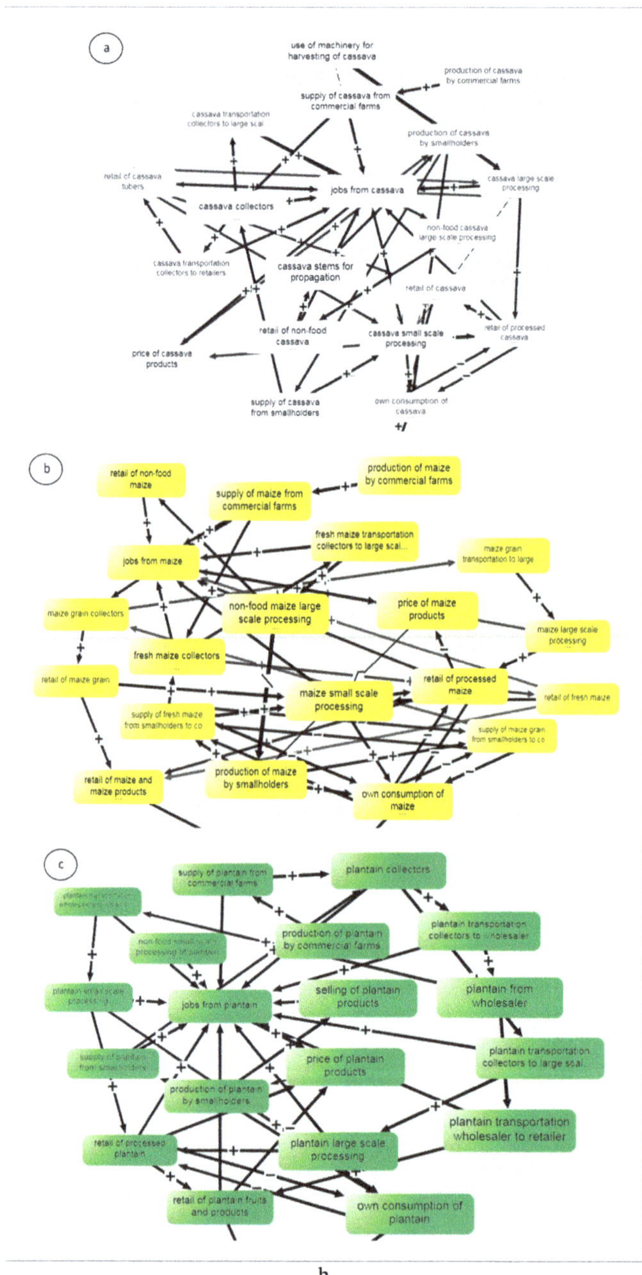

Figure 5. Excerpt schema of Nigerian biomass-based value web with cassava ((**a**); white), maize ((**b**); yellow) and plantain ((**c**); green) factors leading to *Availability*. *Access* (dashed red box) also connects to *food security* in the model but has been filtered out from the display. Two non-crop-specific factors connect the otherwise ramified crop-specific value webs (blue), *increased use of capacity of machines* and *fewer seasonal layoffs of processors*. Factors and connections can be further explored in iMODELER using the link provided in the supplementary material.

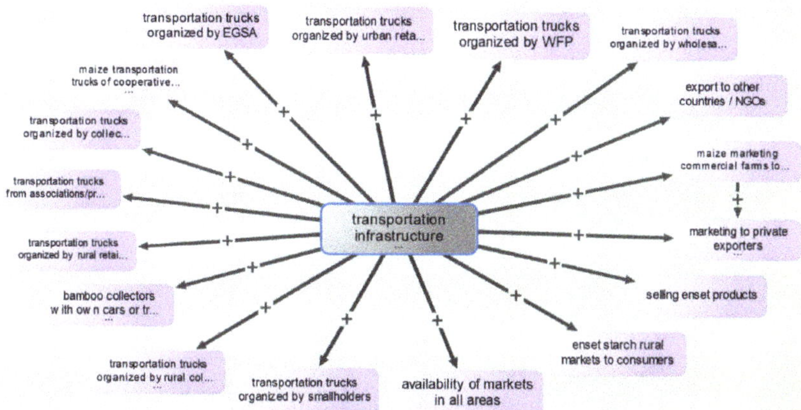

Figure 6. Excerpt of the Ethiopian model with the factor *transportation infrastructure* and its 17 outgoing connections.

3.2. Generic Value Web Model and Policy Trends

The generic model represents the aggregated significant and non-crop-specific factors and connections from the country value webs in a systems model within which each factor's degree of influence on food availability and access can be quantified. Because the generic model is originally derived from stakeholder input, determining the influence of factors and connections using iMODELER provides a description of their current influence within the value webs in relation to food security.

Enough similarity exists in the factors and flows of the country models to generalize, albeit with a significant reduction in detail, leaving the final generic model with 82 factors (including the three target factors *availability* of and *access* to food, as well as *food security*) and 213 connections. Excluding the three target factors, a total of 41 supporting factors, 17 hindering factors, and 21 neutral factors were identified. For both *availability* and *access*, the five supporting factors with the most influence and five supporting factors with the least influence (Table 3), as well as the five hindering factors with the most influence and five hindering factors with the least influence (Table 4), were extracted and interpreted.

Table 3. Supporting factors with the most and least influence on *availability* and *access*.

	Availability	Access
Most influential	(1) streamline processes (2) food aid (3) increase access to information for farmers (4) promote extension services (5) support information services	(1) food aid (2) increase export of commodities (3) implement subsidies (4) improve food processing techniques (5) streamline processes
Least influential	(37) implement subsidies (38) promote partnerships (39) improve unique selling points of processed products (40) improve food processing techniques (41) increase export of commodities	(37) land title holdings (38) increase access to market for smallholders (39) loans to smallholders (40) train farmers to handle food (41) increase use of farm machinery

Several patterns emerge from the model output of supporting factors. First, there are clear differences in the order of factors between *availability* and *access*. Only the factors food aid and streamline processes (i.e., improving the efficiency of exchanges among actors) are currently highly influential for both *availability* and *access*. Food aid was directly linked to *availability* and can act to reduce food prices. Although food aid can also reduce incomes for local market traders [44], the model recognizes a reduction in food prices as an improvement for food security, highlighting the need for contextual model interpretation. The factor *streamline processes* can lead to fewer losses during storage, transport, and retail and is highly embedded in the model, with nine direct connections.

The role of information for farmers is highly influential in the model for *availability*, as it can act to reduce losses and maximize profits by leveraging market prices. For *access*, the strength of reduced prices and increased incomes are illustrated by the factors *increase export of commodities, implement subsidies*, and *improve food processing techniques*. These three factors are some of the least influential for *availability* given that their effects on prices and incomes are only connected to food production through further indirect connections, as represented in the model.

The least influential supporting factors are generally those that are being prioritized within national and regional food security policy in SSA. Increasing exports is perhaps the most pervasive of the factors across national and regional policy documents, particularly with regard to increasing regional trade within Africa [3,14,45–52]. The roles of *improve unique selling points of processed products* along with *improve food processing techniques* are also specifically highlighted in policy briefs [50,53] and reflect the aspiration of greater market integration for smallholders and the desired shift from primarily subsistence to more commercial agriculture across SSA [1,3,45,50,53,54]. Likewise, market integration is often directly tied to the objective of promoting partnerships, particularly as a means of integrating smallholders into value webs [55] by bringing together farmers, agribusiness, and civil society, as laid out in the Malabo Declaration [1].

The least influential factors for *access* are perhaps even more related to current policy efforts, particularly with the push for increased market access and use of farm machinery [1,3,45,50–56]. The African Union Commission's Agenda 2063 report [3] (p. 3) states under action point 13 that "the hand hoe will be banished by 2025 and the sector will be modern … ", and the latest report from the Malabo Montpellier Panel is entitle "Mechanized: Transforming Africa's Agriculture Value Chains" [52]. At a national level, in both Ghana and Nigeria, mechanization is one explicit component of development policies, considered a major constraint to current agricultural productivity [56,57]. To a lesser extent, the roles of land title holdings for economic growth and increasing productivity are recognized [46,58], as well as the importance of farmer training [3,14,50].

Although there are fewer total hindering factors (n = 17), the degree of overlap is still significant between the most influential hindering factors for both *availability* and *access*, with four of the five factors shared (Table 4).

Table 4. Hindering factors with the most and least influence on *availability* and *access*.

	Availability	Access
Most influential	(1) challenge of seasonality	(1) overpriced food
	(2) post-harvest loss	(2) post-harvest loss
	(3) market queens fix prices and markets	(3) challenge of seasonality
	(4) impacts of weather extremes	(4) market queens fix prices and markets
	(5) losses at retail	(5) losses at retail
Least influential	(13) other land uses	(13) mining (legal and illegal)
	(14) corruption	(14) underutilized equipment
	(15) transportation costs	(15) transportation costs
	(16) underutilized equipment	(16) cultural barriers
	(17) overpriced food	(17) corruption

The hindering factors tend to represent direct losses associated with the lack of a physical abundance of food and income generated through the value webs. The only factors that are not shared as highly influential are *impacts of weather extremes* for *availability* and *overpriced food* for *access*, two factors directly implicated in a loss of production and inability to purchase, respectively. *Overpriced food*, similar to the supporting factor *implement subsidies*, is among the least influential for *availability* and most influential for *access*. This output is a result of the factor *implement subsidies* leading to less *food price*, which directly connects to *access* in the generic model. In fact, policies around subsidies in SSA are variable across countries, but there is a general call for moving away from most forms of subsidies as a means of achieving the general regional consensus of increased trade [46]. Their role, however, is still prominent in many SSA countries, with input subsidies in particular being a significant part of agricultural development policies [46].

Contrary to the supporting factors, the hindering factors with the most influence are generally those that are receiving the most recent policy attention. This is appropriate given that the hindering factors identified by experts represent the current situation and the most negatively influential factors should be mitigated. *Post-harvest loss* and *challenge of seasonality* are highly influential across both *availability* and *access*. The role of subsidies for minimizing the effects of overpriced food on consumers spiked across SSA following the food price crisis in 2007/2008, and these policies continue to be prominent across the region [46]. Likewise, post-harvest losses [1] (p. 3) and dealing with climate variability [1] (p. 5) are featured in recent policy documents.

The findings and interpretation were tested by subjecting the most and least supporting (Table 3) and hindering factors (Table 4) to a ranking exercise with stakeholders. When experts were asked to rank the most and least influential factors based on their current *importance* for either supporting or hindering efforts towards improved food security, a number of factors "flipped", either from being highly influential in the model to lowly ranked in the workshop, or vice versa (Table 5).

Table 5. Supporting and hindering factors for *availability* and *access* that are identified as either high influence in the model and of low importance for sustainable food security by expert stakeholders, or of low influence in the model and high importance by expert stakeholders.

	Supporting Factors		Hindering Factors	
	High Model Influence → Low Expert Ranking	Low Model Influence → High Expert Ranking	High Model Influence → Low Expert Ranking	Low Model Influence → High Expert Ranking
AVAILABILITY	food aid streamline processes increase access to information for farmers support information services	implement subsidies promote partnerships improve unique selling point of processed products improve food processing techniques	challenge of seasonality impacts of weather extremes losses at retail	other land uses costs of transportation overpriced food
ACCESS	food aid increase export of commodities implement subsidies	land title holdings increase access to market for smallholders loans to smallholders	challenge of seasonality	costs of transportation

Several highly influential factors in the model were judged to be of relatively low importance by the stakeholders when prioritizing efforts to improve food security. The factor *food aid*, for example, was highly influential in the model as it can directly increase *availability* and *access*, but it received a low rank of importance in the stakeholder exercise. This reflects both the lack of demand for food aid in Ghana and its perception by stakeholders as relatively unimportant for sustainable improvement of food security. This demonstrates the need for interpreting the output factors from the generic model as functions of direct and indirect influence rather than on the basis of the sustainable importance for effectuating change. Factors seen to have a more direct market impact such as *implement subsidies*, *promote partnerships, improve unique selling point of processed products, improve food processing techniques,* and *increase access to market for smallholders* were highly ranked by the stakeholders but of low influence in the model.

The experts agreed to a lack of agency in regard to certain factors, and therefore, a lack of current efforts regarding seasonality, weather, and to a certain degree losses at retail and thus gave related factors low rankings. Land use, transportation and prices were judged and prioritized as modifiable through policy intervention and given higher ranks. It is evident that while the models do not account for feasibility, the experts rely on more contextual and historic knowledge with consideration of behavioral and governmental agency.

4. Discussion

4.1. From Value Chains to Multicrop Value Webs

Highly complex value webs with, however, a lack of integration among crop-specific factors, are two principal findings demonstrated by the country models. The concept of a value web as a more realistic approximation of actual interconnections is supported. This adds to a growing body of evidence that simplistic and linear value chains may be inadequate in the context of actors and exchanges of food and non-food biomass [11,21–26]. The finding is also in line with predictions of increased complexity among actors and exchanges [11].

However, a lack of integration across crop-specific factors in the country models was surprising given the potential gains in efficiency and an increasing domestic and international demand for food and non-food biomass-based products. This suggests either that predictions made by [11,59] may overemphasize the degree of eventual integration or that such integration is yet to occur. It is unclear why this is the case, particularly given the potential benefit of increased efficiency. However, based on stakeholder input and understanding of the value webs in the case study countries, several theories can be posited. Because of the economic importance, reliability, and sheer size of the markets of the crops modelled, the value webs can accommodate a number of actors that are all able to make a profit without moving towards the diversification of biomass-derived income. Taking Ghana as an example, the value webs of cassava and maize have been described [29,31,43] and show a number of interlinked institutions and industries respective to each crop, with significant flows of money and co-dependent actors. While this may be the case with other crops to some extent, there is less total available profit and thus less incentive for specialization. A further explanation involves the crops chosen for the modeling—they all play unique roles in their social-ecological systems and respective biomass economies, with relatively little overlap in terms of products made from multiple crops. Contrarily, the value webs of other crops, for example maize and wheat, can be connected through cereal processors [40]. Despite the current dynamics, the rapidly increasing demand for biomass in SSA, partly driven by new uses in the form of non-food or industrial products, could incentivize further interlinkages in the value webs [9,14,40]. However, growing markets with space for new actors do not necessarily require maximum efficiency for substantial profit, and thus the necessary incentive may only develop over time. Tracking these changes will be crucial for management decisions regarding cascading and circular uses of biomass for sustainable resource use [60] as well as monitoring and evaluation of policy options [31,33], and the country models provide useful baselines in this sense.

Furthermore, although crop-specific factors do not show significant integration, the models demand the consideration of more overarching factors that are often neglected but highly interlinked (e.g., processing capacity and transportation infrastructure). The findings here are not surprising given past research. Poku et al. [21] used a systems mapping approach for the cassava value web in Ghana and similarly found that highly connected factors such as road infrastructure and access to information by means of public services currently have pervasive effects on the cost of food and its availability. The ability of extreme weather and climate variability (particularly regarding seasonality) to undermine agricultural systems is also well understood [61,62] and emerged as an influential factor for food security in the generic model.

4.2. Generic Model Results and Food Security Policy in SSA

Overall, the generic model performed well and is interpretable in terms of mapping out the current landscape of supporting and hindering factors and their current influence as embedded within a larger biomass-based value web system. Although both availability of and access to food are problematic in the case study countries, scores from the Global Food Security Index [37] indicate that affordability, similar to access, is even more problematic than availability in Ghana, Nigeria, and Ethiopia. Based on these scores, it is likely that the supporting factors, particularly for *access* (Table 3), are currently misaligned within the actual value webs. In other words, the factors with the least current influence should be more emphasized in policy to effectuate future change. This is reasonable given that the model dynamics are based on stakeholder input describing the present realities within the countries and extrapolated to SSA. As opposed to attempting to leverage those factors that currently exert more direct influence, sustainable and transformative policy should rather be prioritized and weaker supporting factors bolstered. Therefore, efforts to increase their impact have the potential to improve food security by shifting the current dynamics of biomass-based value webs. Findings generally suggest that current overarching trends regarding efforts to improve food security in SSA are accordant with where resources are best allocated.

This interpretation is further supported by the expert ranking exercise (Table 5). There were fewer changes between model output of influence and expert determination of importance among hindering factors, but those factors that did change positions show the similar trend of representing current policy efforts. While the highly influential factors in the model represent general dynamics, the least influential factors as well as the ranking exercise tend to reflect current policies. Furthermore, the model does not account for feasibility, while the experts rely on more contextual and historic knowledge with a consideration of behavioral and governmental agency.

Several of the hindering factors (Table 4) demonstrate one limitation of generalizing across the diversity of SSA countries. The influence of "market queens", a term from Ghana that represents women leaders who function essentially as cartels to skew market signals and limit market entrance, is high for both *access* and *availability*. Despite its specificity, the factor is included in the generic model to represent the prevalence of both price control and instability within markets across SSA [63]. Similarly, the factor mining (legal and illegal) is not a significant issue in Ethiopia but plays a role in soil and general environmental degradation in Ghana, Nigeria, and other SSA countries [64,65].

4.3. Model Limitations and Future Outlook

The findings, along with the model, should be considered generalized and descriptive of broad overall trends, whereas smaller scale spatial and temporal dynamics are often responsible for the effectiveness of programs and projects to improve food security [66]. While the process of creating the generic model was conducted to best capture trends across SSA, the original data in the form of stakeholder input originally represent Ghana, Nigeria, and Ethiopia. Nevertheless, the agro-ecology of these countries just south of the Sahel allows for potential increased agricultural production, and regional dynamics are therefore crucial for SSA food security.

Along with being generalized, both the model output in terms of influencing factors as well as the expert ranking exercise are inherently based on potentially biased stakeholder perceptions. For instance in Ghana, some research has shown that in the case of the "market queens", their negative reputation is not entirely justified as they can also provide a social safety net [67], whereas their negative influence in the generic model was very high for both *availability* and *access*. In addition, it is possible that the influence of more insidious or slow-acting deleterious effects of the hindering factors in particular are underweighted. For example, neither country model workshop stakeholders nor the experts during the ranking exercise in Ghana weighted factors like *corruption, mining, other land uses*, and *cultural barriers* highly. While it is possible that the effects are relatively minor compared to other factors, it could also be that their undervaluation is a result of their effects being less salient.

Understanding what exactly accounts for the influence of specific factors within the generic model requires analyzing individual causal pathways. One determinant of factor influence within the models is a factor's distance to the target factors (i.e., *availability* and *access*) in terms of number of intermediary connections. Although many intermediary connections with high weights could equate to more influence than few connections with low weights, in the generic model a greater distance on average resulted in reduced factor influence. The average distances in number of connections to the target for the five most influential supporting and hindering factors are 2.2 and 3.2, respectively, while the five least influential are 3.6 and 3.8, respectively. This is theoretically sound, as the hindering factor *overpriced food*, for example, more directly affects *access*, as opposed to mining or process inefficiencies.

If this criterion of directness is a legitimate indicator of influence in actual biomass-based value webs as presented here, one way to increase the influence of the least influential supporting factors would be policy options that more directly link these factors to *availability* and *access*. For example, if the effects of promoting *loans to smallholders* or *land title holdings* can more directly influence *access*, then food security as a whole would be improved.

Determining best practices in this sense requires more focused and spatially explicit research. This is important when considering trade-offs in policy contexts. However, the utility of the models in terms of drawing out specific recommendations is challenging due to their size and complexity. For this purpose, model segments can be focused on or the models adjusted by adding, removing, or changing factors and connections based on context or new data. The biomass-based value webs for Ghana, Nigeria, and Ethiopia, as well as the generic model for SSA, are accessible in iMODELER through links provided as supplementary material, and their continued use and development is encouraged for specific research or policy agendas. Furthermore, all of the supporting and hindering factors in the generic model were identified by stakeholders from the case study countries and should therefore be closely considered in research or policy contexts designed at improving food security.

5. Conclusions

Findings suggest that the shift from a value chain to a value web perspective more accurately represents the actual interconnections of biomass actors and flows in SSA countries. However, there is currently a lack of integration among actors dealing with different crops, with interconnections in the models primarily due to overarching general factors. Results from a generic systems model suggest that current trends in efforts towards improving food security in SSA are at least nominally appropriate. This is based on the current dynamics of biomass-based value webs as identified by stakeholders in Ghana, Ethiopia, and Nigeria and compiled into a generic model for SSA. Specifically, results correspond to the policy trends of promoting increased access to and success within markets for smallholders as well as reducing the impacts of post-harvest losses and climate variability. Notwithstanding model assumptions and limitations, increasing the impact of supporting factors while mitigating those with the most current detrimental influence should make gains in this regard.

Future research should focus on the utility of the models or sub-sections of the models to inform specific research or policy queries. Duplication studies using other crops or scales below the national level should be conducted. These may result in different findings and would thus better position this research in terms of its contribution to understanding SSA biomass-based value webs. Furthermore, such studies could be useful for developing methodological and interpretative best practices. Lastly, the methods described could be replicated in the future to determine how value web dynamics change and the relation of these changes to policy initiatives such as increased mechanization and regional cooperation.

It will become increasingly important to consider a holistic range of factors to better understand complex potential trade-offs and plan for improved food security. The models should be viewed within the spatial-temporal context in which the data were acquired and as providing a baseline for inevitable future changes.

Author Contributions: Conceptualization, M.D., K.N., K.A., and C.T.; Data curation, C.C.A., M.D., and K.N.; Formal analysis, C.C.A. and M.D.; Funding acquisition, M.D.; Investigation, C.C.A., M.D., K.N., K.A., and C.T.; Methodology, C.C.A., M.D., and K.N.; Project administration, M.D.; Resources, M.D. and K.N.; Software, C.C.A. and K.N.; Supervision, M.D.; Visualization, C.C.A.; Writing—original draft, C.C.A.; Writing—review editing, C.C.A., M.D., K.N., K.A., and C.T.

Funding: This research was funded by the German Federal Ministry of Education and Research (BMBF) through the collaborative project "Improving food security in Africa through increased system productivity of biomass-based value webs (BiomassWeb)". This project is part of the GlobE—Research for the Global Food Supply Programme (grant no. 031A258A).

Acknowledgments: The authors would like to thank all of the workshop participants and those who supported the project, as well as the German Federal Ministry of Education and Research (BMBF) for funding this research.

Conflicts of Interest: The authors declare no conflict of interest.

Appendix A

Ghana 1: http://www.imodeler.info/ro?key=COqgfn-SmYdzcQvxN1sgfcw.
Ghana 2: http://www.imodeler.info/ro?key=CQAN5XNQiuCgzAS0wgZiH3g.
Nigeria: http://www.imodeler.info/ro?key=CxLn3VdLkPLgppuYGJAQn7w.
Ethiopia: http://www.imodeler.info/ro?key=CEb9NshM2dTx9RUNcmmxaXg.
Generic Model: http://www.imodeler.info/ro?key=CE7ubGF9etOpqrx2cp8JM2w

References

1. African Union. *Malabo Declaration on Accelerated Agricultural Growth and Transformation for Shared Prosperity and Improved Livelihoods*; African Union: Malabo, Guinea Bissau, 2014; Available online: https://au.int/en/documents/20150617-2 (accessed on 11 May 2018).
2. United Nations. *Transforming Our World: The 2030 Agenda for Sustainable Development*; Resolution Adopted by the General Assembly; United Nations: Rome, Italy, 2015.
3. African Union Commission. *Agenda 2063: The Africa We Want. Popular Version*; African Union Commission: Addis Ababa, Ethiopia, 2015; Available online: https://au.int/en/Agenda2063/popular_version (accessed on 3 April 2018).
4. FAO. *Food and Agriculture: Driving Action across the 2030 Agenda for Sustainable Development*; FAO: Rome, Italy, 2017; Available online: http://www.fao.org/sustainable-development-goals/news/detail-news/en/c/901768/ (accessed on 11 April 2018).
5. United Nations Economic and Social Council. *Progress towards the Sustainable Development Goals: Report of the Secretary-General*; United Nations Economic and Social Council: New York, NY, USA, 2017; Available online: https://sustainabledevelopment.un.org/ (accessed on 26 July 2018).
6. African Union. *Maputo Declaration on Agriculture and Food Security in Africa. Maputo, Mozambique*; African Union: Malabo, Guinea Bissau, 2003; Available online: http://www.nepad.org/resource/au-2003-maputo-declaration-agriculture-and-food-security (accessed on 14 July 2018).
7. Alliance for a Green Revolution in Africa (AGRA). *Africa Agriculture Status Report: Progress towards Agricultural Transformation in Africa*; Alliance for a Green Revolution in Africa: Nairobi, Kenya, 2016; Available online: https://agra.org/aasr2016/ (accessed on 10 September 2018).
8. Kanu, B.S.; Salami, A.O.; Numasawa, K. Inclusive growth: An imperative for African agriculture. *Afr. J. Food Agric. Nutr. Dev.* **2014**, *14*, 3.
9. Allen, T.; Heinrigs, P. *Emerging Opportunities in the West African Food Economy*; OECD Publishing: Paris, France, 2016; Volume 1.
10. Nebe, S. *Bio-Based Economy in Europe: State of Play and Future Potential: Part 2 Summary of Position Papers Received in Response to the European Commission's Public On-Line Consultation*; European Commission: Brussels, Brussels, 2011. [CrossRef]
11. Virchow, D.; Beuchelt, T.D.; Kuhn, A.; Denich, M. Biomass-based value webs: A novel perspective for emerging bioeconomies in Sub-Saharan Africa. In *Technological and Institutional Innovations for Marginalized Smallholders in Agricultural Development*; Springer: Cham, Switzerland, 2016; pp. 225–238. [CrossRef]
12. El-Chichakli, B.; von Braun, J.; Lang, C.; Barben, D.; Philp, J. Policy: Five cornerstones of a global bioeconomy. *Nat. News* **2016**, *535*, 221. [CrossRef]
13. Ahenkan, A.; Osei-Kojo, A. Achieving sustainable development in Africa: Progress, challenges and prospects. *Int. J. Dev. Sustain.* **2014**, *3*, 162–174.
14. Hollinger, F.; Staatz, J.M. *Agricultural Growth in West Africa: Market and Policy Drivers*; FAO and AfDB: Rome, Italy, 2015; Available online: http://www.fao.org/policy-support/resources/resources-details/en/c/447562/ (accessed on 11 June 2018).
15. Morris, M.; Binswanger-Mkhize, H.P.; Byerlee, D. *Awakening Africa's Sleeping Giant: Prospects for Commercial Agriculture in the Guinea Savannah Zone and Beyond*; The World Bank: Washington, DC, USA, 2009. [CrossRef]
16. Burchi, F.; Fanzo, J.; Frison, E. The role of food and nutrition system approaches in tackling hidden hunger. *Int. J. Environ. Res. Public Health* **2011**, *8*, 358–373. [CrossRef]

17. Allee, V. Reconfiguring the value network. *J. Bus. Strategy* **2000**, *21*, 36–39. [CrossRef]
18. Kelly, E.; Marchese, K. Supply chains and value webs. In *Business Ecosystems Come of Age*; Business Trends Series. Available online: https://www2.deloitte.com/content/dam/insights/us/articles/platform-strategy-new-level-business-trends/DUP_1048-Business-ecosystems-come-of-age_MASTER_FINAL.pdf (accessed on 11 May 2018).
19. Smith, R.W.; Broxterman, W.E.; Murad, D.S. *Understanding Value Webs as a New Business Modeling Tool: Capturing & Creating Value in Adhesives*; The Adhesive & Sealant Council: Las Vegas, NV, USA, 2000.
20. Stabell, C.B.; Fjeldstad, Ø.D. Configuring value for competitive advantage: On chains, shops, and networks. *Strateg. Manag. J.* **1998**, *19*, 413–437. [CrossRef]
21. Poku, A.G.; Birner, R.; Gupta, S. Is Africa ready to develop a competitive bioeconomy? The case of the cassava value web in Ghana. *J. Clean. Prod.* **2018**. [CrossRef]
22. Adeyemo, T.A.; Abass, A.; Amaza, P.; Okoruwa, V.; Akinyosoye, V.; Salman, K.K. Increasing smallholders' intensity in cassava value web: Effect on household food security in Southwest Nigeria. In Proceedings of the Conference on International Research on Food Security, Natural Resource Management and Rural Development, Organised by the Humboldt-Universitat, Berlin, Germany, 16–18 September 2015.
23. Scheiterle, L.; Ulmer, A.; Birner, R.; Pyka, A. From commodity-based value chains to biomass-based value webs: The case of sugarcane in Brazil's bioeconomy. *J. Clean. Prod.* **2018**, *172*, 3851–3863. [CrossRef]
24. Loos, T.; Hoppe, M.; Dzomeku, B.; Scheiterle, L. The Potential of Plantain Residues for the Ghanaian Bioeconomy—Assessing the Current Fiber Value Web. *Sustainability* **2018**, *10*, 4825. [CrossRef]
25. Lin, J.; Gupta, S.; Loos, T.K.; Birner, R. Opportunities and Challenges in the Ethiopian Bamboo Sector: A Market Analysis of the Bamboo-Based Value Web. *Sustainability* **2019**, *11*, 1644. [CrossRef]
26. Neumann, K.; Anderson, C.C.; Denich, M. Food Security in Sub-Saharan Africa: An increase of labor productivity vs. an increase of land productivity. In preparation.
27. Tortoe, C.; Quaye, W.; Akonor, P.T.; Oduro-Yeboah, C.; Buckman, E.S. Analysis of Plantain Biomass-Based Value Chain in Selected Regions in Ghana. *Afr. J. Sci. Technol. Innov. Dev.* **2019**. submitted.
28. Mekonnen, Z.; Worku, A.; Yohannes, T.; Alebachew, M.; Kassa, H. Bamboo Resources in Ethiopia: Their value chain and contribution to livelihoods. *Ethnobot. Res. Appl.* **2014**, *12*, 511–524. [CrossRef]
29. Angelucci, F. *Analysis of Incentives and Disincentives for Maize in Ghana*; Technical Notes Series; MAFAP; FAO: Rome, Italy, 2012; Available online: http://www.fao.org/in-action/mafap/resources/detail/en/c/394295/ (accessed on 16 July 2018).
30. Angelucci, F. *Analysis of Incentives and Disincentives for Cassava in Ghana*; Technical Notes Series; MAFAP; FAO: Rome, Italy, 2013; Available online: http://www.fao.org/in-action/mafap/resources/detail/en/c/394295/ (accessed on 21 July 2018).
31. Ayanwale, A.B.; Fatunbi, A.O.; Ojo, M.P. *Baseline Analysis of Plantain (Musa sp.) Value Chain in Southwest of Nigeria*; FARA Research Report; FARA: Accra, Ghana, 2018; Volume 3, p. 84.
32. Kassahun, T. Review of bamboo value chain in Ethiopia. *Int. J. Afr. Soc. Cult. Tradit.* **2014**, *2*, 52–67.
33. Balié, J.; Nelgen, S. *A Decade of Agriculture Policy Support in Sub-Saharan Africa. A Review of Selected Results of the Monitoring and Analysing Food and Agricultural Policies*; MAFAP: Rome, Italy, 2016.
34. Angelucci, F.; Balié, J.; Gourichon, H.; Mas Aparisi, A.; Witwer, M. *Monitoring and Analysing Food and Agricultural Policies in Africa*; FAO: Rome, Italy, 2013.
35. Neumann, K. 'Know why' thinking as a new approach to systems thinking. *Emerg. Complex. Organ.* **2013**, *15*, 81–93. Available online: https://www.questia.com/library/journal/1P3-3115959041/know-why-thinking-as-a-new-approach-to-systems-thinking (accessed on 8 January 2018).
36. The World Bank, World Bank National Accounts Data. *GDP Growth (Annual %)*. *2017*. [Data File]. Available online: https://data.worldbank.org/indicator/ny.gdp.mktp.kd.zg (accessed on 20 July 2018).
37. Economist Intelligence Unit. *Global food Security Index*; Economist Intelligence Unit: London, UK, 2017; Available online: https://foodsecurityindex.eiu.com/ (accessed on 10 May 2018).
38. FAO. *FAO Statistical Yearbook 2014: Africa Food and Agriculture*; FAO: Accra, Ghana, 2014; Available online: http://www.fao.org/economic/ess/ess-publications/ess-yearbook/en/#.W60jMmNoSUk (accessed on 11 June 2018).
39. National Bureau of Statistics. *Nigerian Gross Domestic Product Report: Q4 and Full Year 2017*; National Bureau of Statistics: Abuja, Nigeria, 2017. Available online: http://nigerianstat.gov.ng/elibrary (accessed on 22 July 2018).

40. Ministry of Food and Agriculture (MoFA). *Agriculture in Ghana. Facts and Figures 2015*; Ministry of Food and Agriculture: Statistics Research and Information Directorate (SRID): Accra, Ghana, 2016. Available online: http://mofa.gov.gh/site/# (accessed on 3 April 2018).
41. Neumann, K.; Anderson, C.; Denich, M. Participatory, explorative, qualitative modeling: Application of the iMODELER software to assess trade-offs among the SDGs. *Econ. Open-Access Open-Assess. E-J.* **2018**, *12*, 1–19. [CrossRef]
42. Akinwale, A.A. The menace of inadequate infrastructure in Nigeria. *Afr. J. Sci. Technol. Innov. Dev.* **2010**, *2*, 207–228.
43. Tadesse, G.; Shively, G. Food aid, food prices, and producer disincentives in Ethiopia. *Am. J. Agric. Econ.* **2009**, *91*, 942–955. [CrossRef]
44. OECD/UNDP/AfDB. *African Economic Outlook 2014: Global Value Chains and Africa's Industrialization*; Organisation for Economic Co-operation and Development, United Nations Development Programme, African Development Bank: Tunis, Tunisia, 2014. [CrossRef]
45. United Nations Economic Commission for Africa. *Urbanization and Industrialization for Africa's Transformation: Economic Report on Africa*; United Nations: Addis Ababa, Ethiopia, 2017; Available online: https://www.uneca.org/publications/economic-report-africa-2017 (accessed on 5 May 2018).
46. Alliance for a Green Revolution in Africa (AGRA). *Business Pathways to the Future of Smallholder Farming in the Context of Transforming Value Chains*; International Food Policy Research Institute Africa Agriculture Status Report 2017; AGRA: Nairobi, Kenya, 2017; Available online: https://agra.org/aasr2017/chapter-2/ (accessed on 29 June 2018).
47. African Union Commission. *Implementation Strategy and Roadmap to Achieve the 2025 Vision on CAADP*; African Union: Addis Ababa, Ethiopia, 2014; Available online: http://www.nepad.org/resource/implementation-strategy-and-road-map-achieve-2025-vision-caadp (accessed on 18 July 2018).
48. New Partnership for Africa's Development (NEPAD). *Agriculture in Africa: Transformation and Outlook*; NEPAD: Pretoria, South Africa, 2014; Available online: http://www.nepad.org/resource/agriculture-africa-transformation-and-outlook (accessed on 13 July 2018).
49. United Nations Economic Commission for Africa. *Economic Report on Africa 2013: Making the Most of Africa's Commodities: Industrializing for Growth, Jobs and Economic Transformation*; United Nations: Addis Ababa, Ethiopia, 2013; Available online: https://repository.uneca.org/handle/10855/22095 (accessed on 7 July 2018).
50. United Nations Economic Commission for Africa. *Assessing Regional Integration in Africa VI: Harmonizing Policies to Transform the Trading Environment*; United Nations: Addis Ababa, Ethiopia, 2013; Available online: https://repository.uneca.org/handle/10855/22176 (accessed on 7 July 2018).
51. United Nations Economic Commission for Africa. *Economic Report on Africa 2015—Industrializing through trade. Addis Ababa*; United Nations: Addis Ababa, Ethiopia, 2015; Available online: https://repository.uneca.org/handle/10855/22767 (accessed on 7 July 2018).
52. Malabo Montpellier Panel. *Mechanized: Transforming Africa's Agriculture Value Chains*; Malabo Montpellier Panel: Dakar, Senegal, 2018; Available online: https://www.mamopanel.org/resources/reports-and-briefings/mechanized-transforming-africas-agriculture-value-/ (accessed on 2 July 2018).
53. African Development Bank. *African Development Report 2015: Growth, Poverty and Inequality Nexus: Overcoming Barriers to Sustainable Development*; African Development Bank: Abidjan, Côte d'Ivoire, 2016; Available online: https://www.afdb.org/en/documents/document/african-development-report-2015-growth-poverty-and-inequality-nexus-overcoming-barriers-to-sustainable-development-89715/ (accessed on 19 May 2018).
54. African Development Bank. *African Economic Outlook 2018*; African Development Bank: Tunis, Tunisia, 2018; Available online: https://www.afdb.org/en/knowledge/publications/african-economic-outlook/ (accessed on 21 July 2018).
55. Badiane, O.; Ulimwengu, J. Africa Agriculture Status Report 2017. 2017. Available online: https://agra.org/aasr2017/ (accessed on 28 July 2018).
56. Benin, S. Impact of Ghana's agricultural mechanization services center program. *Agric. Econ.* **2015**, *46*, 103–117. [CrossRef]
57. FAO. *Mechanization for Rural Development: A Review of Patterns and Progress around the World*; Integrated Crop Management: Rome, Italy, 2013; Volume 20.

58. OECD/FAO. *OECD-FAO Agricultural Outlook 2016-2025: Special Focus: Sub-Saharan Africa*; OECD Publishing: Paris, France, 2016. [CrossRef]
59. Von Braun, J. Addressing the food crisis: Governance, market functioning, and investment in public goods. *Food Secur.* **2009**, *1*, 9–15. [CrossRef]
60. Carus, M.; Dammer, L. The Circular Bioeconomy—Concepts, Opportunities, and Limitations. *Ind. Biotechnol.* **2018**, *14*, 83–91. [CrossRef]
61. Intergovernmental Panel on Climate Change (IPCC). *Climate Change 2014—Impacts, Adaptation and Vulnerability: Food Security and Food Production Systems*; Cambridge University Press: Cambridge, UK, 2014; A Report of the Intergovernmental Panel on Climate Change; IPCC: Cambridge, UK, 2014; Available online: http://www.ipcc.ch/report/ar5/wg2/ (accessed on 22 July 2018).
62. Wheeler, T.; Von Braun, J. Climate change impacts on global food security. *Science* **2013**, *341*, 508–513. [CrossRef] [PubMed]
63. Poulton, C.; Kydd, J.; Dorward, A. Overcoming market constraints on pro-poor agricultural growth in Sub-Saharan Africa. *Dev. Policy Rev.* **2006**, *24*, 243–277. [CrossRef]
64. Hilson, G. Small-scale mining, poverty and economic development in sub-Saharan Africa: An overview. *Resour. Policy* **2009**, *34*, 1–5. [CrossRef]
65. Reichl, C.; Schatz, M.; Zsak, G. World mining data. *Miner. Prod. Inter-Natl. Organ. Comm. World Min. Congr.* **2014**, *32*, 1–261.
66. De Graaff, J.; Kessler, A.; Nibbering, J.W. Agriculture and food security in selected countries in Sub-Saharan Africa: Diversity in trends and opportunities. *Food Secur.* **2011**, *3*, 195–213. [CrossRef]
67. Scheiterle, L.; Birner, R. Gender, knowledge and power: A case study of Market Queens in Ghana. In Proceedings of the 2018 Annual Meeting, Washington, DC, USA, 5–7 August 2018; No. 274125. Agricultural and Applied Economics Association: Washington, DC, USA, 2018.

 © 2019 by the authors. Licensee MDPI, Basel, Switzerland. This article is an open access article distributed under the terms and conditions of the Creative Commons Attribution (CC BY) license (http://creativecommons.org/licenses/by/4.0/).

sustainability

Article

Opportunities and Challenges in the Ethiopian Bamboo Sector: A Market Analysis of the Bamboo-Based Value Web

Jessie Lin [1,*], Saurabh Gupta [2], Tim K. Loos [3,4] and Regina Birner [4]

[1] Department of Agricultural Economics and Rural Development, University of Goettingen, 37073 Goettingen, Germany
[2] Centre for Development Policy and Management, Indian Institute of Management, Udaipur 313001, India; saurabh.gupta@iimu.ac.in
[3] Food Security Center, University of Hohenheim, Wollgrasweg 43, 70599 Stuttgart, Germany; timloos@uni-hohenheim.de
[4] Institute of Agricultural Sciences in the Tropics (Hans-Ruthenberg-Institute), University of Hohenheim, Wollgrasweg 43, 70599 Stuttgart, Germany; regina.birner@uni-hohenheim.de
[*] Correspondence: jessie.lin@uni-goettingen.de; Tel.: +49-551-39-20208

Received: 28 September 2018; Accepted: 13 March 2019; Published: 19 March 2019

Abstract: Bamboo is one of the more important natural resources in Ethiopia and contributes to the bioeconomy as a potential source for high-value products. While the country is the largest producer of bamboo in Africa, the existing utilization of the bamboo sector in Ethiopia remains under-developed, with little value addition. This study identifies the current market challenges and opportunities for future developments of the northern Ethiopian bamboo sector, with a focus on the Injibara township. This research adopts the "value web" approach to assess the potentials of different product lines that create the bamboo biomass value web. We utilize qualitative data collection methods, in particular, semi-structured interviews and informal focus group discussions with key stakeholders. Our findings suggest that bamboo farmers in Injibara are constrained by a lack of local demand and market for bamboo products with high-value addition, leading to an absence of product diversification and innovation. Furthermore, there is an overreliance on foreign technology and methods that are poorly matched for local needs. We recommend that policymakers invest in targeted and effective training strategies on bamboo cultivation and processing. Furthermore, farmers can benefit from decreasing their reliance on middle men with cooperatives or contract arrangements.

Keywords: biomass; value web; bioeconomy; bamboo; Ethiopia

---

1. Introduction

Throughout the history of time, the production of food, energy, feed, and other environmental goods has largely depended on both agricultural and natural resources [1]. In many developing countries today, an increasing population, changing diets, and urbanization place immense pressure on these resources [1]. For this reason, increasing attention has been given to what is coined the "bioeconomy." This term can be defined as an economy in which its basic structure for materials, chemicals, and energy originates from regenerated biological materials from plant or animal sources [2]. When properly implemented, a bioeconomy is able to satisfy many conditions for a sustainable economic and social environment [2].

There are many drivers for the increasing development of a bioeconomy, including rising challenges to respond to issues like environmental degradation, energy, and food security [3]. Both the challenge and advantage for many sub-Saharan African countries, such as Ethiopia, is to transform biomass resources into products with multiple usages and make use of that extra share of added value [4]. It is expected that through the process of value addition in biomass resources, opportunities for further employment and an increase in income will arise [4].

This study addresses these issues by exploring the marketing opportunities and constraints in the northern Ethiopian highland bamboo sector. We seek to fill this knowledge gap in this context by analyzing the Ethiopian highland bamboo biomass value web. The organization of this article is as follows. We start with a general background of bamboo and its utilization in Ethiopia. Section two presents the conceptual framework that guides the analysis of this study. Section three details our methodology. The results and main bottlenecks are identified and presented in section four and we give conclusions and policy recommendations for the sustainable development of the northern Ethiopian bamboo sector in section five.

1.1. Background

Ethiopia is one of the poorest countries in the world. It has been constantly plagued with problems such as food insecurity and seasonal vulnerabilities. Although poverty reduction rates have been impressive in the last few decades in this East African nation, around 30 percent of the population still lives below the national poverty line, surviving on less than US$1.25 Purchasing Power Parity (PPP) per day [5]. With a growing population and rising urbanization, Ethiopia, like many other countries in Sub-Saharan Africa, increasingly faces problems related to energy scarcity and food insecurity.

Many rural Ethiopian households depend on non-timber forestry products (NTFP), through their provision of goods and services, as a significant part of their livelihood [6]. However, there has been an annual acute shortage of forestry products due to an expected increase in the demand for lumber and other wood products [7]. This occurrence has brought bamboo into the spotlight as a possible replacement for wood for a variety of usages and purposes. Bamboo is identified as having great potential for poverty reduction, environmental enhancement, and sustainable economic development [8]. Due to population growth and increasing urbanization, forest resources, in particular wood, are facing growing pressure for harvest [9]. Bamboo has increasingly replaced wood for industrial usages throughout the last few decades, thereby lessening the pressure for forest clearings [9].

Ethiopia has the largest resource of bamboo in Africa, estimated at around one million hectares, accounting for 67 percent within the continent and seven percent of the world total [10]. There are two types of bamboo in Ethiopia: highland and lowland bamboo. Highland bamboo, which our study centers on, accounts for around 15 percent of bamboo coverage and is mostly cultivated on a smallholder's own land, whereas lowland bamboo grows in natural groves and forests in the southern parts of Ethiopia. Despite the abundant quantities, bamboo has traditionally been treated as an undervalued resource in Ethiopia in comparison to other forest resources [8]. In contrast, bamboo is a highly valued resource in many Asian countries and has been transformed into more than 1500 products worldwide [10]. Value-addition in bamboo contributes much to the international bamboo market and trade. It is estimated that by 2015, the world market for commercialized bamboo products reached $20 billion US dollars [11].

In 2009, the United Nations Industrial Development Organization (UNIDO) presented the Bamboo Sector Strategy Framework to the government of Ethiopia with the purpose of strengthening the bamboo sector within a five-year period. Several years have passed, yet the status of the sector remains rudimentary. Little has changed for farmers and communities whose livelihoods depend on bamboo. The absence of an enabling environment and ineffective governmental and institutional efforts remain the main challenges to the development of the sector.

1.2. Literature Review

To date, only a few studies have looked into the bamboo sector in Ethiopia. Kelbessa et al. found that bamboo is an important commodity, both environmentally and ecologically, for producing regions [12]. At the same time, it is seen as a "poor man's wood" [7]. The present utilization, marketing, and trade of bamboo products remain under-developed, with minimal contribution to the economy. It is limited to the construction of huts, fencing, furniture, and such [13]. Many households only engage in bamboo production for additional household revenues [12]. The participation of actors in the bamboo value chain remains at a low level of knowledge, skills, and value-addition [7,8]. The current market and demand in Ethiopia for hand-made bamboo products is minimal, and there are few existing enterprises that specialize in processing bamboo furniture with higher value [14].

Despite these limitations, the market for hand-made and highly processed bamboo products has shown positive signs of growth. In a 2014 survey, 85 percent of respondents mentioned a rising demand for bamboo products [7]. There is a big potential for the bamboo market to grow further. These are reasons for advancements and developments in the utilization and marketing opportunities of bamboo in Ethiopia. There have been efforts within the last couple of decades to develop the bamboo sector by numerous international organizations, non-governmental organizations (NGOs), and the government of Ethiopia. Despite these efforts, the performance of the bamboo market remains rudimentary and limited. Little has changed for farmers and communities whose livelihoods depend on bamboo. Due to the natural characteristics and the possibilities for high-value-added production of bamboo, smallholder farmers and producers in Ethiopia can greatly benefit from improved utilization and marketing of bamboo. However, the existing marketing structure of the bamboo sector is informal and there are no connections, coordination, and integration among the product markets and sources at both the local level and at central markets; in addition, the participation of actors in the bamboo value chain remains at a low level of knowledge and skills, with little value addition [7,8].

The aim of this research is to assess the opportunities and constraints of the northern Ethiopian bamboo sector, with a focus on the area of Injibara. We utilize the value web concept to better assess the various dimensions of highland bamboo-based biomass in the north of Ethiopia. We aim to fill the knowledge gap by analyzing the marketing and policy constraints along the biomass value web. This comprises the potentials, relations, and linkages of various stakeholders, including farmers, traders, enterprises, officials, policymakers, and international research and development agencies. The main bottlenecks are identified, and policy recommendations are given for a more sustainable development of the highland bamboo sector.

2. Conceptual Framework

In this section, we present the concepts that guide the study. We first introduce the idea of the biomass value web in the context of bamboo. Then, the conceptual framework for the analysis of this study is presented.

Biomass Value Web

Biomass is defined as a natural substance that is derived from either living or recent living organisms [4]. It can be categorized into food, feed, energy sources, and raw materials [4]. Countries that depend on agriculture can benefit from an increased demand for biomass, especially when there are high levels of value-addition to the biomass raw product in labor-intensive processing sectors [15]. Bamboo is classified as "lignocellulosic biomass"; its composition is higher in inorganic makeup than ash and wood [11]. It contains high levels of biomass productivity and is one of the most sustainable in terms of energy and materials production [11]. This is important for Ethiopia since around 90 percent of the primary energy source comes from biomass [16].

There are many complexities involved when exploring biomass-based economic growth. For many agricultural crops, the value chain approach is suitable. This method details the range of activities involved in bringing a product from its production to consumption [17]. However, in order to address the multiple usages and flow of a biomass resource, such as bamboo, we utilize the biomass value web in order to address these complexities and constraints [4]. The value web provides a holistic approach for the purpose of this study, as it applies a multi-dimensional web perspective to assess the linkages and relationships among the numerous value chains involved and how each is governed [4].

This approach is appropriate for further understanding the interactions amongst the value chains, pinpointing deficiencies in the use of biomass, and identifying the potential for growth in the entire biomass-based value web [4]. To date, there have been studies that have used the value web concept to analyze the opportunities and challenges of the sugarcane case in Brazil [18] and the potential of plantain residues in Ghana [19]. Figure 1 shows a visualization of the value web concept. It has been recreated to incorporate bamboo as a biomass-based product in the value web.

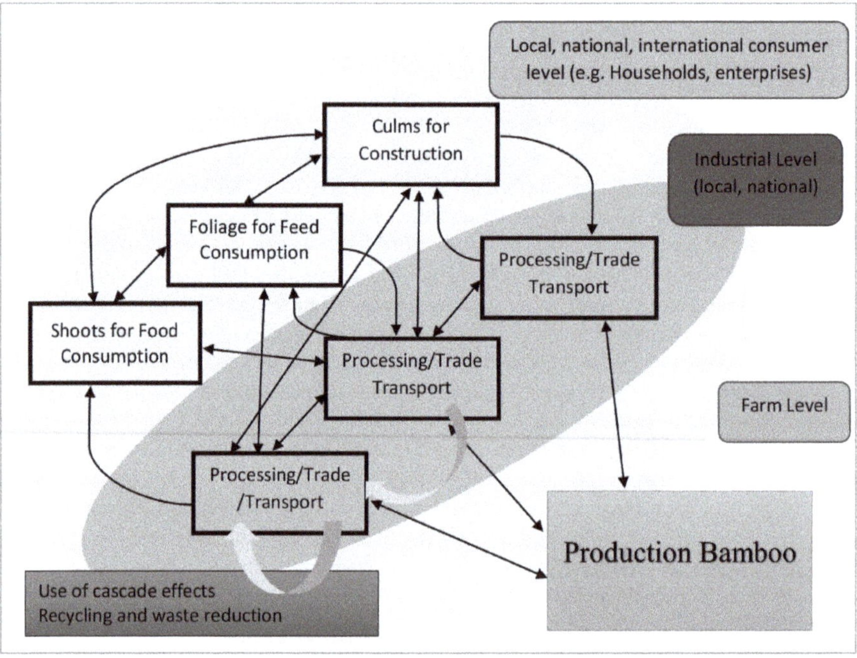

Figure 1. Biomass-based Value Web Concept, recreated for bamboo; Source: adaptation from Virchow et al. [15].

To address and analyze the numerous usages and various challenges of the bamboo sector in northern Ethiopia, we have constructed a conceptual framework, as shown in Figure 2. The framework considers the institutional and marketing aspects of bamboo in this area.

The framework demonstrates the complex process within the bamboo biomass system. First, the production of raw materials goes through the institutional setup of the country before they can be transformed into finished products. The three major outputs mentioned are (1) construction and fencing, (2) furniture and handicrafts, and (3) energy. There are market and institutional challenges within the institutional setup of a country. These aspects influence the cultivation, processing, and trade of bamboo. These three features lead to the production, sale, and utilization of the three main outputs. They also determine the marketing and development of the bamboo sector. Further and more advanced development would upgrade and add value to the three main outputs.

We assess our framework through socio-economic and environmental factors. They take into account the following:

- Socio-economic factors: knowledge regarding the benefits of bamboo, market accessibility, demands for bamboo resources, access to advanced technology, cost of raw culms and such;
- Environmental factors: location of markets, availability of bamboo raw material, seasonal and weather patterns.

These together can identify the entry points to the unexplored potential of bamboo, and the development of the market.

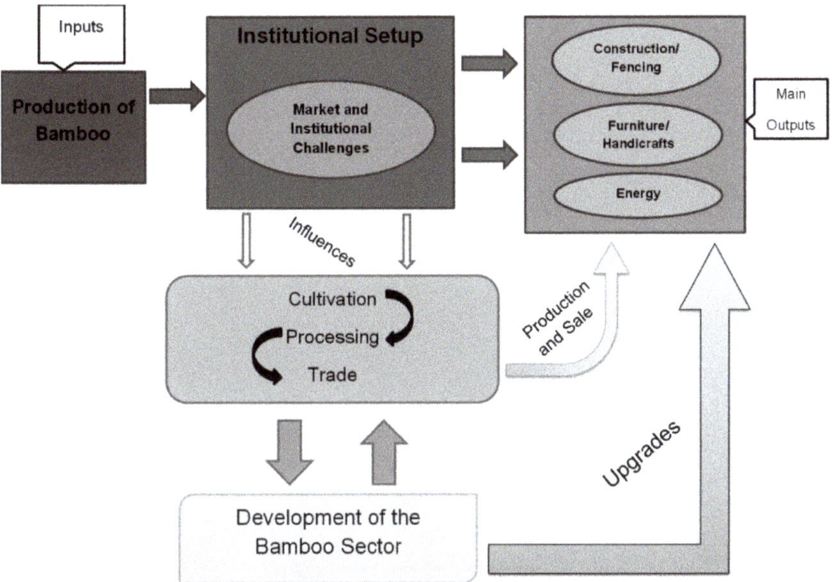

Figure 2. Conceptual Framework of the Marketing Challenges in the Ethiopia Bamboo-based Value Web; Source: own illustration.

3. Methodology

This section presents the research area and data collection methods utilized in this study.

3.1. Research Area

This research explores and investigates key institutions and stakeholders involved in the northern Ethiopian bamboo sector. Interviews and visits with key institutional informants and stakeholders were conducted in the capital city of Addis Ababa. This is where international organizations, research centers, and governmental offices are located. Visits were also made to bamboo workshops, processors, retailers, and factories to better understand the process of marketing, production, and trade, and the diverse products produced, as well as to compare the price differences.

Fieldwork and visits with farmers and small handicraft processors were limited to the town of Injibara and its surrounding regions. The township of Injibara is located in the northern highland of Ethiopia, situated 447 km north of the capital city, Addis Ababa, and 118 km south of Bahir Dar, another major city in Ethiopia. This region receives around 2500 mm of annual rainfall. Due to its high altitude location and climate, the major crops that grow in this area are predominately teff, barley, peas, and potatoes. Injibara is well-connected to nearby towns and markets, with easy access to transportation and well-paved roads for the movements of products from production to consumption [20]. Figure 3 shows a map of the Amhara region. The study area is identified with a red circle.

We have chosen Injibara as the study area since it is one of the major growing regions of highland bamboo in Ethiopia. Bamboo also plays a main economic role in the communities' livelihoods.

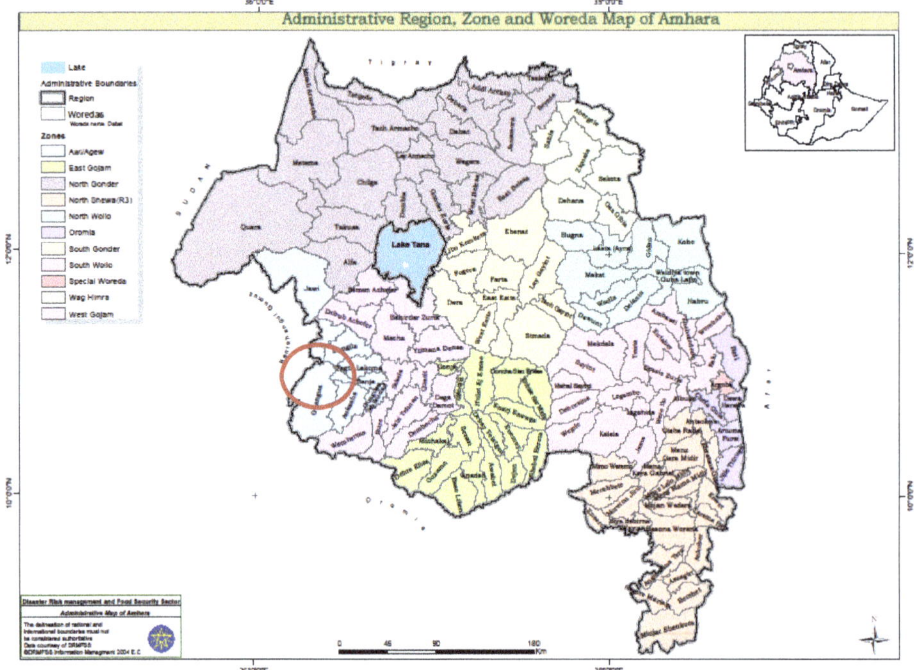

Figure 3. Map of Study Region; Source: DRMFSS (2005).

3.2. Data Collection Method

Qualitative data collection methods were used for the purpose of this study, including semi-structured interviews and informal focus group discussions. In addition, to further complement the bamboo value-web, secondary data were obtained by reviewing published literature, reports by international organizations, and University theses that have been written on the bamboo sector.

Semi-Structured Interviews

Semi-structured interviews were conducted with major institutional actors in the bamboo sector. This interview method mostly involves general questions and becomes the basic structure for other detailed questions which have not been prepared beforehand [21]. Many questions are created on the spot, during the interview, which allows for greater flexibility for both the interviewer and the interviewee [21].

The goal was to explore the different aspects of the bamboo sector with the interviewees in order to gain a better understanding of some specific challenges and other factors that were not foreseen in advance. Various institutions were visited to gain perspectives from different sides. Table 1 lists the institutions that participated in semi-structured interviewees and their respective functions.

Table 1. Institutions that Participated in Semi-Structured Interviews.

Institution	Function
Ministry of Environmental and Forest Research Institute	Governmental
Wood Technology Research Center	Senior Researcher
FEMSEDA	Trainer
GIZ SLM	Deputy Program Manager
Addis Ababa University	Research
Hope College	Research
INBAR	Team Leader
INBAR	Training Coordinator
INBAR	Manager, Development Project
INBAR	National Project Coordinator
ADAL Industrial Plc	Deputy Manager
African Bamboo Factory	General Manager
Ethiopian Tourist Trade Enterprise	Employee

Table 2 lists the number of farmers, small processors, and retailers who took part in semi-structured interviews. It is important to note that many smallholders in Injibara engage in both the farming and processing of bamboo. In this case, they are accounted for in both categories. All of the small processors interviewed in Addis Ababa and Bahir Dar also retail their own products on-site, and are thus also categorized as retailers.

Table 2. Number of farmers, entrepreneurs, and retailers who participated in semi-structured interviews.

Location	Total # Interviewed [1]	# of Small Handicraft Processors	# of Retailers	# of Farmers
Addis Ababa	6	5	6	0
Bahir Dar	3	3	3	0
Injibara and surroundings	14	11	0	9
Total	23	19	9	9

[1] Some interviewees in different categories are counted as one in the total number interviewed.

4. Results

This section highlights the findings of our study. We first present the biomass-based value web to account for the multifunctional aspects of bamboo. Then, we specify the marketing challenges along the value web.

4.1. Bamboo Biomass Value Web Flow Map

Information regarding the bamboo biomass value web was obtained in order to look at the entire Ethiopian sector and as a whole and the actors who are involved. Figure 4 details the different uses of bamboo. As described in the methodology section, our data was compiled through field visits, interviews, and secondary data. We have contrasted the usage of highland and lowland bamboo in the value web, because of their structural differences and their suitability for various utilizations.

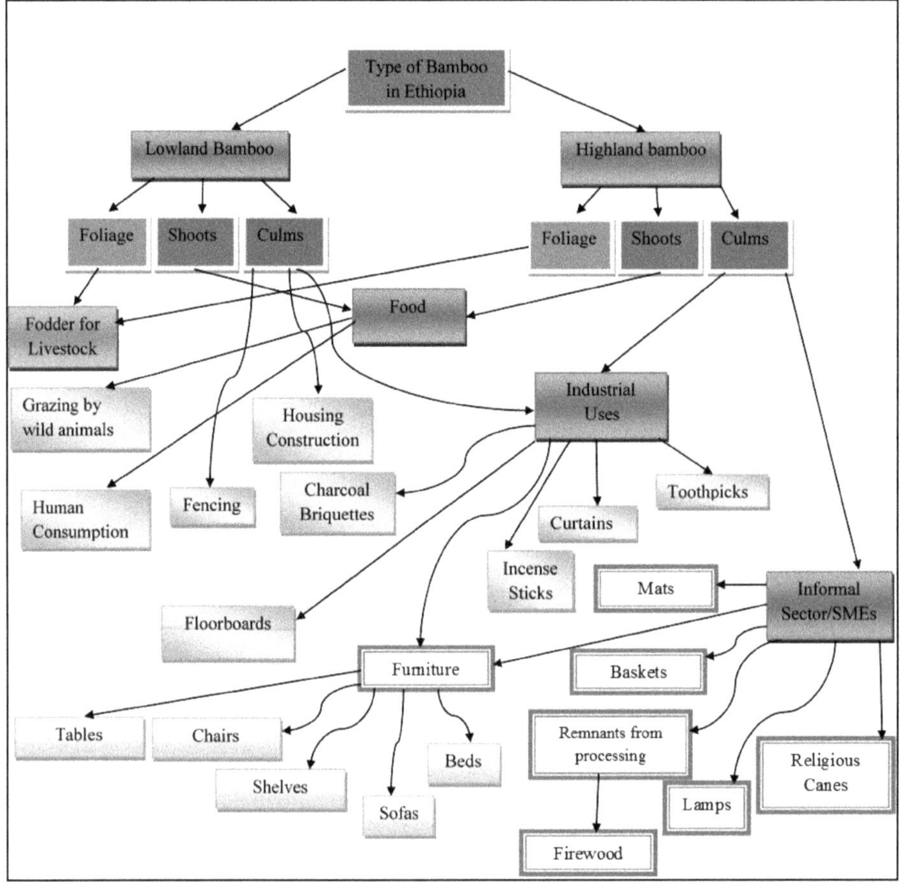

Figure 4. Bamboo-Based Value Web in Ethiopia; compiled from Interviews and Literature Research; Source: own research.

The bamboo value web shows three useable parts of bamboo. They consist of the culm, shoot, and foliage or leaves. There is a clear knowledge gap regarding the extent of usage in the shoots and foliage. The consumption of bamboo shoots by humans remains at a very minimal level and there is a lack of awareness of its edibility and nutritional value. Very few existing studies mention the human consumption of bamboo shoots. An interviewee commented that wild animals consume bamboo shoots in the forests rather than humans (Private Farmer). There is some research concerning the usage of bamboo foliage and leaves as fodder for livestock animals. This was not revealed during the field interviews in this study. The prevalence and details of this usage remain unclear.

The major utilization and potential for economic activities of bamboo come from its culm. Many of the products include items with minimal value addition, such as small handicrafts, basic furniture, mats, and construction. These productions usually involve manual labor and are made by small processors in both cities and rural areas. Products with high-value addition include floorboards, decorative furniture, curtains, and incense sticks. These are produced by factories or by craftsmen who have received training and have access to tools and machinery.

Lowland bamboo predominately grows in nature and has a larger area coverage in Ethiopia. It is predominately used in the southern regions of the country for housing construction and fencing. This type of bamboo is also more suitable for the production of bamboo charcoal briquettes. However, this potential remains largely untapped. Highland bamboo is for the most part homegrown by smallholder farmers as an additional activity on top of agricultural farming activities. Many smallholders use this species to make mats, furniture, and other more delicate handicraft items for commercial purposes. In addition, highland bamboo is used more for furniture than lowland bamboo.

In small productions, the scraps and remnants from bamboo production are either used by farmers or craftsmen themselves as a source of firewood, or they are left at the working site for people in the community for collection. One craftsman commented that he reuses residuals to fix holes and mend other disfigurements on furniture items (Production employee, private firm). The actual extent of residual usage, however, is unclear. At the ADAL factory, the residuals from bamboo production are used to produce charcoal (Production manager, private firm).

There is a lack of knowledge on the management of bamboo throughout the value web. For example, the timing of bamboo harvest is crucial in determining the quality of culms. Many farmers harvest when they need income. In addition, without proper storage, especially in the rainy season, culms become prone to insect attacks. The suitability of bamboo to make products, such as furniture, also differs by the age of the culm. For example, younger bamboo contains a higher starch and glucose content than mature bamboo, leading to pest infestations (University Professor, academia). These occurrences often lead to inferior quality of the end product.

The biomass value web demonstrates the diverse possibilities of the usage and flow of bamboo. This confirms that further efforts in marketing and policy-making must address and assess the usage and flow of bamboo in Ethiopia.

4.2. Marketing Challenges for Small Farmers and Processors

From the semi-structured interviews and focus group discussions, we were able to identify the marketing challenges in the northern Ethiopian bamboo sector. As mentioned, only the highland bamboo species is found in Injibara and its surrounding areas. Unlike lowland bamboo, which grows naturally in the forest in the Assossa region of Ethiopia, highland bamboo is mostly planted, cultivated, and harvested on the homestead by smallholder farmers. In Injibara, not only do smallholders engage in bamboo planting, but many in towns and areas in its vicinity also take part in making rudimentary bamboo products. The products are primarily traditional furniture, such as benches and stools, and also bamboo mats. Only one entrepreneur we encountered produces a different product, bamboo canes, which are used in religious events and ceremonies. Due to the lack of machines and tools required to make the products, the production process relies largely on manual labor. Entrepreneurs in Injibara have a limited market base. Bamboo products are seen as inferior compared to other timber products; therefore, people are unwilling to pay a higher premium. The quality of bamboo products is inconsistent due to the lack of knowledge on the timing of harvest. There are certain times when bamboo should not be harvested during the year. However, because of this lack of knowledge, many farmers harvest bamboo, even when it is not suitable. The government puts the focus on value addition, skill management, and training in the development of the bamboo sector (Government Official), but little attention is given to the cultivation and propagation (University Professor, academic sector). In addition, bamboo needs proper treatment and drying time after harvest to better prevent problems, such as insect infestation. The lack of adequate storage facilities and space exacerbates this problem. These are major challenges that might have been overlooked by organizations that provide merely skills training to the craftsmen.

Both the Ethiopian Federal Micro and Small Enterprises Development Agency (FEMSEDA) and the International Network of Bamboo and Rattan (INBAR) provide skills training to bamboo craftsmen to make more "modern" and advanced furniture with decorations that require higher levels of processing. However, customers, especially those in rural areas, are not willing to pay a high price premium for products made out of bamboo. Since these products have no market, many craftsmen go back to making the basic traditional furniture with little value addition. Even though the profit margin is much lower for traditional furniture, at least they are able to sell a number of them to obtain some sort of income. Furthermore, one could argue that trainings targeted at value-addition are only for show. For example, farmers are trained in facilities with machines and electricity. However, when they return home in rural areas, the equipment is not accessible to them.

Table 3 gives an example of the price differentiation and daily profits between a rural area and a city close by.

Table 3. Price Comparison of a Traditional Bamboo Bench Between Injibara and Bahir Dar (City).

Place	Price per Culm	# of Culms per Bench	Price per Bench	Total Profit	# of Benches Produced per Day	Estimated Profit per Day
Injibara	~25 birr	2	60 to 70 birr	10 to 20 birr	3 to 4	30 to 80 birr
Bahir Dar	~25 birr	2	Up to 120 birr	Up to 70 birr	3 to 4	Up to 280 birr

In contrast to the usages of bamboo in rural townships, increasingly, in Addis Ababa, there are workshops that produce bamboo products with higher value addition. In the capital city, there is a much broader market base, due to foreigners who have higher incomes and higher regards for bamboo products. The Ethiopian Tourist Trading Enterprise (ETTE) is a government-supported institution that engages in bamboo production. There, workers produce highly decorated furniture and small ornamental products. They are sold at much higher prices. When asked, the employee stated that most customers who buy bamboo products are foreigners and not local Ethiopians (Production Employee, public firm). Established and formal means of marketing bamboo products seem to not exist for both the northern countryside and cities. In all three sites visited during field work, most bamboo products were both produced and sold on the same premise. In Injibara, customers walk pass production sites and buy products directly. Some processors in the cities receive orders from their customers. For example, in Addis Ababa, areas near the church of Urael and the Kazanchis area are known for bamboo production. Customers know beforehand that bamboo products are being produced and sold there. The more skilled small- and medium-sized enterprises have a photo album of the type of products available, in addition to a few pieces of furniture on display. A good portion of the production is done by order. Customers can pick out items from the photo album and place their orders. At the same time, there remains a lack of a proper marketing channel for bamboo products. There are no investments in small enterprises and rural handicrafts beyond basic skill training. Even with good quality finished end products, there is no place to sell them. This limits the potential for the sector for further growth and opportunities.

5. Discussion

This section discusses and analyzes the marketing challenges and constraints throughout the bamboo value web as presented in the results section.

5.1. Socio-Economic Challenges

Our results confirm what Kelbessa et al. [12] describe in their study; areas covered with bamboo play an important role in the livelihoods of the local communities. At the same time, there is a lack of awareness regarding the proper cultivation management of highland bamboo at both the farm and market level, as Mulatu and Kindu [22] also note in their research.

As previously noted, the quality of bamboo products largely depends on the harvest time and proper storage of culms. These factors combined can severely impact the quality of the culms and the products that come out of it. Many of the bamboo retailers in Addis Ababa have mentioned customer complaints due to insect problems. This paints a negative image for the already under-valued bamboo products and serves as a major constraint for the development of the bamboo market.

The majority of smallholder farmers and processors lack means for transportation and access to the market. Their only option is through middlemen or traders. They are also the ones who can obtain licenses to sell finished products. This reliance, combined with a lack of product diversification in concentrated cluster areas, gives them very limited bargaining power in terms of pricing for raw bamboo and for finished bamboo products. If one refuses to sell at a lower price, then the middlemen have many different options in the same area. Throughout the value chain, it is the middlemen who have the most to gain. As a result, many smallholders end up receiving income that can only sustain their livelihoods at a bare minimum.

Bamboo products are seen as inferior by the Ethiopian society in general. This is partly exacerbated by the problem of the inappropriate storage of bamboo culms. For products with high value and labor addition, people are unwilling to pay a higher premium. This brings up an important point that skill training has little effect when there is a very minimal demand and market for more expensive bamboo products.

Because of these types of challenges, there is no incentive for local small processors to innovate new bamboo products for the market. For example, we asked why locals do not produce kitchen utensils out of bamboo, as many Asian countries do. The response was that it simply is not in their culture to do so, and no one has ever thought about it (Production Employee, private firm). The absence of a proper market leads to a lack of incentives to innovate.

One of the biggest differences that has been found in terms of pricing of bamboo products is the location of the products being sold. As shown in the results section, the same bamboo bench sold in the city of Bahir Dar can bring in twice as much as income as a bench sold on the roadside of Injibara. The high value-added furniture in the ETTE is sold at a high markup price because they have access to the expatriate market in Addis Ababa. As foreigners tend to prefer bamboo products more than the local population, it will be interesting to see how the bamboo market will be altered as the expat community grows in Addis Ababa.

Another factor is related to the seasonal variabilities in bamboo cultivation, process, and trade. First, because bamboo requires additional time for drying during the rainy season, space becomes a big issue for small processors as they have limited space for storing excess culms. Demand for bamboo products decreases in the rainy season while the price of raw bamboo culms increases. Theoretically, an increase in raw culm price should benefit the bamboo farmers. However, harvest is much more difficult in the rainy season, and several farmers also remarked that they do not cut bamboo during the rainy season. At the same time, the increase in culm pricing strains the small handicraft sectors. For families, whose entire livelihood largely depends on bamboo, the rainy season becomes especially difficult for them.

It is also a major constraint there are no factories set up in places where bamboos are cultivated and harvested. It is currently difficult to set up factories near bamboo forests. Problems, such as a lack of electricity and inconvenience for factory owners, make it unappealing to have an operation on site. In addition, there is a lack of policy support for the investments of these purposes. If bamboo factories were actually set up near the resources, then this would cut out the middlemen problem, with possibly more income going directly to the farmers. Factories would also have more capacities to store and dry the bamboo culms after harvest.

5.2. Main Challenges Throughout the Bamboo Value Web

There is a clear lack of diverse bamboo products on the market. In Injibara, identical products are being produced in clusters. The same types of products that require similar skills are produced in the same area. For example, on the same road that stretches for many kilometers, mats are the only products available. Once in a while, craftsmen receive special product orders from customers. For the most part, traders come and collect the mats to be sold in other markets and cities. Because smaller processors have no means of transporting the finished mats elsewhere, they depend on the traders to provide them with their source of income. If traders do not come to collect the mats, then they have limited access to other means of selling or they do not sell at all. Since everyone is producing the same product, the middlemen have much more bargaining power to negotiate lower prices. If one family is unwilling to sell at a certain price, the traders still have many other people from whom they can collect the mats. In this case, it is hard to foresee how the livelihoods of families can improve dramatically without a means of transport to markets. According to Mulatu and Kindu [22], the experiences of the utilization of bamboo in many countries have resulted in significant financial gains and environmental protection. They note that in order to have sustainable bamboo utilization, there must be "a function of bamboo resource development availability of new technologies and scientific information, production of bamboo products using the technologies, and marketing" (p. 80). A combination of suitable institutions, policies, and strategies must be implemented in order for these to occur [22]. It is clear from this value web analysis that there is much room for improvement in each of these façades.

Policymakers and international institutions involved in training and research should put the focus on bamboo propagation in addition to value-addition. While the utilization of bamboo is an important aspect to develop the sector, it is also crucial for bamboo farmers to obtain adequate cultivation skills and knowledge in order to avoid low quality bamboo for production. Better managed training can facilitate this process.

Governments can assist bamboo processors in Injibara to gain markets by setting up government bazaars. Awareness campaigns can be a useful strategy to make the public more knowledgeable about bamboo products. Setting up cooperatives could be an option for farmers and craftsmen to gain stronger bargaining power in the process of trade and access to markets.

6. Conclusions

This study focused on the current utilization and future potential of the bamboo market in Injibara, Ethiopia, within the framework of the biomass value web in the forthcoming bioeconomy. As our research suggests, there is a set of complex and multifaceted hindrances that has kept the market for bamboo in northern Ethiopia under-developed, despite the many efforts that have been made in the last decade. There have been some improvements and signs for a positive outlook in the future.

For local consumers, there is a general lack of demand and market for products with high-value addition. This is due to the mindset that bamboo products are of low quality. The phenomenon is further exacerbated by the lack of proper cultivation and management.

The value-addition training given by governmental and international organizations has been ineffective. There is neither a sufficient demand nor market for more sophisticated bamboo products. Therefore, the effects of value addition training are minimal. Furthermore, the majority of the given training requires the use of large and expensive machinery that smallholders simply do not own nor have access to. If processors have the skills, but no tools to utilize the given knowledge, then the training given is of little use and value. Finally, there are no established means of bamboo marketing. Other than through word of mouth, government bazaars and such, small bamboo processors have limited access to the market both in the promotion and the transportation of the finished products.

Further research in the bamboo sector to fill a wide knowledge gap is needed. There is currently very little research and literature that exist in the bamboo sector in Ethiopia. The last document that surveyed bamboo in Ethiopia was the LUSO Consult in 1997. Almost 20 years have elapsed and both the sector and the country have developed rapidly. Therefore, a new set of research should explore the current state of bamboo in both the north and south of Ethiopia.

As the findings and analyses in this study demonstrated, the untapped potential of the bamboo biomass is a multi-faceted and complicated problem. The development of the bamboo sector needs to be a holistic approach. The outlook for the future to develop the sector needs to be a long-term investment with an emphasis on every stage, from propagation to production. By far the most difficult aspect would be to educate and have people accept knowledge that they did not know for most of their lives. However, with enough interest and motivation to commit long-term to this development, there is no reason why the bamboo sector cannot develop as a major contribution to the bioeconomy. If that is the case, this would definitely mean an improvement in the many lives and communities that depend on bamboo as a part of their everyday livelihoods and for bamboo to play a major role within the context of the bioeconomy.

Author Contributions: Conceptualization, J.L., S.G., T.L., and R.B.; Methodology, J.L. and S.G.; Investigation, J.L.; Writing—original draft preparation, J.L.; writing—review and editing, J.L., S.G., T.L., and R.B.

Funding: This research was funded by the Foundation Fiat Panis (Ulm) and by the German Federal Ministry of Education and Research (BMBF) within the framework of the GlobE–Research for the Global Food Supply program. It is part of the collaborative project "Improving food security in Africa through increased system productivity of biomass-based value webs" (BiomassWeb); Grant No. 031A258H.

Acknowledgments: The authors are thankful for the assistance of Bamlak Alemu (University of Addis) during the fieldwork activities for the logistical support. We would also like to express our sincerest thanks to all respondents who made time for us and kindly shared information during the interviews and discussions.

Conflicts of Interest: The authors declare no conflict of interest.

References

1. Swinnen, J.; Riera, O. The Global Bio-Economy. *Agric. Econ.* **2013**, *44* (Suppl. 1), 1–5. [CrossRef]
2. McCormick, K.; Kautto, N. The Bioeconomy in Europe: An Overview. *Sustainability* **2013**, *5*, 2589–2608. [CrossRef]
3. Rosegrant, M.W.; Ringler, C.; Zhu, T.; Tokgoz, S.; Bhandary, P. Water and Food in the Bioeconomy: Challenges and Opportunities for Development. *Agric. Econ.* **2013**, *44* (Suppl. 1), 139–150. [CrossRef]
4. Virchow, D.; Beuchelt, T.D.; Kuhn, A.; Denich, M. Biomass-Based Value Webs: A Novel Perspective for Emerging Bioeconomies in Sub-Saharan Africa. In *Technological and Institutional Innovations for Marginalized Smallholders in Agricultural Developmet*; Gatzweiler, F.W., von Braun, J., Eds.; Springer International Publishing: Basel, Switzerland, 2016; pp. 225–238.
5. World Bank Group. *Ethiopia Poverty Assessment 2014*; World Bank Group: Washington, DC, USA, 2015.
6. Melaku, E.; Ewnetu, Z.; Teketay, D. Non-Timber Forest Products and Household Incomes in Bonga Forest Area, Southwestern Ethiopia. *J. For. Res.* **2014**, *25*, 215–223. [CrossRef]
7. Mekonnen, Z.; Worku, A.; Yohannes, T.; Alebachew, M.; Teketay, D.; Kassa, H. Bamboo Resources in Ethiopia: Their Value Chain and Contribution to Livelihoods. *Ethnobot. Res. Appl.* **2014**, *12*, 511–524. [CrossRef]
8. Adnew, B.; Statz, J. *Bamboo Market Study in Ethiopia*; UNIDO: Addis Ababa, Ethiopia, 2007.
9. Phimmachanh, S.; Ying, Z.; Beckline, M. Bamboo Resources Utilization: A Potential Source of Income to Support Rural Livelihoods. *Appl. Ecol. Environ. Sci.* **2015**, *3*, 176–183. [CrossRef]
10. Embaye, K. The Indigenous Bamboo Forests of Ethiopia: An Overview. *AMBIO A J. Hum. Environ.* **2000**, *29*, 518–521. [CrossRef]
11. Poppens, R.; van Dam, J.; Elbersen, W. *Bamboo Analyzing the Potential of Bamboo Feedstock for the Biobased Economy Colofon*; NL Agency: Utrecht, The Netherlands, 2013.
12. Kelbessa, E.; Bekele, T.; Gebrehiwot, A.; Hadera, G. *A Socio-Economic Case Study of the Bamboo Sector in Ethiopia: An Analysis of the Production-to-Consumption System*; PHE Ethiopia Consortium: Addis Ababa, Ethiopia, 2000.
13. Embaye, K. Ecological Aspects and Resource Management of Bamboo Forests in Ethiopia. Doctoral Dissertation, Swedish University of Agricultural Sciences, Uppsala, Sweden, 2003.
14. Desalegn, G.; Tadesse, W. Resource Potential of Bamboo, Challenges and Future Directions towards Sustainable Management and Utilization in Ethiopia. *For. Syst.* **2014**, *23*, 294–299. [CrossRef]
15. Virchow, D.; Beuchelt, T.; Loos, T.K.; Hoppe, M.; Gmbh, C.S.P. The Value Web Approach—So That the South Can Also Benefit from the Bioeconomy. *Rural 21* **2014**, *48*, 16–18.
16. Seboka, Y. Charcoal Production: Opportunities and Barries for Improving Efficiency and Sustainability. In *Bio-carbon Oppor. East. S. Afr.*; UNDP: New York, NY, USA, 2009; Chapter 6; pp. 102–126.
17. Kaplinsky, R.; Morris, M. *A Handbook for Value Chain Research*; IDRC: Ottawa, ON, Canada, 2001; Volume 113.
18. Scheiterle, L.; Ulmer, A.; Birner, R.; Pyka, A. From Commodity-Based Value Chains to Biomass-Based Value Webs: The Case of Sugarcane in Brazil's Bioeconomy. *J. Clean. Prod.* **2018**, *172*, 3851–3863. [CrossRef]
19. Loos, T.K.; Hoppe, M.; Dzomeku, B.M.; Scheiterle, L. The Potential of Plantain Residues for the Ghanaian Bioeconomy-Assessing the Current Fiber Value Web. *Sustainability* **2018**, *10*, 4825. [CrossRef]

20. Endalamaw, T.B.; Lindner, A.; Pretzsch, J. Indicators and Determinants of Small-Scale Bamboo Commercialization in Ethiopia. *Forests* **2013**, *4*, 710–729. [CrossRef]
21. Case, D. *The Community's Toolbox: The Idea, Methods and Tools for Participatory Assessment, Monitoring and Evaluation in Community Forestry*; FAO: Bangkok, Thailand, 1990; pp. 1–146.
22. Mulatu, Y.; Kindu, M. Status of Bamboo Resource Development, Utilisation and Research in Ethiopia: A Review. *Ethiop. J. Nat. Resour.* **2010**, *1*, 79–98.

© 2019 by the authors. Licensee MDPI, Basel, Switzerland. This article is an open access article distributed under the terms and conditions of the Creative Commons Attribution (CC BY) license (http://creativecommons.org/licenses/by/4.0/).

Case Report

Biomass-Based Innovations in Demand Driven Research and Development Projects in Africa

Jatta Raymond [1], Kwapong Nana Afranaa [1,2,*], Nero Bertrand Festus [1,3] and Fatunbi Oluwole [1]

[1] Directorate of Research and Innovation, Forum for Agricultural Research in Africa (FARA), 12 Anmeda Street, Roman Ridge, PMB CT 173 Accra, Ghana; rjatta@faraafrica.org (J.R.); bnero@faraafrica.org (N.B.F.); ofatunbi@faraafrica.org (F.O.)

[2] Department of Agricultural Extension, University of Ghana, P.O. Box LG 68, Legon, GA-492-3175 Accra, Ghana

[3] Department of Land Reclamation and Rehabilitation, Kwame Nkrumah University of Science and Technology, AK-448-4835 Kumasi, Ghana

* Correspondence: nkwapong@ug.edu.gh or nkwapong@faraafrica.org

Received: 28 May 2018; Accepted: 11 July 2018; Published: 27 July 2018

Abstract: The case for demand-driven research and development has received important considerations among governments, donors and programme implementing partners in development planning and implementation. Addressing demand is believed to be a bottom-top approach for designing and responding to development priorities and is good for achieving development outcomes. In this paper, we discuss the concept and application of demand-driven research and development (DDRD) in Africa. We use evidence of six projects implemented under the BiomassWeb Project in Africa. We focus on parameters on level of engagement of stakeholders—whose demand is being articulated, the processes for demand articulation, capacity building and implementation processes, innovativeness of the project, reporting and sustainability of the project. We find that the nature of the institutions involved in articulation and implementation of demand-driven research and development projects and their partnerships influence the impact and reporting of demand-driven projects.

Keywords: demand-driven research; Biomass; innovation; Ghana

---

1. Introduction

There is an increasing recognition of the challenge that African agriculture faces to grow more food to feed its fast growing and urbanizing population in situations of greater uncertainty because of impact of climate change and growing instability associated with land, water and energy shortages. Today, productivity (i.e., cereal yields) is estimated at 1.6 tons/ha in Africa, compared to 6.6 tons/ha in Europe and 3.9 tons/ha globally. The imperative is to deepen the application of science, technology and innovation in all agricultural processes [1]. The ability of science to lead to agricultural transformation depends to a considerable extent on what science is to be applied and for which constituency. This has a bearing on the use of research evidence, policy action, adoption rates as well as sustainable of project and programme outcomes. This brings to the fore the concept and application of demand-driven research and development (DDRD). The argument of demand- driven research is based on a realization that when research is close to the needs of the end users, it can easily be adapted, adopted and used [2]. Demand-driven and locally-driven national agricultural research systems are believed to support better overall institutional capacities, linkages among partners in the sector as well as sustainability of outcomes.

However, the inadequate funding for research often means that donors are the main drivers of research in Africa. Service delivery therefore remains largely supply-driven and organizations fail to effectively contribute to the real goal of providing more efficient and effective quality services for farmers to enhance rural development. These arguments are also similar to that of demand-driven development programmes promoted by donor and development partners. For example, World Bank's lending to demand-driven projects was estimated to rise from $325 million in 1996 to $2 billion in 2003 [2]. Demand-driven interventions are regarded as a mechanism for enhancing sustainability, making development more inclusive and empowering end users. However, demand-driven intervention in some areas also suffer from challenges of elite capture, misapplication of funds, low capacity etc. These issues are applicable to research funding much as they are to mainstream development interventions. This study is significant to the extent that it shows that the processes for demand articulation, the stakeholders involved in the process (articulation and implementation) and their linkages to end users are important in ensuring impact and sustainability of the project outcomes.

In 2016, the Forum for Agricultural Research in Africa (FARA) and the Centre for Development Research (ZEF), University of Bonn initiated demand-driven research and development under the BiomassWeb Project. The demand-driven research was to provide an opportunity for partner organizations within the BiomassWeb project to carry out further research and development activities to reinforce the possible outcomes of on-going BiomassWeb activities, and in addition to use the knowledge gained to improve the livelihoods of the people. The DDRD activities aimed to increase stakeholders' participation and contribute to research and development activities of BiomassWeb project. The DDRD activities must be driven by need or demand from potential beneficiaries. Grants were provided to successful partner organizations. On the basis of evidence on demand, innovativeness, ease of adoption and potential impact, Forum for agricultural research in Africa (FARA) and Center for Development Research (ZEF), University of Bonn shortlisted thirteen projects. The thirteen shortlisted projects were then submitted to external evaluators for their review. Following the external evaluators' review and recommendations, six proposals were selected for funding (see Table 1). Using evidence on the emerging results of six projects implemented in Ethiopia, Ghana and Nigeria, this paper examines the following questions.

1. Are demand driven projects easily adoptable or upscaled?
2. Does the implementation of demand-driven projects necessarily ensure the higher impact and sustainability of outcomes?

Table 1. Selected demand-driven research for development (DDRD) sub projects under the Biomassweb project.

Applicant Affiliation	Title of the Project	Country	Crop
Ministry of Food and Agriculture, Accra, Ghana	Using cassava peels for mushroom cultivation	Ghana	cassava
Council for Scientific and Industrial Research (CSIR) Food Research Institute, Accra, Ghana	Developing biomass-based value chain of plantain and reduce post-harvest losses of plantain through the development of value added products for small scale farmers and processors in two regions in Ghana	Ghana	plantain
International Network for Bamboo and Rattan, Kumasi, Ghana	Exploring the potential of bamboo leave fodder for livestock production in Ghana	Ghana	bamboo
Department of Agricultural and Environmental Engineering, University of Ibadan, Nigeria	Production of bio-plastics and bio-gels from agricultural waste to promote their biomassweb values	Nigeria	cassava, maize, banana
Department of Agricultural and Environmental Engineering, Federal University of Technology, Akure, Ondo State, Nigeria	Mass and energy balance analysis of pneumatic dryers for cassava and development of optimization models to increase competitiveness	Nigeria	cassava
YOM Institute of Economic Development, Ethiopia	Exploring potentials of the bamboo sector for employment and food security in Ethiopia: An institutional analysis of bamboo-based valueweb	Ethiopia	bamboo

2. Materials and Methods

To answer the questions indicated above, our analysis is guided by the "FARA bio-economy innovation-to-impact framework" for developing, testing and refining models for generation, uptake, out-scaling and commercialisation of innovations [3]. The key selling point is about ensuring technology adoption for increased livelihood outcomes through the use of multi-stakeholder innovation platforms for articulation of demand-driven research and for technology development and outreach.

We review the DDRDs based on FARA's Innovation to Impact model framework described below (Figure 1). To operationslize the model our framework is built on three key parameters:

- Partner engagement in project design
- Quality of the research results/outputs
- Sustainability of project outcomes

The use of these parameters allowed us to assess the quality of the deliverables and engagement of partners at each of the project processes from identification of the project activities to communication of research outputs. Under partner engagement, we consider the type of partner e.g., research institutions, NGOs' and the kind of partnership arrangement and funding mechanism. Also, who or which institution is demanding the research and whose demand is the project responding to. This included the research needs, community needs, industry needs, environmental needs, etc. The process of implementation and the level of participation of different stakeholders especially women is also assessed.

The DDRD projects are evaluated based on the demand for the project to address specific socio-economic and environmental challenges through the innovative use of biomass to generate and develop bio-based services and products. The results of the DDRD projects are to be easily up-scaled, adapted, adopted and used. Communication of research output/outcomes and capacity building of actors is considered key to ensure adoption, upscaling and sustainability.

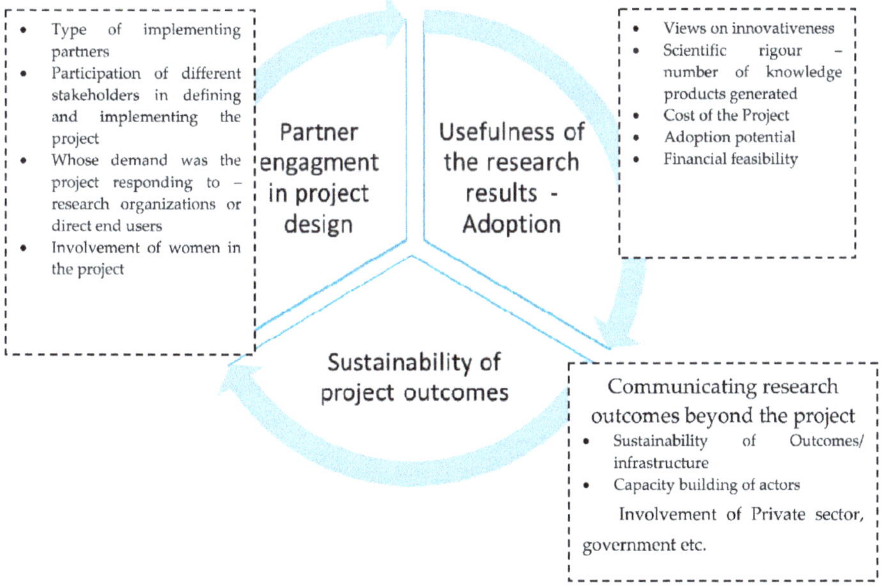

Figure 1. Analytical Framework.

For this study, we use qualitative methodology for the data collection and analysis. Project documents and reports on six implemented DDRD projects (Table 1) were reviewed and content analyzed. We focused on the parameters described in the conceptual framework. This included the partner engagement, demand for project, project goal, implementation process, innovativeness of the project, sustainability and scalability of project, project end-users, capacity building, delivery and adoption of innovative technologies. Focus group discussions and personal interviews were conducted with some of the project beneficiaries. Also key informant interviews were conducted with project managers and principal researchers of the BiomassWeb DDRD projects. BiomassWeb stands for improving food security in Africa through increased systems productivity of biomass-based value webs. BiomassWeb aims at improving the availability of and access to food (food security) in Sub-Saharan Africa through producing, processing and trading of biomass in biomass-based value webs. Biomass-based value webs are complex systems of interlinked value chains in which food and non-food biomass is produced, processed, traded and consumed.

3. Results

The following section provides information on the innovation processes in each of the six projects (Table 1) and the relation between end-users and researchers in the generation and adoption of new technologies in innovative use of cassava biomass for the production of mushroom, addressing postharvest losses of plantain, use of bamboo biomass for feeding livestock, production of bio-plastics and bio-gels from agricultural waste, improved engineering design of pneumatic flash dryers to regulate heat loss and improve on the quality of high quality cassava flour produced, and lastly exploring the potential of the bamboo sector for employment and food security. Discussions are centred on BiomassWeb DDRD project planning, innovations in the use of biomass, capacity building/demonstrations, delivery and adoption of innovative technologies.

3.1. Using Cassava Peels for Mushroom Production, Ghana

3.1.1. Demand Driven Research Planning

The goal of the project was to use cassava peel waste which is generated in the production of Gari/cassava chips, as substrate for the production of mushroom to generate income, improve household nutrition security and minimize environmental degradation. Three major actors were involved in this demand-driven research and development (DDRD) project. The funder—Forum for Agricultural Research in Africa (FARA/BiomassWeb project) and Centre for Development Research (ZEF), the facilitators/implementers (Women in Agricultural Development), and the beneficiaries (extension agents, and farm families of Gomoa). A key observation to note in this DDRD is the fact that the innovation was generated by the researchers (or facilitators/implementers) and supplied or transferred to the local community with funding from the development partner. In this case the donor agency (FARA/ZEF) demands the knowledge and innovation on behalf of the end-users and goes a step further to demand the empowerment of these end-users. Local farm families and the youth of Gomoa were trained on cassava peel composting, bagging the substrates, inoculating the substrate with spores, daily culture of the inoculated substrate/growing mushrooms, harvesting the matured mushrooms, drying and packaging for the market, as well as the marketing of either the mushrooms or the substrate. A mushroom house was constructed near a major Gari processing centre where cassava peels had accumulated and polluted the environment.

3.1.2. Innovation in the Use of Biomass

The idea and know-how to convert cassava peel waste into income generating opportunity and improve environmental quality was the innovative contribution of this DDRD. To operationalize the idea, cassava peels were composted and subsequently used as substrate for the cultivation of mushrooms. Mushroom is an important food in diet of Ghanaians [4]. It is a valuable source of high quality proteins (21–40%) dry weight, and rich in vitamins (B1, B2, B6, B12, C, D) [5–7]. Both the substrate and the mushrooms produced could be sold. This could be sufficient to keep farm families and the youth in business assuming there is ready market and the substrate as well as mushrooms are produced in large quantities. Through this intervention, the cassava peel waste is put to productive use and does not pollute the environment. Figure 2 shows the framework for converting cassava peels, which is a by-product or waste from the Gari/cassava chips processing activity and the waste substrate used as manure for crop production. The peels can also fed to livestock.

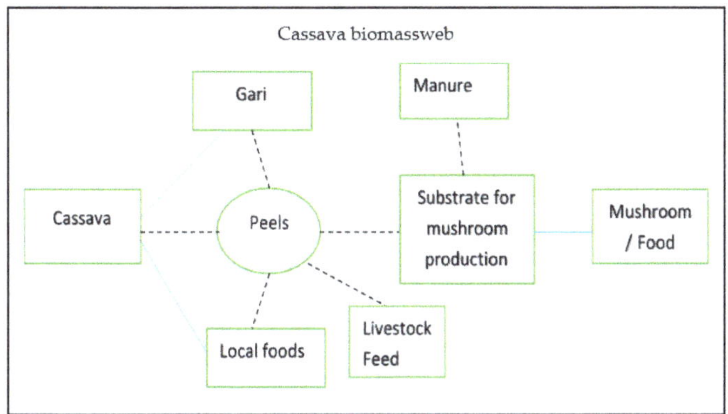

Figure 2. The flowchart illustrating conversion of cassava peel waste into substrate for mushroom cultivation and manure for agricultural production. Cassava peel waste also serves as livestock feed. (Dotted lines show productive uses of waste generated).

3.1.3. Capacity Building

For this project, about thirty (30) local women from a gari/cassava chips processing group and the youth of Gomoa community were trained on how to undertake mushroom cultivation. The training covered compost preparation leading to the generation of substrate, bagging, inoculation with spores (seed), incubation and harvesting of the mushroom. A mushroom house was constructed, and a solar dryer provided. According to the project managers there was the need to construct a structure for the mushroom production to be able to control the temperature which should not exceed 30 degrees Celsius, and also keep the room humid. This is consistent with other studies that have found an increase in temperature of 40–60 degree celcius likely to kill the mycelium in less than 24 h [4,8].

Trainees were also trained on how and where to market mushrooms. Since mushroom is currently produced on a small scale, the local markets and interested individual mushroom consumers were the main market options explored. Training was conducted by a consultant who was hired by the implementers Women in Agricultural Development (WIAD).

3.1.4. Delivery of the Project and Adoption

The concept of converting cassava peel waste to mushrooms for food and income is theoretically easy to adopt. Interactions with participants who were trained revealed they had understood how the concept is operationalized. However, many had the notion that a mushroom house is needed to operationalize this innovation. In which case, they felt financially handicapped to truly benefit from this innovation. To ensure the sustainability of the innovation, there should be (1) small-scale production of substrate from cassava peels using simple home-based materials and subsequently following up with the rest of the processes, and (2) private sector investments in constructing a mushroom house for the mass production of substrate for sale to small-scale mushroom producers.

3.2. Development of Plantain Biomass into Composite Flour for Traditional Foods and Bakery Products, Ghana

3.2.1. Demand Driven Research Planning

Researchers from the Food Research Institute of the Centre for Scientific and Industrial Research (CSIR-FRI) Ghana, identified the need for this project in response to high postharvest losses of plantain and the need to convert plantain biomass into composite flour which could substitute wheat in the production of local foods and provide highly nutritious foods. Local food processors were trained in plantain processing technologies of converting plantain into composite flour for various traditional foods such as plantain fufu and value-added products for making of bakery products such as plantain chips, cakes, pies, bread and doughnuts.

3.2.2. Innovation in the Use of Biomass

Plantain is highly nutritious, provides rich dietary energy, and contains micronutrients such as carotenes, ascorbic acid, as well as minerals such as iron, potassium, zinc, calcium, and phosphorus [9]. Processing plantain into composite flour during plantain peak season when it is readily available on the market and prices are moderate will help reduce postharvest losses which are estimated at 20% at production level and 15% at consumption level [9]. The composite flour can be stored for up to a year, hence can serve as a convenient raw material for making bakery products. The composite flour can serve as a substitute for wheat flour which is a major import commodity in Ghana. Currently Ghana imports about 700,000MT of wheat annually. Usage of plantain composite flour in bakery products will reduce importation of wheat flour. The plantain peels can serve as food for feeding livestock and as raw material for local soap industries (Figure 3).

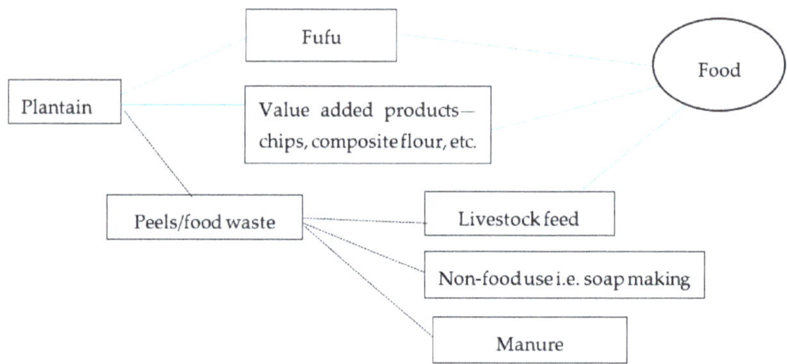

Figure 3. Plantain biomass value web (Dotted lines show productive uses of waste generated).

3.2.3. Capacity Building

About 117 local food processors were trained in converting plantain into composite flour and adding value to plantain products and developing better market linkages for plantain products. The technology is easy to adopt as it involved locally sourced materials and simple processing steps of boiling plantain, milling and drying to get the plantain flour. It does not require a huge start-up capital hence can be started on a small scale. Because the adoption of new technologies is likely to occur when individuals perceive some relevant advantage over an existing innovation or status quo, and such an innovation is compatible with existing practices and not too complex as well as offering observable results [10]. The initial start-up package for this project consisted of 1kg of composite flour supplied to each beneficiary to encourage uptake of the technology. Beneficiaries reported uptake of the technology to have resulted in higher incomes as the addition of the plantain flour increased volumes of products and thereby increased their profits. Also, the project created jobs for some women who otherwise had no source of income. It was not immediately apparent what the level of adoption of the plantain value addition techniques and marketing strategies acquired from the CSIR-FRI team is. It appears too soon to notice measurable levels of adoption of the techniques. Adoption happens over time depending on the individual's decision to take up innovation [11], and the availability of resources, which in this case is seasonal with peak and minor seasons.

3.2.4. Delivery of the Project and Adoption

Widespread adoption and production of plantain composite flour is however hindered by the lack of access to milling machines and mechanized or solar dryer units to commercialize production. Current production level is on small-scale using corn milling machines and sun-drying which is challenged during rainy season. Innovative ways to assist in increasing adoption of the technology and commercializing the plantain composite floor production is therefore essential. This has to do with addressing the underlying challenge of acquiring milling machinery purposely for processing plantain into composite flour. Also, providing solar or mechanical drying units for efficiently drying the flour especially during the raining season which is also the peak season for plantain production. The beneficiaries can be organized into groups to mobilize funds for the purchase of milling machine and mechanical or solar drying unit. The group could go into commercial plantain flour production which members could purchase for their bakery production or food preparations. A private investor could also profit greatly by introducing milling machines and a solar dryers in the peak plantain growing areas. These aside, some easy to adopt strategies such as plantain chips production are already being utilized widely in major urban centers across the country.

3.3. Bamboo Leaf as Fodder for Livestock Production, Ghana

3.3.1. Demand Driven Research Planning

The idea of introducing bamboo as an alternative fodder for livestock feeding was developed by the International Network for Bamboo and Rattan (INBAR), with financial support from FARA and ZEF within the BiomassWeb project. Livestock production in Ghana is limited by access to sustainable feed supply especially during the dry season. The evergreen nature of bamboo and its high nutritive content makes it an ideal fodder especially during the dry season when tree leaves and grasses dry up and are burnt by frequent bush fires. INBAR piloted a bamboo-based agroforestry model in the dry semi-deciduous zone of Ghana to promote the integration of bamboo into indigenous cropping systems to meet socioeconomic needs and provide fodder for livestock. Experimental bamboo feeding trials were set up to explore the consumption patterns and digestibility of bamboo fodder and evaluate growth and health of livestock fed with bamboo leaves either as sole feed or feed supplement.

3.3.2. Innovation in the Use of Biomass

In Ghana, bamboo use as fodder is largely unknown. There is insufficient awareness of bamboo utilization as fodder throughout Ghana. Partey et al. [12] in their study on perception of bamboo leaf as fodder for livestock production found less than 26% of respondents were aware of use of bamboo leaves as fodder. However, the evergreen nature of bamboo and its high nutritive content -i.e., rich crude protein (9–19%) and content of crude fiber (18–34%)—makes it an ideal fodder especially during the dry season [13]. Successful introduction of bamboo as an alternative feed stock will ensure that livestock have fresh fodder all year round, thereby encouraging livestock production. The bamboo stems will provide fuelwood, while the young regenerated shoots can be consumed as food. Bamboo can also be used in alley cropping systems to boost food production while at the same time stabilizing the soils and minimizing environmental impacts including mitigation and adaptation to climate change (Figure 4). INBAR has introduced two bamboo species from India (*Bambusa balcooa* or *beamer bamboo*) and Ethiopia (*Oxythenanthera abyssinica*). Both species are drought and fire tolerant. Fire and droughts are typical of the forest-savannah transition belt and the savannah zone of Ghana.

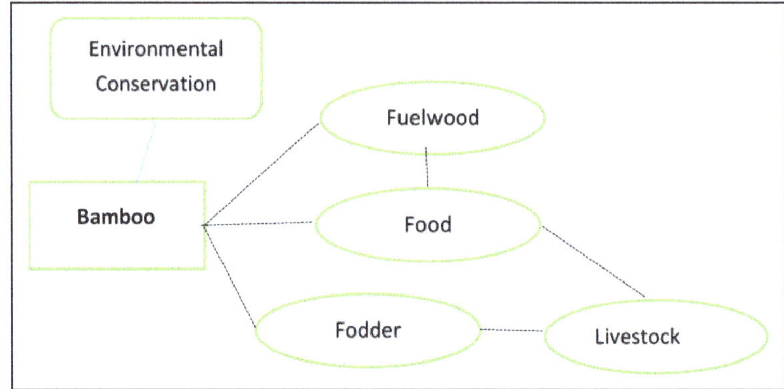

Figure 4. Bamboo agroforestry and bamboo fodder for livestock production (Dotted lines show productive uses of waste generated).

3.3.3. Experimental Set-Up

The bamboo feeding trial experiment was set as a completely randomized design experiment with two replications. Plant fodder for goats was the main treatment and consisted of three feed types: (1) Bamboo leaves, (2) Grass—sugar cane grass or *Saccharum spp.*, and (3) Leaves of *Millettia* species and *Gmelina arborea*. Six goats were used in the experiment, with two assigned to each of the fodder types above. The data was collected on weekly weights of goats, feacal matter, urine and blood samples. The study revealed that bamboo is acceptable to the sheep and was completely consumed when offered ad libitum and therefore using it as a feed supplement can increase feed intake of the basal diets by 40% and increase the weight of sheep by 2.31 kg [14]. The Bamboo crude protein (CP) of 124 g/kg; ash—80 g/kg; neutral detergent fibre (NDF)—464 g/kg compares well with cowpea haulm, a leguminous haulm, with a CP, ash, NDF, and gas production (GP) of 124–268; 89; 419 g/kg, respectively, except the gas production (GP), which is 2 times lower than cowpea haulm [15]. The study however finds that the bamboo used as supplement led to a reduction in the blood parameters measured at the end of the experiment, the RBC (5.8–8.0 L^{-1}), total protein (63–70 g dL^{-1}), and albumin (23–28 g dL^{-1}) compared well with RBC (6.4–9.9 L^{-1}), total protein (63–71g dL^{-1}) when sheep were fed with sorghum stover and dried poultry droppings, except for the albumin which was 1.5 times lower than those reported by the researchers [14]. Despite the comparably higher CP, GP, ash and the positive influence on growth performance, bamboo should be fed alongside with leguminous for ages in an attempt to meet the energy-protein requirement of the animals and also improve the health status of the animals through the supply of minerals and protein [14].

3.3.4. Delivery of the Project and Adoption

For bamboo leaves to be adopted and used as fodder by local subsistence and commercial livestock farmers, there is the need for massive awareness creation and public education at the grassroots level where the need for such fodder will be most required. Key to this process will be the sustainable availability of the bamboo resources. INBAR has undertaken efforts to ensure sufficient bamboo material is available for the provisioning of leaves as fodder by establishing bamboo plantations in selected communities. There are plans to expand the bamboo plantation cover in the country. Hence, there is significant promise that sufficient bamboo fodder will be made available for large scale livestock production in the future. So far more than 50% of farmers in bamboo growing areas in Ghana have demonstrated their willingness to accept bamboo as a livestock feed [14]. The bamboo fodder should however be supplemented by a leguminous fodder to supply both energy, protein and minerals for the animals to meet their nutrient requirement [14].

3.4. Production of Bio-Plastics and Bio-Gels from Agricultural Waste to Promote Their Biomassweb Values, Nigeria

3.4.1. Demand Driven Research Planning

The project idea was conceptualized through collaborative effort by researchers from different institutions both national and international to address the challenge of environmental pollution from plastic waste in Africa using waste from agricultural products such as cassava, banana, maize to produce bio-gels and bio-plastics that are biodegradable. The institutions involved in the demand-driven research are the Federal Institute of Industrial Research Oshodi (FIIRO), International Institute for Tropical Agriculture (IITA), and University of Ibadan. The collaboration among the institutions ensured that different partners pursued various aspect of the research and analysis. Funding for the research was provided by Forum for Agricultural Research in Africa (FARA) and Centre for Development Research (ZEF) under the BiomassWeb project.

3.4.2. Innovation in the Use of Biomass

Plastic waste disposal is a huge environmental burden in Africa and the world, polluting land, water and air. The alternative to synthetic plastic is bio-degradable plastics which is degradable by microorganisms and enzymes such as bacteria, algae and fungi. Such biodegradable plastics can reduce plastic waste in the environment. The research team in this DDRD project explored the use of various agricultural waste products to develop biodegradable plastics and bio-fuels (Figure 5).

The research team was successful in developing a protocol for the production of bio-plastics from starches obtained from cassava peels, acetic acid and glycerol (200:20:10) to form a good bioplastic resin [16]. The bio-plastics produced are highly degradable. Degradation of the bio-plastics at day 3 was approximately 14% after burying in soil and by day 12, approximately 86 percent of the bio-plastic was degraded [16]. The team also successfully produced bioethanol using starch extracted from cassava peel and cellulose from corn cobs using acid hydrolysis and yeast fermentation. The average yield of bioethanol is 757.33ml for cassava peel and 595.56 for corn cobs from 3.55L and 3.35L of filtrates respectively [17]. The properties of bio-ethanol produced from the two samples compared favorably with the commercial ethanol [17].

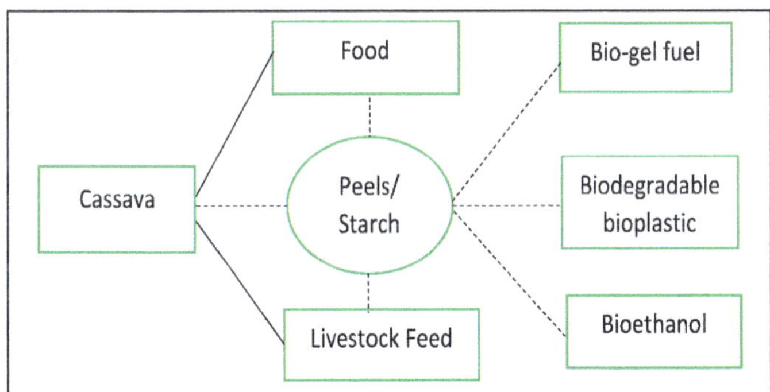

Figure 5. The flowchart illustrating conversion of cassava waste into bio-plastics, bio-ethanol and bio-gels. (Dotted lines show productive uses of waste generated).

3.4.3. Capacity Building

The research team presented their research findings at scientific conferences and workshops to disseminate research findings. A project dissemination workshop was organized to inform policy makers, manufacturers from the plastic industry and other researchers of the novel findings of the study. Two masters' students from the University of Ibadan were involved in the research study and trained on performing the experimental setup.

3.4.4. Delivery of the Project and Adoption/Upscaling

The research findings show evidence of the potential of using waste from agricultural products to address environment pollution. Scaling up bioplastic production will first need an economic feasibility of such production using a pilot scale. Such study will present a real life situation before commercial production [17]. In the case of ethanol gel, processing plants utilizing starch crop like maize, cassava and cellulose for ethanol are already in existence, however, most of them are large scale. There is a need to set-up micro/small scale plants for bio-gel production to ensure lower cost of production and even distribution at affordable price [17].

3.5. Mass and Energy Balance Analysis of Pneumatic Dryers for Cassava and Development of Optimization Models to Increase Competitiveness, Nigeria

3.5.1. Demand Driven Research Planning

This DDRD project implemented by The Federal University of Technology Akure (FUTA) addressed the inefficiencies of locally manufactured pneumatic flash dryer for processing of cassava into High Quality Cassava Flour (HQCF), and further identified new ways to improve the drying performance of the dryers. The project had the objectives of evaluating the performance of pneumatic flash dryer models operated by cassava processors in Nigeria using mass and energy balance analysis, and further modifying the flash dryer models to improve the drying efficiency and to develop a detailed engineering design of an efficient flash dryer model. The research team was composed of researchers from FUTA, Federal Institute of Industrial Research Oshodi (FIIRO), International Institute of Tropical Agriculture (IITA), and Kwara State University Ilorin who collaboratively embarked on the study to provide a solution to the inefficiencies in existing dryers for local manufacturers. Funding for the research was provided by Forum for Agricultural Research in Africa (FARA) and Centre for Development Research (ZEF) under the BiomassWeb project.

3.5.2. Innovation in the Project

The research team evaluated four design models of pneumatic flash dryers based on energy efficiency, specific heat consumption, thermal efficiency, heat losses to the ambient and heat losses via air outlets. The results of the study showed that existing models had a combination of both efficient and inefficient component parts which results in heat loss and low heat energy utilization. The inefficiencies identified on the existing dryers included the absence of insulation on the drying duct which facilitated greater loss, absence of feeder on some of the flash dryer models, improper design of the multiple cyclone which affect proper separation of product from the exhaust air and absence of heat control system on the burners [18].

An improved engineering design of a pneumatic flash dryer was designed by the researchers to regulate the heat loss and improve on the quality of HQCF produced. The modification included the introduction of insulation for reduction of heat losses, instrumentation circuit diagram for heat control system, and an efficient two passes heat exchanger system for maximum heat utilization. A prototype model design of the improved pneumatic flash dryer was developed and fabricated by the research team.

3.5.3. Capacity Building

A dissemination and training workshop was organized by the research team to disseminate information on the improved efficient pneumatic flash dryer for processing of High Quality Cassava Flour (HQCF). Fifty fabricators and engineers from South-West Nigeria participated in the workshop.

Participants at the workshop were informed of the improved model design and the needed improvement on the existing dryers to improve drying efficiency.

3.5.4. Delivery of the Project and Adoption/Upscaling

To transfer the knowledge on manufacturing efficient pneumatic flash dryers to equipment manufacturers, trainings in the development of detailed engineering design and fabrication of the prototype improved flash dryer need to be carried out. The trainings should include hands-on practical session for manufacturers on how to fabricate the improved pneumatic flash dryer model.

3.6. Exploring Potentials of the Bamboo Sector for Employment and Food Security in Ethiopia: An Institutional Analysis of Bamboo-Based Valueweb, Ethiopia

3.6.1. Demand-Driven Research Planning

The idea for implementation of the DDRD project was conceptualized by researchers from the institute of economic development, Ethiopia and the University of Hohenheim, Germany. The primary objective of the research project was to provide holistic insights into the current status and future potentials of the bamboo sector in Ethiopia in order to enhance sustainable livelihoods and employment generation. The researchers surveyed 468 households from two major bamboo growing regional states in Ethiopia—Amhara and Benishangul Gumuz. The research findings are to inform policy decision makers on the potential benefits of promoting the bamboo sector for socio-economic benefits. This research was researcher led, incorporating local people perceptions. The funding for the research study was provided by provided by Forum for Agricultural Research in Africa (FARA) and Centre for Development Research (ZEF) under the BiomassWeb project.

3.6.2. Innovation in the Project

Ethiopian rural households earn a significant part of their livelihood from natural resources mostly forest products [19]. Bamboo is such a natural resource which serves as a source of energy, fodder, food, construction material and handicrafts. Ethiopia's natural bamboo forest, the largest in the African continent, is estimated to be around one million hectares, of which 850,000 hectares are lowland and 350,000 hectares are highland bamboo varieties [19–21]. Even though Bamboo has significant benefits, the bamboo sector has received little attention and its potential contribution to the economy has been underexploited. This situation is similar to that of many African countries including Ghana [12,22,23], which calls for the need creation of awareness among the local people and policy makers to promote the bamboo sector to enhance livelihoods and enhance sustainability of the environment. This DDRD research project was setup to contribute to this agenda.

The findings from the study [19] revealed that poor rural families preferred to engage in bamboo production requiring few resources. Also, market prices of bamboo culms significantly increase the probability of employment in bamboo sector, and the probability of the variation in the income from bamboo of rural households. Bamboo has the potential to ensure food security through provision of higher incomes and better food security of poor rural smallholder farmers. Further the rapid growth of bamboo and short growing cycle makes bamboo a suitable option as income source during food shortages.

3.6.3. Capacity Building

The findings of this study seek to create awareness on the potential benefit of the bamboo in addressing food security, creating employment which will lead to an increase in income and wellbeing in the livelihood of people in communities where bamboo grows.

3.6.4. Delivery of the Project and Adoption/Upscaling

To tap into the full potential of bamboo, a collective action of action of actor in the bamboo biomassweb is required. This includes the engagement of governments, NGOs, local farmers, processors and private sector. There is the need to create awareness on the diverse benefit of bamboo, improved methods of growing and extracting culms, and adoption of improved technologies.

4. Discussion

Demand-driven research process involve several actors and beneficiaries at each stage of the process, with the ultimate benefit of enhancing ownership and increasing applicability of research [24]. Practical operationalization of demand-driven research planning process is limited by the level of perspective of actors on the innovation and capacity to fully participate and operationalize the demand-driven research [24]. For the case studies presented in this paper researchers identified the need for the innovation to address identified challenges and supplied or transferred knowledge and innovation to the end-users. The identification of the challenge was done through surveys, or consultations with the end-users. This was followed by developing proposals to source funds from funding agencies who prioritized the identified research area/ innovation as key to engendering development. Thus, the funding agency (FARA/ZEF) demands the knowledge and innovation on behalf of the end-users.

In contrast, DDRD in the advanced economies are spearheaded by the end-user, in this respect the farmer. The end-users fundamentally identify a research problem, mobilize funds from among end-users, private enterprises/beneficiaries or the public sector (Government) to solve the problem and contracts a researcher or an agency to implement and/or solve the problem. End-users in this case have the opportunity to contribute knowledge and innovations into the DDRD, although more often the final outcome of the DDRD program does not adequately reflect end-users' (farmers) needs due to influence by several actors in the DDRD planning process [25]. This implies that even in the developed economies where DDRDs are end-user demanded in theory, in practice, the influence of other stakeholders in the planning compromise the power of end-users. Active participatory methods must be pursued if end-user power in DDRDs are to be greater priority.

Implementation was done together with the end-users through trainings, hands-on experiences, experimental set-up and surveys. For the innovation generated to be transferred, it should respond to the needs of the beneficiaries or end-users. As Rogers [10] indicates, new innovations are likely to occur in response to economic opportunity or scarcity.

In the case studies presented in this paper (Table 2), the innovations in the use of biomass of plantain responded to reducing postharvest losses, substituting wheat which is imported in the preparation of bakery products, and also creating jobs and increasing incomes [9]. The use of drought tolerant and fire resistant bamboo biomass highlights the multiple uses of the bamboo in feeding livestock, especially in the dry season, inter-alley cropping to boost food production and stabilizing the soils used as fuelwoods because of its potential for energy recovery to be used as energy source compared to other woody biomass [13]. The bamboo shoot served as a food source for locals. However, there is low awareness among end-users even though there is evidence of bamboo contributing to improving food security. This calls for the need to create awareness on the benefit of bamboo and to involve policy makers and other stakeholders in the bamboo valueweb [12,19].The innovation in the use of cassava biomass showed the conversion of cassava peels which are generated waste from cassava processing into substrate for cultivation of mushroom which were sold to raise income and included in household food.

Table 2. Innovation in Demand Driven Research and Development Project and implementation Modalities.

DDRD Project	Implementing Institution	Planning and Scope of Project. How Was the Process Planned or Conceptualized?	Innovation in the Use of Biomass. What Is Innovative about the Process? Number of Research Products Generated	Capacity Building/Experimental Setup. How Was Training/Experimental Set Up Done?	Delivery of Project and Adoption. How Was Innovation Uptake? Sustainability?
Using cassava peels for mushroom production	Ministry of Food and Agriculture, Women in Agricultural Development, (WIAD) Ghana. WIAD is a department of the Ministry of Food and Agriculture. It has National, Regional and District offices across Ghana. It is very active at local level	Researchers identified innovation and facilitated implementation of process for end users (local women and youth)	Conversion of cassava peels which are waste from processing of cassava chips into substrate for production of mushroom to improve household food security, generate income, minimize environmental degradation. A manual on production of mushroom. A video used in farmer training	Researchers trained beneficiaries on usage of innovative technology of using cassava peels in compost as a substrate for mushroom production. Constructing of low-cost building for mushroom with local materials.	Adoption low. A mushroom house needed to be constructed to operationalize this innovation, which comes as a cost. There is the need for more cost-effective approaches using a group based approach with access to microfinance.
Development of plantain biomass into composite flour for traditional foods and bakery products	Centre for Scientific and Industrial Research—Food Research Institution (CSIR-FRI), Ghana. This is an institute of the National Agricultural Research Institute. Its mandate is mainly of research. It works through other units (like WIAD) for technology outreach and adoption.	Researchers identified need to process plantain into composite flour and trained local processors and women in use of the technology.	Plantain processed into composite flour to reduce on postharvest losses and a potential subsite for wheat in bakery products. Seven knowledge products were generated, three peer reviewed articles, manuals, posters	Researchers trained beneficiaries on converting plantain biomass into composite flour, adding value to plantain products, linkage to markets.	Uptake of technology low. Hindered by the lack of access to milling machines and mechanized and solar dryer units. Locally fabricated milling machines specifically for milling of plantain should be developed and also solar drying units. Processors could access and use the facility at a fee. WIAD could mobilize the women into groups to acquire and manage the milling and drying units.
Bamboo leaf as fodder for livestock feeding	International Bamboo and Rattan Organization (INBAR), Ghana. INBAR is an Intergovernmental organization working in Agroforestry—focus on Bamboo and Rattan. Its regional office in West Africa is in Ghana. It works with research organizations, local NGOs and government at the frontline	Researcher led survey and experimental set-up.	Drought tolerant and fire-resistant bamboo used for feeding livestock in the dry season; inter-alley cropping to boost food production and stabilizing the soils, used as fuelwood. A peer reviewed article, a study report	Bamboo feeding trial experiment set as a completely randomized design experiment with two replications	Experimental trials show bamboo feed by as a viable feed supplement for livestock feeding. Need to supplement bamboo feed by a leguminous fodder to supply both energy, protein and minerals for the animals to meet their nutrient requirement. There is the need for education and sensitization on the locals on uses of bamboo.
Production of bio-plastics and bio-gels from agricultural waste to promote their biomassweb values	Federal Institute of Industrial Research Oshodi (FIIRO), International Institute for Tropical Agriculture (IITA), and University of Ibadan. All three institutions are located in Nigeria	Researchers responded to the need to address the challenge of environmental pollution from plastic waste in Africa using waste from agricultural products	Development of a protocol from cassava peel starch for the production of bio-gel fuel and biodegradable plastic. Study reports and peer reviewed articles	Dissemination workshops to share findings with policy makers, manufacturers and researchers. Training of research scientist.	Scaling up bioplastic production will first needs an economic feasibility of such production using a pilot scale. There is need to set-up micro/small scale plants for bio-gel production to ensure lower cost of production and even distribution at affordable price

Table 2. *Cont.*

DDRD Project	Implementing Institution	Planning and Scope of Project How Was the Process Planned or Conceptualized?	Innovation in the Use of Biomass What Is Innovative about the Process? Number of Research Products Generated	Capacity Building/Experimental Setup How Was Training/Experimental Set Up Done?	Delivery of Project and Adoption How Was Innovation Uptake? Sustainability?
Mass and energy balance analysis of pneumatic dryers for cassava and development of optimization models to increase competitiveness	Federal University of Technology Akure (FUTA), Nigeria	Researchers identified the need to design an efficient pneumatic flash dryer to address inefficiencies in existing models.	An improved engineering design of pneumatic flash dryer was designed by the researchers to regulate the heat loss and improve on the quality of HQCF produced. The modification included the introduction of insulation for reduction of heat losses, instrumentation circuit diagram for heat control system, and an efficient two passes heat exchanger system for maximum heat utilization. Study reports and peer reviewed articles are output of the project.	Fifty fabricators and engineers from South-West Nigeria were trained on the design of the improved efficient pneumatic flash dryer.	Uptake of the technology will require hands-on practical session for manufacturers on how to fabricate the improved pneumatic flash dryer model.
Exploring potentials of the bamboo sector for employment and food security in Ethiopia: An institutional analysis of bamboo-based valueweb	YOM Institute of Economic Development (YIED), Ethiopia	Researcher-led survey	A study reports and peer reviewed articles are outputs of the research study.	Survey of 468 households from two major bamboo growing regional states in Ethiopia—Amhara and Benishangul Gumuz.	Need to create awareness on the diverse benefit of bamboo, improved methods of growing and extracting culms, and adoption of improved technologies.

There is the need for more cost-effective approaches using a group-based approach with access to microfinance to scale up the adoption of the using cassava peels for mushroom production. Production of bioplastic and bio-fuels from cassava peel starch is innovative and addresses a serious environmental problem of plastic pollution. For this innovation to be up-scaled there is the need for an economic feasibility of such production using a pilot scale. Also, there is the need to set up micro/small scale plants for bio-gel production to ensure lower cost of production and even distribution at affordable prices. The research team from FUTA developed an improved engineering design of pneumatic flash dryer to regulate the heat loss and improve on the quality of HQCF produced. The modification included the introduction of insulation for reduction of heat losses, instrumentation circuit diagram for heat control system, and an efficient two passes heat exchanger system for maximum heat utilization. Uptake of the technology will require hands-on practical session for manufacturers on how to fabricate the improved pneumatic flash dryer model.

These innovations even though identified by researchers addressed needs of the end-users which necessitated the buy-in of the end-users. It must, however, be noted that even though there was a need for the innovation to address a pressing challenge of the end-users, this was not sufficient. The adoption and uptake of the innovation depended on the technological and technical support received by the end-users and in some of the project the reliable supply of biomass (plantain, bamboo, cassava). For instance, in the case of bamboo biomass, there is the need to expand bamboo plantation cover in the area and across the livestock growing regions in the country. In the use of plantain biomass, operationalization of the innovation was hindered by the availability of technology and machinery. The end-users needed to have local milling machines and solar drying units to effectively take up the technology and upscale production of the plantain composite flour. The case of use of cassava biomass required end-users to have a housing structure for the mushroom production (Table 2). In designing demand-driven research and development projects, the uptake by the end-users is crucial and such technology and infrastructural needs should have been factored into the implementation and funding plan for the full uptake of the technology.

Technical support in the uptake of innovations is also essential in the supply-side of the demand-driven research. This was fulfilled by the researchers providing training or education to end-users on the operationalization and processes of the innovations as well as the relevance and benefit of the innovation. However, the decision to take up an innovation largely depends on the innovation itself addressing an economic opportunity or scarcity, knowledge and understanding of the end-user, the end-users decision to adopt and the end-user actually confirming and implementing the innovation. Time is a key factor in adoption [11]. Adoption of innovation happens over time. The immediate results on the level of uptake of innovation of the demand-driven research could therefore not be fully assessed as the projects implementation phase lasted between three to nine months.

Lessons learnt from the DDRD projects include:

1. Innovative thinking such as in the WIAD, University of Ibadan bioplastic, and biofuel projects can lead to improved environmental conditions by reducing environmental pollution.
2. Waste can be used profitably by carefully process it to reduce the environmental externalities, reduce the land area required for its disposal, and provide income.
3. Extending the shelf life of perishable products and thus minimizing postharvest losses can add value and alleviate poverty.
4. Bamboo has the potential to serve as an alternative feed for livestock farmers especially in the lean season when feed is scarce. This could boost meat production, increase food and nutritional security and alleviate poverty.
5. Uptake of the technology such as improved design of pneumatic flash dryers by the research team from FUTA will require hands-on practical session for manufacturers on how to fabricate the improved pneumatic flash dryer model.

6.   Training and dissemination of research findings and output is essential to generate the desired impact and outcome of the projects.

5. Conclusions

The type of the beneficiary organisation (research organisation and government technical unit) is an important variable to ensuring stakeholder involvement and reporting. The experience implementing the demand-driven projects (i.e. small grants) generated several research outcomes, perhaps much more than the value of the investments. We observe some significant differences in the level of participation of communities and end-users in the various projects. In the case of WAID, we observe strong community participation (albeit low involvement of women) as well as community/district level extension agencies. In the other DDRD projects (INBAR, FUTA, UI YIED, CSIR), we observe relatively less participation of end users in actual implementation of the project. These projects were mainly research based enquiry, implemented by researchers from the institutions. Trainings, surveys and dissemination workshops to some extent provided some level of participation for the end users.

There were significant differences in the knowledge products produced from the DDRD projects, more from the research oragnisations (CSIR, YIED, FUTA, UI and INBAR) compared to the department of the Ministry of Agriculture (WIAD). It is also observed that research institutions often collaborated with other research institutions in implementing the research project. All the project embarked on finding innovative ways of addressing pressing socio-economic and environmental challenges exploring the use of biomass.

In summary,

(1)   Innovations around the use of biomass was led by researchers, funded by a donor who demanded on behalf of end-users to address challenges and create opportunity around the use of biomass.

(2)   Innovation opportunities were generated around the use of biomass in addressing challenges of postharvest losses of plantain, income generation from conversion of cassava peels into mushroom production, use of bamboo for livestock feeding in dry season and for promoting food security, production of bioplastics and bio-fuels from cassava peels, and development of improved efficient model of pneumatic flash dryers.

(3)   However, these innovations were not sufficiently adopted. The adoption/uptake and operationalization depended on the availability of reliable supply of biomass, technological and technical support. Adoption of the innovation around the biomass in the end depends on the end-users understanding the innovation, decision to adopt and actually adopting the innovation.

Author Contributions: All authors contributed equally to the conceptualization and writing of manuscript.

Funding: This research was funded by the German Federal Ministry of Education and Research (BMBF) through the collaborative project "Improving food security in Africa through increased system productivity of biomass-based value webs." This project is part of the GlobE—Research for the Global Food Supply programme (Grant No. 031A258H).

Acknowledgments: This paper presents six demand-driven research and development projects implemented in Ghana within the BiomassWeb project which is coordinated by Forum for Agricultural Research in Africa (FARA), Ghana and the Centre for Development Research (ZEF), Germany. We acknowledge the financial support from the German Federal Ministry of Education and Research (BMBF) supported with funds from the German Federal Ministry for Economic Cooperation and Development (BMZ). We also acknowledge support of all project partners.

Conflicts of Interest: The authors declare no conflict of interest.

References

1. Forum for Agricultural Research in Africa FARA. *Science Agenda for Agriculture in Africa (S3A): "Connecting Science" to Transform Agriculture in Africa*; Forum for Agricultural Research in Africa (FARA): Accra, Ghana, 2014.
2. Mansuri, G.; Rao, V. Community-based and-driven development: A critical review. *World Bank Res. Obs.* **2004**, *19*, 1–39. [CrossRef]
3. Kouévi, T.A.; Fatunbi, O.A. Achieving sustainable impact from development projects through multistakeholders innovation platforms: Lessons from Ghana and Rwanda. In Proceedings of the Innovation Conference—Ghana Proceedings, Accra, Ghana, 27–28 September 2016; pp. 123–131.
4. Atikpo, M.; Onokpise, O.; Abazinge, M.; Louime, C.; Dzomeku, M.; Boateng, L.; Awumbilla, B. Sustainable mushroom production in Africa: A case study in Ghana. *Afr. J. Biotechnol.* **2007**, *7*, 249–253.
5. Atikpo, M.; Onokpise, O.; Abazinge, A.; Awumbilla, B. Utilizing seafood waste for the production of mushrooms. In Proceedings of the Florida Academy of Sciences, Melbourne, FL, USA, 10–11 March 2006; FAS Abstracts 2006 Meeting. Agric. Sci. (AGR): AGR-10. Florida Academy of Sciences Inc.: Melbourne, FL, USA, 2006.
6. Hafiz, F.; Begum, M.; Parveen, S.; Nessa, Z.; Azad, A.K.M. Study of edible Mushroom grown on *Eucalyptus camaldulensis* trunk and under soil of *Albizzia procera*. *Pak. J. Nutr.* **2003**, *2*, 279–282.
7. Matila, P.; Salo-Vaananen, P.; Kanko Aro, H.; Jalava, T. Basis Composition and amino acid contents of Mushrooms cultivated in Finland. *J. Agric. Food Chem.* **2002**, *50*, 6419–6422. [CrossRef]
8. Lelley, J.I.; Janben, A. Interactions between supplementation, fructification surface and productivity of the substrate of *Pleurotus* species. In Proceedings of the First International Conference on Mushroom Biology and Products, Hong Kong, China, 23–26 August 1993; Chang, S., Buswell, J.A., Chiu, S., Eds.; 1993; pp. 85–92.
9. Tortoe, C.; Quaye, W.; Akonor, P.T.; Buckman, E.S. *Improving Food Security in Africa through Increased System Productivity of Biomass-Based Value Webs*; Technical Report, CSIR-FRI/FARA; FARA: Accra, Ghana, 2017.
10. Rogers, E.M. *Diffusion of Innovations*; Free Press: New York, NY, USA, 1995.
11. Nutley, S.; Davies, H.; Walter, I. *Conceptual Synthesis 1: Learning from the Diffusion of Innovations*; ESRC UK Centre for Evidence Based Policy and Practise: Swindon, UK; Research Unit for Research Utilization: St Andrews, UK, 2012.
12. Partey, S.I.; Kwaku, M.; Frik, O.B. *Perception of Bamboo Leaf as Fodder for Livestock Production in Jeduako in the Dry Semi-Deciduous Zone of Ghana*; Technical Report, INBAR/FARA; FARA: Accra, Ghana, 2017.
13. Engler, B.; Schoenherr, S.; Zhong, Z.; Becker, G. Suitability of Bamboo as an Energy Resource: Analysis of Bamboo Combution Values dependent on the Culm's Age. *Int. J. For. Eng.* **2012**, *23*, 114–121. [CrossRef]
14. Partey, S.I.; Kwaku, M. *Bamboo Leaves as Feed Supplement to Sheep Fed a Basal Diet of Pennisetum purpureum (Hematology and Serum Biochemical Analysis)*; Technical Report, INBAR/FARA; FARA: Accra, Ghana, 2018.
15. Partey, S.I.; Kwaku, M. *Nutritional Potential of Bamboo Leaves for Feeding Goats within the Forest Transition and Savannah Zones of Ghana and the Proximate Composition*; Technical Report, INBAR/FARA; FARA: Accra, Ghana, 2018.
16. Raji, A.O.; Asiru, W.B.; Abass, B.; Adefisan, O.O. *Production of Bio-Plastics and Bio-Gels from Waste of Cassava, Maize and Banana to Promote Their Biomassweb Values*; Technical Report, UI/IITA/FIIRO/FARA; FARA: Accra, Ghana, 2018.
17. Asiru, W.B. *Production of Bio-Gel and Bio-Plastic from Agricultural Wastes*; Contribution to Online BiomassWeb DGroup Discussion on Demand Driven Research Projects in Africa; FARA: Accra, Ghana, 2018.
18. Adegbite, S.A.; Olukunle, O.J.; Olalusi, A.P.; Adebayo, A.; Asiru, W.B.; Awoyale, W. *Mass and Energy Balance Analysis of Pneumatic Dryers for Cassava and Development of Optimization Models to Increase Competiveness in Nigeria*; Technical Report, FUTA/IITA/FIIRO/FARA; FARA: Accra, Ghana, 2018.
19. Bamlaku, A.; Loos, T.K.; Habtamu, D.; Saurabh, G. *Potential of Bamboo for Employment Creation and Food Security Enhancement in Ethiopia*; Technical Report, YIELD/FARA; FARA: Accra, Ghana, 2018.
20. Zenebe, M.; Adefires, W.; Temesgen, Y.; Mehari, A.; Demel, T.; Habtemariam, K. Bamboo Resources in Ethiopia: Their value chain and contribution to livelihoods. *Ethno Bot. Res. Appl.* **2014**, *12*, 511–524.
21. International Network for Bamboo and Rattan (INBAR). *Study on Utilization of Lowland Bamboo in Benishangul Gumus Region, Ethiopia*; INBAR: Beijing, China, 2010.

22. Kwapong, N.A. Bamboo for land restoration in Ghana. In *FAO and INBAR 2018. Bamboo for Land Restoration*; INBAR Policy Synthesis Report 4; INBAR: Beijing, China, 2018.
23. FAO; INBAR. *Bamboo for Land Restoration*; INBAR Policy Synthesis Report 4; INBAR: Beijing, China, 2018.
24. Klerks, L.; Leeuwis, C. Operationalizing Demand Driven Agricultural Research: Institutional Influences in a Public and Private System of Research Planning in the Netherlands. *J. Agric. Educ. Ext.* **2009**, *15*, 161–175. [CrossRef]
25. Klerks, L.; Leeuwis, C. Institutionalizing end-user demand steering in Agriculture R & D: Farmer levy of R & D in the Netherlands. *Res. Policy* **2008**, *37*, 460–472.

© 2018 by the authors. Licensee MDPI, Basel, Switzerland. This article is an open access article distributed under the terms and conditions of the Creative Commons Attribution (CC BY) license (http://creativecommons.org/licenses/by/4.0/).

sustainability

Article

Assessment of Ghana's Comparative Advantage in Maize Production and the Role of Fertilizers

Lilli Scheiterle * and Regina Birner

Institute of Agricultural Sciences in the Tropics (Hans-Ruthenberg-Institute), University of Hohenheim, Wollgrasweg 43, 70599 Stuttgart, Germany; Regina.Birner@uni-hohenheim.de
* Correspondence: Lilli.Scheiterle@uni-hohenheim.de

Received: 19 August 2018; Accepted: 9 November 2018; Published: 13 November 2018

Abstract: Maize is one of the most important cereal crops produced and consumed in West Africa, but yields are far under their potential and the production gap leads to growing import bills. After the structural adjustment program, fertilizer subsidies again became a popular intervention to increase yields in most African countries. Ghana introduced fertilizer subsidies in 2008, with high government expenses. This study assesses the competitiveness of Ghanaian maize production and the significance of socio-economic and management variables in determining high yields in northern Ghana. Household survey data and secondary data were applied in a Policy Analysis Matrix (PAM) to test private and social profitability of the fertilizer subsidy policy. Additionally, a probit model is used to determine the characteristics that contribute to higher yields. The results suggest that production systems with Ghana's above-average yields of 1.5 Mt/ha are profitable at household level and contribute to its economic growth, whereas production systems below this threshold report negative social profits and depend on government intervention. However, fertilizers did not increase the likelihood of a household to fall in the category of high-output production system, whereas the use of improved seeds and herbicides does. In conclusion, the analysis highlights the importance of additional measures, especially the use of supporting inputs as well as management practices, to increased maize productivity.

Keywords: maize; Policy Analysis Matrix; comparative advantage; probit; Ghana

---

1. Introduction

Addressing persistent yield gaps remains a challenge in West Africa. Achieving higher yields is not only important for food security, improved diets, and higher income but it is also a precondition to foster crop use diversification for development of both food and non-food bio-based products [1]. Agriculture is an important sector in Ghana, crop production accounts for about 23% of Ghana's gross domestic product (GDP) and employs one fourth of the country's households [2]. Ghana is informally divided between the south and north. The northern part of the country suffers from poorly developed infrastructure whereas the south has more natural resources, such as minerals, it calls upon a better educated labor force and has a more developed infrastructure with access to the Gulf of Guinea and the lake Volta. Compared to the bimodal rains in the south, the north suffers from unfavorable climatic conditions for crop production due to the single rainy season and high temperatures, nevertheless agriculture it is still the main source of employment (over 60%) and maize is the main crop grown, the Northern region in particular, ranks fifth in the country average maize production (three years average, 2013–2015) [2]. In Ghana, maize is broadly appreciated as a stable crop and it is grown in all agro-ecological zones. More than 50% of rural households cultivate it, traditionally under rain fed conditions. In addition, also 16% of urban households are involved in its production [3]. However, the yield gap is severe, especially in the north. Average maize yields fluctuate between 1.2 and 1.9 metric tons (Mt) per hectare (ha), whereas on-station and on-farm trials suggest that yields average between 4 and 6 Mt/ha of maize are attainable in the country [4]. Agricultural production below production potential, especially for maize yields, is a common phenomenon in most African countries. Generally, it is stated that low yields correlate with low input use and poor technology adoption [5]. Over the past decade, mainly after the successful Malawian fertilizer program between 2005 and 2007, fertilizer subsidy programs have become one of the most popular agriculture policies in Africa. The vision for the program is to increase fertilizer use by smallholder farmers, to improve productivity and to contribute to food security [6]. However, there is no evidence of the efficiency of this policy. The study aims to assess the impact of a fertilizer subsidy program on the private and social profitability of maize production in the more neglected area of the country.

Ghana represents a good case study country as it has operated the fertilizer subsidy program (FSP) since 2008 to counter the high fertilizer cost of 2007. At inception, the FSP, was implemented through a voucher system, which was not crop-specific and not wealth-dependent, that enabled farmers to have access to fertilizer at 50% of the retailer's price. In 2010, the mode of delivery changed to the 'waybill system', which enabled all farmers to access the subsidized prices, but implied high costs for the government [7,8].

As the yield gap and the increasing demand for maize, led to high import bills the government has invested greatly in the FSP. The average expenditure for the program is about 20% of the total expenses of the Ministry of Food and Agriculture (MoFA), reaching up to 53% in 2012 [9]. Import bills in some years exceeded the amount spent on the fertilizer subsidies. In the best-case scenario, the FSP would address the yield gap and the country could be self-sufficient on maize production. Given the high cost of imports despite the FSP, the question arises whether the country has a comparative advantage in maize production or if it would be economically more efficient to import entirely the maize for domestic consumption.

This study empirically investigates whether Ghana has a comparative advantage in maize production under current world prices and domestic policies. To pursue this objective, we apply the Policy Analysis Matrix (PAM) to evaluate the effectiveness of government FSP on the maize production system. As a budget-based method, it measures the effects of a policy on private and social revenues, costs, and profits. Furthermore, the aim is to estimate, based on households' socio-economic characteristics and management factors, the probability that a farm is classified as one with an above-average maize yield. To do so, we applied a probit model to identify the factors that contribute to higher outputs, especially to analyze the role of fertilizers. Ultimately, the study aims to assess whether the country has a comparative advantage to produce maize in the north and how fertilizers contribute to the probability of a farmer to achieve above-average production maize

The rest of the paper is organized as follows: Section 2 provides a description of the research method. The empirical results are presented in Section 3 and in Section 4 provide the discussion of the results. Concluding remarks and summary of the papers main findings are presented in Section 5.

2. Methodology

The following section will provide detailed information on how the data was collected and the research methods applied to assess the efficiency of resource use and the role of single inputs in the above-average production systems.

2.1. Data Collection

The household data being analyzed in this research is from a comprehensive household survey conducted from May to July 2011 in three different regions in northern Ghana: Sissala East (Upper West), Tolon-Kumbungu, Yendi (Northern Region) and Nkoranza (Brong Ahafo). The questionnaire captured data on maize production of 2010. A two-stage clustered sampling procedure was followed. The sampling is based on a work by Winter-Nelson and Aggrey-Fynn (2008), which was based on data from 2007 before inception of the FSP [10]. This approach enabled us to compare the two studies. First, the four districts in the northern sector were selected; then, a sub-cluster was taken from a list of all communities belonging to the district, and at this stage three were randomly selected. In each sub-cluster of communities, a random selection of maize farmers was made. This system yielded 175 datasets on maize-producing households. The farmers within the districts share common characteristics, such as relatively scarce use of inputs, non-irrigated cultivation, and limited mechanization.

For analysis, we divided the dataset at the threshold of 1.5 Mt/ha as it represents the average maize production in the country and we distinguish between high and low producing systems. This allowed to investigate the differences between the systems, and to draw specific conclusions.

Prices, taxes, and levies needed to compute the social costs (both in the crop budget and in the PAM) were acquired from documents and reports from the Ministry of Trade and Industry (MoTI), Customs Exercise and Preventive Service (CEPS) the custom division of the Ghana Revenue Authority (GRA). Additional information on extension services and crop production's features and challenges were recorded during interviews with the MoFA and at Savanna Agriculture Research Institute (SARI).

For triangulation purposes, data on fertilizer, agrochemicals, and transport costs have been recorded from the local markets during informal talks.

2.2. Conceptual Framework

The data was used to construct the PAM and a probit model. For an increased accuracy, the effective plot sizes were measured by GPS devices. The two types of output systems (divided by above or below national average maize production) were considered separately. The framework allows assessing which of the systems is profitable at private and social prices (including family labor costs). Additionally, the probit model enables to estimate the determinants of the high maize productivity farms.

The PAM approach was developed by Monke and Person (1989) to address three principal issues: First, most important for ministries of agriculture, it represents a tool to determine how agricultural policies affect farming profits. Second, how policies impact the economic efficiency of the analyzed system and how this pattern can be, eventually, influenced by public investment. The third issue is related to the allocation of future funds, with the main concern of increasing social profits by improving crop yields and reducing social costs. Additionally, the PAM can be used to estimate the impacts of alternative policies on economic performance and income levels of the researched systems. The PAM includes several theoretical assumptions and empirical simplifications, but the results are theoretically consistent and valuable for policy makers [11].

Therefore, the PAM ultimately contributes to "identify how a given agricultural policy or investment would be likely to affect the general performance of the economy and the private income of farmers" [10] (p. 9). Because of the importance of this topic from a point of view of policy effectiveness, a probit model was applied to estimate the partial effect of the determinants of high-yield maize farmers. Determinants are suggested by the literature and observation from the field; the model included both socio-economic regressors as well as plot-specific input factors.

2.3. Policy Analysis Matrix

The PAM is a general equilibrium and policy-orientated simulation model, which can disaggregate crop production activities by revenues and costs (tradable and non-tradable inputs) [9]. Table 1 shows the how the matrix is constructed. It is used to measure the contribution of a specific agricultural system to a famer's private income and to the overall economy. This approach enables to compare and identify policies that contribute to increase household and national income [11]. The advantage of this approach is that it quantifies the economic efficiency of a given production system and the effects of policies on production technologies. The PAM uses farms budgets, to estimate separately the effects of micro and macro policies, market failures and distortions along the different stages in the production chain. Therefore, the outcome of the matrix facilitates assessing the policies that support the development of new technologies [12].

Table 1. The Policy Analysis Matrix (PAM).

	Revenues	Cost		Profits
		Tradable Inputs	Non-Tradable Inputs	
Private Prices	A	B	C	D
Social prices	E	F	G	H
Divergence	I	J	K	L

Source: Monke and Person 1989. D = A − B − C Private Profits; H = E − F − G Social Profits; I = A − E Output Transfers; J = B − F Input Transfers; K = C − G Factor Transfers; L = D − H = I − J − K Net Transfers.

The first row represents revenues and costs of private prices and reflects what farmers face in the existing market. Costs are divided into two categories: tradable inputs (e.g., fertilizer and fuel) and domestic factors of production, which are generally considered non-tradable in nature (e.g., land, labor, and capital). The costs include the effect of policies e.g., tax, subsidies. Ultimately private profits are calculated through the difference between revenue and both tradable and non-tradable inputs. This data was collected in using the household questionnaire.

The second row represents values adjusted from the first row (private prices), using economic prices. As proxy for the economic prices, world market prices adjusted to their import and export parity prices are used. Opportunity costs are used to estimate the domestic factors of production. For this purpose, we made use of the secondary data collected from statistical departments of the ministries and research institutions in Ghana. The sign of the social profit is important to determine the production system's performance. A negative sigh indicates that social costs exceed the costs of import and the sector relies on economic intervention to function.

The third and last row is calculated by subtracting the values of the social prices from the private prices. It shows the effect of distorting policies and market failures on economic efficiency. It includes the value of the output transferred from society to individuals (A − E). The same method can be applied for transfer of tradable and non-tradable inputs (B − F) and C − G). Economic efficiency can be measured by social profits (H), which is calculated by subtracting the costs of tradable and non-tradable inputs (F + G) from social revenues (E). Social values are calculated, in the case of exported goods in F.O.B. (free on board) prices and import goods in C.I.F. (cost, insurance, freight) prices. The private profitability indicates the competitiveness of a given commodity at its current technology, input costs, output prices and policy transfers. If the value is higher than 0, it implies a comparative advantage.

2.4. Additional Indicators for Policy Analysis

The PAM additionally enables to compute indicators of distortion and comparative advantage. The Effective Protection Coefficient (EPC) is measured as a ratio between the value added in the domestic market and the value added in the international market prices ((A − B)/(E − F)). A value greater than 1 indicates a positive commodity policy (e.g., subsidy to farmer); if the EPC value is less than 1, it means that negative incentives to farmers (e.g., taxes) are applied. Cost Benefit Ratio (CBR) is a broader measure of economic efficiency and indicates private profitability ((F + G)/E). CBR < 1 implies private profit. The Domestic Resource Cost (DRC) measures the comparative advantage or the economic profitability of crop production. The case of social costs for land cannot be assessed because of lacking information on alternatives, as the DRC can be calculated with respect to labor and capital only. The DRC is the ratio of the value of non-tradable inputs to the value added in economic terms (G/(E − F)); furthermore, it is used as a proxy to measure social profits. It indicates the cost of the non-tradable inputs that must rise to get one more unit of value added in economic terms. The lower the value (lower than 1), the greater the country's comparative advantage in its commodity production. The Nominal Protection Coefficient for Outputs (NPC_O) and tradable inputs (NPC_I) are ratios of the private value to the social value, respectively for revenues and the tradable inputs. The Nominal Protection Coefficient (NPC) is used to determine how well government policies give incentives to grow specific crops. If the NPC (A/E) of a crop is greater than 1, the domestic price is higher than the price on the international market, which incentivizes farmers in the country to produce the crop [11].

2.5. The Probit Model

The probit model was applied to analyze the household survey data to estimate the factors that determine household's above-average yields. The regressors included socio-economic characteristics of the household such as experience in farming, level of education and size of household; other characteristics include wealth variables such as the number of cattle and total land owned. Furthermore, variables related to field management were included: use of improved seed, the quantity of fertilizer and herbicides applied, labor, and the size of the plots. The probit model is expressed as:

The probit model in its general form is given below:

$$Yi = Xi\beta i + \mu i$$

In this case, the Yi is the dependent variable (farmers producing more than 1.5 Mt/ha), Xi is the vector of explanatory variables that contribute to the plot yield above the national average, whereas μi is the error term.

The specific probit model used is specified as:

$$Yi = \beta_0 + \beta_1 FarmExp + \beta_2 Edu + \beta_3 HhSize + \beta_4 NCattle + \beta_5 FarmSize + \beta_6 Ext + \beta_7 ImpSeed + \beta_8 Fert + \beta_9 Herb + \beta_{10} Labor + \beta_{11} PloSize + \beta_{12} FarmGroup + \beta_{13} District1 + \beta_{14} District2 + \beta_{15} District3 + \mu i$$

The socio-economic variables in the empirical model are denoted as follows. FarmExp denotes the experience in farming of the household head and Edu her or his education level. It was hypostasized that with more experience and a higher level of education, farmers would have a better understanding of farm practices and input use. As determinants related to the household's wealth, we included NCattle, which denotes the number of cattle owned, and FarmSize, as it was assumed that better-off farmers are more likely to have better access to resources and therefore achieve higher yields. Ext denotes access to extension services by the household as farmers are more likely to have higher yields as they receive information on practices and input use. ImpSeed refers to the use of seeds other than the ones farmers saved from the previous harvest. Part of the success of the Green Revolution lies in the joint use of seeds, agrochemicals, and fertilizers; therefore, we included this input as regressors. Fert stands for fertilizer and Herb for herbicide. Labor was also included as input variable. As more labor is supplied in a single plot, for example for weeding, the farm will more likely have higher maize yields. PlotSize represents the size of the plot measured by GPS device. Plot size is very much discussed in literature as a factor of productivity and can have either a negative or a positive effect of yield. FarmGroup denotes the participation of the household head in a farm-based organization (FBO). A farmer part of an FBO is expected to have better access to information on market prices, distribution systems and resource use, which can influence the yield performance. Additionally, we controlled for district-specific, non-observed factors (e.g., market access, specific soil, and weather conditions).

3. Results

The following section details the results from different analytical methods to assess the comparative advantage in the two production systems and identify the role of inputs and other farm specific factors.

The crop budget is the first important result of the analysis, production revenues and cost of inputs, thus the determination of the farm profits. Table 2 represents the outcome of the crop budget, based on the household (HH) questionnaire, illustrating the allocation of household resources and the output in the two systems. Net revenues should be interpreted as returns to land and management skill, since management and land are not charged in the budget. Therefore, net revenue is not "profit" but rather the returns to having land and knowing how to allocate labor and inputs on it. The last row of Table 2 shows household (HH) revenue not accounting family wages. In this case, all systems allocate their resources efficiently. If we consider the opportunity cost of family labor, the outcome highly drops.

Table 2. Crop budget of high and low-output system (values in GHS/ha unless otherwise specified).

		High	Low
Total Grain (Kg/Ha)		2149	938
Tradable Inputs		152	86
Non-Tradable Factors	HH labor	142	106
	Wage labor	68	30
Own Capital (Tools and Small Implements)		43	33
Service And Non-tradable Inputs		116	70
Total Costs		521	324
Total Revenue		855	373
Net Revenue		334	49
Net Revenue (Without HH Labor)		476	155

Based on crop budget and the secondary data collected from government and research institutions on transport, taxes, levies, subsidies, and other cost we computed the PAM. Tables 3 and 4 present the results for two output systems. Private profits are outcomes of the crop budget and reflect the difference between revenues and costs at current market prices (first row, letter D). Whereas social prices (second row, letter H) are used to measures the efficiency and competitive advantage of maize production. The positive outcomes in the high production system indicate efficient use of economic resources, suggesting that production costs are lower than import costs at the current policies and technology adoption. The situation changes in the below average system.

Table 3. Average of PAM values in GHS/ha of 121 farmers in the low-output systems.

Low	Revenues	Input Costs	Factor Cost	Profits
Private	389 (A)	107 (B)	218 (C)	64 (D)
Social	334 (E)	107 (F)	217 (G)	−10 (H)
Divergences	55 (I)	0 (J)	1 (K)	54 (L)

Table 4. Average of PAM values in GHS/ha of 54 farmers in the high-output systems.

High	Revenues	Input Costs	Factor Cost	Profits
Private	890 (A)	186 (B)	334 (C)	369 (D)
Social	746 (E)	191 (F)	339 (G)	234 (H)
Divergences	144 (I)	−5 (J)	−5 (K)	155 (L)

Private and social prices are import price plus inland transport costs, adjusted for processing losses. The private prices are validated against market prices. In the second row, social outputs (E) are valued at C.I.F. prices since they are treated as exportable, inputs (F) are valued according to F.O.B. prices since they are imported goods, and international prices are used since the products are traded at world prices.

The divergence, in the third row, between observed private (actual market) and estimated social prices (efficiency) are explained by market failures or policies [13]. Two possible policies influence the divergence observed in input transfers (J) and output transfers (I) between reported and international market prices: either commodity-specific or exchange rate policies. The slight overvaluation of the currency indicates a small divergence in the input transfers (J), although neglectable. All tradable inputs are calculated by separating each component of the intermediate inputs into factor costs and tradable input categories. Output transfers (I) are relatively high compared to input transfers (J). This factor (I) indicates the market price minus the efficiency valuation of maize; the divergence can be attributed to distorting policies, in particular to import and sales tax on goods, since market failures are difficult to identify empirically. Factor transfers (K) are the difference between all factors of production (C) and their social cost (G): the effects of distorting policies on output or factor markets are very common in developing countries. Net transfers (L) are an important result of the PAM and show the extent of inefficiency of the system; policy can be aimed to reduce the degree of distortion. The positive net transfers (L) suggest that the net effect of policy intervention is increasing production at household level in all systems.

3.1. Protection and Competitiveness Coefficients

Protection and competitiveness coefficients of output systems are summarized in Table 5.

Table 5. Protection and competitiveness coefficients derived from PAM.

	EPC	**CBR**	**DRC**	**NPC$_O$**	**NPC$_I$**
High	1.27	0.58	0.61	1.19	0.98
Low	1.24	0.84	0.96	1.16	1.00

The Effective Protection Rate (EPC) indicates the effect of protection policies on the agricultural system, combining the effect of commodity price policies. In our case, government's fertilizer subsidies result in a net positive incentive for maize production. Therefore, producers are protected by policy intervention on value added processes. This finding is confirmed by the values of the NPC$_O$ and NPC$_I$. Both inputs and outputs are protected by commodity (price) policies. The CBR indicates private profitability. The values of the CBR in the low and high production systems suggest that 0.84 and 0.58 GHS are needed to generate 1 GHS of output in the respective systems. The indicators of comparative advantage are DRC and social profits (H). For both output systems the DRC is less than one, indicating that the systems are economically efficient. However, DRC indicates the cost of domestic factors incurred to obtain one unit of added value in economic terms. The value of 0.61 indicates a higher comparative advantage of maize production of farms in the high-output system, whereas the low-output systems are barely in the range of economic efficiency which is reflected by the negative value (letter F) in Table 3. The NPC$_O$ coefficients greater than one, which indicates agricultural policy is protecting the output price at domestic level, raise farm-gate prices to a higher level than the world maize price. The NPC$_I$ values close to one indicate that input costs in these systems are only slightly lower and equal (low-output) to the international price.

3.2. Empirical Probit Model Estimates for High Production Plots

From the previous section one can conclude that both systems are profiting from the FSP, but the high-output system can make better use of its resources. To identify the factors that contribute to the above-average output of the plots a probit model was computed. The estimates derived are presented in Table 6. The results indicate the regressors, which positively or negatively contributed to the probability of being part of the high-output maize plots, while controlling for variations (e.g., soil and weather) with district dummies. Farmer's experience did significantly affect the yield level, since farmers management skills are supposed to increase with experience. The variable on farmer's age was omitted from the model specification since highly correlated with the farming experience variable.

Table 6. Plot specific factors determining the categorization into the high-output systems.

	Robust Std. Err.	Marginal Effects
Farming experience (years)	0.003	0.006 *
Education HH head (years)	0.057	0.106
Household size	0.005	0.000
No. Cattle	0.037	0.043
Use of improved seeds (dummy)	0.094	0.225 *
Extension service (dummy)	0.073	0.087
Fertilizer (kg/ha)	0.000	0.000
Herbicides (L/ha)	0.013	0.029 *
Labor (person-days/ha)	0.000	0.000
Size of maize plot (ha)	0.034	−0.090 *
Total farm size (ha)	0.006	0.006
Farmers group (dummy)	0.078	−0.042
Observations	175	
Wald Chi2	44.53 ***	
Log pseudolikelihood	84.50	

Significant at 10% level *, at 1% level ***.

Plots with improved seeds had a 22% higher likelihood to yield above average. In this case, we considered improved seeds as those that were purchased in agro-input dealers' shops in a certified package. The variety that is the most available in the Ghanaian seed market is the open-pollinated variety *Obatampa*, and a large share of farmers mentioned it as the variety that they purchased. Furthermore, the use of hybrids is very limited in Ghana, about 3% [14].

The variable addressing the use of herbicides, expressed in liters per hectare, increased the likelihood of a plot to be classified as an above-average output plot. Weeds are a major problem, and if not addressed, fertilizers may support their growths and inhibit maize production. It is also true that weeding is a time-consuming activity and labor is limited during peak seasons. Therefore, herbicide use is a welcoming and efficient alternative to manual labor.

Differently than expected, fertilizer did not have a significant effect and smaller sized plots decreased the likelihood of being part of the high production system. We tested the model against the uncentred variance inflation factor (VIF) test; single regressors all show values below 7, and an overall mean of 3.43, which indicates no multicollinearity issues.

4. Discussion

This study assessed the comparative advantage of maize production in northern Ghana, and the contribution of single inputs, especially fertilizer application, to high-output plots. The results showed that a set of agricultural and macroeconomic policies are supporting competitiveness of maize production as import substitution, this is particularly true for high-output-systems.

The PAM shows that positive private profits are achieved by farmers, indicating cost effectiveness of the production systems in the short to medium term. However, high-output farmers use scarce resources more efficiently. The low-output farmers instead depend more on government intervention at the margin and barely have a comparative advantage in maize production. The negative divergences between private and social costs of domestic factors (K) in the high production systems indicates a reduction of cost to individual farmers, a sign of interventions that lowers the cost of capital and labor. The positive values of the divergences of private and social profits (L) suggest that the net effect of policies on maize production increases profitability in both systems analyzed. The data suggest that, Ghana might not be able to export but is still better off with the domestic production than importing, since C.I.F prices are high compared to F.O.B. prices, respectively 407 and 66 $/Mt in the survey year 2010.

The ratios of the protection and competitiveness coefficients are summarized in Table 5. The value of the nominal protection coefficient on tradable outputs, NPCo, is 16 and 19 points higher than one, indicating that the inland farm-gate price is higher than the world trade price due to import duty on maize. The NPC_I of less than one indicates that the price of tradable inputs is lower than the international market price, suggesting that policies in Ghana are reducing the cost of tradable inputs. This leads to a positive input transfer to the agricultural system, which is confirmed by the government FSP. These two effects, output price and tradable input price, are combined in the EPC. An EPC greater than one indicates how much the observed value differs from what it would be without policy effects, in this case, the value added in private prices and the world prices. The policy transfers from output and input transfers are about 25% (EPC is 27 and 24 points above 1 in the high and the low system, respectively) greater than private profits would be without policy interventions. The DRC is an indicator of efficiency closely related to the social profits row (E, F, G and H). In the low-output system, the value is closer to one, indicating that the value of domestic resources used in production is higher. This suggests that at the current level of technology and input management, especially in the low-output system, scarce resources are not used efficiently. The policies in place should aim at achieving a comparative advantage for the low-output farm system, since they constitute the larger share of the farms. Production levels below the national average are socially expensive and the import price is lower than the production price. The lower value of DRC of the high-output system indicates a higher comparative advantage.

Overall, the PAM result shows that domestic maize production is barely socially profitable for the low-output farms. With the introduction of the FSP, small landholders had access to initially economically inaccessible inputs. Low-output farmers using fertilizers now depend on government intervention. Furthermore, factor transfers indicate small support of policy intervention in the capital market, which is not very well developed [15], it is expected that lower cost of capital and contribute to private and social profits.

These findings differ from the results of the study by Winter-Nelson and Aggry-Fynn undertaken from November 2007 to February 2008, before the introduction of the FSP. Private input costs in the low-production system are twice as much as in the previous study, e.g., factor costs. The main difference found was that farmers in the previous study did not apply any mineral fertilizer to their maize fields. The higher cost is related to the greater use of fertilizers, a shift that might be related to the governmental FSP introduced in August 2008. With the new policy, smallholder farmers had access to initially economically inaccessible inputs. This massively increased their expenditure. The use of fertilizers presumably also affected labor allocation. Repeated application of fertilizer and the more intense weeding needed due to increased weed growth explain the increased labor costs. Despite increased fertilizer application, maize yield did not improve.

Unlike in the previous study, low-output farmers using fertilizers now depend on government intervention. Factor transfers (K) indicate small support of policy intervention in the capital market (lower cost of capital would increase private and social profits). With values similar to the study of 2008, even with substantially higher private costs, net transfers are positive, and the net effect of policies increased profitability of all systems to the same level as in 2007. Our results also indicate that the income situation did not change due to the fertilizer subsidies program, since farmers spent more to purchase tradable input, but did not obtain higher yields. Leading to the conclusion that the potential of mineral fertilizer is not yet fully realized. The strategy should aim not only to increase the amount of fertilizer used but also invest in extension service to improve management skills that foster production. As the probit model showed other factors contribute more to the probability of a plot to fall in the high-output system than fertilizer. Furthermore, the timing of applications is of primary importance to achieve its full potential of the inputs. It is also known that improved maize varieties adapted to the harsh conditions perform better and would increase fertilizer efficiency. Higher producing systems showcase the importance of using improved maize varieties. Traditionally, farmers save seeds from the previous harvest for the next growing season; in this study only 22% of the land cultivated with maize was sown with purchased seeds. This reduces input costs and problems with an insufficient seed supply. Furthermore, only 7 out of 52 farmers recalled that the extension service officer provided information on seed varieties. The main constraints against adoption of improved seeds mentioned by farmers were lack of financial means (74%), the perception that the quality of recycled seeds is enough (26%), and lack of knowledge about the existence of improved seeds (10%). The results are reflected in a recent study, in the same region, which reports that only 3% of the farmers grow hybrids [16]. The MoFA district office was reported to be one of the few places where seeds could be purchased, aside from the input dealers but, both rarely had them available. The same problem was observed for fertilizers, which hinders appropriate application. Poor input availability is a major reason for low productivity [17]. Furthermore, farmers prefer the well-known, open-pollinated varieties (OPVs), especially *Obatampa* which was released in 1992 and is since than the most popular variety [18]. With proper management this variety has the potential to yield up to 4.6 Mt/ha as shown in both on-station and on-farm trials [18]. However, varieties such as *Sanzal-sima*, known as one tolerant to drought and parasitic weed *Striga* sp., and *Mamaba*, a hybrid maize variety, were not known by the households even though they have very good yield performances in northern Ghana [19]. Besides the price (50% more than for *Obatampa*), hybrids are considered risky since its characteristics are lost when replanted and they are in greater need for inputs. Farmers investing in improved seeds may increase their yield to 2.3 Mt/ha and those combining them with fertilizers achieve up to 3.4 Mt/ha also in dry areas; however initial costs and risks are higher [20]. The findings from the probit model fall in line with the literature, as the use of improved seeds was significant in high-output farms. Only 26 farmers used to buy seeds of which 21 fall in the high producing category. However, the use of fertilizer was not significant to explain the likelihood of being part of the high-yield plots. This is surprising and questions the importance of the FSP. Farmer's experience, the use of herbicides and improved seeds were significant, which illustrates that better farm management and complementing inputs need as much attention as fertilizers alone. In Ghana, the extension service coverage is very low (1 officer for about 1000 farmers) and the information material available is very limited [21]. This explains why the extension service variable in the probit did not have a significant effect on agricultural productivity. The low outreach of the service is very limited, only 26% of the interviewed households had received extension service in the previous two years. This reflects the literature on Ghana which shows that only 12% of farmers had adopted a new technique in the previous two years, indicating that farmers face considerable challenges to innovate [22]. Low access to physical resources for transport and trainings, low salary, as well as lack of training and incentives are affecting the quality of the extension officer's service delivery. Extension service and the education system are key policy instruments to improve productivity in agriculture [23].

Farmers mentioned only two types of fertilizer SoA (Sulphate of Ammonia) and NPK (Nitrogen, Phosphorus and Potassium), yet little was known about their composition, efficient application methods, and timing. Correct timing of mineral fertilizer application is important to ensure maximum efficiency and reduce runoff and leaching. Cost-efficient measures to improve soil fertility can also be achieved through soil conservation practices, such as mulching to increase soil organic matter, reduce kinetic energy of runoff and increase soil moisture. Fertilizer use must be supported by training to avoid wastage and negative externalities. Major constraints in crop production perceived by the farmers interviewed include other aspects, such access to financial services, access to extension services and pest control. The lack of information on improved seeds, insecticides, and pesticides, as well as lack of a well-developed seed supply system clearly affects adoption rate [24,25]. Though some farmers claimed that they do not need seeds or that improved seeds are not profitable, price is an important factor preventing adoption [26], as fertilizer price is.

The FSP targets only one variable, but the results confirm that other factors are equally important to improve biomass production and food security in the study region. The adoption of improved maize varieties, proper timing of fertilizer application, good agricultural practices to improve soil fertility and a functioning extension service that provides information on management practices are pivotal to the close the yield gap in northern Ghana. The World Bank underlines the need to offer major inputs (fertilizers, agrochemicals, and improved seed) as a package and to improve the currently fragmented distribution network [27]. The use of complementary inputs and management practices that consider soil parameters and include local limitations are likely have the potential to increase yields more than if the focus is only on fertilizers [28,29]. It is also true that farmers face resource constrains (e.g., financial and labor) to follow the recommendations of fertilizer application (e.g., timing advices and application modalities) [30].

Political forces and priorities play a part in the allocation of government's resources. An empirical study examined the effect of districts' political characteristics on fertilizer voucher allocation in 2008. The program first targeted districts where the ruling party had lost support in the previous elections and more intensively those districts that registered high percentage losses. This type of allocation was not efficiency-based but politically orientated, resulting in fewer vouchers to poorer farmers. According to this study, the three regions where we conducted our survey rank in the top four positions for average number of vouchers available per 1000 farmers [7]. Governance challenges of the FSP are found to still affect targeting of smallholder farmers in Ghana [9].

5. Conclusions

This research uses the PAM approach to analyze the comparative advantage of two systems of maize production in the northern Ghana. The agricultural sector is important to Ghana's economy, and as stated by many authors, agricultural growth in early stages of development has the greatest impact on overall economic performance and poverty reduction.

The data show that even the northern regions of Ghana, where the environmental conditions are harsh the high-output system has the comparative advantage in maize production. The lack of access to information on crop management and improved technologies prevents small-scale farmers from realizing their crops full potential. This is coupled with a weak financial and credit system and poor infrastructure. However, the use of fertilizer, under the current conditions, did not increase the likelihood of producing more than the national average, whereas other measures have a higher impact on maize yield and should be considered by policy makers to bridge the productivity gap. This is an important finding, especially looking at the controversial policy issue debated in many African countries.

In conclusion, maize production systems have the potential to increase productivity to ensure food security and meet the challenges of biomass use in the bioeconomy. However, the policy measures need to address more than just the increased use of fertilizer. The combined access to information and new technologies, as well as the improvement of soft and hard infrastructure to reach out to a larger number of rural communities, are considered to be pivotal for the development of the agricultural sector, especially in the northern Ghana.

Author Contributions: L.S. and R.B. conceptualization, design and analysis; L.S. investigation and writing.

Funding: The research conducted for this paper was supported by the International Food Policy Research Institute (IFPRI) and by the German Federal Ministry of Education and Research (BMBF). The BMBF funds are part of the collaborative project "Improving food security in Africa through increased system productivity of biomass-based value webs" (BiomassWeb); the project is part of the GlobE—Research for the Global Food Supply programme (Grant No. 031A258H).

Acknowledgments: The authors are thankful to Alex Winter-Nelson, University of Illinois, USA, Shashi Kolavalli from IFPRI, Ghana, and to Felix Asante from the University of Ghana, Legon, for their input during conceptualization and design of the study, and their support during field research. We would like to express our gratitude to Yaa Adwinmea Attipoe and Harald Zeller for their assistance during data collection. We would also like to convey our heartfelt appreciation to the communities and respondents from the local institutions who kindly contributed their valuable time towards the data collection. Our sincerest thanks go to the reviewers for their comments and suggestion on the earlier version of this paper.

Conflicts of Interest: The authors declare no conflict of interest.

References

1. Virchow, D.; Beuchelt, T.D.; Kuhn, A.; Denich, M. Biomass-Based Value Webs: A Novel Perspective for Emerging Bioeconomies in Sub-Saharan Africa. In *Technological and Institutional Innovation for Marginalized Smallholders in Agricultural Development*; Gatzweiler, F.W., von Braun, J., Eds.; Springer International Publishing: Basel, Switzerland, 2016; pp. 225–238.
2. Ministry of Food and Agriculture (MoFA). *Agriculture in Ghana: Facts and Figures 2015. Statistics, Research and Information Directorate*; MoFA: Accra, Ghana, 2016.
3. Quiñones, E.J.; Diao, X. *Assessing Crop Production and Input Use Patterns in Ghana–What Can We Learn from the Ghana Living Standards Survey (GLSS5)*; Ghana Strategy Support Program (GSSP) Working Paper No. 0024; International Food Policy Research Institute: Accra, Ghana, 2011.
4. FAOSTAT. Available online: http://www.fao.org/faostat/en/#home (accessed on 14 September 2018).
5. FAO. *Ghana Country Fact Sheet on Food and Agriculture Policy Trend*; APDA—Food and Agriculture Policy Decision Analysis: Rome, Italy, 2015.
6. Benin, S.; Johnson, M.; Abokyi, E.; Ahorbo, G.; Jimah, K.; Nasser, G.; Owusu, V.; Taabazuing, J.; Albert, T. *Revisiting Agricultural Input and Farm Support Subsidies in Africa: The Case of Ghana's Mechanization, Fertilizer, Block Farms, and Marketing Programs*; IFPRI Discussion Paper 01300; International Food Policy Research Institute: Accra, Ghana, 2013.
7. Banful, A.B. Old Problems in the New Solutions? Politically Motivated Allocation of Program Benefits and the 'New' Fertilizer Subsidies. *World Dev.* **2011**, *39*, 1166–1176. [CrossRef]
8. Houssou, N.; Andam, K.S.; Collins, A.-A. *Can Better Targeting Improve the Effectiveness of Ghana's Fertilizer Subsidy Program? Lessons from Ghana and Other Countries in Africa South of the Sahara*; IFPRI Discussion Paper 1605; International Food Policy Research Institute: Accra, Ghana, 2017.
9. Resnick, D.; Mather, D. *Agricultural Inputs Policy under Macroeconomic Uncertainty: Applying the Kaleidoscope Model to Ghana's Fertilizer Subsidy Programme (2008–2015)*; Discussion Paper 01551; International Food Policy Research Institute: Accra, Ghana, 2016.
10. Winter-Nelson, A.; Aggrey-Fynn, E. *Identifying Opportunities in Ghana's Agriculture: Results from a Policy Analysis Matrix*; Ghana Strategy Support Program (GSSP) Background Paper No. 12; International Food Policy Research Institute: Accra, Ghana, 2008.
11. Monke, E.; Pearson, S. *The Policy Analysis Matrix for Agricultural Development*; Cornell University: Ithaca, NY, USA, 1989; ISBN 0-8014-9551-2.
12. Shapiro, I.B.; Staal, J.A. The Policy Analysis Matrix Applied to Agricultural Commodity Markets. In *Price, Products and Market: Analyzing Agricultural Markets in Developing Countries*; Lynne Rienner Pub. Inc.: Boulder, CO, USA, 1995; pp. 73–96.
13. Ogbe, O.A.; Okoruwa, V.O.; Jelili, S.O. Competitiveness of Nigerian Rice and Maize Production Ecologies: A Policy Analysis Approach. *Trop. Subtrop. Agroecosyst.* **2011**, *14*, 493–500.
14. Tripp, R.; Ragasa, C. *Hybrid Maize Seed Supply in Ghana*; Ghana Strategy Support Program (GSSP) Working Paper No. 40; International Food Policy Research Institute: Accra, Ghana, 2015.
15. Nair, A.; Fissha, A. *Rural Banking: The Case of Rural and Community Banks in Ghana*; Agriculture and Rural Development Discussion Paper 48; World Bank: Washington, DC, USA, 2010.
16. Morris, M.L.; Tripp, R.; Dankyi, A. *Adoption and Impacts of Improved Maize Production Technology: A Case Study of the Ghana Grains Development Project*; Economics Program Paper 99-01; CIMMYT: Mexico City, Mexico, 1999.
17. Feder, G.; Richard, E.J.; Zilberman, D. Adoption of Agricultural Innovations in Developing Countries: A Survey. *Econ. Dev. Cult. Chang.* **1985**, *33*, 255–298. [CrossRef]
18. Ragasa, C.; Dankyi, A.; Acheampong, P.; Wiredu, A.N.; Chapoto, A.; Asamoah, M.; Tripp, R. *Patterns of Adoption of Improved Rice Technologies in Ghana*; Ghana Strategy Support Program (GSSP) Working Paper, No. 36; International Food Policy Research Institute: Accra, Ghana, 2013. [CrossRef]
19. Wiredu, A.N.; Gyasi, K.O.; Abdoulaye, T.; Sanogo, D.; Langyintuo, A. *Characterization of Maize Producing Households in the Northern Region of Ghana Country Report—Household Surve Y Characterization of Maize Producing Households in the Northern Region of Ghana*; CSIR-SARI: Tamale, NY, USA, 2008.

20. WABS Consulting Ltd. Draft Report: Maize Value Chain Study in Ghana: Enhancing Efficiency and Competitiveness. Available online: http://www.valuechains4poor.org/file/Maize_Value%20Chain_WAB_Dec_08.pdf (accessed on 16 August 2018).

21. Sam, J.; Osei, S.K.; Dzandu, L.P. Evaluation of Information Needs of Agricultural Extension Agents in Ghana. *Inf. Dev.* **2017**, *33*, 463–478. [CrossRef]

22. Belden, C.; Birner, R.; Asante, F.; Horowitz, L. Agricultural Extension in Ghana Results of a Survey of Agricultural Extension Agents in Six Districts. 2010. Available online: https://gssp.files.wordpress.com/2010/05/agricultural_extension_draft.pdf (accessed on 16 August 2018).

23. Binam, N.J.; Tonye, J.; Nyambi, G.; Akoa, M. Factors Affecting the Technical Efficiency among Smallholder Farmers in the Slash and Burn Agriculture Zone of Cameroon. *Food Policy* **2008**, *29*, 531–545. [CrossRef]

24. Doss, C.R.; Morris, M.L. How Does Gender Affect the Adoption of Agricultural Innovations? The Case of Improved Maize Technology in Ghana. *Agric. Econ.* **2001**, *25*, 27–39. [CrossRef]

25. Poku, A.-G.; Birner, R.; Gupta, S. Why Do Maize Farmers in Ghana Have a Limited Choice of Improved Seed Varieties? An Assessment of the Governance Challenges in Seed Supply. *Food Secur.* **2018**, *10*, 27–46. [CrossRef]

26. Tahirou, A.; Sanogo, D.; Langyintuo, A.; Bamire, S.A.; Olanrewaju, A. *Assessing the Constraints Affecting Production and Deployment of Maize Seed in DTMA Countries of West Africa*; IITA: Ibadan, Nigeria, 2009; ISBN 978-131-342-0.

27. World Bank. *Awakening Africa's Sleeping Giant: Prospects for Commercial Agriculture in the Guinea Savannah Zone and Beyond*; World Bank: Washington, DC, USA, 2009; ISBN 978-0-8213-7941-7.

28. Marenya, P.P.; Barrett, C.B. State-Conditional Fertilizer Yield Response on Western Kenyan Farms. *Agric. Appl. Econ. Assoc.* **2009**, *91*, 991–1006. [CrossRef]

29. Scheiterle, L.; Häring, V.; Birner, R.; Bosch, C. Soil, striga or subsidies? Determinants of maize productivity in northern Ghana. *Agric. Econ.*. Unpublished work.

30. Van Asselt, J.; Di Battista, F.; Kolavalli, S.; Udry, C. *Agronomic Performance of Open Pollinated and Hybrid Maize Varieties Results from on-Farm Trials in Northern Ghana*; Ghana Strategy Support Program (GSSP) Working Paper 44; International Food Policy Research Institute: Accra, Ghana, 2018.

© 2018 by the authors. Licensee MDPI, Basel, Switzerland. This article is an open access article distributed under the terms and conditions of the Creative Commons Attribution (CC BY) license (http://creativecommons.org/licenses/by/4.0/).

sustainability

Article

Options for Sustainable Intensification of Maize Production in Ethiopia

Amit Kumar Srivastava [1,*], Cho Miltin Mboh [1], Babacar Faye [1], Thomas Gaiser [1], Arnim Kuhn [2], Engida Ermias [2] and Frank Ewert [1]

[1] Institute of Crop Science and Resource Conservation, University of Bonn, Katzenburgweg 5, D-53115 Bonn, Germany; cmboh@uni-bonn.de (C.M.M.); babafaye@uni-bonn.de (B.F.); tgaiser@uni-bonn.de (T.G.); frank.ewert@uni-bonn.de (F.E.)
[2] Institute for Food and Resource Economics, University of Bonn, Nussallee 21, D-53115 Bonn, Germany; arnim.kuhn@ilr.uni-bonn.de (A.K.); ermias.engida@ilr.uni-bonn.de (E.E.)
* Correspondence: amit.srivastava@uni-bonn.de; Tel.: +49-228-73-2876

Received: 23 September 2018; Accepted: 17 March 2019; Published: 21 March 2019

Abstract: The agricultural intensification of farming systems in sub-Saharan Africa is a prerequisite to alleviate rural poverty and to improve livelihood. In this modelling exercise, we identified sustainable intensification scenarios for maize-based cropping systems in Ethiopia. We evaluated Conventional Intensification (CI) as continuous maize monocropping using higher Mineral Fertilizer (MF) rates with and without the incorporation of Crop Residues (CR) in the soil. We also evaluated the effect of groundnut in rotation with the maize-based cropping system with the current Farmer's Practice + Rotation (FP + Rotation) and increased MF application rates (CI + Rotation) combined with CR incorporation. The results suggest that, under CI, there was a positive effect of MF and CR. The incorporation of only CR in the field increased the maize yield by 45.3% compared to the farmer's yield under current MF rates. CR combined with higher MF (60 kg N ha^{-1} + 20 kg P ha^{-1}) increased the yield by 134.6%. Incorporating CR and MF was also beneficial under rotation with groundnut. The maize yields increased up to 110.1% depending upon the scenarios tested. In the scenario where CR was not incorporated in the field, the maize yield declined by 21.9%. The Gross Economic Profit suggests that groundnut in rotation with maize is advantageous across Ethiopia in terms of the net return with a few exceptions.

Keywords: sustainability; intensification options; maize; groundnut; crop residue; crop model

1. Introduction

Sub-Saharan Africa (SSA) is the region at greatest food security risk because, by 2050, its population is likely to increase 2.5-fold and the demand for cereals will approximately be tripled [1]. Also noted is that SSA's self-sufficiency (a ratio between domestic production and total consumption or demand) in staple cereals is among the lowest compared to other subcontinents, indicating the current levels of cereal consumption already depend on substantial imports.

The main reasons for the low self-sufficiency include soil nutrient depletion, soil erosion, and erratic or low precipitation [2]. The prevailing practice of low-input agriculture is not only providing little outputs but also detrimental to soils [3]. These issues, combined with continuous cereal-based cropping systems without sufficient nutrient inputs to the soil, have led to large-scale declines in soil fertility and persistently poor crop yields on smallholder farms [4]. The challenge of meeting the demands of a growing population can be met by bringing new land into cultivation [5] but that is not a sustainable solution, as often the suitable lands are already in use [6] and land naturally is a limited resource. Nitrogen (N), as one of the most limiting nutrients in agriculture, is a key component in the proper functioning of cropping systems. However, a dependence on synthetic N fertilizers has adverse economic and environmental consequences, such as the nitrate pollution of groundwater, the atmospheric pollution from ammonia, and a contribution to global warming due to nitrous oxide emissions [7]. There are situations in arable systems where the introduction of legumes has both economic and environmental advantages, especially when grain legumes achieve high prices as human food. Grain legumes fix atmospheric nitrogen gas (N_2) and can contribute to the N economy of fields, provide other rotational benefits to subsequent crops, produce in situ high-quality organic residues with a high N concentration and a low C to N ratio, and thereby contribute to integrated soil fertility management [8]. Moreover, ecosystem services, including reduced nitrous oxide emissions and nitrate leaching and increased biodiversity, are currently not awarded through payments, so they are not considered in a farmer's economic calculations. Farmers seldom consider the long-term benefits, focusing instead on single years. This leads to an underestimation of the services provided by legumes. The valuation of such services requires an assessment at the cropping-system scale [9].

To meet the increasing food demands in SSA and to protect the environmental quality simultaneously in a sustainable manner, it is necessary to optimize agronomic management practices to enhance the nitrogen and water use efficiency [10]. Quantifying the rotational effect of grain legumes on subsequent crops is important for understanding the adoption potential of legume technologies as well as their impact on the sustainability of production [11].

Various approaches have been proposed to overcome soil nutrient limitations such as (i) conventional intensification mainly based on the increased use of mineral fertilizer [12], (ii) using legumes in rotation with the main crop or intercropping systems [3], or (iii) a mix of both by rotation with legumes and supplementary mineral N supply [13]. The mixed approach is widely being promoted in agricultural development programs for small-scale farming in SSA [14], while conventional intensification is the approach mostly taken currently in large-scale farming [2]. We are unaware of any recent studies in Ethiopia that synthesize the effect of legumes in rotation with the main crop evaluating the options of sustainable intensification.

Therefore, here, we use a biophysical modelling framework combined with a cost-benefit analysis to identify sustainable intensification scenarios for maize-based cropping systems in Ethiopia through investigating (i) the long-term crop yield response and (ii) the change in crop available N and organic carbon over time.

2. Materials and Methods

2.1. Study Area and Simulation Units

Ethiopia lies within the tropics between $3°24'$ and $14°53'$ N and $32°42'$ and $48°12'$ E with an estimated arable land area of 15.1 million hectares [15]. The climate of the country is diverse, ranging from semiarid desert in the lowlands to humid and warm (temperate) in the southwest. The mean annual rainfall distribution ranges from a maximum of more than 2000 mm over the Southwestern highlands to a minimum of less than 300 mm over the Southeastern and Northwestern lowlands. The mean annual temperature also varies widely, from lower than 15 °C over the highlands to above 25 °C in the lowlands [16]. The simulations were done at the 1 km grid cell level, where cropland (Figure 1) and soil data are available (details about soil data is under Section 2.6.1). The long maturing cycle maize variety (see details in Section 2.3) was used in the simulations in Agroecological Zones (AEZs) 1 and 2 (Figure 1) where the length of major crop growing season is more than 160 days, elsewhere (AEZ 3) a medium maturing cycle variety was used in the simulations. The simulated yield from all the simulation units over 7 years (2004–2010) within each administrative zone was averaged to obtain a representative value for a specific year to compare them with the observed yield.

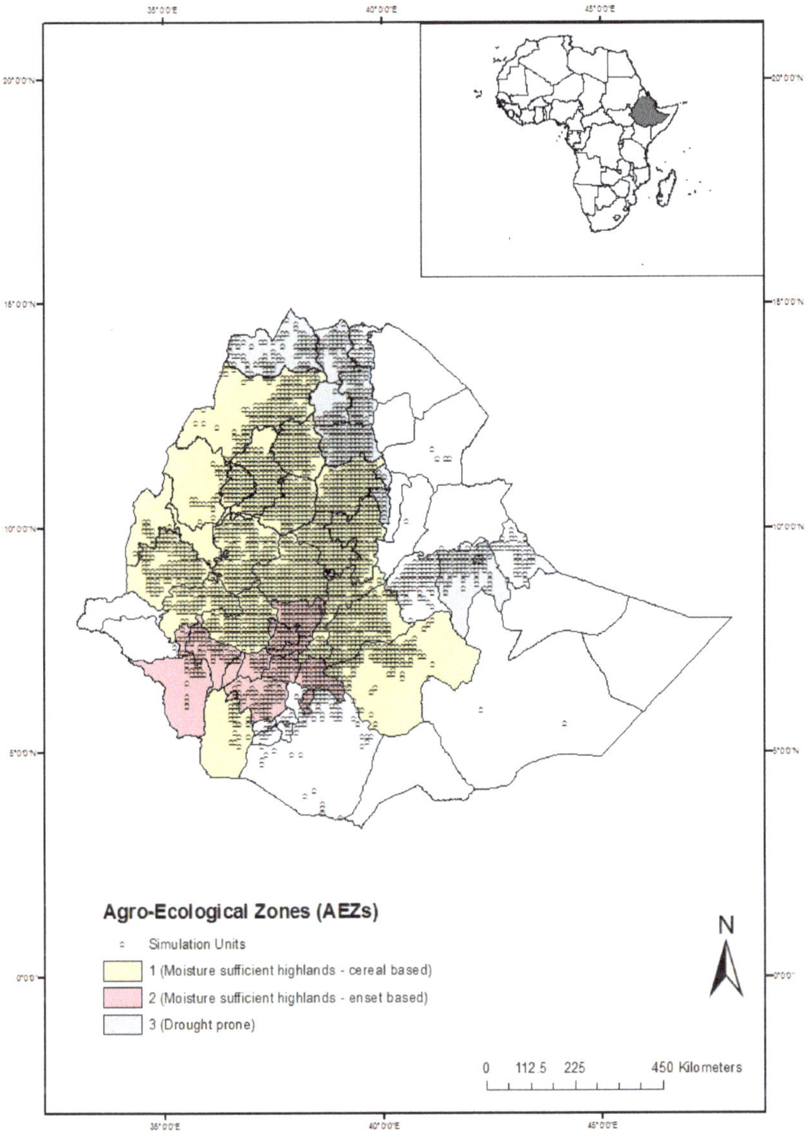

Figure 1. A map of Ethiopia showing the simulation units and the Agroecological zones.

2.2. Model Setup and Description

LINTUL5 is a biophysical model that simulates plant growth, biomass, and yield as a function of climate, soil properties, and crop management using experimentally derived algorithms. LINTUL5 has been widely used in various studies at the field, country, and continental scale [17–20]. Additionally, the crop model used in the current study has been used in earlier studies, showing its ability to simulate the growth and development of Groundnut (variety "Fleur11") [21] and the crops in rotations [22,23]. The effect of legume crops on the subsequent crops in the rotation is simulated as a fraction of the crop N uptake by biological fixation which was fixed as a parameter in the crop parameter file [21].

The applied version of LINTUL5 simulates the potential crop growth (limited by solar radiation only) under well-watered conditions; ample nutrient supply; and the absence of pests, diseases, and weeds [24]. Biomass production is based on intercepted radiation according to Lambert–Beer's law and light use efficiency. The produced biomass is partitioned among various crop organs (leaves, stems, storage organs, and roots) according to the partitioning coefficients defined as a function of the development stage of the crop. The phenology is simulated by the accumulation of thermal time above a defined base temperature. Photosynthesis and the total crop growth rate are calculated by multiplying the intercepted light and radiation use efficiency (RUE). The total crop growth, root-shoot partitioning, and leaf area expansion are further influenced by water stress. Water stress occurs when the available soil water is between a defined critical point and a wilting point or higher than the field capacity (water-logging). The critical point is a crop specific value which is calculated according to Reference [25] and depends on crop development, soil water tension, and potential transpiration. Water, nutrients (NPK), temperature, and radiation stresses restrict the daily accumulation of biomass, root growth, and yield. The stress indices are calculated daily for the water and nutrient limitations and range from 0.0 to 1.0. The estimation of the daily increase in crop biomass considers, on a given day, the maximum stress index among water, nitrogen, phosphorus, and potassium stress. Water stress occurs when available water in the soil is below the crop-water demand. The same holds for nitrogen stress, that is, when the crop available nitrogen in the rooted soil profile is lower than crop nitrogen demand. To simulate a continuous cropping system, the model was embedded into a general modeling framework, SIMPLACE (Scientific Impact Assessment and Modelling Platform for Advanced Crop and Ecosystem Management) [22]. The SIMPLACE<LINTUL5-SLIM-SoilCN> solution of the modeling platform was used in this study. SLIM is a conceptual soil water balance model subdividing the soil in a variable number of layers, substituting the two-layer approach in Lintul5. The crop residue effect on the crop yield was implemented in the crop model. Before the harvest of annual crops, some crop parameters define the death rate of roots and the start of the root senescence. The dead roots are transferred to the so-called "root litter pool" in the SoilCN SimComponent. At harvest, after removing the marketable part of the crop (in this case grains of legume and maize), the above and below ground residues are rooted into the respective litter pools of the SoilCN SimComponent. In both cases, carbon in the residues are routed into the carbon fraction for the respective litter pool and the amount of N in the residues is routed into the nitrogen fraction of the respective litter pool. The decomposition, mineralization, and humification of the litter pools in each soil layer are mainly triggered by the soil moisture, soil temperature, clay content, and maximum decomposition rates as shown in Reference [26]. The effect of crop residue retention on the crop is mainly through the maintenance of a certain soil organic nitrogen content and subsequently higher average mineral nitrogen concentrations in the soil, thus improving average N supply to the crop.

2.3. Dataset for Maize Model Calibration

In this study, two sets of hybrid maize cultivar-related parameters, namely BH660 (a long maturing cycle variety) and BH540 (a medium maturing cycle variety), (Table 1) were calibrated against the experimental data (yield and phenology) under rain-fed conditions collected from the Melko (Jimma Agricultural Research Centre), located on 7°39'56.4" latitude north and 36°46'56.4" longitude east in Ethiopia for the years 2008 to 2012. The fertilizer application rate used in the experiments was 23 kg ha^{-1} of urea and 217 kg ha^{-1} DAP (Di-Ammonium Phosphate) at planting and 150 kg ha^{-1} urea after 35 days of planting. According to Reference [27], both BH660 and BH540 are the most popular and widely grown maize varieties in the country, covering major maize producing areas. The Maize (*Z. mays*) crop parameter dataset (provided with the LINTUL5 model and Reference [17]) was used as a starting point to establish a new parameter set for these maize varieties.

2.4. Dataset for Groundnut Model Calibration

The field experiments were conducted in Bambey located at 14°42′ N and 16°29′ W and in Nioro located at 13°45′ N and 15°46′ N in Senegal during the dry and rainy seasons of 2014 and dry season of 2015 [28]. The peanut cultivar selected was Fleur11, known to be an early (90 days) maturity cultivar. Phenology observations were taken approximately every seven days to determine the parameters such as the day of emergence, the day of flowering, the beginning of peg, the beginning of pod formation, the beginning of seed, and physiological maturity as described in References [29,30]. The total dry matter was determined in leaves, stems, and pods on a weekly basis. At final harvest, the biomass and seed yield were determined in each plot in an area of 3.9 m^2 (1.95 m × 2 m). Most of the parameters used in the Lintul5 model are default values reported in Reference [24]. However, as no published studies exist with LINTUL5 for peanut or other legumes, some parameters values were adjusted based on the literature and from field measurements. Some parameter values were manually adjusted during the calibration process in order to adapt them to local conditions. The parameters of the model are given in Table 1.

2.5. Statistical Tools Used

As a measure of accuracy to compare the statistical data and simulated values, the following objective functions were used [31]:

a.   The mean relative error (*MR*) as

$$MR = \frac{1}{n}\sum_{i=1}^{n}\frac{(y_i - x_i)}{x_i} \tag{1}$$

b.   The mean residual error (*ME*) as

$$ME = \frac{1}{n}\sum_{i=1}^{n} y_i - x_i \tag{2}$$

where n is the sample number, x is the observed, and y is the simulated value. A value of 0 for *ME* indicates no systematic bias between the simulated and measured values. The *MR* gives an indication of the mean magnitude of the error in relation to the observed value. Small values indicate little difference between the simulated and measured values.

c.   Root mean square error (*RMSE*) as

$$RMSE = \sqrt{\frac{1}{n}\sum_{i=1}^{n}(s_i - o_i)^2} \tag{3}$$

where S_i is the simulated yield, O_i is the observed yield, and n is the total number of observations.

The daily climatic data were recorded at the meteorological stations of Jimma situated at 1718 meters above sea level and provided by the National Meteorological Agency (NMA) of Ethiopia. The required soil data for the crop model were provided by the National Soil Testing Center of Ethiopia (Table A2 in Appendix A).

Table 1. The crop parameters of LINTUL5 used in the study of maize (varieties BH660 and BH540) and Groundnut (variety Fleur11).

Name	Description	Unit	Value for Maize Varieties	Value for Groundnut Variety
			BH660/ BH540	Fleur 11
Crop parameters				
TSUM1	Temperature sum from emergence to anthesis	°C day-1	1000/ 680	422
TSUM2	Temperature sum from anthesis to maturity	°C day-1	990/760	1285
TBASEM	Lower threshold temperature for emergence	°C	8.0	10.0
TEFFMX	Maximum effective temperature for emergence	°C	30.0	30.0
TSUMEM	Temperature sum from sowing to emergence	°C	56.0	25.0
RUE-0.0	Radiation use efficiency at development stage 0	g MJ-1	3.2	2.6
RUE-1.0	Radiation use efficiency at development stage 1.0	g MJ-1	2.5	2.6
RUE-1.50	Radiation use efficiency at development stage 1.50	g MJ-1	2.2	2.4
RUE-2.0	Radiation use efficiency at development stage 2.0	g MJ-1	2.0	2.0
SLATB-0.0	Specific leaf area at development stage 0	m2 g-1	0.03	0.022
SLATB-1.0	Specific leaf area at development stage 1.0	m2 g-1	0.02	0.021
SLATB-2.0	Specific leaf area at development stage 2.0	m2 g-1	0.02	0.018
LAI critical	Critical leaf area beyond which leaves die due to self shading	m2 m-2	4.0	4.0
RGRLAI	Maximum relative increase in LAI	ha ha-1 day-1	0.02	0.018
ROOTDI	Initial rooting depth	m	0.1	0.1
ROOTDM	Maximum rooting depth	m	2.0	0.6
RRDMAX	Maximum rate of increase in rooting depth	m	0.012	0.012
TDWI	Initial total crop dry weight	kg ha-1	5.0	6.0

2.6. Datasets Used at the National Level

2.6.1. Climate and Soil Data

The climate data at the national scale was made available from the National Aeronautics and Space Administration (NASA), Goddard Institute of Space Studies (https://data.giss.nasa.gov/impacts/agmipcf/agmerra/), AgMERRA [32] and consists of the daily time series over the 1980–2010 period with a global coverage of the climate variables required for agricultural models (i.e., the minimum and maximum temperature, solar radiation, precipitation, and windspeed). These datasets are produced by combining state-of-the-art reanalyses (NASA's Modern-Era Retrospective analysis for Research and Applications, MERRA [33] and NCEP's Climate Forecast System Reanalysis, CFSR [34], with observational datasets from in situ observational networks and satellites. The dataset is stored at $0.25° \times 0.25°$ horizontal resolution (approx. 25 km). The values for relevant soil parameters for each soil layer down to the maximum soil depth (sand, clay, bulk density, Soil available water, and organic carbon) were extracted from the soil property maps of Africa at a 1 km $\times$ 1 km resolution (http://www.isric.org/data/soil-property-maps-africa-1-km) (Figure 2). Other parameters such as soil water at field capacity, wilting point and saturation point, and the Van–Genuchten parameters were computed [35]. These soil parameter values were used as an input to the soil water balance module of the crop model used in simulations at a regional scale (Figure 2).

2.6.2. Crop Yield and Fertilizer Application Data

The Maize and Groundnut yields (Mg ha^{-1}) and fertilizer application (Nitrogen and Phosphorus) rates in maize over seven years (2004–2010) at the administrative zone level have been collected from the Central Statistical Agency, Ethiopia. These values have been averaged over cropland and used for the model calibration at the national scale. However, no fertilizer was applied in the Groundnut simulations.

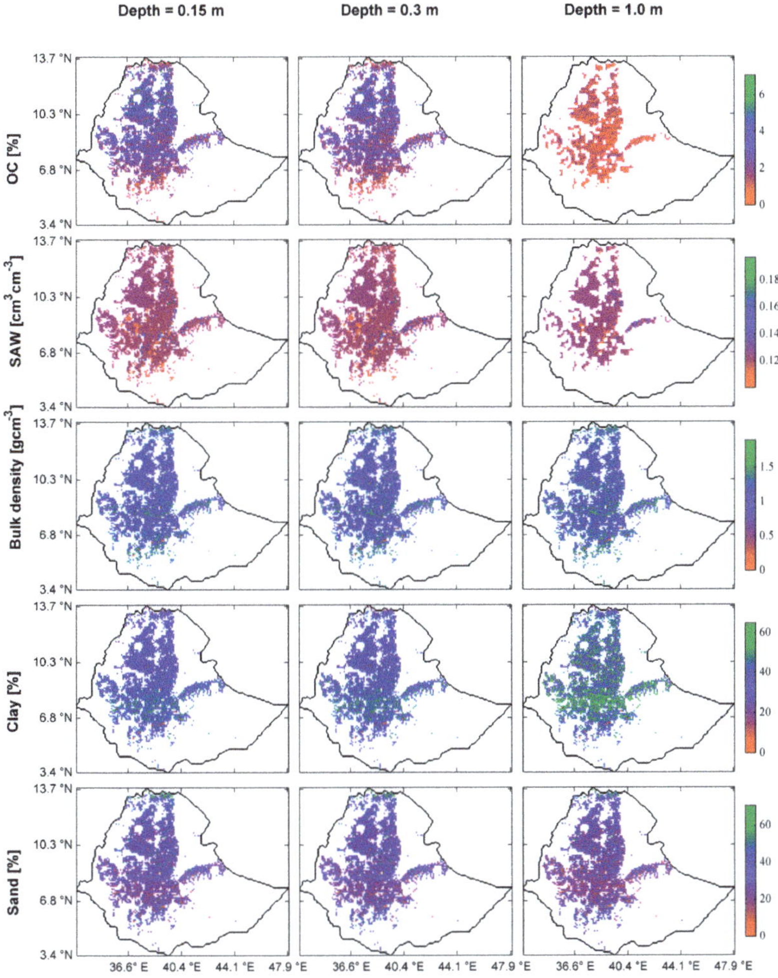

Figure 2. The soil physical properties (sand, clay, bulk density, soil available water (SAW), and soil organic carbon (OC) at three soil depths (i.e., 0.15, 0.3, and 1.0 m) of Ethiopia covering the cropland.

2.7. Intensification Scenarios

Groundnuts are also a source of cash income in Ethiopia that contribute significantly to food security and alleviate poverty [36]. As a legume, groundnuts improve soil fertility by fixing nitrogen and thereby increasing the productivity of the semiarid cereal cropping systems [37]. Therefore, we are exploring different crop management scenarios (Table 2) in the maize production system by using supplementary a mineral fertilizer supply, by recycling crop residues, and by rotating with N fixing Groundnuts. For the simulation of the intensification scenarios, the simulation period was 7 years with a preceding spin-up phase of 7 years to make sure that the soil organic matter pools of the model solution were approximating the equilibrium level of the respective treatment.

Table 2. The crop management scenarios for Maize production explored in Ethiopia.

Scenario name	Scenario description	N+P fertilizer input	Crop Residue Management	
Farmer's Practice (FP)	Maize in major season	MF1	Crop residue removed from the field	(MF1 - CR)
	Maize in major season	MF1	Crop residue added in the field	(MF1 + CR)
Conventional Intensification (CI)	Maize in major season	MF2	Crop residue removed from the field	(MF2 - CR)
	Maize in major season	MF2	Crop residue added in the field	(MF2 + CR)
FP + Rotation	Groundnut in minor season	No mineral fertilizer applied to Groundnut	Crop residue removed from the field	(MF1 - CR)
	Maize in major season	MF1		
	Groundnut in minor season	No mineral fertilizer applied to Groundnut	Crop residue added in the field	(MF1 + CR)
	Maize in major season	MF1		
CI + Rotation	Groundnut in minor season	No mineral fertilizer applied to Groundnut	Crop residue removed from the field	(MF2 - CR)
	Maize in major season	MF2		
	Groundnut in minor season	No mineral fertilizer applied to Groundnut	Crop residue added in the field	(MF2 + CR)
	Maize in major season	MF2		

where FP = Farmer's Practice; CI = Conventional Intensification; MF1 = current fertilizer rates; MF2 = 60 kg N ha^{-1} + 20 kg P ha^{-1}; CR = Crop Residue.

3. Results

3.1. Model Calibration and Evaluation

The observed and simulated days of maturity under the fertilized production treatment agreed well. In the case of variety BH660, the model overestimated the Day of Maturity (DOM) by one day compared with the observed value, whereas for variety BH540, the observed and simulated DOM matched exactly (Table 3). The model overestimated the day of anthesis by 4 days for both the maize varieties (Table 3). The simulated average grain yield of both the maize varieties was comparable to the corresponding observations where the model overestimated the grain yield of variety BH660 by 4% and underestimated the grain yield of variety BH540 by −1.6% (Table 3), whereas plotting all the observation points of variety BH660 (n = 3) and BH540 (n = 5) with the corresponding simulated values resulted in high root mean square values (RMSE) (Figure 3). When applied at the 35 administrative zones in Ethiopia, the average simulated yields of the districts were in the range of the observed yields averaged over 7 years (Figure 4) with an RMSE of 0.64 Mg ha^{-1}. However, the RMSE varied from 0.4 Mg ha^{-1} to 1.4 Mg ha^{-1} across the administrative zones where all the observed yield values were plotted against the simulated values for 7 years (2004–2010) (Figure 5).

The discrepancy observed in the simulated yield could have been due to the soil parameters used from the ISRIC (International Soil Reference and Information Centre) database, which refer to soil samples that are representative of large areas. Available soil data does not likely represent long-term cultivated, nutrient-depleted soils. Regarding Groundnut, at both test sites, the model underestimated the day of anthesis by 1 day (Table 3), whereas the day of maturity was overestimated by 1 day at Niaro and at Bambey; the simulated day of maturity matched the observed (Table 3). The average simulated grain yield at both sites was overestimated by 14.4% and 15% respectively (Table 3), whereas plotting all the observation points in Niaro (n = 3) and Bambey (n = 3) with the corresponding simulated values resulted in a higher variability and higher root mean square values (RMSE) (Figure 3). When applied at the 10 administrative zones in Ethiopia, the average simulated yields of the districts were in the range of the observed yields averaged over 7 years (Figure 6) with a root mean square error (RMSE) of 0.29 Mg ha^{-1}, whereas the RMSE varied from 0.97 Mg ha^{-1} to 0.40 Mg ha^{-1} across the administrative zones when all the observed yield values were plotted against the simulated yield values (Figure 7).

Table 3. The observed and simulated Grain yield, Day of Anthesis (DOA), Day of Maturity (DOM), mean residual error (ME), and mean absolute error (MR) for the maize variety BH660 and BH540 at Jimma in Ethiopia and for the Groundnut (variety Fleur11) at Nioro and Bambey in Senegal.

Variety	Site	Treatment	DOA	DOA	DOM	DOM	Grain Yield	Grain Yield	ME	MR
			Observed	Simulated	Observed	Simulated	Observed (Mg ha-1)	Simulated (Mg ha-1)	Grain Yield (Mg ha-1)	Grain Yield (%)
BH660	Jimma	Fertilized	70	74	160	161	10.0	10.4	0.4	4.0
BH540	Jimma	Fertilized	71	75	150	150	6.2	6.1	−0.1	−1.6
Fleur11	Niaro	Without fertilizer	102	101	166	167	2.03	2.28	0.25	14.4
Fleur11	Bambey	Without fertilizer	110	109	175	175	1.81	2.14	0.32	15.0

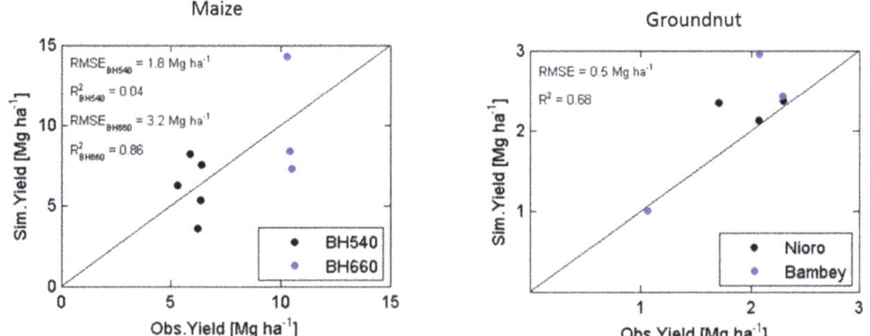

Figure 3. A 1:1 plot of the simulated (water and nutrient-limited) and observed (farmer's yield) two varieties of maize (BH540 n = 5 and BH660 n = 3) and groundnut variety "Fleur11" at two sites Nioro (n = 3) and Bambey (n = 3). RMSE = Root Mean Square Error and R^2 = Coefficient of Determination.

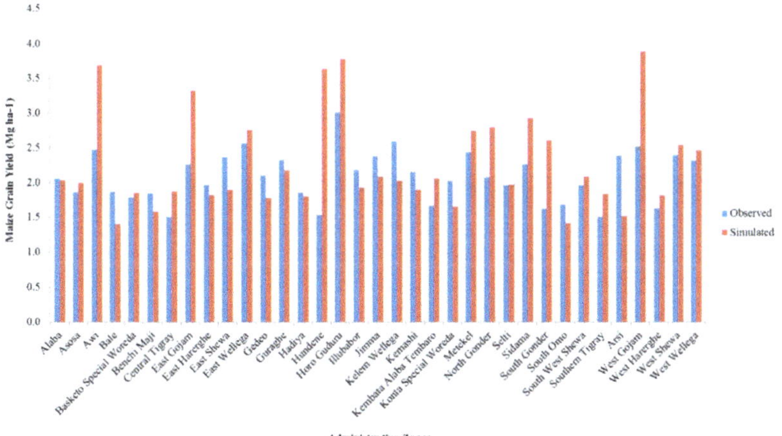

Figure 4. The simulated (water and nutrient-limited) versus observed (farmer's yield) maize yield averaged over 7 years (2004–2010) across 35 administrative zones in Ethiopia with Root Mean Square Error (RMSE) = 0.64 Mg ha^{-1}.

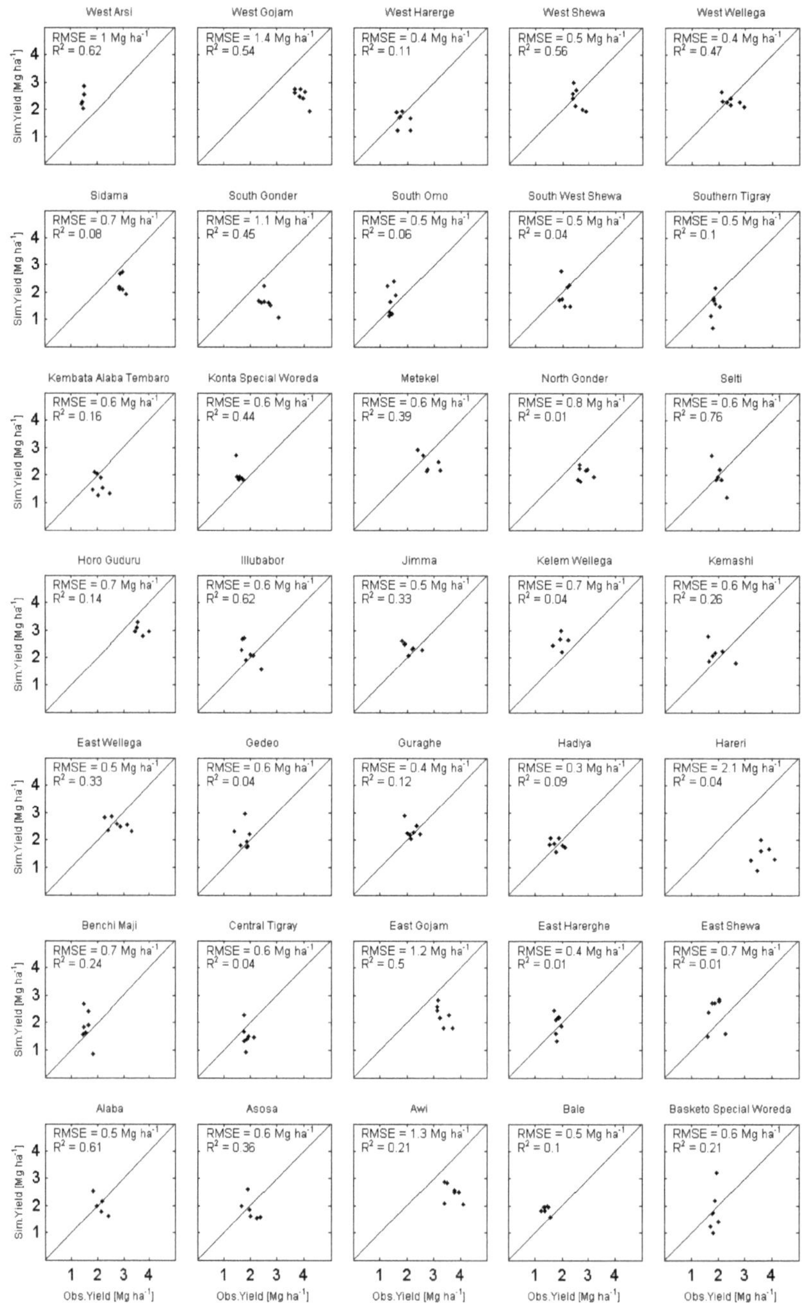

Figure 5. The simulated (water and nutrient-limited) versus observed (farmer's yield) maize yield across 35 administrative zones in Ethiopia across the 7 years (2004–2010) with the Root Mean Square Error (RMSE) and Coefficient of Determination (R^2).

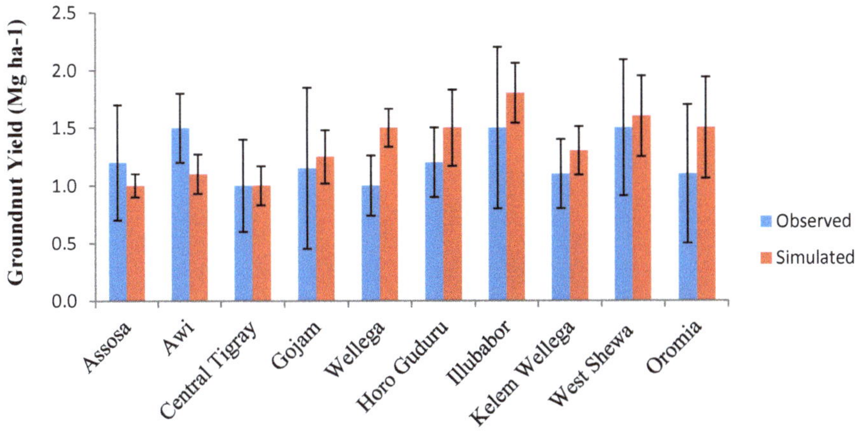

Administrative Zones

Figure 6. The simulated (water and nutrient-limited) versus observed (farmer's yield) groundnut yield averaged over 7 years (2004–2010) across 10 administrative zones in Ethiopia with Root Mean Square Error (RMSE) = 0.29 Mg ha^{-1}.

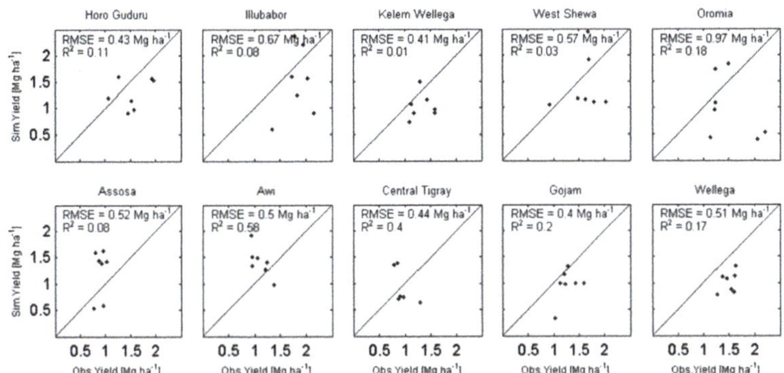

Figure 7. The simulated (water and nutrient-limited) versus observed (farmer's yield) groundnut yield across 10 administrative zones in Ethiopia with the Root Mean Square Error (RMSE) and Coefficient of Determination (R^2) values.

3.2. Effects of Integrating Crop Residue and Mineral Fertilizer into the Cropping System

Under the conventional intensification options explored in the current study, the simulation results show the positive effect of mineral fertilizer application and incorporation of the crop residue into the soil. The incorporation of only the crop residue in the field can lead to an increase in the maize yield by 45.3% compared to the farmer's yield under current mineral fertilizer application rates, whereas the crop residues combined with an increased amount of mineral fertilizer resulted in an increased yield of about 134.6% (Figure 8).

The effect of incorporating crop residues in combination with mineral fertilizer was also beneficial under the option where maize was grown in rotation with Groundnut (CI + rotation). The yields were increased up to 110.1% depending upon the scenarios tested (Figure 8). However, in the scenario where crop residues were not incorporated in the field (FP + rotation-CR), the maize yield was reduced by 21.9% (Figure 8).

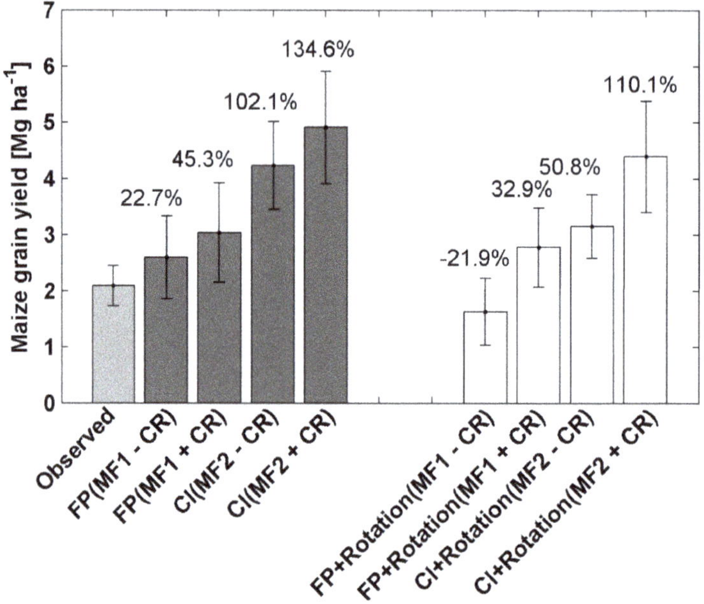

Figure 8. The Maize grain yield (in Mg ha^{-1}) under different intensification scenarios for maize production systems explored in Ethiopia: The bars are a standard deviation, and the numerical values placed above them are the percentage difference in Maize grain yield with the observed (farmer's yield) yield values. FP = Farmer's Practice; CI = Conventional Intensification; MF1 = current fertilizer rates; MF2 = 60 kg N ha^{-1} + 20 kg P ha^{-1}; CR = Crop Residue.

3.3. Effect of Management Scenarios on Nitrogen Uptake

There was a significant positive effect in the incorporation of crop residues and higher doses of mineral fertilizer application on nitrogen uptake by the plant ($p < 0.005$) in all the intensification scenarios explored in the current study (Figure 9). As expected, the highest Nitrogen uptake (138.8 kg ha^{-1}) was estimated in the conventional intensification where crop residues plus higher doses of mineral fertilizer (60 kg N ha^{-1} + 20 kg P ha^{-1} + CR) were applied in the field. The lowest nitrogen uptake (40.5 kg ha^{-1}) was estimated in the scenario where Groundnut was grown in rotation with Maize without incorporating the crop residues back in the field (FP + rotation-CR).

3.4. Economic Profitability Calculations

Based on the fact that the short rainy season comes first and then the long rainy seasons ("Belg" and "Meher" respectively) follow, the rotation scenario intends to introduce the cultivation of groundnut in the "Belg" season followed shortly by maize in the "Meher" season. The introduction of such a crop rotation within a single production year in non-irrigated areas could provide a number of agronomic advantages in addition to improving the soil fertility as a result of the nitrogen-fixing capacity of groundnut. It could also improve the income of farm households as a result of the production of groundnut as an additional crop.

In this section, the economic advantages of the crop rotation scenario are calculated in comparison with the conventional scenario. As clearly shown in Figure 8, the only difference between the two scenarios is the cultivation of groundnut in the minor or "Belg" season and the difference in the yield rate for maize. Since the scope of this paper is to identify the possible advantages of the rotation method over the conventional, it is appropriate to deal only with the profitability of groundnut production and market value of any yield difference in maize across corresponding simulations in the two scenarios.

Based on data from the World Bank Ethiopian socioeconomic survey [38], groundnut production and transportation costs are calculated per hectare (see Table A1 in Appendix A for a detailed explanation). Since groundnut is currently produced in only a few areas in Ethiopia, the number of observations is low, which makes the calculation of these costs at the zone level difficult. The national average figures are calculated instead for all the variables needed here. The total production costs per hectare are calculated to be 6625.3 birr ("birr" is the basic monetary unit of Ethiopia), and the transportation cost is 472 birr based on the national average of 1.63 tons of output per hectare. Based on the price of 9160 birr per ton, the average yield results in a total farm sales revenue of 14,909.7 birr per hectare from groundnut production. This is calculated to be of 7812.4 birr gross economic profit (GEP) per hectare. (Note: Since land is usually rented for a minimum of a year, there is no need to consider land rental costs for groundnut cultivation within the crop rotation method.)

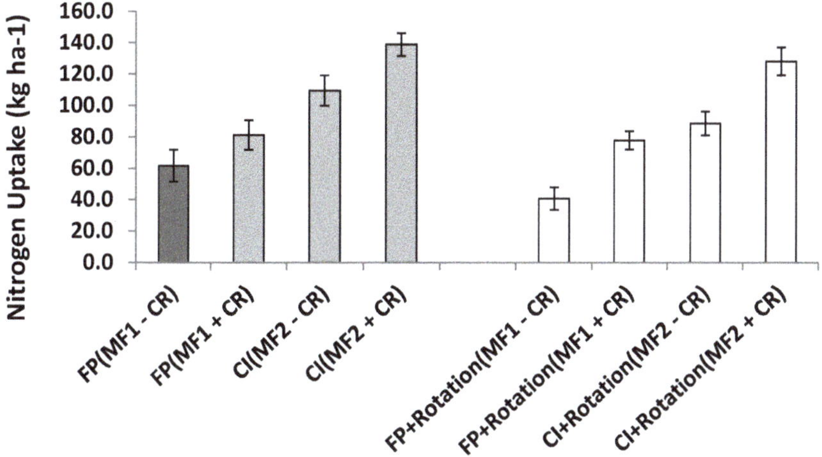

Figure 9. The nitrogen uptake (in kg ha^{-1}) under different intensification scenarios for maize production systems explored in Ethiopia: The bars are the values of standard deviation. FP = Farmer's Practice; CI = Conventional Intensification; MF1 = current fertilizer rates; MF2 = 60 kg N ha^{-1} + 20 kg P ha^{-1}; CR = Crop Residue.

4. Discussion

The objective of this study was to quantify the long-term crop yield response and change in soil organic carbon and crop available N over time for conventional intensification with or without groundnut included in the rotation. Our results demonstrate that the value of crop available nitrogen in soil profiles have increased from 5 kg ha^{-1} under current farmer's practices (i.e., FP, MF1-CR) to 30 kg ha^{-1} under the rotation scenario (i.e., CI, MF2 + CR) including groundnut in the rotation (Figure 10). However, to take advantage of the benefits of groundnut on nitrogen availability to the maize crop, the release of N from above- and below-ground residues must be synchronous with maize N uptake.

Various researchers have reported the lower beneficial effect of crop rotation on maize yield with increasing N fertilization, especially when legumes were included [39,40]. In our study, the inclusion of groundnut as a rotation crop did result in a lower maize yield gain under higher fertilization compared to the yields under conventional intensification scenarios (Figure 8) despite the fact that the incorporation of legume N, e.g., through the incorporation of groundnut biomass, led to large amounts of mineral N being available in the soil (Figure 11a). Compared with the cereals, legume residues are relatively rich in N with a lower C:N ratio, and these characteristics favour the rapid decomposition and release of N to subsequent crops [11,41]. The reason for a lower maize yield when including groundnut in the rotation could be due to the mismatch of N release from groundnut residues and the development of maize roots and maize N demand. Thus, maize was not able to take advantage of the higher amount of crop available N (Figure 11a). In addition, groundnut consumes soil water during the short rainy season (belg) which can lead to less total crop available water over the rooting depth (CAWT) for the subsequent maize crop compared to the conventional intensification scenario without any crop in the short season (see Table A3 in Appendix A). The same findings were also reported in Western Kenya by Reference [42].

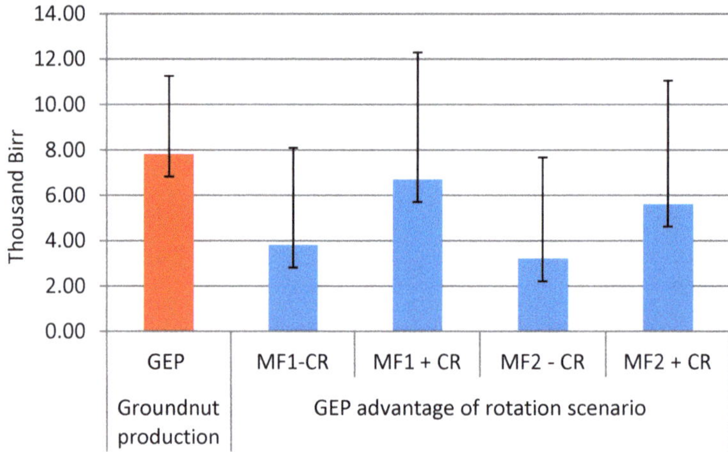

Figure 10. The average gross economic profit (GEP) (Birr ha^{-1} year^{-1}) over Ethiopia from groundnut production and the average annual GEP advantage of rotation scenario over the conventional (in '000 Birr).

The observation of total organic carbon in the maize monoculture production system (i.e., conventional intensification) and legume-cereal rotation (Figure 11b) did not show a significant difference, which goes in accordance with the findings of earlier studies in SSA [43,44]. However, the plausible way through which the incorporation of grain legumes into the cereal-based system could enhance soil C contents is through the enhanced productivity of a subsequent cereal crop or through the intercropping systems with an enhanced total biomass production providing high-quality residues to enhance the buildup of soil organic matter [45]. Therefore, when compared to the respective conventional intensification scenario, the rotation scenarios have a slightly higher soil organic matter content.

Livestock is an important component of the farming systems in Ethiopia [46], and the use of maize and legume residues as feed is common. Residues of certain legumes, e.g., that of groundnut, are more likely to be removed for feed. The use of residues as feed creates a potential trade-off between the use of legume residues for soil fertility improvement or for feed [42]. In the latter case, the ability of the farmer to return manure from their livestock to their fields and to handle feed and manure well during storage and transport strongly affects the carryover rates of nutrients and carbon and the impacts on soil fertility [47].

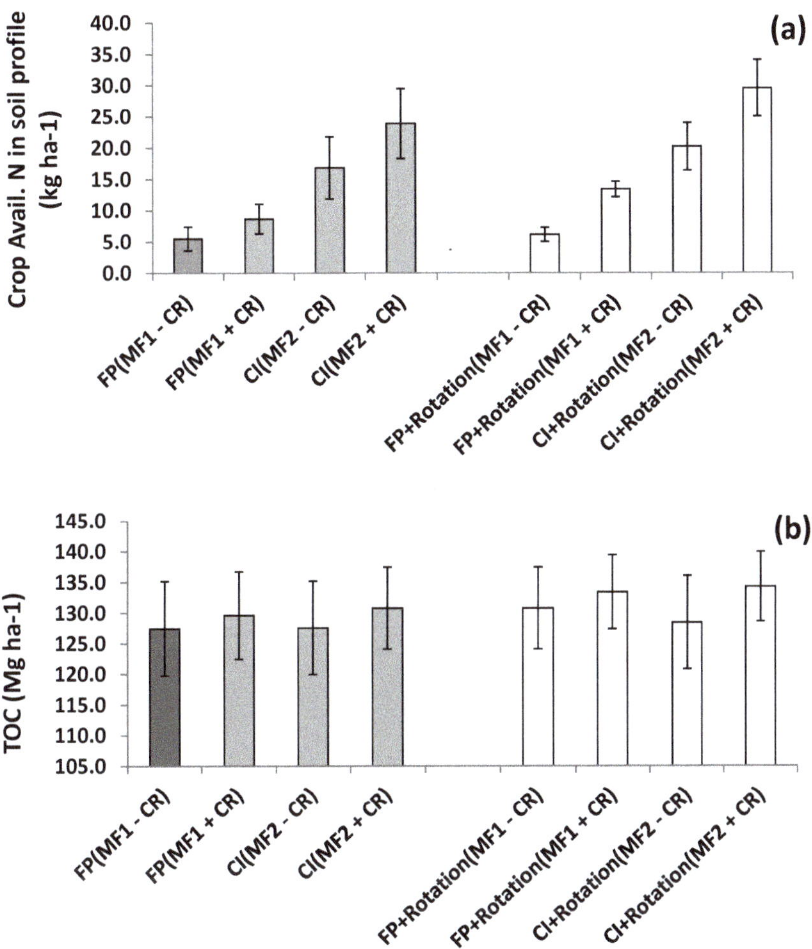

Figure 11. (**a**) The crop available Nitrogen in soil profile (in kg ha^{-1}) and (**b**) the Total Organic carbon (TOC; in Mg ha^{-1}) under different intensification scenarios for maize production systems explored in Ethiopia: The bars are the values of standard deviation. FP = Farmer's Practice; CI = Conventional Intensification; MF1 = current fertilizer rates; MF2 = 60 kg N ha^{-1} + 20 kg P ha^{-1}; CR = Crop Residue.

Regarding the cost-benefit analysis with respect to the introduction of groundnut in rotation with maize, the results show that the inclusion of groundnut could be a profitable venture for farmers, although, averaged over Ethiopia, a rotation with groundnut has a negative impact on the maize yield (Figure 8). The blue bars in Figure 10 show the net GEP increase of the four rotation scenarios compared to the four scenarios without groundnut when averaged over the cropped area in the whole country. However, when looking at the administrative zone level results (Table 4, the profitability of the rotation scenario has a large variation which is due to the variation in the calculated yield levels for groundnut. The very poor groundnut yield levels in zones like Asosa, Awi, Metekel, North and South Gonder, and West Gojam cannot compensate the loss of return from the reduced maize yields, leading to an overall lower economic return of the rotation system. Thus, the introduction of groundnut in crop rotation with maize in areas like these is disadvantageous, although the soil organic matter content may be slightly increased. However, in most of the zones in Ethiopia, the rotation scenarios have economic and ecological advantages.

Table 4. The average annual Gross economic profit (GEP) advantage across Ethiopia (in 35 administrative zones) from groundnut production in rotation scenarios over the conventional (in '000 Birr).

Zone Name	MF1 - CR	MF1 + CR	MF2 - CR	MF2 + CR
		Rotation Scenarios		
Alaba	7.4	11.3	6.7	9.9
Asosa	−4.5	−3.1	−5.4	−4.7
Awi	−2.5	−0.4	−3.9	−0.9
Bale	7.5	11.5	7.8	9.8
Basketo Special Woreda	9.1	13.9	8.2	12.6
Benchi Maji	7.9	12.9	7.2	11.3
Central Tigray	−1.0	0.5	1.8	3.2
East Gojam	1.2	2.6	0.0	1.9
East Harerghe	6.8	8.6	6.7	8.3
East Shewa	2.4	4.2	0.4	0.8
East Wellega	3.2	6.2	1.8	4.3
Gedeo	9.8	16.2	9.4	15.0
Guraghe	6.2	9.0	6.0	8.1
Hadiya	7.6	12.3	6.8	10.5
Hundene	0.9	0.9	0.9	0.7
Horo Guduru	1.6	4.6	0.8	3.9
Illubabor	6.1	10.2	4.9	8.3
Jimma	8.9	13.8	7.7	12.5
Kelem Wellega	−0.3	1.7	−2.0	−1.4
Kemashi	2.9	5.5	1.3	3.1
Kembata Alaba Tembaro	9.1	13.5	8.3	12.2
Konta Special Woreda	8.4	13.6	8.2	12.2
Metekel	−3.4	-1.9	−4.6	−3.2
North Gonder	−3.9	-2.8	−4.2	−3.1
Selti	5.8	9.8	5.1	8.4
Sidama	7.4	10.7	6.1	9.9
South Gonder	−1.5	0.2	−2.1	0.0
South Omo	7.4	11.8	8.9	11.9
South West Shewa	5.3	7.4	4.8	6.0
Southern Tigray	2.4	3.2	4.1	5.0
Arsi	8.6	12.4	8.1	11.1
West Gojam	−1.8	−0.4	−3.1	−1.0
West Harerghe	3.2	5.3	2.6	3.3
West Shewa	4.3	6.2	3.5	5.0
West Wellega	1.1	3.3	−0.4	1.6

5. Conclusions

The conventional intensification option explored in the current study shows the positive effect of mineral fertilizer application and the incorporation of the crop residue back in the field. The incorporation of crop residue alone in the field can lead to an increase in maize yield by 45.3% compared to the farmer's yield under current mineral fertilizer application rates, whereas the crop residues combined with an increased amount of mineral fertilizer resulted in an increased yield of about 134.6%. The same phenomenon was observed where Maize (in the long season) was grown in rotation with Groundnut (in the short season). The yields were increased by up to 110.1% depending upon the scenarios tested. However, in the scenario where crop residues were not incorporated in the field, the maize yield was reduced by 21.9%. The cost-benefit analysis suggests that the incorporation of groundnut in maize-based systems could provide additional income to the farmers when averaged over Ethiopia; with some local exceptions in some administrative zones, the introduction of groundnut in the rotation turned out to be economically disadvantageous, although they slightly increased soil organic carbon over time compared to maize monoculture.

Author Contributions: All authors substantially contributed to the conception, writing, and revisions of the paper. A.K.S. was the lead author.

Funding: The work was carried out under the project BiomassWeb within the GlobeE research programme. Funding by the Federal Ministry of Education and Research (BMBF) of Germany is highly acknowledged.

Acknowledgments: We are grateful to the anonymous reviewers for their valuable suggestions and comments which helped to improve the manuscript.

Conflicts of Interest: The authors declare no conflict of interest.

Appendix A

Table A1. The detailed calculations of the Gross economic profit for Ground Nut cultivation (Source: Authors' calculation based on data from the World Bank Ethiopia Socioeconomic Survey (ESS, 2014).)

Description	
Production cost (Birr per ha)	6625.26
Labor cost (Birr per ha)	6093.7
Wage rate (Birr per day)	31.17
Total labor day (per ha)	195.5
Seed cost (Birr per ha)	531.56
Quantity of seed (Kg per ha)	58.03
Seed price (Birr per Kg)	9.16
Fertilizer cost (Birr per ha)	0
Quantity Fertilizer (Kg per ha)	0
Pesticide + herbicide cost (Birr per ha)	0
Quantity of pesticide + herbicide (Kg per ha)	0
Total revenue (Birr per ha)	14909.732
Price (Birr per Mg)	9160
Quantity of output (Mg per ha)	1.628
Total transportation cost (Birr per ha)	472.12
Quantity of output (Mg per ha)	1.6277
Transportation cost (Birr per Mg)	290
Gross economic profit (GEP) (Birr per ha)	7812.35

Table A2. The sand, silt, clay, and organic carbon (OC) content used as an input for the model calibration at Jimma (in Ethiopia) and Bambey and Niori (in Senegal).

Sites	Depth (cm)	OC (%)	Sand (%)	Silt (%)	Clay (%)
Jimma	0–5	2.0	30.0	31.0	39.0
	5–40	1.5	15.5	14.5	71.0
	40–70	1.0	15.0	9.0	76.0
	70–110	0.8	15.0	6.5	78.0
	110–200	0.6	17.5	6.5	76.0
Bambey	0–10	0.2	95.0	1.5	2.8
	10–20	0.1	92.7	1.7	3.8
	20–30	0.1	91.8	1.6	5.1
	30–40	0.2	91.2	1.7	6.2
	40–50	0.3	89.9	1.8	7.0
	50–60	0.1	89.7	1.7	7.2
	60–70	0.1	91.5	2.2	6.5
	70–80	0.1	91.1	2.6	6.2
	80–90	0.1	90.4	2.0	6.1
	90–100	0.1	87.6	1.3	5.9

Table A2. *Cont.*

Sites	Depth (cm)	OC (%)	Sand (%)	Silt (%)	Clay (%)
	0–10	0.5	92.8	4.3	4.5
	10–20	0.4	88.7	9.7	3.2
	20–30	0.4	86.5	10.3	3.4
	30–40	0.4	86.6	9.2	3.1
Nioro	40–50	0.3	84.3	12.0	4.0
	50–60	0.2	80.7	13.9	4.6
	60–70	0.2	77.6	3.6	18.4
	70–80	0.2	74.0	4.1	21.4
	80–90	0.2	73.1	11.0	14.8
	90–100	0.2	72.5	4.3	23.9

Table A3. The total crop available water over the actual rooting depth (CAWT) in mm under different intensification scenarios for the maize production systems explored in Ethiopia. Abbreviations: FP = Farmer's Practice; CI = Conventional Intensification; MF1 = current fertilizer rates; MF2 = 60 kg N ha^{-1} + 20 kg P ha^{-1}; CR = Crop Residue.

	FP		CI		FP + Rotation		CI + Rotation	
	MF1 - CR	MF1 + CR	MF2 - CR	MF2 + CR	MF1 - CR	MF1 + CR	MF2 - CR	MF2 + CR
CAWT (mm)	23.7	26.0	31.7	33.6	14.3	20.6	22.3	25.6

References

1. Van Ittersum, M.K.; van Bussel, L.G.J.; Wolf, J.; Grassini, P.; van Wart, J.; Guilpart, N.; Claessens, L.; de Groot, H.; Wiebe, K.; Mason-D'Croz, D.; et al. Can sub-Saharan Africa feed itself? *Proc. Natl. Acad. Sci. USA* **2016**, *113*, 14964–14969. [CrossRef] [PubMed]
2. Folberth, C.; Yang, H.; Gaiser, T.; Abbaspour, K.C.; Schulin, R. Modeling maize yield responses to improvement in nutrient, water and cultivar inputs in sub-Saharan Africa. *Agric. Syst.* **2013**, *119*, 22–34. [CrossRef]
3. Akinnifesi, F.K.; Ajayi, O.C.; Sileshi, G.; Chirwa, P.W.; Chianu, J. Fertiliser trees for sustainable food security in the maize-based production systems of East and Southern Africa. A review. *Agron. Sustain. Dev.* **2010**, *30*, 615–629. [CrossRef]
4. Sanginga, N. Role of biological nitrogen fixation in legume based cropping systems; a case study of West Africa farming systems. *Plant Soil* **2003**, *252*, 25–39. [CrossRef]
5. Van Oort, P.A.J.; Saito, K.; Tanaka, A.; Assagba-Amovin, E.; Van Bussel, L.G.J.; van Wart, J.; de Groot, H.; van Ittersum, M.K.; Cassman, K.G.; Wopereis, M.C.S. Assessment of rice-sufficiency in 2025 in eight African countries. *Glob. Food Secur.* **2015**, *5*, 39–49. [CrossRef]
6. Hall, A.J.; Richards, R.A. Prognosis for genetic improvement of yield potential and water-limited yield of major grain crops. *Field Crop Res.* **2013**, *143*, 18–33. [CrossRef]
7. Bouwman, L.; Goldewijk, K.K.; van Der Hoek, K.W.; Beusen, A.H.W.; van Vuuren, D.P.; Willems, J.; Rufino, M.C.; Stehfest, E. Exploring global changes in nitrogen and phosphorus cycles in agriculture induced b livestock production over the 1900–2050 period. *Proc. Natl. Acad. Sci. USA* **2013**, *110*, 20882–20887. [CrossRef]
8. Vanlauwe, B.; Bationo, A.; Chianu, J.; Giller, K.E.; Merckx, R.; Mokwunye, U.; Ohiokpehai, O.; Pypers, P.; Tabo, R.; Shepherd, K.; et al. Integrated soil fertility management: Operational definition and consequences for implementation and dissemination. *Outlook Agric.* **2010**, *39*, 17–24. [CrossRef]
9. Reckling, M.; Hecker, J.-M.; Bergkvist, G.; Watson, C.; Zander, P.; Stoddard, F. A cropping system assessment framework—Evaluating effects of introducing legumes into crop rotations. *Eur. J. Agron.* **2016**, *76*, 186–197. [CrossRef]
10. Chen, X.; Cui, Z.; Fan, M.; Vitousek, P.; Zhao, M.; Ma, W.; Wang, Z.; Zhang, W.; Yan, X.; Yang, J. Producing more grain with lower environmental costs. *Nature* **2014**, *514*, 486–489. [CrossRef] [PubMed]

11. Franke, A.C.; van den Brand, G.J.; Vanlauwe, B.; Giller, K.E. Sustainable intensification through rotations with grain legumes in Sub-Saharan Africa: A review. *Agric. Ecosyst. Environ.* **2018**, *261*, 172–185. [CrossRef] [PubMed]

12. Crawford, E.W.; Jayne, T.S.; Kelly, V.A. Alternative Approaches for Promoting Fertilizer Use in Africa. In *Agriculture and Rural Development Discussion Paper 22*; The World Bank: Washington, DC, USA, 2006.

13. Sanchez, P. Tripling crop yields in tropical Africa. *Nat. Geosci.* **2010**, *3*, 299–300. [CrossRef]

14. Palm, C.A.; Smukler, S.M.; Sullivan, C.C.; Mutuo, P.K.; Nyadzia, G.I.; Walsh, M.G. Identifying potential synergies and trade-offs for meeting food security and climate change objectives in Sub-Saharan Africa. *Proc. Natl. Acad. Sci. USA* **2010**, *107*, 19661–19666. [CrossRef] [PubMed]

15. World Bank Group. Available online: https://data.worldbank.org/indicator/AG.LND.ARBL.ZS?view= chart (accessed on 20 March 2019).

16. Regassa, S.; Givey, C.; Castillo, G.E. The Rain Doesn't Come on Time Anymore: Poverty, Vulnerability, and Climate Variability in Ethiopia. *Oxfam Policy Pract.* **2010**, *6*, 90–134.

17. Srivastava, A.K.; Mboh, C.M.; Gaiser, T.; Zhao, G.; Ewert, F. Climate change impact under alternate realizations of climate scenarios on maize yield and biomass in Ghana. *Agric. Syst.* **2017**. [CrossRef]

18. Trawally, D.; Webber, H.; Agyare, W.A.; Fosu, M.; Naab, J.; Gaiser, T. Effect of heat stress on two maize varieties under irrigation in Northern Region of Ghana. *Int. J. Biol. Chem. Sci.* **2015**, *9*, 1571–1587.

19. Eyshi Rezaei, E.; Siebert, S.; Ewert, F. Impact of data resolution on heat and drought stress simulated for winter wheat in Germany. *Eur. J. Agron.* **2015**, *65*, 69–82. [CrossRef]

20. Zhao, G.; Webber, H.; Hoffmann, H.; Wolf, J.; Siebert, S.; Ewert, F. The implication of irrigation in climate change impact assessment: A European wide study. *Glob. Chang. Biol.* **2015**, *21*, 4031–4048. [CrossRef] [PubMed]

21. Faye, B.; Webber, H.; Diop, M.; Mbaye, M.; Owusu-Sekyere, J.D.; Naab, J.B.; Gaiser, T. Potential impact of climate change on peanut yield in Senegal, West Africa. *Field Crops Res.* **2018**, *219*, 148–159. [CrossRef]

22. Yin, X.G.; Kersebaum, K.C.; Kollas, C.; Manevski, K.; Baby, S.; Beaudoin, N.; Ozturk, I.; Gaiser, T.; Wu, L.H.; Hoffmann, M.; et al. Performance of process-based models for simulation of grain N in crop rotations across Europe. *Agric. Syst.* **2017**, *154*, 63–77. [CrossRef]

23. Kollas, C.; Kersebaum, K.C.; Nendel, C.; Manevski, K.; Nüller, C.; Palosuo, T.; Armas-Herrera, C.M.; Beaudoin, N.; Bindi, M.; Charfeddine, M.; et al. Crop rotation modelling—A European model intercomparison. *Eur. J. Agron.* **2015**, *70*, 98–111. [CrossRef]

24. Wolf, J. *User Guide for LINTUL5: Simple Generic Model for Simulation of Crop Growth under Potential, Water Limited and Nitrogen, Phosphorus and Potassium Limited Conditions*; Wageningen University: Wageningen, The Netherlands, 2012; pp. 1–63.

25. Allen, R.G.; Pereira, L.S.; Raes, D.; Smith, M. *Crop Evapotranspiration: Guidelines for Computing Crop Water Requirements*; Irrigation & Drainage Paper 56; FAO: Rome, Italy, 1998; p. 301.

26. Corbeels, M.; Mcmurtrie, R.E.; Pepper, D.; O'Connell, A.M. A process-based model of nitrogen cycling in forest plantations: Part I. Structure, calibration and analysis of the decomposition model. *Ecol. Model.* **2015**, *187*, 426–448. [CrossRef]

27. Jaleta, M.; Yirga, C.; Kassie, M.; de Groote, H.; Shiferaw, B. Knowledge, Adoption and Use Intensity of Improved Maize Technologies in Ethiopia. In Proceedings of the 4th International Conference of the African Association of Agricultural Economists, Hammamet, Tunisia, 22–25 September 2013.

28. Faye, B.; Webber, H.; Gaiser, T.; Diop, M.; Owusu-Sekyere, J.D.; Naab, J.B. Effects of fertilization rate and water availability on peanut growth and yield in Senegal (West Africa). *J. Sustain. Dev.* **2016**, *9*, 111–131. [CrossRef]

29. Boote, K.J. Growth stages of peanut (*Arachis hypogaea* L.). *Peanut Sci.* **1982**, *9*, 35–49. [CrossRef]

30. Meier, U. (Ed.) Stades Phénologiques des mono-et Dicotylédones Cultivées BBCH Monographie, 2nd ed. Available online: http://www.iki.bund.de/fileadmin/damuploads/veroeff/bbch/BBCH-Skala_franz% C3%B6sisch. (accessed on 20 March 2019).

31. Papula, A. *Mathematik fur Chemiker*, 2nd ed.; Enke-Verlag: Stuttgart, Germany, 1982; p. 405.

32. Ruane, A.C.; Goldberg, R.; Chryssanthacopoulos, J. AgMIP climate forcing datasets for agricultural modeling: Merged products for gap-filling and historical climate series estimation. *Agric. For. Meteorol.* **2015**, *200*, 233–248. [CrossRef]

33. Rienecker, M.M.; Suarez, M.J.; Gelaro, R.; Todling, R.; Bacmeister, J.; Liu, E.; Bosilovich, M.G.; Schubert, S.D.; Takacs, L.; Kim, G.K.; et al. MERRA: NASA's Modern-Era Retrospective Analysis for Research and Applications. *J. Clim.* **2011**, *24*, 3624–3648. [CrossRef]

34. Saha, S.; Moorthi, S.; Pan, H.L.; Wu, X.; Wang, J.; Nadiga, S.; Tripp, P.; Kistler, R.; Woollen, J.; Behringer, D.; et al. The NCEP Climate Forecast System Reanalysis. *Bull. Am. Meteorol. Soc.* **2010**, *91*, 1015–1057. [CrossRef]

35. Rawls, W.J.; Ahuja, L.R.; Maidment, D.R. *Handbook of Hydrology*; McGraw-Hill: New York, NY, USA, 1993; p. 1442.

36. Chala, A.; Mohammed, A.; Ayalew, A.; Skinnes, H. Natural occurrence of aflatoxins in groundnut (*Arachis hypogaea* L.) from eastern Ethiopia. *Food Control* **2013**, *30*, 602–605. [CrossRef]

37. Smart, M.G.; Shotwell, O.L.; Caldwell, R.W. Pathogenesis in Aspergillus ear rot of maize: Aflatoxin B1 levels in grains around wound inoculation sites. *Phytopatology* **1994**, *80*, 1283–1286. [CrossRef]

38. Ethiopia Socioeconomic Survey (ESS). *World Bank Ethiopia Socioeconomic Survey*; ESS: Addis Ababa, Ethiopia, 2014.

39. Nevens, F.; Reheul, D. Crop rotation versus monoculture; yield, N yield and ear fraction of silage maize at different levels of mineral N fertilization. *Wagening. J. Life Sci.* **2001**, *49*, 405–425. [CrossRef]

40. Riedell, W.E.; Schumacher, T.E.; Clay, S.A.; Ellsbury, M.M.; Pravecek, M.; Evenson, P.O. Cornand soil fertility responses to crop rotation with low, medium or high inputs. *Crop Sci.* **1998**, *38*, 427–433. [CrossRef]

41. Anyanzwa, H.; Okalebo, J.R.; Othieno, C.O.; Bationo, A.; Waswa, B.S.; Kihara, J. Effects of conservation tillage, crop residue and cropping systems on changes in soil organic matter and maize–legume production: A case study in Teso District. *Nutr. Cycl. Agroecosyst.* **2010**, *88*, 39–47. [CrossRef]

42. Ojiem, J.O.; Franke, A.C.; Vanlauwe, B.; de Ridder, N.; Giller, K.E. Benefits of legume-maize rotations: Assessing the impact of diversity on the productivity of smallholders in Western Kenya. *Field Crops Res.* **2014**, *168*, 75–85. [CrossRef]

43. Franke, A.C.; Berkhout, E.D.; Iwuafor, E.N.O.; Nziguheba, G.; Dercon, G.; Vandeplas, I.; Diels, J. Does crop-livestock integration lead to improved crop productivity in the savanna of West Africa? *Exp. Agric.* **2010**, *46*, 439–455. [CrossRef]

44. Bationo, A.; Ntare, B.R. Rotation and nitrogen fertilizer effects on pearl millet, cowpea and groundnut yield and soil chemical properties in a sandy soil in the semi-arid tropics, West Africa. *J. Agric. Sci.* **2000**, *134*, 277–284. [CrossRef]

45. Samake, O.; Stomph, T.J.; Kropff, M.J.; Smaling, E.M.A. Integrated pearl millet management in the Sahel: Effects of legume rotation and fallow management on productivity and *Striga hermonthica* infestation. *Plant Soil* **2006**, *286*, 245–257. [CrossRef]

46. Birara, E.; Zemen, A. Assessment of the Role of Livestock in Ethiopia: A Review. *Am. Eurasian J. Sci. Res.* **2016**, *11*, 405–410.

47. Rufino, M.C.; Rowe, E.C.; Delve, R.J.; Giller, K.E. Nitrogen cycling efficiencies through resource-poor African crop-livestock systems. *Agric. Ecosyst. Environ.* **2006**, *112*, 261–282. [CrossRef]

 © 2019 by the authors. Licensee MDPI, Basel, Switzerland. This article is an open access article distributed under the terms and conditions of the Creative Commons Attribution (CC BY) license (http://creativecommons.org/licenses/by/4.0/).

 *sustainability*

Article

Household Welfare Implications of Better Fertilizer Access and Lower Use Inefficiency: Long-Term Scenarios for Ethiopia

Ermias Engida Legesse [1,*]**, Amit Kumar Srivastava** [2]**, Arnim Kuhn** [1] **and Thomas Gaiser** [2]

[1]   Institute for Food and Resource Economics, University of Bonn, Nussallee 21, 53115 Bonn, Germany
[2]   Institute of Crop Science and Resource Conservation, University of Bonn, Katzenburgweg 5, D-53115 Bonn, Germany
*   Correspondence: ermias.engida@ilr.uni-bonn.de or ermiasengida@gmail.com; Tel.: +49-228-73-2326

Received: 20 September 2018; Accepted: 17 July 2019; Published: 20 July 2019

Abstract: High population growth in Ethiopia is aggravating farmland scarcity, as the agrarian share of the population stays persistently high, and also creates increasing demand for food and non-food biomass. Based on this fact, this study investigates welfare implications of intensification measures like interventions that improve access and use efficiency to modern farming inputs. Using a dynamic meso-economic modeling framework for Ethiopia, ex-ante scenarios that simulate a) decreased costs of fertilizer use and b) elevated efficiency of fertilizer application for all crops are run for a period of 20 years. Fertilizer-yield response functions are estimated (based on results from an agronomic crop model and actual survey data) and embedded into the economic model in order to get realistic marginal returns to fertilizer application. This is our novel methodological contribution in which we introduce how to calculate input use inefficiency based on attainable yield levels from agronomic crop model and actual yield levels. Simultaneous implementation of these interventions lead to annual yield increases of 8.7 percent for an average crop farmer compared to the current level. Increased fertilizer application is also found to be profitable for an average farmer despite price reduction for crops following increased market supply. As a result of price and income effects of the interventions, all household types exhibit welfare gain. Non-farming households, being net consumers, enjoy lower costs of living. Rural farming households enjoy even higher welfare gain than non-farming households because they consume a higher share from crop commodities that become cheaper, and because their farming profits increase.

Keywords: CGE; fertilizer-yield-response; productivity; welfare; Ethiopia

---

1. Introduction

Rapid population growth is likely to cause increasing farmland scarcity in Sub-Saharan Africa. While rising population density and farmland scarcity typically causes a transition from fallow-based systems to permanent cultivation [1], more intensive cultivation, uncontrolled deforestation of the natural vegetation cover for farmland expansion, high stocking rates, and farming practices with little concern for conservation and poor soil management practices may lead to protracted environmental and soil degradation which can significantly reduce soil fertility [2]. This results in low and stagnating yields, and it becomes increasingly difficult to maintain households' livelihoods from the land. This is at odds with the necessity of substantial growth in agricultural output that will have to originate from higher yields. Among other measures, farmers will have to raise their use of supplementary plant nutrients from organic and inorganic fertilizers [2].

In developing countries like Ethiopia where the majority of the labor force is engaged in agriculture, the growth of the farming sector is particularly inclusive and has been shown to contribute significantly to poverty reduction. For every 1 percent growth in agricultural output, poverty was reduced by 0.9 percent during the 2000-2011 period in Ethiopia [3]. Higher productivity through adoption of improved agricultural technologies is considered as the main pathway for solving food insecurity problems and escaping poverty traps. Agricultural technologies such as mineral fertilizers can help improve household welfare directly by raising incomes of adopters, and also indirectly by creating employment and lowering food prices [4,5]. Duflo et al. [6] experimentally show that, when used appropriately, mineral fertilizer is highly profitable with mean annual returns of 36 percent over a season. Minten & Barrett [7] found that doubling rice yields in Madagascar, as a result of higher rates of adoption of improved agricultural technologies, was associated with a reduction of the number of food insecured by 38 percent and a 1.7 months shorter lean period.

Despite these potential benefits, fertilizer use intensity is still at low levels, and the efficiency of fertilizer use in raising output per unit of land is significantly lower in Sub-Saharan Africa than in Asia [8]. Several arguments are brought forward in the economics literature about possible reasons for the low fertilizer use in developing countries like Ethiopia. There is poor access to modern farming inputs which emanates from both supply and demand side constraints. Weak infrastructure and non-conducive policy environment together with institutional problems lead to relatively high input costs and the absence or late arrival of supplies. Rashid et al. [9], and Zerfu et al. [10] observe that fertilizers are more expensive in Africa as compared to other developing regions, like Asia. This is because ocean freight costs in Asia are lower due to economies of scale, and domestic transport costs are much higher in Africa than in Asian countries. According to Rashid et al. [9], in Ethiopia transaction (plus transportation) costs up to input distribution centers constitute 11-23 percent of the price at the port. Transport costs alone account for 64–80 percent of these price differentials. Studies on the most remote areas in Ethiopia estimate transport costs from input distribution centers to the farms to be as high as 50–80 percent of the transport costs from port to the distribution centers [11]. Regarding demand side constraints, lack of liquidity/credit and lack of information and/or knowledge are among the major factors for low fertilizer adoption. Minten and Barrett [7] show that lower illiteracy levels in Madagascar are generally associated with significantly higher modern input adoption rates. Demand side constraints discourage the use of chemical fertilizers through reducing the willingness and ability of farmers towards fertilizer adoption and highly contribute to the inefficient use of inputs as well.

Technical use inefficiency reduces fertilizer response and its profitability which discourages adoption. Studies find low yield response to fertilizer application in Ethiopia [12,13]. Nesrane et al. [14] estimated an average level of farming efficiency for smallholder farmers in Ethiopia to be 0.46 during 1994–2009, indicating that an average farmer produces less than half of the value of output produced by the most efficient farmer using the same technology and inputs. Technical efficiency of input use depends upon the quality and appropriateness of inputs, the timing of their delivery to farmers, the availability of complementary resources (for example, improved seed and fertilizer together), availability and quality of extension services, agro-ecological conditions, and farmers' technical skill or competence in using the inputs [15]. Bold et al. [16] find modern farming inputs purchased in local markets in Uganda to be of low quality. They find that 30 percent of nutrient is missing in fertilizer, and hybrid maize seed contains less than 50 percent authentic seeds. Lack of education and past experience in modern input use is the other major factor for fertilizer use inefficiency. Yu and Nin Pratt [17], and Beshir et al. [18] argue that farmers face high knowledge costs related to the adoption of new technologies which can effectively increase the adoption barrier and hence significantly slow down the diffusion of new technologies. They also indicate that extension services can help cut the adoption cost and input use inefficiency.

This study aims to quantify the output and welfare effects of increased fertilizer use as a result of reduced observed costs and improved fertilizer use efficiency (i.e., lower unobserved cost). A number of studies analyze this issue based on micro-level survey data or partial equilibrium models in which it is hard to control for or identify all impact pathways (see Duflo et al. [6]; Verkaart et al. [19]; Minten and Barrett [7]). The topic needs analysis of a broad range of effects as the policy interventions may generate spillovers that benefit non-recipients or may compete for resources and so indirectly affect other programs or economic agents. Very few studies manage to capture this broad range of effects with an economy-wide modeling approach (see Caria et al. [20] and Arndt et al. [21]). Though the use of economy-wide model is appropriate, we observe two drawbacks in incorporating marginal returns to fertilizer use in Caria et al. and Arndt et al.. First, these studies typically apply a linear fertilizer-yield relationship which ignores the principle of diminishing marginal returns to variable inputs; and second, they mention that fertilizer use efficiency rates are key determinants of expected benefits but did not empirically treat this issue in their modeling approaches. In this study we calculate levels of inefficiency in fertilizer use for different crops based on maximum attainable fertilizer-yield response levels from agronomic crop models and maximum actual fertilizer-yield response levels from survey data. These are incorporated into the production function in the economic model. The maximum attainable yield response at various fertilizer application rates is expressed by a quadratic functional form which allows a more realistic derivation of marginal economic returns. Thus, our contribution is threefold. First, unlike previous studies that use linear relationships, we introduce to the economic model an agronomically more realistic estimation of marginal returns to fertilizer use; second, we incorporate input use inefficiency which is key determinant factor of actual returns; and third, based on these novel methodological contributions, we manage to quantify micro and macro level impacts of higher rates of fertilizer application and improved use efficiency.

To enable the analysis of a broad range of effects, a dynamic Computable General Equilibrium (CGE) model for Ethiopia is used. The model is subject to specific modifications that allow illustrating the effect of increasingly efficient fertilizer use on crop yields. The rest of this paper is organized as follows. Section 2 gives background information on the current state of the Ethiopian crop sub-sector, fertilizer adoption, and household welfare. Detailed explanation on the empirical methods used in this paper for both agronomic and economic analysis follows in Section 3, together with details of the specific policy simulations which are run. Section 4 presents analysis of the results, and Section 5 concludes.

2. Background

Agriculture is the dominant sector of the Ethiopian economy. Though its GDP share declined in recent years, its importance to the economy is still significant, as it contributes around 40 percent of the GDP and employs 77 percent of the total workforce. Agricultural employment is also growing; for instance, by 2.5 percent annually on average during the period 2005 to 2013 [22,23]. The sector contributed 80.8 percent of total commodity exports during 2004/05–2013/14 [22,24]. The heterogeneity in topography, climatic conditions, and soil types enable the country to grow a wide variety of crops. According to 2015/16 AgSS (nationally representative Agriculture Sample Survey of the Central Statistics Agency of Ethiopia) data [25], cereal crops cover the biggest share with 71 percent of the total cropland and 68 percent of total crop output. Among the eight major cereal crops cultivated in Ethiopia, teff (*Eragrostis tef*), maize (*Zea mays*), sorghum (*Sorghum bicolor*), and wheat (*Triticum aestivum*) are the biggest four both in area coverage (85.2 percent) and total cereal output (87.2 percent).

2.1. Productivity and Agronomic Practices in the Crop Sub-sector

Referring to AgSS data, from 2004/5 to 2015/16 [25–36], total crop production significantly increased at a faster rate than crop area expansion - production more than doubled (124 percent) during this period whereas the area covered expanded by just 27.3 percent. This implies that yields have grown consistently over this decade. Despite this growth, obtained yield levels are still relatively low in comparison to world average. For instance, in 2004 Ethiopian maize yield was less than a quarter of that in Egypt and a fifth of that in the USA. By 2013, the gap had narrowed, although still considerable, with Ethiopian maize yield reaching 44 percent of Egypt's and a third of USA's [22].

Ethiopia's agriculture is characterized by a low-input, low-value, subsistence-oriented, rain-fed crop sub-sector. It is dependent on a highly erratic rainfall regime and vulnerable to frequent weather fluctuations and drought episodes that often lead to harvest failures. In addition, the shortage of arable land in the densely populated highlands leads to severe and protracted soil degradation and nutrient mining, threatening to lower the fertility of cropland in the longer run that is further exacerbated by critically low levels of human and physical infrastructure [37]. Modern input use is still at a very low level. For instance, based on data from the 2015/16 AgSS [25], only 55.7 percent of the entire cropland cultivated by smallholder farmers is treated with some type of fertilizer including natural fertilizer like manure. Cereal crops like teff, maize and wheat enjoy relatively higher fertilizer application, being 76, 73 and 84.4 percent of their area, respectively. However, the application rate is low even for these cereals. Moreover, the share of cropland cultivated with improved seeds is still not higher than 8 percent, with the only exceptions being maize and wheat which got due attention from local breeding institutions. Improved seed application was almost inexistent for teff and sorghum in 2015/16 (2.1 and 0.2 percent respectively). Only 23.6 percent of the total cropland received pesticide treatment. It is significantly higher for cereals, where 55 and 48 percent of wheat and teff area got pesticide treatment in 2015/16, whereas that of maize and sorghum was 9 and 12 percent. In addition, 30 percent of the total cropland was reported to have benefited from extension services, with 38 percent for cereals. Irrigation is another promising modern farm management practice that is practically non-existent in Ethiopia's smallholder farming – irrigation coverage is 1.2 percent for all crops in general and 0.7 percent for cereals in particular which confirms the sector's complete dependence on rainfall. In summary, the crop sub-sector in Ethiopia, with special emphasis on the major cereal crops, is still traditional with a low level of application of modern inputs and modern agronomic practices.

Regarding farm structural change, individual farm plot holdings are continuously diminishing in size as a result of rapid growth in the farming population. The already scarce supply of farmland is constantly tightening, and farmsteads and plot allocation have become more and more fragmented in the course of time. For instance, as reported in AgSS (2004/05-2015/16), the average land holding size declined by 12.7 percent over this period. This phenomenon raises the need for modern farm practices and inputs.

2.2. Use and Access to Modern Inputs

Currently, the use of fertilizer is increasing all over Ethiopia. According to the CSA [25–36,38], the area of fertilized land has doubled in the last 13 years. In line with this, the amount of fertilizer used all over the country has increased more than threefold during the same period which indicates that the application rate has also increased, although slowly (Figure 1). The application rate on average increased from 66 kilograms per hectare in 2003/04 to 104 kilograms per hectare in 2015/16, considering fertilized cropland only. If applied to the entire cultivated cropland, this would be 59 kilograms per hectare. In terms of coverage, chemical fertilizer was applied to only 40 percent of Ethiopia's cropland in 2003/04, which has increased to 55 percent in the very recent past. Despite the fact that the application rate and coverage gradually increased, adoption of fertilizer as a technological option is still at a low level. Data from the World Bank's Ethiopian socioeconomic survey 2015/16 shows that 56 percent of the respondent households never used chemical fertilizer on their farm plots in any instance (Table A1).

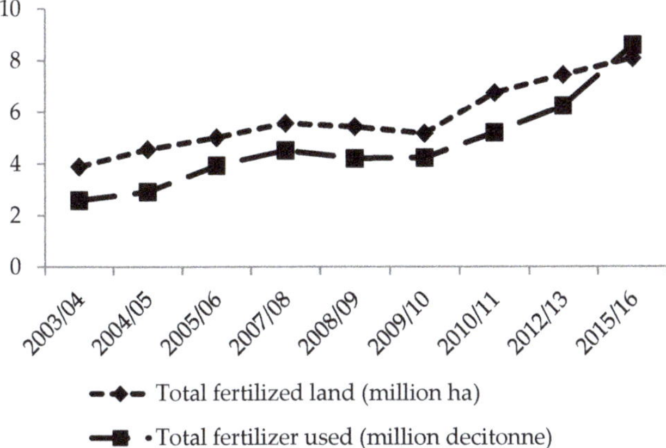

Figure 1. Trend in fertilizer application (2003/04 – 2015/16).

So far Ethiopia has exclusively been using imported chemical fertilizer. The country has no fertilizer manufacturing or blending factory. Currently, there is a government-backed initiative to commence domestic chemical fertilizer production in four regions of the country.

2.3. Household Welfare Situation

Ethiopia is among the poorest countries in the world. Based on UNDP 2017 Human Development Index report [39], Ethiopia stands at 173rd place out of 189 countries with an HDI of 0.463 which is below the world and Sub-Saharan African averages. However, according to the World Bank's recent poverty assessment report for the years 2000–2011 [3], significant decrease in poverty rates and progress in wellbeing has been observed. Ethiopian households experienced a decade of remarkable improvement in wellbeing and the number of poor inhabitants living on less than USD 1.25 PPP a day fell from 56 percent to around 30 percent. This progress is not without its challenges, as poverty remains widespread and the very poorest have not seen improvements—on the contrary, even a worsening—of consumption since 2005, which poses a challenge to achieving shared prosperity in Ethiopia. As the majority of poor households are agricultural producers, the reduced costs of fertilizer use that is analyzed in this study is expected to improve their incomes, while non-farm households and agricultural households in the poorest decile that produce very little (less than three months of consumption than other poor households, and were more likely to report suffering from food price shocks than any other group) may benefit from a more abundant food supply following improved crop productivity.

3. Materials and Methods

3.1. The Model

The model used in this study is a recursive dynamic extension of the standard CGE model by the International Food Policy Research Institute (IFPRI). In general a CGE model covers all economic sectors of a country or region and ensures macro-economic consistency. It captures all economic agents' interactions which enables us to analyze both direct and indirect effects. A recursive-dynamic version CGE model, in particular, is used for multi-annual projections that are driven by trends in economic drivers such as population growth and technical productivity. It is based on the assumption that the behavior of economic agents is characterized by adaptive expectations: economic agents make their decisions on the basis of past experiences and current conditions [40]. This is an alternative that captures developing countries' realities better than inter-temporal dynamic models that are explained by forward-looking expectations.

The model is solved recursively which means that it solves one period at a time with a series of static one-year equilibrium solutions that are linked between periods by stock variables such as population or production factors. The static model assumes that producers maximize profit subject to costs governed by the specific production function employed. A multi-stage production function is used. First, factors of production are made to combine using a constant elasticity of substitution (CES) with elasticity levels adopted from Dorosh and Thurlow (2009) [41]. The optimal amount of factors is ruled based on their relative prices. The value-added composite is then combined with fixed share intermediates using a Leontief specification. Profit maximization drives producers to sell these products in domestic or foreign markets substituted through a constant elasticity of transformation (CET) function with an assumption of imperfect transformability. The domestically marketed domestic output is an imperfect substitute for imported goods ruled by a CES Armington specification. These trade elasticities are adopted from Global Trade Analysis Project (GTAP) database.

Ethiopian households are assumed to be the sole owners of factors of production (labor, land, livestock, and capital) and maximize their income by allocating these factors of production across activities. Supply and demand for these factors have to equilibrate based on closure rules which affect returns to factors and thus, incomes of the households. All factors of production except capital are made to face endogenous demand from the different activities and have fixed (or exogenous) supply, which is the sum total of the demands from activities. The balance between the two is secured by the economy-wide wage rate. This makes them mobile across the activities. Factor capital, on the other hand, is assumed to be fully employed and activity specific. Demand for capital is exogenously set to grow with the amount of investment in the economy, while the supply is an endogenous sum total of the demands. Activity-specific wage rates are made flexible to balance the demand and supply for factor capital. In addition to factor returns, these households get a smaller portion of their income from government transfers and remittances from abroad. Their demand for goods and services, on the other hand, is represented by a linear expenditure system (LES) with income elasticities estimated by Dorosh and Thurlow [41] for Ethiopia based on data from the 2004/05 Household Income, Consumption and Expenditure Survey (HICES). The model assumes the households maximize utility subject to budget constraint and save a fixed share of their income. They are also subject to direct taxation at a fixed rate. The total revenue collected from this and other types of taxes (import tariff and sales tax) represents government income. Transfer from the rest of the world in terms of aid and borrowing augments this revenue. Assuming real government consumption held constant, government budget adjusts to price changes. Thus, positive or negative government savings bridge any mismatch between the revenues and expenditures.

The external balance of Ethiopia with the rest of the world is maintained using a flexible foreign saving and fixed exchange rate regime. Whenever the demand for foreign currency exceeds the supply, it is assumed to be covered by increased foreign savings and it decreases whenever the opposite happens. This current account closure is believed to better represent the current managed floating exchange rate regime in Ethiopia.

The savings pool collects money from domestic (households and government) and foreign sources and finances the economy's investment demand. The total amount of the savings is assumed to drive the amount of investment in the economy, which is the closure rule in order to maintain balance between savings and investment.

The model is designed to solve only for relative prices and real variables of the economy. To achieve this and anchor the absolute price level, a normalization rule has been applied. The consumer price index (CPI) is chosen as the numéraire, so all changes in nominal prices and incomes in simulations are relative to the weighted unit price of households' initial consumption bundle (i.e., a fixed CPI).

Regarding the specification of the dynamic model, the process of capital accumulation is modeled endogenously; in every period of the model run, the capital stock continues updating with the total amount of new investment and depreciation. New capital is distributed among sectors based on each sector's initial share of aggregate capital income. Population growth is exogenously imposed on the model based on separately calculated growth projections. It is assumed that a growing population raises subsistence demand of household consumption. Additionally, exogenous growth rates are applied on labor force and Total Factor Productivity (TFP) in every period of the model run.

3.2. Data

The Social Accounting Matrix (SAM) on which this model is calibrated was first developed by the Ethiopian Development Research Institute (EDRI) for a 2005/06 snapshot of the Ethiopian economy and later updated for 2009/10 by Engida et al. [42]. The SAM was updated with the following procedure. The dynamic CGE model is used to simulate the growth of the Ethiopian economy based on actual economic developments from 2005/06–2009/10. The resulting solution is a new, balanced SAM for 2009/10. The projected 2009/10 SAM and GDP were then converted to current prices. Actual value added shares of activities and actual aggregate demand components of 2009/10 (from national accounts) were then used to adjust value added by sector in the projected 2009/10 SAM. This resulted in an unbalanced SAM, which was then balanced using a cross entropy program.

The SAM currently in use is an aggregation with 18 agricultural activities and 16 other activities of which 7 are industrial and the remaining 9 are in the service sector. Every activity produces and serves the economy with its respective commodity. In the SAM there are different factors of production; labor – disaggregated by skill level, land, livestock, and capital. The SAM contains 8 different types of institutions from which 6 are representative households which are disaggregated by their occupation and income class (i.e., farming/non-farming and poor/non-poor), and a government and rest of the world (RoW). A direct and two different indirect (sales and import tariff) tax types are considered, and there is a saving-investment account as well.

Regarding Ethiopian economic structure, let's look at some relevant figures derived from the SAM. As seen in the Annex, Table A2, the agriculture sector in general, and the cereal sub-sector in particular intensively use factor inputs, especially land and labor with marginal contribution from intermediate inputs. Moreover, the biggest share of the intermediate inputs in these activities is composed of agricultural commodities, indicating traditional agricultural practices with low application of modern inputs like chemical fertilizers, pesticides, herbicides and improved seeds from the industry sector. Additionally, the non-agriculture sector satisfies an insignificant portion of its intermediate input demand from agricultural commodities. This clearly shows the weak linkage between the two sectors. The fact that agriculture outputs are mostly (73 percent) consumed unprocessed by the households, with a smaller share being exported (6.5 percent) and fed to the non-agriculture sector (16.2 percent) (see Table A3), shows that growth in the agriculture sector has a weak impact on the rest of the economy. Factors' distribution among sectors, as seen in Table A4, also supports this argument. Factors mostly engaged in agriculture, land and unskilled labor, are rarely engaged in the non-agriculture sector which mostly employs factors like capital and skilled labor.

3.3. Endogenizing Crop Productivity

3.3.1. Simplified Yield Response Function

The impact of increasing rates of typically used mineral fertilizer on Rain water use efficiency (WUE) and Radiation use efficiency (RUE) of maize grain yield and stover biomass productivity was estimated across the Agro-Ecological zones (AEZs) using the crop model LINTUL5 which is later embedded into a general modeling framework. LINTUL5 is a bio-physical model that simulates plant growth, biomass, and yield as a function of climate, soil properties, and crop management using experimentally derived algorithms. The applied version LINTUL5 simulates potential crop growth (limited by solar radiation only) under well-watered conditions, ample nutrient supply and the absence of pests, diseases, and weeds. To simulate a continuous cropping system, the model was embedded into a general modeling framework, SIMPLACE (Scientific Impact Assessment and Modelling Platform for Advanced Crop and Ecosystem Management) [43]. The SIMPLACE<LINTUL5-SLIM-SoilCN

solution of the modeling platform was used in this study. SLIM is a conceptual soil water balance model subdividing the soil in a variable number of layers, substituting the two layer approach in Lintul5 [44]. In this modeling framework, water, nutrients (NPK), temperature, and radiation stresses restrict the daily accumulation of biomass, root growth, and yield.

Spatial resolution was at the 1 km grid cell level, where cropland and soil data are available. Long maturing cycle maize variety was used in the simulations in AEZ1 and AEZ2 where the length of crop growing season is more than 160 days, elsewhere (AEZ 3) a medium maturing cycle variety was used in the simulations. The simulated yield from all the simulation units within each administrative zone was averaged to obtain a representative value for a specific year to compare with the observed yield.

Two sets of parameters for hybrid maize cultivars, namely BH660 (a long maturing cycle variety) and BH540 (a medium maturing cycle variety) were calibrated against experimental data (yield and phenology) under rain-fed conditions collected from Melko (Jimma Agricultural Centre) for the years 2008 to 2012. Fertilizer application rates used in the experiments were 23 kg ha^{-1} of urea and 217 kg ha^{-1} DAP (Di-Ammonium Phosphate) at planting and 150 kg ha-1 urea after 35 days of planting. According to Jaleta et al. [45], both BH660 and BH540 are the most popular and widely grown maize varieties in the country. The maize crop parameter dataset (provided with the LINTUL5 model and Srivastava et al. [46]), was used as a starting point to establish a new parameter set for these maize varieties. Further details on the crop modeling effort can be found in Srivastava et al. [47].

Results show a strong effect of the application rate of mineral fertilizer on maize yield and stover biomass across the AEZs. The national average of maize grain yield under different application rates of the mentioned mix of fertilizers (DAP and urea in a 1:0.9 ratio) is illustrated by Figure 2. The data points allow for a quadratic approximation which is differentiable with respect to fertilizer use and can thus be used to derive economically optimal fertilizer use rates under alternative input and output price constellations. This yields a fertilizer yield response function for maize. (Table A5 displays parameters and results of the quadratic approximation function that is integrated into the CGE model)

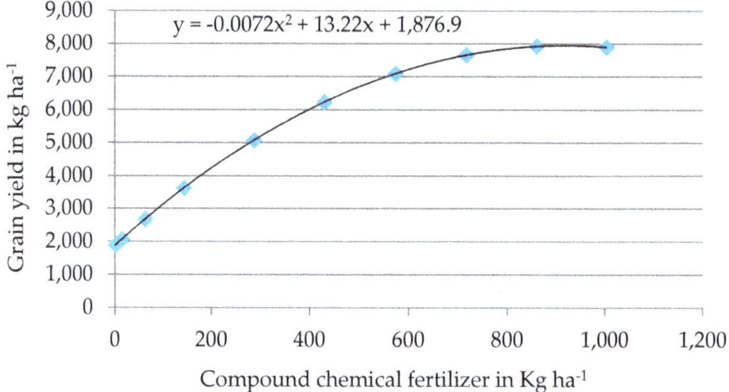

Figure 2. LINTUL5-simulated maize yields, and quadratic approximation function (Ethiopian national average results) at different use levels of mixed chemical fertilizer.

The results show that the maximum yield level to be obtained in Ethiopia is far above the current maize yield, which is due to low fertilizer use. At given price levels for fertilizer and maize, the profit function that can be obtained from the yield response function has its optimal fertilizer use level per hectare far above the observed levels. The reason might be unobserved costs in fertilizer use in Ethiopia. Thus, to use this yield function in an economic model, the first order condition (FOC) of the profit function has to be calibrated to current use levels, which is explained in Section 3.3.2.

Though we have a crop model only for maize, the other crops in the SAM are considered as well in order to not have maize as the only crop where flexible fertilizer use is linked to productivity.

Unlike the estimation strategy applied for maize, actual survey data is used to econometrically estimate fertilizer yield response function for the other crops in the SAM (teff, wheat, barley (*Hordeum vulgare*), sorghum, pulses, oilseeds, vegetable and fruits, 'enset' (false banana), and cash crops like coffee, chat, flower and sugar). We obtain the required data from three rounds (2009/10, 2010/11 and 2011/12) AgSS dataset. Obviously, fertilizer use efficiency of field experiments and actual farmers' practices cannot be comparable. In order to narrow the gap, we only use farmers that are relatively best in terms of modern agronomic practices – that are under extension coverage and use improved seeds in addition to chemical fertilizer. The fertilizer-yield response function that we estimated for the other crops (i.e., except maize) is displayed in Figure A1. As explained above for maize, at given price levels for the crops and fertilizer, the profit functions that can be obtained from the yield response functions have their optimal fertilizer use levels per hectare far above the observed average levels. These differences exhibit the level of unobserved costs in fertilizer use of the average farmers as compared to the best ones.

3.3.2. Integrating Yield Response into the CGE Model

The principal approach to incorporate a fertilizer yield response function to the CGE model is to create a 'fertilizer module', i.e., a set of equations that a) calculates yields, b) identifies optimal fertilizer use levels per hectare based on a profit function approach, and c) finally ensures that changes in nominal quantities and prices obtained in this module are translated into equivalent changes in quantities and prices in the CGE model. Moreover, minor changes to a very limited number of equations of the original CGE model have to be made.

First, the yield ($Qyperha_{a,c}$) is defined as a function of fertilizer per hectare by a quadratic model:

$$Qyperha_{a,c} = fertq_{a,c} \cdot Qabsfert^2_{a,c} + fertlin_{a,c} \cdot Qabsfert_{a,c} + fertconst_{a,c} \tag{1}$$

where: a stands for set of activities with fertilizer-yield function and c for set of commodities

$fertq_{a,c}$ is the quadratic parameter in the yield response function

$Qabsfert_{a,c}$ is the quantity of fertilizer input per hectare

$fertlin_{a,c}$ is the linear parameter in the yield response function

$fertconst_{a,c}$ is the constant term in the yield response function

The parameters of the above quadratic function ($fertq_{a,c}$, $fertlin_{a,c}$, and $fertconst_{a,c}$) are shown in Figures 2 and A1. Now it has to be ensured that changes in output (QA_a) per unit of land use ($QF_{f,a}$) in the CGE model compared to its base value are equal to changes of simulated yields per hectare ($Qyperha_{a,c}$) relative to its base level ($Qyperha0_{a,c}$):

$$\frac{QA_a}{QF_{f,a}} = \frac{QA0_a}{QF0_{f,a}} \cdot \frac{Qyperha_{a,c}}{Qyperha0_{a,c}} \tag{2}$$

where: f stands for set of factors but now specifically represents factor land

$QF_{f,a}$ is a simulated unit of land use

$QF0_{f,a}$ is a unit of land use at the base

The next step is defining a fertilizer demand function that expresses marginal revenues and costs of fertilizer use.

Adding prices for the crops and fertilizer ($Poutput_{a,c}$ and $Pfert_c$), the profit per hectare of producing commodity c from activity a ($\pi_{a,c}$) can be expressed as:

$$\pi_{a,c} = Poutput_c \cdot Qyperha_{a,c} - Pfert_c \cdot Qabsfert_{a,c} \tag{3}$$

As yield per hectare ($Qyperha_{a,c}$) is a function of fertilizer input per hectare ($Qabsfert_{a,c}$), profit per hectare can be re-written by combining Equation 1 and 3. Differentiating it for the decision variable $Qabsfert_{a,c}$ and solving for the fertilizer price ($Pfert_c$) yields:

$$Pfert_c = Poutput_c \cdot (2fertq_{a,c} \cdot Qabsfert_{a,c} + fertlin_{a,c}) \tag{4}$$

The problem is now that if we apply recent average prices (of the crops and fertilizer) in this equation, we would get a much higher fertilizer application rate as observed for Ethiopia. That means there must be unobserved cost elements of fertilizer that were not accounted for, and which we have to add to $Pfert_c$ in order to arrive at observed fertilizer application levels. This fixed calibration factor is thus calculated as the difference between marginal profitability of fertilizer application and marginal fertilizer price. Thus the complete fertilizer demand function is now given as:

$$Pfert_c + fertcalib_{a,c} = Poutput_c \cdot (2fertq_{a,c} \cdot Qabsfert_{a,c} + fertlin_{a,c}) \tag{5}$$

$Fertcalib_{a,c}$ can be interpreted as an indicator for unobserved costs in fertilizer marketing and use, and could thus be varied in policy simulations or long-term scenarios of better marketing and use efficiency.

We have created a new equation with the first order condition (FOC) above, and we have created two new variables contained in it, the absolute prices for the crop outputs ($Poutput_c$) and fertilizer ($Pfert_c$) that have to be defined in related equations. In these, changes in absolute prices are determined by changes in relative prices in the CGE part of the model where markets determine price changes for the crops and fertilizer:

$$\frac{Poutput_c}{Poutput0_c} = \frac{PQ_C}{PQ0_C} \tag{6}$$

$$\frac{Pfert_c}{Pfert0_c} = \frac{PQ_C}{PQ0_C} \tag{7}$$

where: PQ_C is the simulated commodity price in the CGE model

$PQ0_C$ is the commodity price in the CGE model at the base

The next step is to enable the CGE model to change its fertilizer input use in accordance with the fertilizer module. By default, single intermediate inputs in the CGE model are a fixed share ($ica_{c,a}$) of total intermediate inputs per activity, i.e., a Leontief demand function.

For variable fertilizer use, this restrictive function has to be relaxed and also altered to better reflect crop production processes. First, the $ica_{c,a}$ input-output coefficients for fertilizer input use are made variable for crops with yield functions while leaving the other inputs' shares in these crops fixed. Generally, input use in such crops is no longer related to the total quantity of inputs ($QINTA_a$ in the CGE model), but rather to land use, which is the standard way to describe a production technology in agronomy, meaning that

$$QINT_{c,a} = ica_{c,a} \cdot QF_{f,a} \tag{8}$$

where: c stands for fertilizer, a for crop activities with fertilizer-yield function, and f for factor land

For these same cropping activities, total input use ($QINTA_a$) is not a fixed proportion of output anymore but simply the sum of all inputs used ($QINT_{a,c}$).

$QINTA_a$ then enters the equation defining total output as a function of factor use (value added) and input use. With a subset of $ica_{c,a}$ being variable, changes of fertilizer use per unit of land use in the physical fertilizer module can now be translated into input use per unit of land in the CGE model:

$$\frac{ica_{c,a}}{ica0_{c,a}} = \frac{Qabsfert_{a,c}}{Qabsfert0_{a,c}} \tag{9}$$

The last necessary step is to relax the fixed 'yield parameter' $fprd_{f,a}$ that is part of the CGE model equation defining value-added creation through factor use. This parameter serves as an equivalent to the yield per hectare from the fertilizer module and therefore has to be made variable for those activity-commodity pairs for which fertilizer yield response function is available. Counterfactual changes in $fprd_{f,a}$ are then equivalent to changes in crop yields from the yield response functions.

3.4. Simulation Scenarios

Three different scenarios are simulated that run for a period of 20 years from 2011 to 2030: a baseline scenario for reference, and two counterfactual scenarios that simulate decreased costs of fertilizer use and elevated efficiency of application.

Scenario level 1: Baseline scenario (BASE) is the scenario in which relevant trends of the recent past in Ethiopian economy are applied.

The baseline scenario works as a benchmark which aims at continuing recent trends in model drivers over the simulation period. Based on CSA AgSS data for several years, total cropland supply is exogenously made to increase at a decreasing rate as shown in Figure 3. Moreover, based on two rounds of labor force survey data from the CSA 1999 and 2014 [48,49], the supply of agricultural labor (including unskilled labor) is made to increase by 3.19 percent annually while we apply a higher growth rate (6.14 percent) for non-agricultural labor (skilled and semi-skilled).

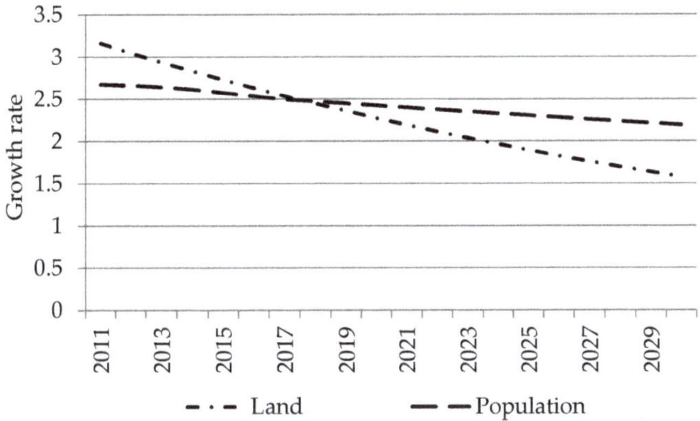

Figure 3. Land and population growth trends (2011-2030).

The other growth drivers in the model are population growth and Total Factor Productivity (TFP). The overall population in the country is made to increase at a decreasing rate for the next 20 years. Using 18 years data (2000–2017) from FAO data portal [50] we calculated and applied the growth trajectory shown in Figure 3 to the model households. The other essential element of economic growth is growth in activities' TFP. The growth rates applied in the model are calculated in a way that enables the economy to continue growing with the average growth rate reported for the last decade (i.e., 8.6 percent annually). Actual sectoral shares are also considered and growth rates for the other growth drivers are already applied. Thus, on average 5.7, 4.6 and 3.9 percent annual TFP growth rates are applied to agriculture, industry, and service sector activities respectively.

Scenario level 2: Better access to inputs and alternative pace of technological progress

In this scenario there are two simulations: one (F_ACCESS1) which experiments improved fertilizer accessibility through a decline in transaction (including transportation) cost, while the other (F_ACCESS2), on top of this, considers reduced use inefficiency through a reduction in unobserved cost that hinders Ethiopian farmers to optimize fertilizer use. The two simulations are similar to the BASE except for these additional interventions.

As briefly explained in Section 1, infrastructural and institutional predicaments in the country have led to high transaction costs. Referring to Rashid et al. [9], fertilizer transaction (plus transportation) cost to input distribution centers is estimated to be 11 - 23 percent (or 15.3 percent on average) of the landed cost. Additional 5 percent transportation cost is added which is an assumption about the total cost from the distribution center to the farm gate. All these calculations come up with a transaction+transportation cost of 20.3 percent. This is maintained in the SAM first and then it is made to decrease by 5 percent every year for both simulations in the second scenario (Table 1). This effectively means around a percent reduction in fertilizer retail price every year. As the price gets cheaper compared to the initial one, the profit maximizing producer is modeled to increase its fertilizer application per hectare. This is well described in the derived fertilizer demand function (see Equation (5) in Section 3.3.2). The optimal level of fertilizer per hectare is then used to calculate total fertilizer physical quantity based on the total area of land for each crop. That gives us the total intermediate demand for fertilizer.

Table 1. Experimental parameters and their respective shocks in the simulations.

Experimental Parameters	Applied on	Annual Change		
		BASE	F_ACCESS1	F_ACCESS2
Transaction + transportation cost (percentage of the total value sales)	Fertilizer commodity	-	−5%	−5%
Unobserved transaction cost / use inefficiency (Birr per hectare)	All crop activities	-	-	−1%

This intervention is with an assumption of a minimum effort by the government to improve the infrastructural and institutional barriers in fertilizer provision which is the public sector's role towards better input accessibility for the advancement of the agriculture sector.

Moreover, as shown in Table 1, a 1 percent reduction in the unobserved cost (or use inefficiency) is applied every year in F_ACCESS2 simulation on top of the intervention on transaction cost. As explained above in Section 3.3, the profit function obtained from the physical yield function postulates an optimal fertilizer use level per hectare far above the observed (actual) levels. We termed the positive profit, at suboptimal levels of fertilizer use (the parameter *fertcalib*$_a$ in the CGE model), as unobserved transaction costs or fertilizer use inefficiency. There are different reasons for this far sub-optimal use. The reasons are well explained in the introduction section.

This scenario assumes an achievement of a 1 percent reduction in unobserved cost as a result of public sector's effort towards addressing the reasons for the observed use inefficiency, for instance through improved access to credit for inputs, facilitating frequent contact with extension agents or provision of any other knowledge dissemination mechanisms that can improve farmers access to and understanding of modern inputs. Obviously, this intervention would have, in absolute terms, different level of effects for different crop types based on the difference in the initial level of unobserved cost (*fertcalib*$_{a,c}$). Those crops with bigger initial unobserved cost would benefit bigger from a percent reduction in unobserved cost applied on all crops.

The above mentioned interventions to improve the agriculture sector in Ethiopia need significant investment, especially from the government side. Even though we do not have the intention to do a thorough analysis on the cost-benefit comparison at the national level in this paper, it is worthwhile to mention quotes on the required public expenditure in order to undertake such interventions and attain the intended progress in the agriculture sector. Moreover, we also calculate the extent to which the possible benefits could cover the intended expenditure. According to the estimation in the Comprehensive Africa Agriculture Development Programme (CAADP) framework, an allocation of at least 10 percent of public expenditure to the agriculture sector is needed for the overall rural and agricultural development envisioned in Africa [51]. Specifically, if we take agriculture extension service in Ethiopia; it has traditionally been financed and provided almost entirely by the public sector.

Thus, these programs represent a significant public investment, roughly estimated to be 2 percent of total annual government expenditure [52].

4. Results and Discussion

As explained above, the interventions in this modeling work are applied to a number of crops. Discussing the results of each of these is time consuming and less important. Hence, we present results on the overall crops in general, and teff and maize in particular as they are the two most important crops in Ethiopia. Teff and maize hold the 1st and 2nd place in terms of crop output and land coverage. From consumption side, teff is mainly used for making 'enjera', the main national dish in Ethiopia (as well as Eritrea). Regarding maize, since it is the cheapest cereal crop, the poor (especially the rural poor) consume it mostly. It's crucial for national food security.

4.1. Supply (production) Effects and Economic Growth

Let us start the analysis with model results on food supply as the counterfactual scenarios are designed to improve productivity in crop production. The reduction in transaction (plus transportation) cost leads to an average of 0.5 percentage points annual decline in fertilizer retail price compared to the base line situation. This intervention together with the decline in unobserved costs increases the use of fertilizer. As a result, the experimental simulations show a positive response in yields and total output.

As shown in Table 2, a 5 percent annual reduction in transaction costs on fertilizer acquisition (i.e., F_ACCESS1) motivates higher fertilizer application. In the BASE run, an average Ethiopian crop farmer applies 56.8 kg of chemical fertilizer on a hectare plot which would rise to 61.5 kg as a result of the intervention. The improved soil fertility, consequently, would give an average 2.2 tons (i.e., additional yield of about 38.2 kg) per hectare crop each year. Looking at individual crops, the intervention encourages teff and maize farmers increase their fertilizer application rate at the BASE (i.e., 38 and 116 kg per hectare respectively) to 40.1 and 122.0 kg per hectare. As a result teff and maize farmers' yield levels rise by 1.2 and 2.0 percent of the BASE and reach 1.18 and 3.35 tons per hectare.

Table 2. Average annual chemical fertilizer application and yield level (kg per hectare).

	BASE		F_ACCESS1		F_ACCESS2	
	DAP+urea	Output	DAP+urea	Output	DAP+urea	Output
Teff	38.0	1162.7	40.1	1177.2	50.6	1246.1
Maize	116.0	3284.1	122.0	3350.1	149.7	3641.6
All crops	56.8	2156.7	61.5	2194.9	76.3	2344.3

On the other hand, if unobserved costs in fertilizer use could be reduced by 1 percent every year (i.e., F_ACCESS2), then the average farmer would annually apply 76.3 kg chemical fertilizer per hectare (i.e., additional of 34.2 percent of the BASE), which would lead to 8.7 percent increase in yield level per hectare. Teff and maize farmers would gain higher yield levels; 1.3 and 3.6 tons, which are higher than the BASE by 7.2 and 10.9 percent as a result of application of additional 12.5 and 33.7 kg fertilizer per hectare on teff and maize plots respectively.

Based on this achievement in yield and the resulted factor re-distribution in the model, the country's total crop production per annum would increase in F_ACCESS1 and F_ACCESS2 by 1.9 and 7.7 percent higher than it increases in the current growth trend (i.e., BASE). This implies a 0.8 and 3.2 million tons of annual crop supply additional to the current achievement level. In teff and maize production F_ACCESS2 would come up with the maximum effect - an additional of 5.7 and 5.5 percent of the current annual production level per annum, respectively. This means around 377 and 385 thousand metric tons of teff and maize.

Looking at the dynamics, yields of all crops, including teff and maize, increase at a decreasing rate. Even though this trend holds for all the simulations, both counterfactual scenarios, especially

F_ACCESS2 results in significantly stronger effects on the yield trend. As seen in Figure 4, the differential between the BASE and the counterfactual simulations is increasing throughout the period for all crops. The cumulated effects of the introduced shocks in both simulations make the activities produce consecutively higher amounts of output on a given plot. Crop yield grows by about 40 kg ha^{-1} on average per annum because of increased fertilizer use as a result of decreased acquisition cost (i.e., F_ACCESS1). Improving fertilizer accessibility and use efficiency simultaneously (i.e., F_ACCESS2) obtains the highest productivity gain for all crops: yield increases on average by about 197 kg ha^{-1}. The dynamics in both fertilizer application and yield gain show that reducing the inefficiency for modern input use is highly critical and makes the effort to improve input adoption even more effective.

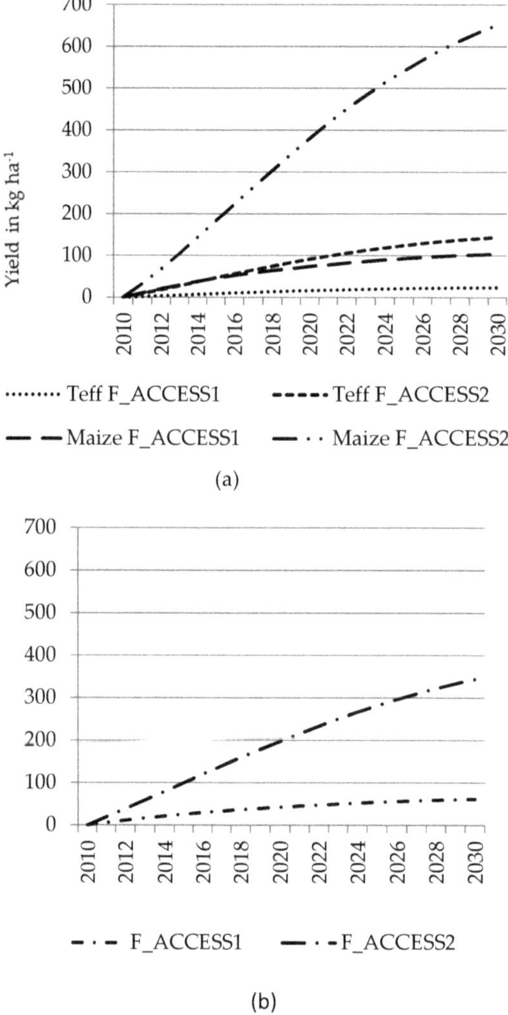

Figure 4. Dynamics in yield (2010-2030) (deviations between BASE and experimental simulation trajectories) for a) teff and maize and b) all crops.

One benefit of the productivity improvements in domestic crop production might be a reduced import dependency of food staples. For instance, due to frequent droughts (especially in the drought-prone areas of east and south-east parts of Ethiopia) and rising food prices, the government of Ethiopia relies on wheat imports for relief and market stabilization. Thus, according to FAOSTAT [50], Ethiopia's wheat imports increased more than three-fold in the last twenty-five years. For instance, from 2008–2013 the country imported an average of about half a billion USD worth of wheat each year, putting a burden on Ethiopia's foreign currency reserve. The additional output obtained as a result of the simulated interventions could contribute to easing this deficit by boosting domestic production of staple grains. Annual wheat import would reduce as a result of the counterfactual interventions by 8.7 percent of the current annual import volume which means the country could save a maximum of a quarter of its import from 2008–2013. Moreover, looking at the country's import of crop commodities, the baseline interventions would increase import demand for all the simulations. However, the counterfactual interventions would reduce crop imports by 3 and 9 percent of the baseline performance annually. Regarding fertilizer import, as both the counterfactual interventions would significantly boost the demand for fertilizer, it would directly be reflected on the country's import bill. Fertilizer import would annually increase in F_ACCESS1 and F_ACCESS2 by 7.7 and 32 percent of the current annual import level.

The interventions introduced in the farming sector have a strong impact on macroeconomic aggregates as well. Aggregate value added is significantly increasing for all the simulations. However, when comparing to the BASE, as can be seen from Figure 5, the additional GDP obtained as a result of the counter factual interventions increases at an increasing rate consecutively. The effect gets stronger when efficiency is improved in fertilizer use together with a higher application rate. The fact that all the counter factual interventions are applied on the crop sub-sector, which contributes around 32 percent of the total GDP, makes interventions with strong implications on this sub-sector strongly reflect on the national GDP. However, the weak linkage between the agriculture and non-agriculture sectors (well discussed in Section 3.2) significantly reduces the implication of agriculture growth to the non-agriculture economy. For instance, following the interventions in F_ACCESS2, crop sub-sector GDP increases by 2 percentage points with the agriculture GDP increasing by 1.4 percentage points, while the non-agriculture GDP rises by only 0.01 percentage points in comparison with the BASE. Moreover, the fact that the country is not exporting most of the crops that receive the interventions, the success obtained in increasing production brings only 0.03 and 0.09 percentage points increase in volume of export. Thus, almost the entire additional production is sold to the domestic market that implies decreasing crop prices that fall on average by 0.33 percentage points in comparison with the BASE. This has adverse effect on (net food seller) farmers' income.

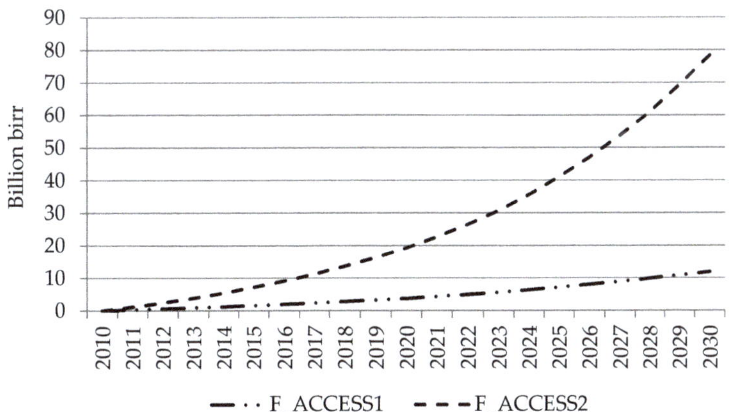

Figure 5. Dynamics in the aggregate value addition (2010-2030), comparison with the BASE (in billion birr).

4.2. Effects on Household Income

All the simulations bring about significantly increasing income changes for both farming and non-farming households. However, compared to the BASE scenario, the additional productivity gain resulting from the two counterfactual simulations leads to a small impact on income for both household types. As highlighted above, the shocks involved in the two counterfactual simulations relatively decrease crop prices, which have an adverse effect on the income of net seller farming households. Moreover, as shown in Table A4 in the annex, non-farming households basically receive most of their income from skilled labor and capital employment in the non-agriculture sector which is marginally affected by the relevant interventions. If we look at the households by their income classification, F_ACCESS1 and F_ACCESS2 enable poor farmers to obtain incomes that are 0.07 and 0.25 percentage points higher, respectively, as compared to the BASE. Non-poor farming households receive 0.06 and 0.26 percentage points higher incomes. Non-farming households similarly enjoy a slight increase in income of 0.04 percentage points as a result of F_ACCESS1 and 0.25 and 0.26 percentage points for the poor and non-poor counter parts respectively as a result of F_ACCESS2.Looking at the aggregate effect doesn't show us the impact on the specific household, so that effects on the profitability of crop farming in particular need to be discussed. It is possible to calculate income effects through profitability of the additional fertilizer applied. A commonly used measure of profitability is the value-cost ratio (VCR) (Equation (10)). Here it is applied to the model results (i.e., additional chemical fertilizer used per holder and the additional output obtained as a result - both in kg to calculate the profitability of additional fertilizer used per holder's plot relative to BASE). This ratio is the value of the additional production due to application of the additional fertilizer to the total cost of the additional fertilizer applied. The additional output from an average crop plot that received the additional fertilizer is X. P is the price of crop output (farm gate price). Price of fertilizer is P_f, and Q_f is the additional amount of fertilizer applied to that particular plot, then the VCR is given as follows:

$$\mathbf{VCR} = \frac{\mathbf{P} \times \mathbf{X}}{\mathbf{P_f} \times \mathbf{Q_f}} \tag{10}$$

Looking at the results from Table 3, the VCR levels mean that each kg of additional fertilizer applied on the crop plots is found to be highly profitable to the holder. As can be understood from the formula, a VCR of 1 is the threshold to profitability. However, it is commonly argued that a VCR of at least 2 is needed for fertilizer to be profitable in Africa. A high threshold level is recommended for Africa just to compensate for the higher probability of adverse conditions that greatly influence profitability, as for instance infestations and weather risks [9].

An average teff farmer in Ethiopia with a 0. 460 ha plot obtains an additional 32.9 and 186.4 Birr profit from applying an additional 1 and 5.76 kg of chemical fertilizer in F_ACCESS1 and F_ACCESS2, respectively. Similarly, an average maize farmer gets from 26.2 to 137 Birr profit from his quarter of a hectare plot as a result of the application of additional 1.5 to 8.4 kg chemical fertilizer in the two simulations. Looking at the overall crop sub-sector, an average crop farmer with 0.219 ha plot enjoys 18.5 and 228.8 Birr profit applying an additional 1 and 4.3 kg chemical fertilizer. As discussed above, the VCR levels from Table 3 show that each Birr spent to purchase additional fertilizer for teff and maize in particular and all crops production in general are very profitable. These results indicate that improved fertilizer accessibility and reduced use inefficiency could make a significant difference regarding fertilizer application and profitability. Thus, as compared to BASE, crop farmers are better off in terms of profitability. This positive effect could be higher with greater fertilizer application.

Table 3. VCR calculations for additional fertilizer application on an average Ethiopian crop farming plot.

Fertilizer price Birr per kg	F_ACCESS1			F_ACCESS2		
	14.76			14.76		
Teff price Birr per kg	7.10			7.09		
Maize price Birr per kg	2.94			2.94		
All crops average price Birr per kg	4.00			4.00		
	Teff	Maize	All crops	Teff	Maize	All crops
Additional DAP+urea used per holder (kg)	0.96	1.49	1.02	5.76	8.35	4.26
Additional output (kg)	6.63	16.38	8.38	38.30	88.64	41.15
Additional spending for fertilizer (birr)	14.18	21.94	15.09	85.08	123.22	62.94
Additional birr obtained from sales	47.08	48.17	33.55	271.52	260.27	291.70
Additional profit in birr	32.89	26.24	18.46	186.44	137.05	228.76
VCR	3.32	2.20	2.22	3.19	2.11	4.63

Note: All the acreage and price figures basically used in this calculation are subject to changes from the model results.

In an attempt to look at the welfare gains and losses to market participants from changes in market conditions as a result of the interventions, aggregate value added is calculated for each intervention. Let us take the last intervention (for the sake of a straight forward calculation of its budgetary cost) and assume the 1 percent annual reduction in unobserved cost for crop farmers is attainable through improved access to better quality extension service. As a result, this intervention would result in a total welfare gain of about 66.6 billion Birr in the country in 2030 which grows from 0.8 billion Birr in 2011 with a 27.5 percent average annual growth rate. Based on data from National Bank of Ethiopia NBE [53] and estimation by Spielman et al. [52] average annual total public expenditure on extension service is calculated to be 3.65 billion Birr between 2009/10 and 2016/17. We also made a projection until 2030 based on recent years' growth trend. The current and projected public expenditure on extension service is then calculated to be 82 percent of the expected welfare gain obtained from increased efficient fertilizer use as a result of improved provision of extension service. (See Figure 6) The welfare gain is encouraging and could sufficiently cover expenditure to further improve the agriculture extension service.

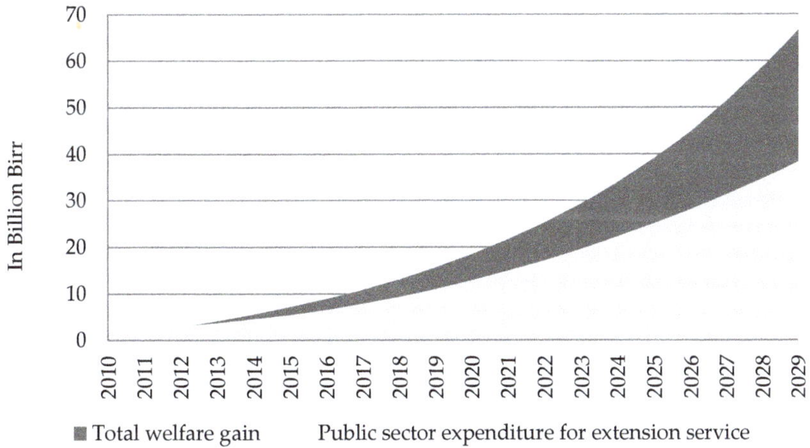

Figure 6. Trajectories for public expenditure on extension service and the expected total welfare gain.

4.3. Effects on Household Consumption

When discussing changes in household welfare as a result of productivity interventions, a distinction between net producers and net consumers of crops has to be made. Subsequent price drops have adverse effects on net producers, whereas they are benefits to net consumers. A broader perspective is needed to deal with this issue, as households have diverse sources of income and lower crop prices even could have positive effects on the producer households' consumption expenditure. Income improvements (see Section 4.2) and price reductions are obtained from the model results. Average crop price annually decreases at a maximum of 0.03 and 0.28 percentage points relative to the BASE in F_ACCESS1 and F_ACCESS2, respectively. Thus, households that are highly dependent on these commodities, such as rural farming households which consume 35 percent of their total consumption from crop commodities, would exhibit higher welfare improvements as a result of cheaper consumption than households that consume less of these commodities (in our case, non-farming households whose consumption bundle comprises of only 15 percent crop commodities). Referring to Figure 7, farming households enjoy higher consumption increases in the experimental simulations than non-farming households. The price decrease under F_ACCESS2 enables farming households to enjoy a higher welfare gain than net consumers even though both enjoy an equal rise in income. However, price and income changes work together to make all household categories better-off as a result of the counter factual interventions. Non-farming households' welfare improves as well as a result of reduced cost of living and rise in income.

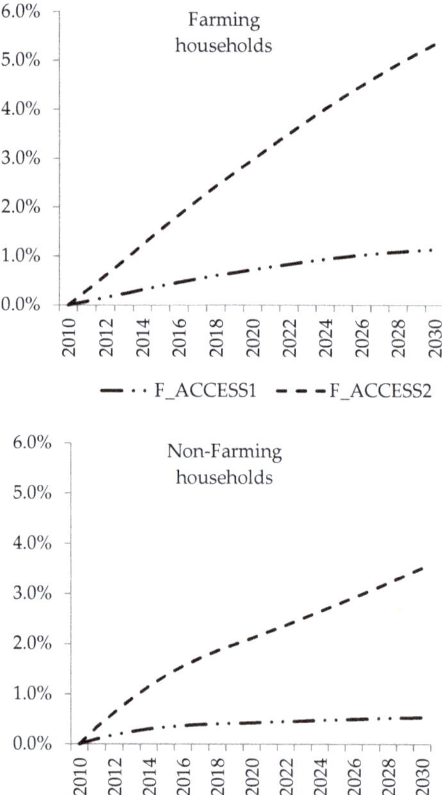

Figure 7. Dynamics in simulation results on households' consumption (2010-2030) (percentage deviation between BASE and simulation trajectories).

5. Conclusions

The overarching objective of this study was to examine the economy-wide implications of improved access to and use efficiency of modern farming inputs, particularly fertilizer with a major emphasis on the resulting household welfare. The study is motivated by the situation in Ethiopia in which cropland scarcity is bound to be a serious challenge for the future growth in food supply. Using a dynamic CGE model for Ethiopia and estimated fertilizer-yield response functions we investigate macro and micro effects of improved fertilizer application.

In comparison with current trends, model results show positive effects in the two experimental simulation scenarios: 1. reduction in fertilizer transaction costs to make fertilizer more accessible, and 2. annual reduction in unobserved transaction cost of fertilizer use, on top of the first one, as a result of the public sector's effort to address the reasons for fertilizer use inefficiency. Both experimental simulations are applied to all crops. In both simulations, the interventions would come up with a significant increase in yields and total output both at the aggregate and household level. Based on the importance of cereals in the national food basket, 46.3 percent according to Brhane et al. 2011 [54], this significant improvement in total domestic production has huge impacts on the national food security and stabilizing the domestic cereal market. It could also ease the country's BoP deficient that is currently worsening as a result of increased food and fertilizer imports, among others. The results also suggest that the increased fertilizer application is highly profitable for an average crop farmer with 0.219 hectare plot. If the entirety of households in the economy is considered, effects of the new interventions on crop price and household income together enables them to enjoy higher welfare, especially farming households when higher fertilizer amount is applied efficiently. This is especially meaningful for the farming population in Ethiopia, who frequently experience food insecurity.

Moreover, the aggregate welfare effect indicates that these interventions are profitable in general. Assuming a reduction in fertilizer use inefficiency attainable through improved access to extension services, the total welfare gain that would be achieved is found to be equivalent to 135 percent of the current annual public expenditure on extension service. Putting this into perspective, we can see how economically viable the intended intervention is. The welfare gain from the intervention on the crop sub-sector could sufficiently justify complete financing of the total cost with extra money that can finance an increase of 35 percent of the entire current public expenditure on farm extension services.

Thus, in order to reap these potential benefits, reductions in transaction costs of fertilizer acquisition could be achieved in different ways. Improving the rural road network would reduce the cost markup from harbor to farm. Additionally, reducing bureaucratic obstacles in fertilizer procurement could avoid belated input application, which decreases the use inefficiency. Finally, improved accessibility to agriculture extension agents and creating possible alternative income sources, especially off-farm, would significantly ease adoption barriers and cut the high knowledge costs related to the adoption of new technologies. The comparison between the two counterfactual simulations shows that improved efficiency regarding modern input use is highly critical and makes the effort to improve input adoption even more effective.

Though the interventions have strong positive effects on the economy, increased fertilizer import has adverse effect on the country's balance of payment (BoP). Moreover, the fact that most of the crops that enjoy the interventions are non-tradables, the increased production ends up in domestic price decrease which has adverse effect on net seller farmers' income. Thus, as a policy recommendation, if Ethiopia introduces most of its crops to the export market, it could be of double advantage; 1. It could serve the BoP deficit resulting from increases in fertilizer import, and 2. It provides the farmers with an opportunity to get better price.

Policy makers should focus on strengthening the agriculture – industry linkage in order to make growth in the agriculture sector transmit and trigger growth in the non-agriculture sector. The possible pathways are 1. Through commencement of domestic fertilizer production and supply which is advantageous for faster and timely distribution, lower price provision and easing BoP burden of fertilizer imports. 2. Through promoting expansion of the food processing industry sub-sector. As the urban centers in Ethiopia get more and more populous and modernized, the demand for ready-made (or packed) foods is increasing. There are few industries operating at micro and small scale trying to satisfy this demand. Policy focus on these kinds of industries could benefit the economy with strong linkage between the agriculture and non-agriculture sector and households with cheaper costs of living.

Author Contributions: Conceptualization, A.K. and E.E.L.; Economic Modelling, A.K. and E.E.L.; Crop Modelling, A.K.S. and T.G.; Investigation, A.K. and E.E.L.; Analysis, E.E.; Writing-Review & Editing, A.K. and A.K.S.

Funding: This research was funded by the German Federal Ministries of Education and Research (BMBF) and of Cooperation and Development (BMZ), grant number: 031A258D.

Acknowledgments: The authors are grateful to Mekdim Dereje and Helina Tilahun for sharing CSA AgSS data and GIS data.

Conflicts of Interest: The authors declare no conflict of interest.

Appendix A

Table A1. Major fertilizer suppliers to the farmer.

Who Are Your Major Suppliers of Fertilizer?	Percent
Government Organization	14.26
Private Organization	0.91
Merchants	3.79
Cooperatives	23.37
Other (Specify)	1.64
Never used fertilizer	56.03
Total	100

Table A2. I-O table and production technology for the 2009/10 Ethiopian economy.

	Cereals	Agriculture Sector	Non-Agriculture Sector
Intermediate inputs	11.54%	5.75%	53.88%
Agriculture commodities	7.38%	3.41%	7.41%
Non-agriculture commodities	4.16%	2.34%	46.47%
Fertilizer	3.63%	1.66%	-
Factor inputs	88.5%	94.2%	46.12%

Table A3. Demand for commodities in the economy by users.

	Intermediate Input Demand	Household Consumption Demand	Export Demand	Investment Demand	Public Sector Demand	Transaction Demand
Cereals	13.06%	86.87%	0.07%	0.00%	0.00%	0.00%
Agriculture commodities	19.23%	73.15%	6.53%	1.1%	0.0%	0.0%
Non-agriculture commodities	31.47%	31.30%	5.99%	13.26%	4.97%	13.01%

Table A4. Factor returns by sectors and their contribution to households' income.

	Income from Factors		Factor Returns from Sectors	
	Farming Households	Non-Farming Households	Agriculture	Non-Agriculture
Skilled labor	15.6%	40.0%	0.00%	100.00%
Unskilled labor	33.0%	16.2%	88.52%	11.48%
Land	24.8%	0.0%	100.00%	0.00%
Capital	26.6%	43.8%	3.58%	96.42%

Table A5. LINTUL5-simulated maize yields, and quadratic approximation function (Ethiopian national average results) at different use levels of mixed fertilizer, and coefficients of the quadratic approximation function.

DAP-Urea Mixed Fertilizer Application in kg/ha	Simulated Yield in kg/ha from LINTUL5	Quadratic Approximation of LINTUL5-Simulated Yields	Deviation
1	1914.24	1890.09	−24.15
62	2687.11	2670.12	−16.99
143	3565.92	3623.61	57.69
287	5024.44	5074.57	50.13
430	6299.13	6229.78	−69.35
573	7153.87	7089.23	−64.64
717	7634.57	7652.93	18.36
860	7848.98	7920.88	71.90
1003	7949.89	7893.07	−56.82

Constant (*fertconst*): 1876.9; Linear coefficient (*fertlin*): 13.22; Quadratic coefficient (*fertq*): −0.0072

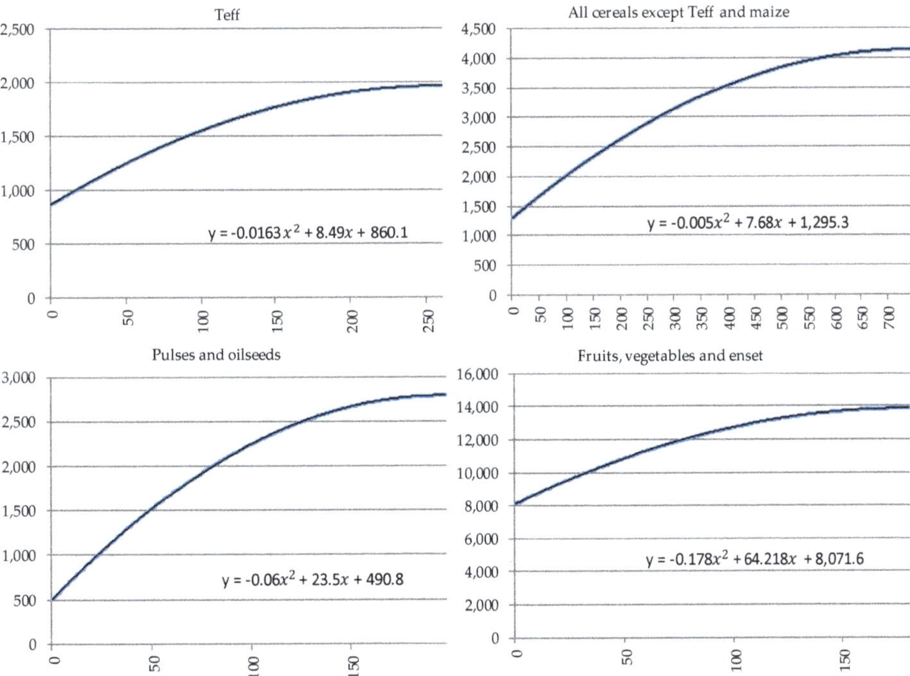

Figure A1. Different use levels of mixed chemical fertilizer and yield levels based on quadratic yield response functions for different crop types (Grain yield kg ha^{-1} on the vertical axis and composite chemical fertilizer kg ha^{-1} on the horizontal axis).

References

1. Boserup, E. *The Conditions of Agricultural Growth: The Economics of Agrarian Change under Population Pressure*; G. Allen and Unwin LTD: London, UK, 1965; pp. 104–105.
2. Demeke, M.; Kelly, V.; Jayne, T.S.; Said, A.; Le Vallee, J.C.; Chen, H. *Agricultural Market Performance and Determinants of Fertilizer Use in Ethiopia. Grain Market Research Project*; Working Paper No.10; Ministry of Economic Development and Cooperation: Addis Ababa, Ethiopia, 1998.
3. World Bank. Ethiopia Poverty Assessment 2014. Poverty Global Practice Africa Region. Report No. AUS6744. Available online: http://documents.worldbank.org/curated/en/131011468247457565/pdf/AUS67440REVISE0sessment0pub02020015.pdf (accessed on 3 March 2019).
4. Kassie, M.; Shiferaw, B.; Muricho, G. Agricultural Technology, Crop Income, and Poverty Alleviation in Uganda. *World Dev.* **2011**, *39*, 1784–1795. [CrossRef]
5. Becerril, J.; Abdulai, A. The impact of improved maize varieties on poverty in Mexico: A propensity score matching approach. *World Dev.* **2010**, *38*, 1024–1035. [CrossRef]
6. Duflo, E.; Kremer, M.; Robinson, J. How High are Rates of Return to Fertilizer? Experimental Evidence from Kenya. *Am. Econ. Rev.* **2008**, *98*, 482–488. [CrossRef]
7. Minten, B.; Barrett, C.B. Agricultural technology, productivity, and poverty in Madagascar. *World Dev.* **2008**, *36*, 797–822. [CrossRef]
8. Jayne, T.S.; Chamberlin, J.; Headey, D.D. Land Pressures, the Evolution of Farming Systems, and Development Strategies in Africa: A Synthesis. *Food Policy* **2014**, *48*, 1–17. [CrossRef]
9. Rashid, S.; Tefera, N.; Minot, N.; Ayele, G. *Fertilizer in Ethiopia: An Assesment of Policies, Value Chain, and Profitability*; Discussion Paper 1304; International Food Policy Research Institute: Washington, DC, USA, 2013.
10. Zerfu, D.; Larson, D. *Incomplete Market and Fertilizer Use: Evidence from Ethiopia*; Policy Research Working Paper, WPS5235; World Bank: Washington, DC, USA, 2010.
11. Minten, B.; Koru, B.; Stifel, D. The Last Mile(s) in Modern Input Distribution: Pricing, Profitability, and Adoption. *Agric. Econ.* **2013**, *44*, 629–646. [CrossRef]
12. Yu, B.; Nin-Pratt, A.; Funes, J.; Asrat, S. Cereal Production and Technology Adoption in Ethiopia. In Proceedings of the 8th International Conference of the Ethiopian Economic Association, Addis Ababa, Ethiopia, 25 June 2010.
13. Asrat, S.; Bizuneh, G.; Taffesse, A.S. Trends and Determinants of Cereal Productivity in Ethiopia. In Proceedings of the 8th International Conference of the Ethiopia Economic Association, Addis Ababa, Ethiopia, 22 June 2010.
14. Nisrane, F.; Berhane, G.; Asrat, S.; Bizuneh, G.; Taffesse, A.S.; Hoddinott, J. *Sources of Inefficency and Growth in Agricultural Output in Subsistence Agriculture: A Stochastic Frontier Analysis*; ESSP2 Working Paper 19; International Food Policy Research Institute: Washington, DC, USA, 2011.
15. Dorward, A.; Chirwa, E. The Malawi agricultural input subsidy programme: 2005/06 to 2008/09. *Int. J. Agric. Sustain.* **2011**, *9*, 232–247. [CrossRef]
16. Bold, T.; Kaizzi, K.; Svensson, J.; Yanagizawa-Drott, D. *Low Quality, Low Returns, Low Adoption: Evidence from the Market for Fertilizer and Hybrid Seed in Uganda*; Working Paper Series rwp15-033; Harvard University, John F. Kennedy School of Government: Cambridge, MA, USA, 2015.
17. Yu, B.; Nin-Pratt, A. Fertilizer Adoption in Ethiopia Cereal Production. *J. Dev. Agric. Econ.* **2014**, *6*, 318–337. [CrossRef]
18. Beshir, H.; Emana, B.; Kassa, B.; Haji, J. Determinants of Chemical Fertilizer technology Adoption in North Eastern Highlands of Ethiopia: The Double Hurdle Approach. *J. Res. Econ. Int. Financ.* **2012**, *1*, 39–49.
19. Verkaart, S.; Munyua, B.G.; Mausch, K.; Michler, J.D. Welfare Impacts of Improved Chickpea Adoption: A Pathway for rural Development in Ethiopia? *Food Policy* **2017**, *66*, 50–61. [CrossRef] [PubMed]
20. Caria, A.S.; Tamiru, S.; Bizuneh, G. *Food Security Without Food Transfers? A CGE Analysis for Ethiopia of the Different Food Security Impacts of Fertilizer Subsides and Locally Sourced Food Transfers*; IFPRI Discussion Paper 01106; International Food Policy Research Institute: Washington, DC, USA, 2011.
21. Arndt, C.; Pauw, K.; Thurlow, J. The Economy-Wide Impacts and Risks of Malawi's Farm Input Subsidy Program. *Am. J. Agric. Econ.* **2016**, *98*, 962–980. [CrossRef]

22. Bachewe, F.N.; Berhane, G.; Minten, B.; Taffesse, A.S. *Agricultural Growth in Ethiopia (2004–2014): Evidence and Drivers*; ESSP Working Paper 81; International Food Policy Research Institute: Washington, DC, USA, 2015.

23. Martins, P. *Structural Change in Ethiopia: An Employment Perspective*; Policy Research Working Paper, WPS 6749; World Bank: Washington, DC, USA, 2014.

24. National Bank of Ethiopia (NBE). Annual Report 2013/2014. Annual Report. 2014. Available online: http://www.nbe.gov.et/pdf/annualbulletin/Annual%20Report%202013-2014/Annex%202013-14newst.pdf (accessed on 25 November 2017).

25. Central Statistics Agency (CSA). *Agriculture Sample Survey 2015/16 (2008 E.C.). Statistical Bulletin, Report on Area and Production of Major Crops (Private Peasant Holdings, Meher Season)*; CSA: Addis Ababa, Ethiopia, 2016.

26. Central Statistics Authority (CSA). *Agriculture Sample Survey 2004/05 (1997 E.C.). Statistical Bulletin, Report on Area and Production of Major Crops (Private Peasant Holdings, Meher Season)*; CSA: Addis Ababa, Ethiopia, 2005.

27. Central Statistics Agency (CSA). *Agriculture Sample Survey 2005/06 (1998 E.C.). Statistical Bulletin, Report on Area and Production of Major Crops (Private Peasant Holdings, Meher Season)*; CSA: Addis Ababa, Ethiopia, 2006.

28. Central Statistics Agency (CSA). *Agriculture Sample Survey 2006/07 (1999 E.C.). Statistical Bulletin, Report on Area and Production of Major Crops (Private Peasant Holdings, Meher Season)*; CSA: Addis Ababa, Ethiopia, 2007.

29. Central Statistics Agency (CSA). *Agriculture Sample Survey 2007/08 (2000 E.C.). Statistical Bulletin, Report on Area and Production of Major Crops (Private Peasant Holdings, Meher Season)*; CSA: Addis Ababa, Ethiopia, 2008.

30. Central Statistics Agency (CSA). *Agriculture Sample Survey 2008/09 (2001 E.C.). Statistical Bulletin, Report on Area and Production of Major Crops (Private Peasant Holdings, Meher Season)*; CSA: Addis Ababa, Ethiopia, 2009.

31. Central Statistics Agency (CSA). *Agriculture Sample Survey 2009/10 (2002 E.C.). Statistical Bulletin, Report on Area and Production of Major Crops (Private Peasant Holdings, Meher Season)*; CSA: Addis Ababa, Ethiopia, 2010.

32. Central Statistics Agency (CSA). *Agriculture Sample Survey 2010/11 (2003 E.C.). Statistical Bulletin, Report on Area and Production of Major Crops (Private Peasant Holdings, Meher Season)*; CSA: Addis Ababa, Ethiopia, 2011.

33. Central Statistics Agency (CSA). *Agriculture Sample Survey 2011/12 (2004 E.C.). Statistical Bulletin, Report on Area and Production of Major Crops (Private Peasant Holdings, Meher Season)*; CSA: Addis Ababa, Ethiopia, 2012.

34. Central Statistics Agency (CSA). *Agriculture Sample Survey 2012/13 (2005 E.C.). Statistical Bulletin, Report on Area and Production of Major Crops (Private Peasant Holdings, Meher Season)*; CSA: Addis Ababa, Ethiopia, 2013.

35. Central Statistics Agency (CSA). *Agriculture Sample Survey 2013/14 (2006 E.C.). Statistical Bulletin, Report on Area and Production of Major Crops (Private Peasant Holdings, Meher Season)*; CSA: Addis Ababa, Ethiopia, 2014.

36. Central Statistics Agency (CSA). *Agriculture Sample Survey 2014/15 (2007 E.C.). Statistical Bulletin, Report on Area and Production of Major Crops (Private Peasant Holdings, Meher Season)*; CSA: Addis Ababa, Ethiopia, 2015.

37. World Bank. *Background Report: Four Ethiopias, A Regional Characterization; Assessing Ethiopia's Growth Potential and Development Obstacles. Economic Memorandum*; World Bank: Addis Ababa, Ethiopia, 2004.

38. Central Statistical Authority (CSA). *Agricultural Sample Survey 2003/2004 (1996 E.C.). Statistical Bulletin, Report on Area and Production of Major Crops (Private Peasant Holdings, Meher Season)*; CSA: Addis Ababa, Ethiopia, 2004.

39. UNDP. Human Development Reports 2017. Available online: http://hdr.undp.org/en/composite/HDI (accessed on 15 March 2019).

40. Lofgren, H.; Harris, R.L.; Robinson, S. *A Standard Computable General Equilibrium (CGE) Model in GAMS*; Discussion Paper 75; International Food Policy Research Institute, Trade and Macroeconomics Division: Washington, DC, USA, 2001.

41. Dorosh, P.; Thurlow, J. Implications of Accelerated Agricultural Growth on Household Incomes and Poverty in Ethiopia: A General Equilibrium Analysis. ESSP2 Working Paper 002. Available online: http://ebrary.ifpri.org/utils/getfile/collection/p15738coll2/id/130942/filename/131153.pdf (accessed on 4 May 2019).

42. Engida, E.; Tamru, S.; Tsehaye, E.; Debowicz, D.; Dorosh, P.A.; Robinson, S. *Ethiopia's Growth and Transformation Plan: A Computable General Equilibrium Analysis of Alternative Financing Options*; ESSP Working Paper 30; International Food Policy Research Institute: Washington, DC, USA, 2011.

43. Gaiser, T.; Perkons, U.; Küpper, P.M.; Kautz, T.; Uteau-Puschmann, D.; Ewert, F.; Enders, A.; Krauss, G. Modeling biopore effects on root growth and biomass production on soils with pronounced sub-soil clay accumulation. *Ecol. Model.* **2013**, *256*, 6–15. [CrossRef]

44. Addiscott, T.M.; Whitmore, A.P. Simulation of solute leaching in soils of differing permeabilities. *Soil Use Manag.* **1991**, *7*, 94–102. [CrossRef]

45. Jaleta, M.; Kassie, M.; Shiferaw, B. Tradeoffs in crop residue utilization in mixed crop–livestock systems and implications for conservation agriculture. *Agric. Syst.* **2013**, *121*, 96–105. [CrossRef]

46. Srivastava, A.K.; Mboh, C.M.; Gaiser, T.; Zhao, G.; Ewert, F. Climate change impact under alternate realizations of climate scenarios on maize yield and biomass in Ghana. *Agric. Syst.* **2018**, *159*, 157–174. [CrossRef]

47. Srivastava, A.; Gaiser, T.; Ewert, F.; Mboh, C.M.; Engida, E.; Kuhn, A. Effect of Mineral Fertilizer on Rain Water and Radiation use Efficiencies for Maize Yield and Stover Biomass Productivity in Ethiopia. *Agric. Syst.* **2019**, *168*, 88–100. [CrossRef]

48. Central Statistics Authority (CSA). *Statistical Report on The 1999 National Labour Force Survey*; Statistical Bulletin No. 225; CSA: Addis Ababa, Ethiopia, 1999.

49. Central Statistical Agency (CSA). *Analytical Report on The 2013 National Labour Force Survey*; Statistical Bulletin; CSA: Addis Ababa, Ethiopia, 2014.

50. Food and Agriculture Organization (FAO). FAOSTAT. Available online: http://www.fao.org/faostat/en/#country/238 (accessed on 4 December 2017).

51. New Partnership for Africa's Development (NEPAD). *Comprehensive Africa Agriculture Development Programme*; New Partnership for Africa's Development (NEPAD): Midrand, South Africa, 2003; ISBN 0-620-30700-5.

52. Spielman, D.J.; Mekonnen, D.K.; Alemu, D. Seed, Fertilizer, and Agriculture Extension in Ethiopia. In *Food and Agriculture in Ethiopia: Progress and Policy Challenges*; Poul, D.A., Rashid, S., Eds.; University of Pennsylvania Press: Philadelphia, PA, USA, 2012; pp. 84–122. ISBN 9780812245295. [CrossRef]

53. National Bank of Ethiopia (NBE). 2016/17 Annual Report. Available online: http://www.nbe.gov.et/pdf/annualbulletin/NBE%20Annual%20report%202016-2017/NBE%20Annual%20Report%202016-2017.pdf (accessed on 3 September 2018).

54. Berhane, G.; Paulos, Z.; Tafere, K.; Tamiru, S. *Foodgrain Consumption and Clorie Intake Patterns in Ethiopia*; ESSP 2. Working Paper 23; International Food Policy Research Institute: Washington, DC, USA, 2011.

 © 2019 by the authors. Licensee MDPI, Basel, Switzerland. This article is an open access article distributed under the terms and conditions of the Creative Commons Attribution (CC BY) license (http://creativecommons.org/licenses/by/4.0/).

 *sustainability*

Article

Making Contract Farming Arrangements Work in Africa's Bioeconomy: Evidence from Cassava Outgrower Schemes in Ghana

Adu-Gyamfi Poku [1,*], **Regina Birner** [1] **and Saurabh Gupta** [2]

[1]   Institute of Agricultural Sciences in the Tropics (Hans-Ruthenberg-Institute), University of Hohenheim, Wollgrasweg 43, 70599 Stuttgart, Germany; regina.birner@uni-hohenheim.de
[2]   Center for Development Management, Indian Institute of Management Udaipur, Udaipur District, Balicha, Rajasthan 313002, India; saurabh.gupta@iimu.ac.in
*   Correspondence: adu-gyamfi.poku@uni-hohenheim.de

Received: 16 March 2018; Accepted: 15 May 2018; Published: 17 May 2018

Abstract: This paper uniquely focuses on rapidly-developing domestic value chains in Africa's emerging bioeconomy. It uses a comparative case study approach of a public and private cassava outgrower scheme in Ghana to investigate which contract farming arrangements are sustainable for both farmers and agribusiness firms. A complementary combination of qualitative and quantitative methods is employed to assess the sustainability of these institutional arrangements. The results indicate that ad hoc or opportunistic investments that only address smallholders' marketing challenges are not sufficient to ensure mutually beneficial and sustainable schemes. The results suggest that firms' capacity and commitment to design contracts with embedded support services for outgrowers is essential to smallholder participation and the long-term viability of these arrangements. Public-private partnerships in outgrower schemes can present a viable option that harnesses the strengths of both sectors and overcomes their institutional weaknesses.

Keywords: contract farming; contract design; cassava; bioeconomy; Ghana

---

1. Introduction

Rapid agro-industrialisation in sub-Saharan Africa is leading to the development of high-value supply chains for agro-food systems [1]. Additionally, there is increasing demand for feed and other biomass-based raw materials, such as fuel and fibre crops, in Africa's emerging bioeconomy [2]. This has meant a transition towards modernised procurement systems even for agricultural commodities that have traditionally been dominated by spot market exchanges between small scale farmers and traders [3]. Consequently, contract farming (CF)—the institutional arrangement wherein processors enter into formal or informal contractual agreements with farmers to produce and supply them with agricultural commodities—has increasingly been embraced by agribusiness firms in developing countries as an efficient approach for coordinating supply chain activities [4–7].

CF, however, remains a highly-contested institutional arrangement in terms of poverty alleviation and rural development. Empirical evidence from developing countries on smallholder participation in CF and its impact presents mixed results. Some studies have found that smallholders actively participate in CF schemes and earn higher income as a result. These farmers were found to have benefitted from better access to inputs and new technology leading to improved farm productivity [5,6,8–10]. Conversely, other studies have reported evidence of smallholder exclusion, high default rates, and various forms of opportunistic behaviour by firms, such as delayed payments and a lack of compensation for crop losses in CF schemes [11–13]. These divergent findings shed light on how essential contract design is to the performance and impacts of CF schemes. However, there is a paucity of studies, particularly in the

African context, examining which institutional arrangements and contract conditions are sustainable for both farmers and agribusiness firms in these schemes. Addressing this knowledge gap is even more pertinent considering the increased competition for the procurement of multi-purpose crop biomass in rapidly developing domestic agricultural value chains. Such domestic value chains have previously been considered as less than suitable for CF arrangements [14]. Therefore, in the wake of numerous failed contract farming arrangements in Africa [15], this paper empirically investigates the role of contract design in facilitating sustainable contract farming arrangements between farmers and agribusinesses, particularly for a staple crop. The study does not directly address the impacts of CF, but rather focuses on investigating which CF arrangements are sustainable for both farmers and agribusiness firms. Sustainability in this context refers to the long-term viability of CF arrangements for all the actors involved. The paper uses the empirical example of two cassava outgrower schemes (state-operated and privately-operated) in Ghana.

Cassava is a major staple crop in Ghana which has the advantage of being able to produce economic yields even under marginal production conditions. Cassava accounts for approximately 50% of all root and tuber production in the country and is second only to maize in terms of area planted [16]. It is, therefore, considered as a primary food security crop. This has made it a preferred crop among small-scale, resource-poor farmers [17]. Cassava is an annual crop that is mainly consumed in the form of cooked fresh roots and domestically processed products traded on the open market. However, it is slowly shedding its image as a "poor man's crop" and overcoming its seasonal marketing challenges due to the economic potential and increasing industrial applications of cassava biomass for food, feed, and energy [18]. Therefore, there has been increasing commercial use of the crop due to rising urban demand for processed cassava products and increased recognition of its industrial potential in the emerging bioeconomy [19]. This has led to the emergence of medium and large scale processors using various contractual arrangements to source cassava from farmers. Cassava roots are not as perishable as fruits and vegetables, which need to be harvested at a specific time to avoid losses. The roots can remain unharvested for some time after maturity. However, unlike grains that can be stored after harvest, cassava roots must be processed or consumed shortly after harvesting. Cassava roots begin to deteriorate within 24 to 36 hours after harvest [20]. This necessitates efficient procurement systems.

CF in Ghana thus far has been dominated by a range of public and private large-scale production arrangements of horticultural and tree crops mainly for export [21,22]. Many of these arrangements have been characterised by contract conditions that allow agribusiness firms to maximise their short-term returns at the cost of the long-term sustainability of the schemes. They, therefore, only operate for a few years before collapsing [23]. This study uniquely highlights the role CF is playing in strengthening the link between biomass production and utilisation as the agricultural sector in sub-Saharan Africa gradually transitions from a food-supplying to a biomass-supplying sector. Therefore, smallholders growing traditional staple crops are increasingly presented with the opportunity of participating in CF schemes for the first time. Poorly designed contracts may, however, expose farmers to additional risks and exploitation by larger agricultural actors. These contractual arrangements have hardly been analysed in the literature.

A complementary combination of qualitative and quantitative methods is employed to assess the suitability and sustainability of these institutional arrangements. First, in-depth interviews and focus group discussions are used to elucidate the contract design features of the schemes, as well as both the firms' objectives and constraints, and farmers' perceptions of these features. Second, probit analysis based on primary survey data is used to determine the contract design features that influence farmers' decisions to participate in each scheme. The paper draws on this comparative case study analysis to examine forms of CF that will promote the long-term sustainability of viable firm-farmer contract arrangements. The study demonstrates that ad hoc or opportunistic investments that only address smallholders' marketing challenges are not sufficient to ensure mutually beneficial and sustainable CF schemes in fast developing domestic value chains. There is the need for direct firm investment in supporting outgrower operations. Therefore, even for a staple crop like cassava, that does not traditionally have an intense cultivation pattern, embedded support services such as input supply and technical assistance are critical to smallholder participation and the long-term success of outgrower schemes. Public-private partnerships may be the best avenue to sustainably providing these conditions.

The rest of the paper is organised into four sections: Section 2 briefly highlights prevalent contract design features of CF in developing countries. The data collection and analytical approach used in the study are described in Section 3. In Section 4, the empirical results are presented. Section 5 discusses the empirical findings, and Section 6 provides the conclusion.

2. Firm-Farmer Contract Relations

The relationship between the agribusiness firm and the farmer in contract production can be conceptualised as a four stage process [5]. A firm first chooses a procurement location; offers farmers a contract; farmers decide whether or not to accept the offered contract; finally, both the firm and farmers choose whether or not to honour the terms of the contract based on how equitable and sustainable the established relationship is for both parties. Agricultural contracts essentially differ based on their intent to transfer decision-rights and risks between the farmer and the contractor. A distinction can be made between three types of such contracts [24]: Market specification contracts where there is a pre-production agreement by both parties on the conditions governing future sale of the produce; resource providing contracts where in conjunction with marketing arrangements the buyer supplies the farmer with key inputs; production management contracts where the farmer additionally agrees to adhere to precise production methods and input regimes. Most contemporary agricultural contracts incorporate various elements of these contract typologies [25]. This invariably implies trade-offs in terms of coordination, motivation and the transaction costs associated with the design of contractual arrangements [26]. The following sections abstract prevalent contract design features that govern crop production in developing countries.

2.1. Output Arrangement

The nature of a CF arrangement is central in assuring farmers of a marketing outlet and firms of the supply of essential raw materials. Contracts can either take the form of an informal oral agreement or a formal written agreement. Written contracts provide superior enforcement possibilities and typically specify pricing, roles and responsibilities, quality and quantity requirements, and conflict resolution mechanisms [5]. However, most CF arrangements in developing countries remain as simple verbal agreements predicated on social capital such as reputation and relationship-specific incentives [27]. Such informal arrangements are less costly for agribusiness firms and provide both parties with the option to opt-out of the arrangement. Some case studies have found that such self-enforcing agreements can work effectively [11,28,29]. However, there remains insufficient evidence on smallholders' preference for either oral or written agreements.

Another key aspect of CF arrangements is the pricing mechanism which is meant to insure farmers against the uncertainty of spot market price volatility. The pricing alternatives used in CF range from fixed pricing, variable or incentive based pricing to formula pricing [30,31]. Empirical evidence from several studies reveals the common use of fixed contract prices in developing countries [8,9]. Therefore, firms tend to bear the marketing risk while smallholders bear the production risk. Farmers effectively accept to pay a risk premium in the event that spot market prices rise above the contract price. Nonetheless, this pricing option generally increases the firm's risk exposure. Agribusiness firms can, however, employ risk management strategies that are not available to smallholders. This gives them a higher risk tolerance to market price fluctuations [32]. An important, and yet often overlooked, aspect of pricing arrangements is the extent to which farmers find the actual price determination mechanism equitable.

2.2. Quality Standards

Agribusiness firms' quality requirements are a major motivating factor for CF [33]. Contractual arrangements typically include either pre-specified minimum quality standards [34,35] or variable quality standards for farmers' produce [31]. Minimum quality standards may be suited to firms targeting single-supply channels while variable quality levels may be appropriate for firms with different marketing outlets. From the farmers' perspective, minimum quality standards offer little incentive for improving quality, although there is a higher risk of complete rejection of produce. Variable quality standards on the other hand expose farmers to potential quality measurement error or bias [36]. Indeed, contractors may be tempted to falsify quality testing in order to reduce the price paid to farmers [37]. Beyond price differentiation based on quality, current case study evidence on contract farming does not provide much insight into the influence the credibility of quality verification procedures has on farmer participation, particularly for crop production.

2.3. Input Arrangement

The interlinkage of input and output markets is a fundamental element of CF in developing countries [38]. Smallholders often have limited access to inputs and technical assistance as input markets are not well developed and the state tends to lack the capacity to adequately provide these services [27,39]. Contracts regularly include seasonal inputs provided on credit, technical assistance, as well as crop delivery arrangements for smallholders [5,11,40]. Such interlocking contracts confer lending advantages on agribusiness firms through monitoring of input use and control over crop management decisions that might jeopardise farmers' output quality or input repayment [13]. As a staple crop widely grown with low input use in sub-Saharan Africa, it is unclear whether input provision will effectively incentivise contract production of cassava among smallholders.

2.4. Contract Enforcement

Conflicts between agribusiness firms and farmers in CF arrangements often arise due to misunderstandings related to the operational aspects of agreements and contract non-compliance [41,42]. Beyond firm-farmer dialogue or third party mediation of disputes, the main contract enforcement mechanism at the disposal of agribusiness firms in the case of oral arrangements is the termination or non-renewal of the contract with non-compliant farmers [8]. Written contacts present both parties with the additional option of sanctions such as legal redress for contract breach. However, in developing countries legal institutions are often absent or ineffective in ensuring contract enforcement [43,44]. In any case, smallholders typically lack the capacity to pursue legal action against firms. Even so, there is deficient empirical evidence on the extent to which farmers consider the means of contract enforcement in deciding whether to participate in CF.

The study aims to contribute to the CF literature by investigating which combination of output and input arrangements, quality standards and contract enforcement mechanisms are sustainable for both farmers and agribusiness firms. Farmers' evaluation of the contract design features that govern CF arrangements as revealed by their participation decision is an integral aspect of this empirical analysis.

3. Data and Methods

This section presents a description of the data collection methods, followed by the method of analysis and profile of the outgrower schemes.

3.1. Data Collection

At the time of data collection for this paper, five cassava outgrower schemes were identified throughout Ghana. The schemes ranged from informal oral arrangements to different forms of written agreements. The schemes studied in this paper are the two largest cassava outgrower schemes in Ghana. The first scheme is operated by a state owned agro-processing firm located in the Awutu Senya district of the Central region while the second scheme is run by a private agribusiness firm located in the Ho Municipal district of the Volta region (see Figure 1). Data collection was done in three stages for each scheme from July 2015 to December 2015, starting with the state-run scheme and subsequently with the privately operated scheme. First, in-depth interviews were conducted with government officials, management personnel and staff of the two selected processing companies, some of their off-takers, as well as with ten purposively-sampled outgrowers in each scheme who have supplied the companies from the inception of the schemes. This was to fully understand how the schemes operate and how they may have evolved over time.

Secondly, two focus group discussions were carried out for each scheme. The first set was done with ten purposively-selected outgrowers (five males and five females) of each scheme identified from the companies' supply ledgers for 2015. This refers to the period from January 2015 to September 2015 when the focus group discussions were conducted. In both cases this was followed by focus group discussions with ten non-participating smallholder farmers from the same communities as the outgrowers. These focus groups also had the same gender profile of five males and five females each. Some of these farmers had previously taken part in the schemes and opted out. This was to gain contextual insight into farmers' understanding and experiences in the schemes. Specifically, the groups were asked to elaborate on their evaluation of the output and input arrangements, quality standards and contract enforcement mechanisms of the schemes. All the interviews and focus group discussions were audio recorded with the expressed permission of the respondents. The qualitative data collection methods of the study are summarised in Table 1.

In the final stage, a pre-tested questionnaire designed on the basis of the first two stages of data collection was administered to a total of 315 famers using a multistage sampling process. For the state-led outgrower scheme, the supply ledger for 2015 was used to identify the four highest supplying communities. Proportional random sampling based on supply was used to select 100 outgrowers from these communities to participate in the survey. Fifty (50) non-participating cassava growing farmers were similarly sampled from these communities from lists provided by community leaders. Subsequently, for the privately-run outgrower scheme, the supply ledger for 2015 revealed 65 active outgrowers all of whom were selected for the survey. These outgrowers were distributed across five communities. Twenty (20) non-participating farmers growing cassava were randomly sampled from each of these communities from lists provided by community heads. All sampled farmers also met the firm's criteria of having a minimum of two acres of farmland.

Non-participant farmers had a high level of awareness and understanding of the contract terms and conditions of both schemes. This was due to extensive community outreach by the firms and farmers' interaction with outgrowers. Furthermore, some of the sampled non-participant farmers had previously been outgrowers of the schemes. Overall, 70% of the farmers who participated in the survey had landholdings of five acres or less. The questionnaire collected a wide range of information on respondents' socio-economic characteristics as well as their experiences and perceptions of the contractual details of the schemes. The interviewers overtly presented themselves as researchers with no affiliation to either agribusiness firm. Neither firm was involved in the selection process of the respondents. All responses were kept confidential.

3.2. Method of Analysis

The study employs the probit model to analyse the survey data. The model is used to estimate the factors that influence a given farmer's decision to participate in the outgrower scheme in each case. Considering the discrete nature of a farmer's decision of whether or not to participate in CF arrangements, binary choice models such as the probit and logit models are most suitable [45]. In most applications, the choice between the probit and logit models does not make much difference. However, the probit model was selected for this study because it can account for non-constant error variances in more advanced econometric settings [46].

Consistent with the objectives of the study, a separate probit model was estimated for each outgrower scheme to account for the different production and marketing arrangements in the respective study areas. The regressors include socio-economic characteristics such as gender, education, farming experience, farm size and off-farm employment, having tested for the possibility of other socio-economic explanatory factors. Additionally, the importance of the contract design features to farmers' participation decision is included in the models as dummy variables. Specifically, farmers were asked whether particular contract design features were important in their participation decision (dummy takes a value of 1) or not (dummy takes a value of 0). This approach was used to validate the qualitative information collected on farmer perceptions.

The probit model is expressed as:

$$Y_i = X_i\beta_i + \mu_i \tag{1}$$

where Y_i is the dependent variable (a farmer's decision of whether or not to participate), X_i is the vector of explanatory variables that influence a farmer's decision of whether or not to participate in the outgrower scheme, β_i is the coefficients of the explanatory variables, and μ_i is the error term capturing all unmeasurable effects that influence a farmer's participation decision. Specifically, the empirical probit model is specified as follows:

$$Y_i = \beta_0 + \beta_1 NoC + \beta_2 PA + \beta_3 PS + \beta_4 PQ + \beta_5 Inputs + \beta_6 Assist + \beta_7 Delivery + \beta_8 Conflict$$
$$+ \beta_9 Sanct + \beta_{10} Edu + \beta_{11} Gend + \beta_{12} FarmExp + \beta_{13} FarmSize + \beta_{14} OffFarm + \mu_i \tag{2}$$

where NoC denotes the nature of the contract for outgrowers, PA signifies the pricing arrangement, PS represents the payment system, PQ denotes the product quality specification, Inputs represents the input supply arrangement, Assist denotes the technical assistance arrangement, Delivery represents the crop delivery arrangement, Conflict denotes the conflict resolution procedure of the arrangement and Sanct signifies the sanctions to be meted out breach of contract. The a priori assumptions of these variables in the context of smallholder farming, such as cassava production in Ghana, have been addressed in Section 2.

In terms of the socioeconomic variables in the empirical model, Edu denotes the number of years of education of the farmer. It is expected that the likelihood of participation in contract farming increases with more years of education due to a better understanding of the contract terms, as well as the benefits of contract farming. Gend represents the gender of the farmer. Cassava production in Ghana is a male-dominated activity [18]. A major contributing factor to this social dynamic is the fact that women are disadvantaged in terms of access to productive resources, such as farmland. Therefore, it is expected that male farmers are more likely to participate in cassava contract farming. FarmExp denotes farming experience. A farmer may become more or less averse to the risks of contract farming based on the amount of farming experience. Thus, this variable can either have a positive or negative effect on a farmer's participation decision. FarmSize represents the total farm size of a farmer. As larger farm sizes are an indicator of wealth and influence, it is expected that farmers with larger farm sizes will be more likely to participate in contract farming. OffFarm denotes off-farm employment. A farmer with off-farm employment has a diversified risk portfolio with multiple income streams. Therefore, the expectation is that a farmer with off-farm employment will be more likely to participate in contract farming.

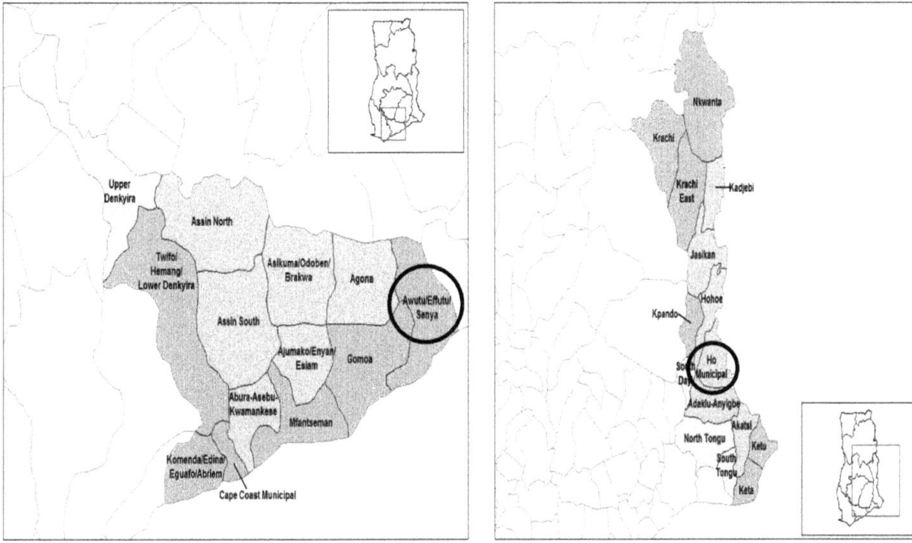

Public Scheme   Private Scheme

Figure 1. Locations of case study outgrower schemes in Ghana; Source: Ghana Districts Repository.

Table 1. Summary of qualitative data collection.

Method	Public Scheme	Private Scheme
In-depth interviews (with stakeholders)		
Company management & staff	7	8
Government officials	5	2
Breweries	1	1
Food processing companies	-	2
Bakeries	-	1
Paperboard manufacturing companies	-	2
Plywood manufacturing companies	-	3
Outgrowers	10	10
Focus Group Discussions (with farmers)	2	2

3.3. Profile of the Schemes

The state-owned agro-processing firm being studied was commissioned in 2003 as part of a Presidential Special Initiative (PSI) aimed at developing industrial cassava starch production for both the domestic and international market and, by so doing, also better the socioeconomic conditions of smallholders in the area. However, the firm was plagued with high operational costs and technological challenges which culminated in it temporarily shutting down on two occasions. This necessitated a more focused and sustainable operational strategy. In 2012, the firm signed an exclusive supply agreement with Guinness Ghana Breweries Limited (GGBL) to supply food-grade starch from cassava. The brewery uses the starch to produce one of its brands of beer for the domestic market. This is in line with an excise tax break announced by the government in 2012 for local content beers (the Customs and Excise Act 855 is an excise duty concession on a sliding scale for breweries using greater than 30% local raw materials for the manufacture of excisable goods). The state firm sources half of its feedstock from its firm-managed farm and the other half from about 400 outgrowers. These outgrowers are all members of a farmers' association that represents their interests when dealing with the firm. There is a verbal agreement in place between the firm and its outgrowers based on trust and understanding. The outgrowers do not receive support from the firm in the form of technical assistance, credit, or inputs. Supplied cassava roots must meet the firm's set quality standards to be accepted. The firm generates additional revenue from selling the by-products (the cassava peels and pulp) from the starch production process to local piggeries as feed.

The privately-owned agribusiness firm was established in 2006. The firm produces high-quality cassava flour (HQCF) and industrial grade flour which is sold to Accra Breweries Limited (a local subsidiary of SABMiller Plc), food processing companies, bakeries, paperboard manufacturing companies, and domestic plywood manufacturing companies. There were also advanced plans for the firm to commence bio-ethanol and biogas production from cassava biomass at the time of data collection for the study. The firm sources approximately 70% of its feedstock from its firm-managed farm and 30% from outgrowers. At the time of the survey there were 65 outgrowers. The outgrowers are all members of a representative farmers' association. There is a standard seasonal written contract between the firm and each of its outgrowers. The firm provides technical assistance and optional inputs in measured quantities to the outgrowers. The cost of inputs provided is deducted from the value of the delivered cassava roots with no interest. The cassava roots must be a healthy approved variety to be accepted. The firm also sells the cassava peels and pulp from its production process as animal feed. The main design features of both outgrower schemes are summarised in Table 2. Both agro-processing firms are located in areas where cassava is the predominant crop grown and the spot market is the only other marketing channel for farmers' produce.

Table 2. Salient outgrower scheme design features for case study firms.

Scheme Design Features	Public Firm *	Private Firm **
Output Arrangement		
Nature of the contract	Oral contract communicated through the outgrower farmers' association	Written contract with individual outgrowers
Farmer selection criteria	No selection criteria	A minimum of two acres of farmland contingent on farm inspection by the firm
Contract duration	Optional seasonal arrangement	Binding seasonal arrangement
Supply quota	No specified quantity of produce to be supplied	No specified quantity of produce to be supplied
Outgrower quota	Unrestricted number of outgrowers	Maximum of 100 outgrowers
Pricing arrangement	Fixed seasonal price per ton whereby the amount paid is determined by a weighing bridge	Fixed seasonal price per ton whereby the amount paid is determined by the number of delivered tractor trailer loads
Payment procedure	Weekly invoice payment system for delivered produce	Cash payments upon delivery of produce
Quality Standards		
Product quality specification and verification	A minimum of 15% starch content of a recommended variety verified through laboratory testing of samples prior to delivery	Healthy appearance of an approved variety verified by physical inspection of produce upon delivery
Input Arrangement		
Input supply arrangement	No input supply arrangement	Supply of free planting material and agro-chemicals in measured quantities, the cost of which is deducted from final payment
Technical assistance	No technical assistance	Technical assistance is provided
Crop delivery arrangement	Specified delivery date without firm-provided transportation services	Transportation of the produce from farm to factory by the firm
Contract Enforcement		
Conflict resolution procedure	No established conflict resolution procedure	Mediation between farmer association executives and the firm
Sanctions	No sanctions	Fines and Legal action

* Based on firm and outgrower interviews, and direct observation; ** Based on written contract agreements, firm and outgrower interviews, and direct observation.

4. Results

This section begins with a descriptive analysis of the socio-economic characteristics of the sampled farmers, as well as an account of the contract design features in both schemes from the survey carried out. This is followed by an analysis of the firms' motivations for scheme design features based on the in-depth interviews with staff of the firms. Finally, the results of the probit analysis of the determinants of participation in the schemes are presented.

4.1. Descriptive Analysis

4.1.1. Socio-Economic Characteristics

A comparison of the sampled groups of participating and non-participating farmers in both outgrower schemes is presented in Table 3. In the state-run outgrower scheme, both groups exhibited similar individual and household level characteristics in terms of age, household size, farming experience, land ownership, and off-farm employment. However, outgrowers were found to have significantly higher levels of education on average. Although both groups were male dominated which reflected the gender profile of the scheme and cassava production in the area, there was a significantly larger number of males among the sampled outgrowers. With regards to income, there was not a significant statistical difference between the two groups. Outgrowers reportedly earned 18% higher total farm income compared to non-participating farmers for the year under consideration. Similarly, outgrowers earned 37% more non-farm income than non-participating farmers. With regard to the production characteristics, farmland showed the only significant difference between the two groups. Outgrowers had an average farmland size of almost six acres which was double that of non-participating farmers. Furthermore, as can be seen in Table 3, non-participating farmers had a higher average gross margin per acre for cassava production than outgrowers in the state-run scheme. However, there was not a significant statistical difference between the two groups' gross margins. The gross margin per acre for cassava production is calculated by subtracting total variable costs per acre (wage labour and agro-chemical costs) from the revenue per acre obtained for cassava production.

In the privately operated scheme, there was not a significant difference between the two groups in terms of average age, household size, farming experience, land ownership and off-farm employment. Outgrowers were, however, significantly more educated than the non-participating farmers. There were also significantly more males participating in the scheme than among non-participating farmers. Outgrowers notably earned 50% higher total annual farm income than non-participating farmers. Non-participating farmers on the other hand earned marginally higher non-farm income, but the difference was not statistically significant. In terms of the production characteristics, outgrowers had an average of six acres of farmland, while non-participating farmers had an average of four acres of farmland. Outgrowers were also found to have significantly higher yield for cassava production. As shown in Table 3, outgrowers in the private scheme had a higher gross margin per acre for cassava production than non-participating farmers. The difference was statistically significant.

It must be noted that the difference in the average set of prices between the two case study areas, as shown in Table 3, reflects the fact that the state scheme is located in a peri-urban area with a relatively higher cost of living. The private scheme on the other hand is located in a rural area with a lower cost of living (the Ghana Statistical Service [47] provides a source of reference for the cost of living across Ghana).

Table 3. Socio-economic characteristics of sample farmers.

Variables	Public Firm			Private Firm		
	Participant Farmers ($n = 100$)	**Non-Participant Farmers** ($n = 50$)	**Equality Test**	**Participant Farmers** ($n = 65$)	**Non-Participant Farmers** ($n = 100$)	**Equality Test**
Individual Characteristics						
Age (years)	44.05 (11.52)	43.94 (10.29)	0.06	49.85 (11.78)	47.47 (12.05)	1.25
Household size (persons)	5.58 (2.39)	5.52 (2.18)	0.15	5.65 (1.80)	5.12 (2.50)	1.47
Educational level (years)	8.4 (3.30)	6.56 (4.27)	2.91 ***	9.14 (2.94)	7.45 (3.73)	3.08 ***
Gender (% of males)	78	60	2.31 **	78	62	2.22 **
Farming experience (years)	18.71 (11.92)	19.03 (11.18)	0.16	21.26 (13.02)	20.99 (11.43)	0.14
Land ownership (% of owners)	21	32	1.47	71	78	1.05
Off-farm employment (% yes)	53	50	0.35	62	52	1.21
Farm income (GH¢ '000)	7.22 [a] (13.40)	5.92 [a] (13.64)	0.55	8.22 [a] (17.51)	4.15 [a] (4.65)	2.21 **
Non-farm income (GH¢ '000)	2.30 [a] (4.55)	1.46 [a] (2.39)	1.22	2.05 [a] (2.41)	2.40 [a] (5.77)	0.46
Production Characteristics						
Farm size (ac)	5.67 [b] (6.23)	3.05 [b] (2.92)	2.83 ***	6.18 [b] (3.20)	4.28 [b] (3.22)	3.70 ***
Distance to market (km)	1.89 (1.87)	2.46 (2.35)	1.62	2.91 (1.75)	2.86 (1.89)	0.19
Family Labour (mandays/ac)	3.37 (7.86)	4.14 (9.86)	0.52	0.75 (3.35)	0.98 (4.11)	0.38
Wage Labour (mandays/ac)	26.46 (27.76)	25.47 (32.38)	0.19	33.25 (35.98)	26.69 (28.53)	1.30
Cassava yield (ton/ac)	5.20 (3.11)	5.72 (2.11)	1.05	8.40 (1.53)	5.62 (1.97)	9.63 ***
Gross Margins for cassava						
Price (GH¢/ton)	220	222		120	150	
Revenue (GH¢/ac)	1144	1269.84		1008	843	
Wage Labour (GH¢/ac)	470.95	496.02		335.66	300.42	
Fertiliser (GH¢/ac)	14.60	10.27		9.61	10.25	
Herbicides (GH¢/ac)	20.23	18.58		48.93	8	
GROSS MARGINS (GH¢/ac)	638.22	744.97	1.26	613.80	524.33	1.91 **

Standard deviations are presented in parentheses; * Significant at the 10% level; ** Significant at the 5% level; *** Significant at the 1% level; [a] Exchange rate October 2015: 4 GH¢/$.
[b] Farm size is calculated as the sum of a farmer's respective acreages in the event that he/she cultivates more than one plot.

4.1.2. Contract Design Features

In the state outgrower scheme, the average number of years of participation was four years among sampled outgrowers (Table 4). Outgrowers confirmed that their arrangement with the state firm was based on verbal commitments with no written proof. More than half (56%) of the outgrowers of the public firm reported receiving their invoices to be cashed at the rural bank within two weeks after delivering their produce, while 21% complained of delayed payments within a month of produce delivery. The firm has mainly attributed this problem to intermittent financial challenges caused by delays in purchases from their sole off-takers. A small number (6%), however, reported receiving their invoices at the time of sale as delivery coincided with weekly disbursements during periods when the firm had sufficient cash flow. Figure 2 shows the contract price offered by the public firm as compared to the variable spot market prices non-participating farmers in the survey received for their cassava roots. In terms of quality standards, 19% of the sampled outgrowers reported some of their produce (from at least one of their farmlands) being rejected for not meeting the product quality specification. Seven percent of the outgrowers reported having conflicts with the firm about opportunistic behaviour stemming from delayed payments and a lack of trust in the quality verification system following rejection of some of their produce. One outgrower reported a violation of contract terms after the firm approved delivery of the produce, but later reneged on the purchase, citing technical challenges and the unreliability of the farmer's contact information to be accordingly notified.

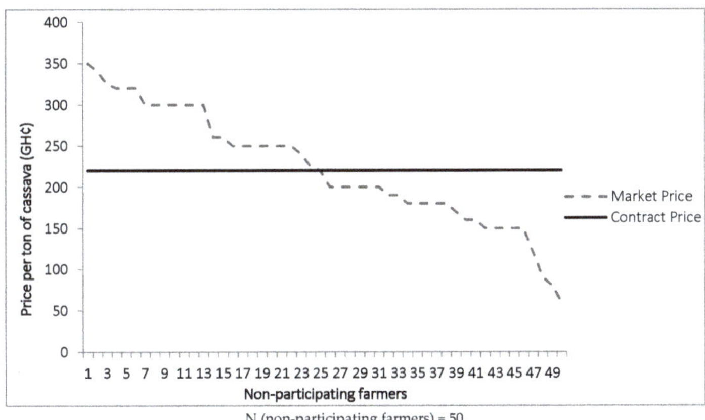

N (non-participating farmers) = 50.

Figure 2. Contract price for state scheme and open market prices for cassava roots.

In the private outgrower scheme, the average number of years of farmer participation was three years (Table 4). All the outgrowers confirmed that they each had a generic signed written contract with the firm which outlined the terms and conditions of the arrangement. Almost all the farmers (90%) confirmed cash payment at the time of produce delivery. The rest received their money within a week of delivering their produce. This was mainly due to a large number of coinciding deliveries whereby the firm ran out of available cash for immediate payments. Most farmers received their payments in the days that followed. Figure 3 shows that the contract price offered by the private firm was low as compared to the variable prices non-participating farmers in the survey received for their cassava roots on the open market. Only two outgrowers reported some of their produce being rejected mainly for being physically damaged. With regards to the input supply arrangement, majority of the outgrowers (94%) reported the use of firm-provided inputs. The standard supply of inputs includes 20 bundles of stem cuttings per acre (for free) and 0.75 pounds of acid equivalent of glyphosate per acre. Outgrowers also had the opportunity to access funds for farming activities as part of this arrangement. However, all outgrowers received technical assistance from the firm. This mostly entailed monthly visits from technical staff of the firm. Technical assistance included guidance on better agronomic practices such as planting in rows. The valuation of farmers produce was a source of conflict for some of the outgrowers (9%). The outgrowers complained that the firm would deliberately overload tractor trailers with cassava roots when transporting their produce, knowing that the amount paid to farmers is contingent on the number of trailers rather than the weight of the produce. Three of the outgrowers also complained of the firm violating the terms of the contract by deducting amounts higher than the cost of firm-provided inputs from their final payments. The firm in these instances insisted the farmers had miscalculated the cost of the inputs they had been advanced.

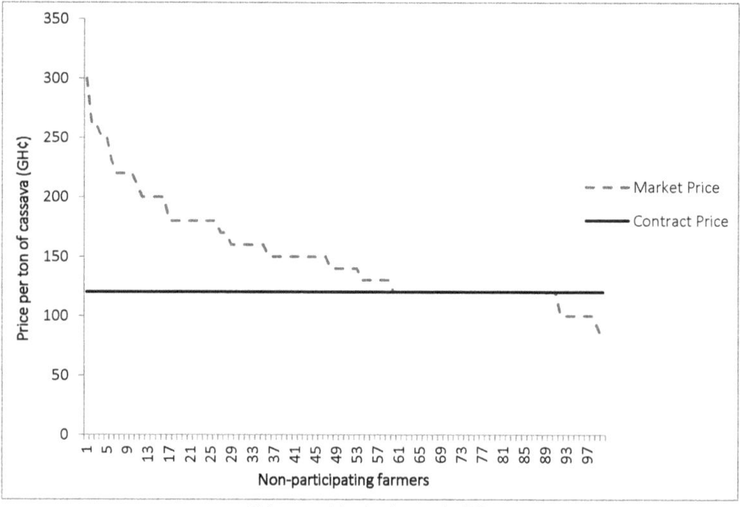

N (non-participating farmers) = 100.

Figure 3. Contract price for private scheme and open market prices for cassava roots.

Table 4. Contract characteristics.

Outgrower Scheme Features	Public Firm Outgrowers	Private Firm Outgrowers
Participation (mean years)	3.83	2.63
Output Arrangement		
Nature of the contract		
Oral/informal	100%	-
Written/formal	-	100%
Time of Payment		
At the time of delivery	6%	90%
Within 1 week of sale	17%	10%
Within 2 weeks of sale	56%	-
Within one month of sale	21%	-
Quality Standards		
Rejection of any produce	19%	3%
Outgrower Scheme Features	**Public Firm Outgrowers**	**Private Firm Outgrowers**
Input Arrangement		
Use of firm inputs	-	94%
Technical assistance from firm	-	100%
Number of visits		
Once per month	-	94%
Several times per month	-	6%
Once per season	-	-
Other	-	-

Table 4. *Cont.*

Contract Enforcement		
Conflict with the firm		
Opportunistic behaviour	7%	9%
Violating contract terms	1%	5%
Number of Observations	100	65

4.2. Firm Motivations for Scheme Design Features

In-depth interviews with the management personnel of both firms revealed that while the private firm made a profit in the year under consideration, the public firm registered losses in its operations. Insufficient supply of cassava roots from outgrowers to supplement the firm's own cassava production was a major contributing factor to this underperformance. In this regard, analysis of the interviews with the firms also revealed their motivations for the various design features of the schemes. This captured some of the key differences between the schemes as shown below.

First, the fixed pricing arrangement of the public firm is set based on firm negotiations with the executives of the outgrower association and representatives of GGBL. The firm takes into account farmers' cost of production, transportation cost and market price trends in the catchment area as well as the price for starch offered by GGBL. The fixed price increases the firm's exposure to price risk but also allows for planned budgeting. The private agribusiness firm also uses a fixed pricing arrangement. However, because costs such as crop delivery and technical assistance are paid for by the firm, the seasonal price mainly reflects the market value of the firm's main product. These motivations are reflected in the following quotations from the in-depth interviews:

"We set the price with their [outgrowers] executives and Guinness, so they know what goes into it. Unlike the unstable price on the open market, this gives the farmers a sense of financial certainty and allows both of us to plan well. […]. We use a weighing bridge to determine the tonnage supplied by the outgrowers so there is no error or misunderstanding." (Production Manager, public firm)

"We give the farmers a lot of support so the price we pay is mainly based on the going rate for the flour we sell, the HQCF. We look at our selling price and try to be reasonable with the farmers in the price we offer. […]. Each tractor trailer we use in transporting the roots weighs 2.5 tonnes with a full load. That is how we know the quantity they [outgrowers] supply." (Production Manager, private firm)

In terms of quality standards, the public firm's minimum starch requirement of 15% has been set in line with the firm's exclusive supply arrangement with GGBL. Accordingly, farmers who approach the firm with the intention of supplying cassava are given a list of recommended varieties to grow. These improved varieties are intended to give farmers the highest probability of meeting the quality specification. The private firm on the other hand sells to different markets with different quality requirements. As such, the firm accepts cassava roots of variable quality, provided outgrowers supply healthy roots of an improved variety. These reasons are expressed in the following statements:

"Previously we didn't demand any specific variety. But Guinness demands a high level of quality. That is why we now have specific varieties we recommend. In fact, 15% is still on the low side for us. At least 20% would have been ideal but we also accept that that would be difficult for a lot of the farmers to meet." (General Manager, public firm)

"Because we have different uses for the cassava, once they [outgrowers] grow an improved variety we have supplied them with we accept it. The only thing we look out for is that the roots are healthy and not damaged." (Production Manager, private firm)

In reference to inputs, the public firm does not have a supply arrangement with outgrowers. This is mainly due to the resource constraints of the firm and the size of the outgrower scheme. The private firm on the other hand provides outgrowers with stem cuttings and herbicides to ensure high output of the raw material supply. These motivations are reflected in the following interview excerpts:

"We can't afford to supply inputs. They [outgrowers] are too many. And supplying them with fertiliser or herbicides also comes with responsibility of monitoring how they use them to make sure they don't divert them, and we don't have the manpower to do that." (General Manager, public firm)

"We want the outgrowers to treat cassava as a cash crop. So we supply them with planting material and chemicals. This way, they can get more money and we can get more roots from them. [...]. We go round and make sure every farmer uses the chemicals correctly because it is an investment we are making." (Outgrower Coordinator, private firm)

Concerning contract enforcement, the public firm has not seen the need to establish a conflict resolution procedure or sanctions given the informal nature of the arrangement. Conversely, the private firm resorts to fines for input diversion and legal action against outgrowers for contract violations, such as side-selling. The firm views this as the most effective means of discouraging contract breaches by farmers. These reasons are supported by the following statements:

"Our arrangement is simple and straightforward so there are no sanctions. There is also no conflict resolution process in place per se. [...]. Well, sometimes farmers complain about delays in payment because they always want instant cash but they are aware that that is how our system works. And they always get their money." (Production Manager, Public firm)

"If we catch them [outgrowers] diverting inputs they pay for it with interest which we [the firm] decide on. [...]. We have caught some of them side-selling and we have terminated their contracts and taken them to court to pay us back. Since we started doing that side-selling has gone down." (Outgrower Coordinator, Private firm)

Information on the evolution of the schemes provided by the firms was consistent with that of the outgrowers who were interviewed. The public company initially provided outgrowers with inputs, technical assistance, and instant cash payments until it run into financial and technical difficulties. The public scheme has been running in its current form since the company's supply agreement with GGBL in 2012. The contract arrangements under the private outgrower scheme on the other hand have not changed since its inception in 2006. These interviews provide insight into the significance of the probit model results for both schemes.

Two separate models were estimated for the state and private outgrower schemes due to the difference in their marketing arrangements and management structures, as revealed by their contract designs. Thus, merging them in a pooled dataset would fail to provide robust findings of how independent variables affect the participation decision of farmers in the schemes.

4.3. Empirical Probit Model Estimates for the State Outgrower Scheme

The estimates derived from the probit model for the state outgrower scheme are presented in Table 5. The results indicate the factors that either positively or negatively affected smallholder participation in the scheme. The oral contract with the firm increased the likelihood of farmer participation in the outgrower scheme. The state firm has been operating in the study area for an extended period of time and has had an outgrower scheme from its inception. According to information from the focus group discussion with outgrowers, farmers have a high level of trust in the relationship based on the firm's reputation and recurrent transactions.

The pricing arrangement also increased the likelihood of participation. This arrangement insures risk-averse farmers against volatile spot market prices for cassava. The outgrowers' farmer association enjoys significant bargaining power in negotiations to set the seasonal price as a satisfactory price is the firm's only guarantee of outgrower participation. Beyond that, the firm provides an objective and acceptable means of valuing farmers' produce through the use of a weighing bridge, leaving little margin for error.

Conversely, the payment procedure decreased the likelihood of participation. The weekly invoice payment system often led to delays in payment. While there does not appear to be any risk of non-payment by the firm, inefficiency in the payment system has been a major source of dissatisfaction for outgrowers and a deterrent to smallholder participation. Additionally, the payment system which required farmers to receive their money through a bank increased their transaction cost.

The fixed product quality specification of the state firm also decreased the likelihood of participation. The minimum starch requirement increased the risk of complete rejection of farmers' produce due to information asymmetry as farmers have no means of knowing or verifying the starch content of their produce.

The lack of an input supply arrangement and a transportation arrangement similarly decreased the likelihood of participation in the outgrower scheme. As a traditional staple crop, cassava is generally viewed by most farmers as a crop that does not require a lot of inputs. However, the firm provides farmers with a more reliable marketing outlet for large quantities of cassava than spot markets. This has provided an incentive for farmers to improve their productivity through the use of improved planting material and agro-chemicals. Furthermore, although transportation is factored into the seasonal price, farmers complained that access to reliable transport services for bulk delivery was hard to come by at affordable prices.

Interestingly, the individual characteristics of farmers did not significantly influence the decision to participate in the scheme. This highlighted the fact that the various contact conditions of the outgrower scheme were the main determining factors in farmers' participation decision. The age variable was found to be highly correlated to the farming experience variable and was, therefore, omitted from the model estimation. The reliability of the model estimation was confirmed using the uncentred variance inflation factor (VIF) test. The result showed a mean of 2.34 which indicated that there was not a problem of multicollinearity.

Table 5. Factors influencing participation in the state outgrower scheme.

Variable	Coefficient	Robust Std. Err.	Marginal Effect
Oral contract	1.394 ***	0.316	0.203
Fixed pricing arrangement	0.969 ***	0.359	0.136
Weekly payment system	−1.128 ***	0.367	−0.171
Fixed product quality specification	−1.103 ***	0.418	−0.164
Lack of input supply	−0.926 ***	0.330	−0.139
Lack of technical assistance	−0.167	0.451	−0.039
Lack of crop delivery arrangement	−1.048 ***	0.342	−0.143
Lack of conflict resolution procedure	−0.485	0.439	−0.062
Education	0.040	0.039	0.012
Gender	0.352	0.313	0.033
Farming experience	−0.010	0.015	−0.001
Farm size	0.036	0.040	0.005
Off-farm employment	−0.392	0.348	−0.048
Constant	0.398	0.572	
Observations	150		
Wald chi2	52.82 ***		
Log pseudolikelihood	−40.93		
Pseudo R^2	57.13%		

* Significant at the 10% level; ** Significant at the 5% level; *** Significant at the 1% level State scheme; VIF, uncentred, 2.34.

4.4. Empirical Probit Model Estimates for the Private Outgrower Scheme

Table 6 reports the probit model estimates for the private outgrower scheme. The results show that the written contract with the firm increased the likelihood of farmer participation in the outgrower scheme. Indeed, as contractual arrangements become more complex particularly with input supply and specialised production practices, written contracts become more beneficial to both firms and farmers. Conversely, the pricing arrangement decreased the likelihood of farmer participation. Price negotiations between the firm and outgrower association appear to be a formality rather than a collaborative process as the firm seems to focus solely on its profit margin. Furthermore, the biased valuation of farmers' produce using the number of trailers delivered, rather than a weighing system, has been a source of contention between outgrowers and the firm.

The system of on-the-spot cash payments used by the private firm increased the likelihood of farmer participation. Smallholders tend to prefer immediate cash payments following delivery of their produce to satisfy their current household consumption requirements. Similarly, firm-provided inputs and technical assistance at the various stages of production also increased the likelihood of farmer participation. Smallholders believed such support would help improve their productivity significantly.

The private firm's provision of crop delivery services gave outgrowers access to dependable transport services that is well-synchronised with the firm's demand for their produce. This increased the likelihood of smallholder participation as an important post-production feature of the scheme, especially given the highly perishable nature of cassava. Conversely, sanctions meted out for contract violations in the form of fines and legal redress in the private outgrower scheme decreased the likelihood of farmer participation as farmers found these sanctions to be excessively harsh.

Similar to the state outgrower scheme, it was found that individual farmer characteristics did not significantly influence the decision to participate in the outgrower scheme. Again, the age variable was also found to be highly correlated to the farming experience variable and was omitted from the estimation. There was no concern of multicollinearity in the model as implied by the mean uncentred VIF test result of 2.62.

Table 6. Factors influencing participation in the private outgrower scheme.

Variable	Coefficient	Robust Std. Err.	Marginal Effect
Written contract	0.597 **	0.268	0.095
Fixed pricing arrangement	−0.940 ***	0.296	−0.144
Instant cash payments	0.699 ***	0.279	0.109
Variable product quality specification	0.308	0.297	0.054
Input supply arrangement	0.996 ***	0.261	0.152
Technical assistance	0.909 ***	0.273	0.152
Crop delivery arrangement	0.836 ***	0.322	0.147
Conflict resolution procedure	0.659	0.390	0.101
Sanctions	−0.871 **	0.284	−0.134
Education	0.052	0.045	0.013
Gender	0.526	0.345	0.079
Farming experience	0.017	0.014	0.003
Farm size	0.082	0.055	0.011
Off-farm employment	0.260	0.268	0.044
Constant	−3.041	0.585	
Observations	165		
Wald chi2	90.70 ***		
Log pseudolikelihood	−49.74		
Pseudo R^2	55.04%		

* Significant at the 10% level; ** Significant at the 5% level; *** Significant at the 1% level Private scheme; VIF, uncentred, 2.62.

5. Discussion

CF arrangements have great potential to simultaneously increase smallholders' productivity and overcome marketing challenges. In this regard, smallholder participation in CF arrangements is widely viewed by policymakers as important for poverty reduction and rural development. However, the growing case study evidence on the impacts of CF in Africa pays little attention to the critical role contract design plays in the sustainability of these arrangements between agribusiness firms and farmers. The current analysis used a comparative case study approach to highlight contract conditions that will promote the sustainability and viability of outgrower schemes in the burgeoning cassava sub-sector in Ghana. The study uniquely focused on rapidly developing domestic value chains in the emerging bioeconomy. The findings add empirical weight to the argument that state-led contract farming schemes are generally not an effective government mechanism for overcoming market failures that inhibit the commercialisation of agricultural production by smallholders. Indeed, the continued existence of inefficient state-led schemes often signals the lack of an enabling environment for the private sector to effectively take over these functions.

The interviews conducted for the study reveal that the public firm's outgrower scheme constitutes a low investment informal CF model. The firm does not invest resources in outgrowers' cassava production and as such does not incur monitoring costs. Correspondingly, the quantitative results show that farmer participation is positively influenced by the oral and informal nature of the agreement. Farmers have the option to opt-out of the arrangement at any point in time. This eliminates the issue of side-selling and allows farmers to take advantage of periods of high local spot market prices. The consequence, however, is limited control over the quantity and quality of produce supplied which increases the risk of the firm not meeting the specific needs of its off-taker. The firm must compete with other buyers who may offer higher prices. This is reflected in a pricing arrangement that farmers find favourable in terms of their collective bargaining power and the valuation of their produce. However, the empirical evidence suggests that the scheme, which initially aimed to improve the socioeconomic conditions of smallholders in the area, is ultimately not beneficial to either the firm or outgrowers. The firm receives an insufficient supply of cassava roots from outgrowers. Outgrowers' productivity and revenue from cassava production also do not appear to have increased through the arrangement.

The private firm's outgrower scheme represents a relatively more capital intensive and formalized CF arrangement as revealed by the interviews conducted with company staff. Due to the provision of inputs, technical assistance, and crop transportation, the firm retains more control over the quality and volumes of outgrowers' output. This makes for more efficient sourcing of cassava roots. Consistently, the quantitative results show that these contract features along with instant cash payments positively influenced smallholder participation in the scheme. Interestingly, even though the firm offers a low contract price with a contestable price determination mechanism, they appear to be able to effectively enforce the contract and control side-selling. This goes contrary to the argument that agricultural commodities with well-developed local markets are not suitable for contract farming because they are associated with a high risk of pervasive side-selling [48,49]. The firm is able to earn a profit using the CF arrangement. Outgrowers also have high farm productivity and earn comparatively higher returns from cassava production.

Contrary, to the results past studies [9,31,50], individual characteristics such as education, farming experience and farm size which are considered critical to farming efficiency did not significantly influence the decision to participate in the schemes. This emphasized the importance of contract design to farmers' participation decision as revealed by the study's results. It must, however, also be noted that the use of different statistical models and data collection techniques may present a nuanced picture. Future studies may benefit from the collection of longitudinal data for richer analysis of farmers' participation decision.

The findings of the study further demonstrate that firm investment in supporting farm production is critical to the success of outgrower schemes. Cassava may not be an input-intensive crop. Nonetheless, there is still the need for embedded support services in outgrower schemes. Farmers desire arrangements that address both production and marketing challenges. Abebe et al. [36] found that smallholders' decision to participate in CF is even more dependent on input market uncertainty than on output market uncertainty. In order for the public firm to improve the economic viability of its model, the scheme must facilitate the adoption of improved technologies to stimulate increased farm productivity among outgrowers through an input supply arrangement. The seed market for cassava in Ghana is missing. Public sector extension agents are the primary source of supply for vegetative propagules (stem cuttings) of improved varieties that have been developed by the research system. These stem cuttings are often distributed to influential farmers with the expectation that they will, in turn, be disseminated to smallholders. This is often not the case as most smallholders do not get access to improved cultivars. Alene et al. [51] reported that the area planted to improved cassava varieties in Ghana only increased from 25% to 36% between 1998 and 2009. It is, therefore, a challenge for many smallholders to grow varieties recommended by the firm. As Wiggins and Sharada [52] observed, smallholders are also susceptible to purchasing adulterated agro-chemicals because they are often more affordable. Likewise, access to reliable transport services for bulk delivery of produce is often inaccessible at affordable prices. The lack of suitable access routes to farms often means that drivers charge higher prices or refuse to transport produce. Firm involvement in providing such post-harvest logistical support is critical given the perishability of the crop, quality requirements and the poor state of the existing road infrastructure in rural areas. An alternative to the firm providing such services directly could be arrangements with intermediaries such as aggregators or lead farmers who have closer proximity to smallholders and could facilitate bulking and crop delivery.

Notably, the bargaining position of the firms in their respective value chains is also a determining factor of the contract features of the outgrower schemes. This is particularly important for quality standards, which can often be one of the most contentious issues in contract arrangements, as reported by Henson et al. [53]. The differentiated marketing strategy of the private firm enabled a system of variable quality standards which reduced the risk of complete rejection of farmers' produce. Comparatively, the fixed product quality specification of the public firm is a direct consequence of its exclusive supply agreement with GGBL. This uncertainty over complete rejection discouraged participation in the public scheme as it largely eliminated the incentive of a guaranteed market. Transparency in quality assurance systems is, therefore, imperative to maintaining smallholders' trust in CF arrangements, especially as they adjust to stricter quality requirements in increasingly competitive value chains for high-value products. Barrett et al. [5] found that firms appear more likely to fabricate quality testing or speciously reject perishable commodities on the grounds of quality when supply is guaranteed from a large pool of smallholders. Indeed, a study by Torero and Viceisza [54] showed that third party quality testing improved farmers' trust in the validity of results as this was perceived to be a more objective system.

Overall, the increased use of cassava biomass associated with commercialisation of the sub-sector in Ghana's emerging bioeconomy has necessitated more organised sourcing arrangements. Government policies like the PSI on cassava starch and the tax incentive policy for local content beers have also served as catalysts to these institutional arrangements. Results from the public scheme reveal that although cassava is a staple crop that has traditionally been cultivated with minimal inputs, ad hoc or opportunistic investments that merely provide a marketing outlet for smallholders are not sufficient to ensure the success of outgrower schemes. The evidence highlights the tendency of public sector schemes to be bureaucratic and lack financial autonomy. Private sector ventures, on the other hand, tend to adopt authoritarian management styles and can also be prone to opportunistic behaviour in contract arrangements with smallholders. Therefore, public-private partnerships in cassava outgrower schemes present a viable and sustainable remedy that harnesses the strengths of both sectors and overcomes their institutional weaknesses. These arrangements would allow schemes to take advantage of government support such as grants and input subsidy programs while also benefitting from the private sector's financial autonomy, systems of accountability and highly-trained and specialised staff. It is the recommendation of this paper that this approach should be pursued by policymakers. Ultimately, smallholder participation and the sustainability of cassava outgrower schemes in Ghana's emerging bioeconomy are contingent on a fully-integrated and comprehensive farm-to-market approach within a conducive enabling environment for agricultural contracting.

6. Conclusions

The study's findings highlight the point of divergence between the low investment model of the state outgrower scheme which has led to insufficient supply from outgrowers and the more capital-intensive arrangement of the private firm that benefits from the productive capacity of smallholders. The state outgrower scheme, initially established to improve smallholders' socioeconomic conditions, offers farmers some favourable contract conditions. However, a lack of embedded support services has not enabled outgrowers to increase their productivity and revenue from cassava production in the scheme. CF arrangements must, therefore, address both production and marketing challenges to be sustainable and mutually beneficial to farmers and firms.

As competitive value chains continue to develop in Africa's evolving agricultural sector, there is the need for equitable and transparent contract design features, as well as direct firm investment in supporting farm production activities within an enabling environment. Public-private partnerships can provide these necessary conditions for ultimately unlocking the potential of CF in Africa's bioeconomy.

Author Contributions: A.-G.P., R.B., and S.G. conceived and designed the study; all authors analysed the data; and A.-G.P. wrote the paper.

Acknowledgments: The authors are thankful to the German Federal Ministry of Education and Research (BMBF) for funding this research through the collaborative project "Improving food security in Africa through increased system productivity of biomass-based value webs." This project is part of the GlobE—Research for the Global Food Supply programme (Grant No. 031A258H). The research conducted for this paper was also supported by a scholarship from the German Academic Exchange Service (DAAD), which is gratefully acknowledged. We would like to express our gratitude to Felix Asante from the University of Ghana, Legon, for his support during the field research. We would also like to thank the communities and staff respondents from the outgrower schemes who kindly contributed their valuable time and perceptions towards the data collection.

Conflicts of Interest: The authors declare no conflict of interest.

References

1. Henson, S.; Reardon, T. Private agri-food standards: Implications for food policy and the agri-food system. *Food Policy* **2005**, *30*, 241–253. [CrossRef]
2. Timilsina, G.R.; Beghin, J.C.; van der Mensbrugghe Mevel, S. The impacts of biofuels targets on land-use change and food supply: A global CGE assessment. *Agric. Econ.* **2012**, *43*, 315–332. [CrossRef]
3. Reardon, T.; Barrett, C.B. Agroindustrialization, globalization, and international development: An overview of issues, patterns, and determinants. *Agric. Econ.* **2000**, *23*, 195–205.
4. Schipmann, C.; Qaim, M. Supply chain differentiation, contract agriculture, and farmers' marketing preferences: The case of sweet pepper in Thailand. *Food Policy* **2011**, *36*, 666–676. [CrossRef]
5. Barrett, C.B.; Bachke, M.E.; Bellemare, M.F.; Michelson, H.C.; Narayanan, S.; Walker, T.F. Smallholder participation in contract farming: Comparative evidence from five countries. *World Dev.* **2012**, *40*, 715–730. [CrossRef]
6. Bellemare, M.F. As You Sow, So Shall You Reap: The Welfare Impacts of Contract Farming. *World Dev.* **2012**, *40*, 1418–1434. [CrossRef]
7. Saenger, C.; Qaim, M.; Torero, M.; Viceisza, A. Contract farming and smallholder incentives to produce high quality: Experimental evidence from the Vietnamese dairy sector. *Agric. Econ.* **2013**, *44*, 297–308. [CrossRef]
8. Warning, M. The social performance and distributional consequences of contract farming. An Equilibrium Analysis of the Arachide de Bouche Program in Senegal. *World Dev.* **2002**, *30*, 255–263. [CrossRef]
9. Minten, B.; Randrianarison, L.; Swinnen, J.F.M. Global Retail Chains and Poor Farmers: Evidence from Madagascar. *World Dev.* **2009**, *37*, 1728–1741. [CrossRef]
10. Rao, E.J.O.; Qaim, M. Supermarkets, Farm Household Income, and Poverty: Insights from Kenya. *World Dev.* **2011**, *39*, 784–796. [CrossRef]
11. Key, N.; Runsten, D. Contract farming, smallholders, and rural development in Latin America: The organization of agroprocessing firms and the scale of outgrower production. *World Dev.* **1999**, *27*, 381–401. [CrossRef]
12. Singh, S. Contracting out solutions: Political economy of contract farming in the Indian Punjab. *World Dev.* **2002**, *30*, 1621–1638. [CrossRef]
13. Simmons, P.; Winters, P.; Patrick, I. An analysis of contract farming in East Java, Bali, and Lombok, Indonesia. *Agric. Econ.* **2005**, *33*, 513–525. [CrossRef]
14. TechnoServe and IFAD (International Fund for Agricultural Development). *Outgrower Schemes-Enhancing Profitability*; Technical Brief; International Fund for Agricultural Development: Rome, Italy, 2011.
15. Oya, C. Contract Farming in Sub-Saharan Africa: A Survey of Approaches, Debates and Issues. *J. Agrar. Chang.* **2012**, *12*, 1–33. [CrossRef]

16. Ministry of Food and Agriculture (MoFA). *Agriculture in Ghana: Facts and Figures 2014. Statistics, Research and Information Directorate*; MoFA: Accra, Ghana, 2015.

17. Polson, R.A.; Spencer, D.S.C. The technology adoption process in subsistence agriculture: The case of cassava in Southwestern Nigeria. *Agric. Syst.* **1991**, *36*, 65–78. [CrossRef]

18. Kleih, U.; Phillips, D.; Wordey, M.T.; Komlaga, G. *Cassava Market and Value Chain Analysis: Ghana Case Study. C: AVA Project, Natural Resources Institute*; University of Greenwich and Food Research Institute: Accra, Ghana, 2013.

19. Koyama, N.; Kaiser, J.; Ciugu, K.; Kabiru, J. *Market Opportunities for Commercial cassava in Ghana, Mozambique, and Nigeria*; Delberg Global Development Advisors and Grow Africa: Nairobi, Kenya, 2015.

20. Iyer, S.; Mattinson, D.S.; Fellman, J.K. Study of the Early Events Leading to Cassava Root Postharvest Deterioration. *Trop. Plant Biol.* **2010**, *3*, 151–165. [CrossRef]

21. Baumann, P. *Equity and Efficiency in Contract Farming Schemes: The Experience of Agricultural Tree Crops*; Working Paper 139; Overseas Development Institute: London, UK, 2000.

22. Amevenku, F.; Yeboah, K.K.; Obuoshie, E. *AgWater Solutions Project Case Study: An Assessment of the IFTC Outgrower Scheme in Ghana*; International Water Management Institute: Accra, Ghana, 2012.

23. Harou, A.; Walker, T. *The Pineapple Market and the Role of Cooperatives*; Working Paper; Cornell University: Ithaca, NY, USA, 2010.

24. Mighell, R.L.; Jones, L.A. *Vertical Coordination in Agriculture*; US Department of Agriculture, Economic Research Service, Farm Economics Division: Washington, DC, USA, 1963.

25. Hueth, B.; Ligon, E.; Dimitri, C. Agricultural Contracts: Data and Research Needs. *Am. J. Agric. Econ.* **2007**, *89*, 1276–1281. [CrossRef]

26. Bogetoft, P.; Olesen, H. Ten rules of thumb in contract design: Lessons from Danish agriculture. *Eur. Rev. Agric. Econ.* **2002**, *29*, 185–204. [CrossRef]

27. Bijman, J. *Contract Farming in Developing Countries: An Overview*; Working Paper; Wageningen University: Wageningen, The Nederland, 2008.

28. Masakure, O.; Henson, S. Why do small-scale producers choose to produce under contract? Lessons from nontraditional vegetable exports from Zimbabwe. *World Dev.* **2005**, *33*, 1721–1733. [CrossRef]

29. Guo, H.; Jolly, R.W.; Zhu, J. Contract Farming in China: Perspectives of Farm Households and Agribusiness Firms. *Comp. Econ. Stud.* **2007**, *49*, 285–312. [CrossRef]

30. Alexander, C.; Goodhue, R.E.; Rausser, G.C. Do Incentives for Quality Matter? *J. Agric. Appl. Econ.* **2007**, *39*, 1–15. [CrossRef]

31. Miyata, S.; Minot, N.; Hu, D. Impact of Contract Farming on Income: Linking Small Farmers, Packers, and Supermarkets in China. *World Dev.* **2009**, *37*, 1781–1790. [CrossRef]

32. Guo, H.; Jolly, R.W. Contractual arrangements and enforcement in transition agriculture: Theory and evidence from China. *Food Policy* **2008**, *33*, 570–575. [CrossRef]

33. Eaton, C.; Shepherd, A.W. Contract Farming: Partnerships for Growth. In *FAO Agricultural Services Bulletin*; FAO: Rome, Italy, 2001.

34. Dolan, C.; Humphrey, J. Governance and Trade in Fresh Vegetables: The Impact of UK Supermarkets on the African Horticulture Industry. *J. Dev. Stud.* **2000**, *37*, 147–176. [CrossRef]

35. Berdegué, J.A.; Balsevich, F.; Flores, L.; Reardon, T. Central American supermarkets' private standards of quality and safety in procurement of fresh fruits and vegetables. *Food Policy* **2005**, *30*, 254–269. [CrossRef]

36. Abebe, G.K.; Bijman, J.; Kemp, R.; Omta, O.; Tsegaye, A. Contract farming configuration: Smallholders' preferences for contract design attributes. *Food Policy* **2013**, *40*, 14–24. [CrossRef]

37. Minot, N.; Ronchi, L. *Contract Farming: Risks and Benefits of Partnership between Farmers and Firms. Public Policy for the Private Sector Viewpoint 102736*; The World Bank Group: Washington, DC, USA, 2014.

38. Dorward, A.; Kyyd, J.; Poulton, C. (Eds.) *Smallholder Cash Crop Production under Market Liberalisation: A New Institutional Economics Perspective*; CAB International: Wallingford, UK, 1998.

39. Little, P.D.; Watts, M.J. (Eds.) *Living under Contract: Contract Farming and Agrarian Transformation in Sub-Saharan Africa*; The University of Wisconsin Press: Madison, WI, USA, 1994.

40. Winters, P.; Simmons, P.; Patrick, I. Evaluation of a Hybrid Seed Contract between Smallholders and a Multinational Company in East Java, Indonesia. *J. Dev. Stud.* **2005**, *41*, 62–89. [CrossRef]

41. Glover, D.; Kusterer, K. *Small Farmers, Big Business——Contract Farming and Rural Development*; Macmillan: London, UK, 1990.

42. Grosh, B. Contract Farming in Africa: An Application of the New Institutional Economics. *J. Afr. Econ.* **1994**, *3*, 231–261. [CrossRef]
43. Gow, H.R.; Streeter, D.H.; Swinnen, F.M. How private contract enforcement mechanisms can succeed where public institutions fail: The case of Juhocukor a.s. *Agric. Econ.* **2000**, *23*, 253–265. [CrossRef]
44. Bellemare, M.F. Agricultural extension and imperfect supervision in contract farming: Evidence from Madagascar. *Agric. Econ.* **2010**, *41*, 507–517. [CrossRef]
45. Scott, L.; Freese, J. *Regression Models for Categorical Dependent Variables Using Stata*; StataCorp LP: College Station, TX, USA, 2006.
46. Greene, W.H.; William, H. *Econometric Analysis*, 6th ed.; Pearson Education, Inc.: Upper Saddle River, NJ, USA, 2007.
47. Ghana Statistical Service. *Ghana Living Standards Survey Round 6 (GLSS 6): Poverty Profile in Ghana (2005–2013)*; Ghana Statistical Service: Accra, Ghana, 2014.
48. World Bank. *An Analytical Toolkit for Support to Contract Farming*; World Bank Group: Washington, DC, USA, 2014.
49. Minot, N. Contract Farming in Developing Countries: Patterns, Impact, and Policy Implications. In *Food Policy for Developing Countries: Case Studies*; Pinstrup-Andersen, P., Fuzhi, C., Eds.; Cornell University: New York, NY, USA, 2007.
50. Maertens, M.; Velde, K.V. Contract-farming in Staple Food Chains: The Case of Rice in Benin. *World Dev.* **2017**, *95*, 73–87. [CrossRef]
51. Alene, A.D.; Abdoulaye, T.; Rusike, J.; Manyong, V.; Walker, T. The Effectiveness of Crop Improvement Programmes from the Perspectives of Varietal output and Adoption: Cassava, Cowpea, Soybean and yam in Sub-Saharan Arica and Maize in West and Central Africa. In *Crop Improvement, Adoption, and Impact of Improved Varieties in Food Crops in Sub-Saharan Africa*; Walker, T., Alwang, J., Eds.; CAB International: Wallingford, UK, 2015.
52. Wiggins, S.; Keats, S. *Leaping and Learning: Linking smallholders to Markets in Africa*; Agriculture for Impact, Imperial College and Overseas Development Institute: London, UK, 2013.
53. Henson, S.; Masakure, O.; Boselie, D. Private food safety and quality standards for fresh produce exporters: The case of Hortico Agrisystems, Zimbabwe. *Food Policy* **2005**, *30*, 371–384. [CrossRef]
54. Torero, M.; Viceisza, A. *Potential Collusion and Trust Evidence from a Field Experiment in Vietnam*; Discussion Paper 01100; International Food Policy Research Institute: Washington, DC, USA, 2011.

© 2018 by the authors. Licensee MDPI, Basel, Switzerland. This article is an open access article distributed under the terms and conditions of the Creative Commons Attribution (CC BY) license (http://creativecommons.org/licenses/by/4.0/).

 *sustainability*

Article

Local Agroforestry Practices for Food and Nutrition Security of Smallholder Farm Households in Southwestern Ethiopia

Omarsherif Jemal [1],*, Daniel Callo-Concha [1] and Meine van Noordwijk [2,3]

[1] Centre for Development research (ZEF), University of Bonn, Genscheralle 3, 53113 Bonn, Germany; d.callo-concha@uni-bonn.de
[2] World Agroforestry Centre (ICRAF), Southeast Asia Regional Programme, P.O. Box 161, Bogor 16001, Indonesia; m.vannoordwijk@cgiar.org
[3] Plant Production Systems Group, Wageningen University, P.O. Box 430, 6700 AK Wageningen, The Netherlands
* Correspondence: omarsherifmm@gmail.com; Tel.: +251-911-410-478

Received: 20 June 2018; Accepted: 27 July 2018; Published: 2 August 2018

Abstract: Food and nutrition security (FNS) rests on five pillars: availability, access, utilization, stability, and sovereignty. We assessed the potentials of local agroforestry practices (AFPs) for enabling FNS for smallholders in the Yayu Biosphere Reserve (southwestern Ethiopia). Data was collected from 300 households in a stratified random sampling scheme through semi-structured interviews and farm inventory. Utility, edibility, and marketability value were the key parameters used to determine the potential of plants in the AFPs. Descriptive statistics, ANOVA, and correlation analysis were employed to determine the form, variation, and association of local AFP attributes. Homegarden, multistorey-coffee-system, and multipurpose-trees-on-farmlands are the predominant AFPs in Yayu. Multipurpose-trees-on-farmlands are used mainly for food production, multistorey-coffee-system for income-generation, and homegarden for both. The 127 useful plant species identified represent 10 major plant utility groups, with seven (food, fodder, fuel, coffee-shade, timber, non-timber-forest-products, and medicinal uses) found in all three AFPs. In total, 80 edible species were identified across all AFPs, with 55 being primarily cultivated for household food supply. Generally, household income emanates from four major sources, multistorey-coffee-system (60%), homegarden (18%), multipurpose-trees-on-farmlands (13%), and off-farm activities (11%). Given this variation in form, purpose, and extracted benefits, existing AFPs in Yayu support the FNS of smallholders in multiple ways.

Keywords: food and non-food benefit; homegarden; multipurpose tree on farmland; multistorey coffee system; multi-functionality; traditional agroforestry; Yayu Biosphere Reserve

---

1. Introduction

In the last four decades, agroforestry has been promoted as an option to address poverty and food insecurity, as well as to enhance the adaptability of small-scale farmers to social-ecological hazards [1–5]. For the former, its potential relies on its contribution to the strengthening of the five pillars of food and nutrition security (FNS): availability, access, utilization, stability, and sovereignty (Figure 1). Examples in local contexts are: (i) the presence of perennial staple food species in the system, like *Ensete ventricosum*, *Musa* spp., *Moringa stenopetala*, or *Manihot esculenta*, which ensure the availability of food [6–9]; (ii) the presence of species that secure cash to farming households, which directly enhance their access to market-based foods, as is the case of *Coffea arabica* or *Theobroma cacao* [10,11]; (iii) the utilization of dimension often enhanced via a diversity of species that offer scarce nutrients, e.g., fruit, leaves, or nuts, as well as the availability of fuel for cooking [11,12]; (iv) an increase in the resilience to and reduction in household social-ecological vulnerability, by the diversity of constituting species and their interactions [5,10,11]; and (v) providing options and choices in the means to grow and/or purchase foods items according to the household's needs in all seasons [9,13,14].

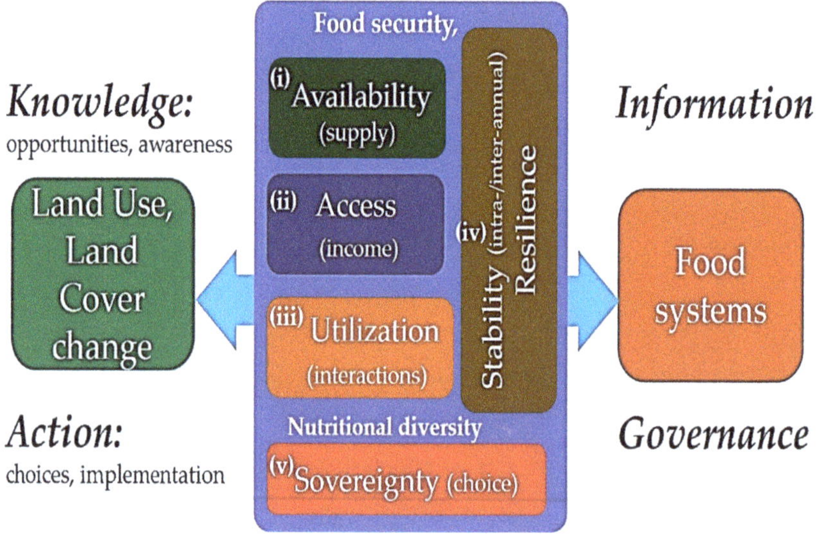

Figure 1. Aspects of food security at the interface of land use and food systems.

Agroforestry practices (AFPs) vary in their composition, structure, and function, depending on the biophysical, ecological, social, economic, and cultural conditions under which they occur. Hence, taking into account this site-specificity is key before attempting any upscaling. Mbow et al. [5] emphasized that although agroforestry has a considerable potential to improve food security, " . . . not all agroforestry options are viable everywhere". Therefore, before recommending or promoting any agroforestry practice for a certain place or community, it is crucial to characterize and recognize its features, attributes, and performance in view of intra- and inter-year variability.

In Ethiopia, a wide variety of local/traditional AFPs exist, with potential to contribute to the FNS security of the practitioner communities. Different forms of homegardens are reported across the country [15–18]. Croplands with scattered trees of *Faidherbia albida* are the oldest form of indigenous agroforestry parklands, omnipresent in central and eastern Ethiopia [4,19,20]. The Enset-Coffee gardens, *Coffea arabica* planted intermingled with *Enset ventricosum*, practiced by the Gedeo people, are well-known to support millions of livelihoods in the most densely populated areas of the country [15,21]. The traditional coffee system dominated AFPs in southwestern Ethiopia are a further example of a well-established traditional AFP [22]. In Yayu, 60 to 80% of rural households rely on these AFPs as the main source of their livelihood [23]; the area was considered food secure in the last 15 years, despite an increasing human population [24–29]. In spite of the potential of the local AFPs to contribute toward FNS, empirical data on this topic are scarce.

For instance, during the last two decades, almost all studies conducted in Yayu mostly focused on the ecology, biodiversity, and conservation of traditional coffee systems [30–36], giving lesser attention to their potential for food and nutrient provision to local communities. Therefore, this study aimed to explore the potential of local AFPs of Yayu to contribute to the FNS of rural households. Specific objectives of the study were: (1) the identification of predominant AFPs and their purpose; (2) the plant species composition of each practice; (3) the ethnobotanical knowledge of uses of the various species encountered; and (4) their respective potential for food provision and cash acquisition, as steps towards (5) an appraisal of the five pillars of food security.

2. Materials and Methods

2.1. Study Area

Yayu is located in the Illubabor zone of the Oromiya state, southwestern Ethiopia, between 8°10′–8°39′ N and 35°30′–36°4′ E (Figure 2). The area was registered by UNESCO in 2010 as a biosphere reserve for the in situ conservation of wild *Coffea arabica*. The Yayu Coffee Forest Biosphere Reserve comprises three concentric zones, i.e., the core area as well as buffer and transition zones, covering about 28,000, 22,000, and 118,000 ha, respectively [22] (Figure 1). The climate is hot and humid; the mean annual temperature is 20.5 °C, with average monthly values between 18.46 and 21.25 °C [35]. The area exhibits a unimodal rainfall pattern with a mean annual precipitation of 2100 mm [32]. Dominant soil groups include nitosols, acrisols, vertisols, and cambisols [37].

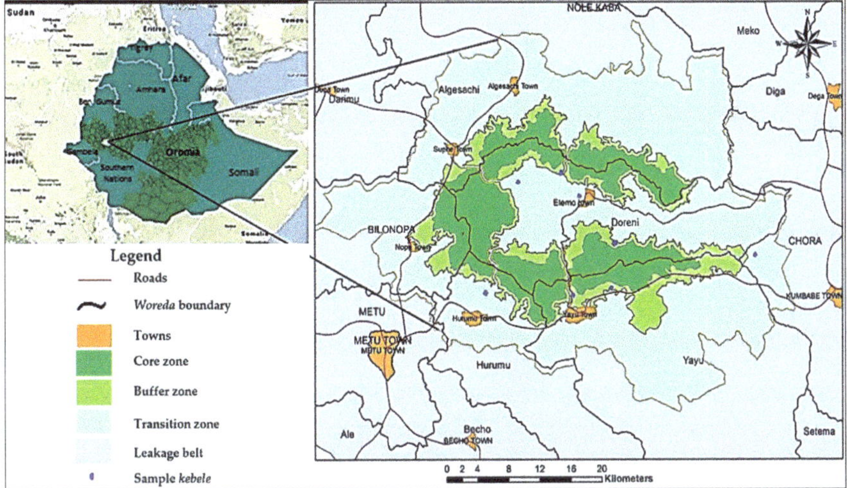

Figure 2. Location of the study area and sampled *Kebele*, with reference to the three zones of the Yayu Coffee Forest Biosphere Reserve and roads. Adapted from Reference [22].

The vegetation cover of the area is relatively well conserved, comprising large areas of forests intermingled with wild coffee plants. The adequate integration of human utilization and environmental conservation makes the system one of the best performing traditional agroforestry systems of Ethiopia [38,39]. However, these coexist with other farming systems, such as annual crop fields, farmlands with scattered trees, homegardens, woodlots, grazing lands, and fallows [22,40]. Land uses exhibit a mosaic arrangement, within which forests, agricultural lands, wetlands, and grazing lands as the most prominent. The major forest uses are undisturbed natural forest, semi-forest coffee systems, fully managed forest for coffee production, and old secondary forests [22,40]. Farming is carried out in both around the farmer homestead and away from it [22,40].

2.2. Sampling Strategy

Prior to sample site selection, potential sources of errors were listed to control their effect on the results of the study. Mainly, the relative distances of farming households to the forest core zone and to market facilities were assumed to be the most influential factors [8]. Thus, the sample stratification was based on these factors. A *kebele* is the smallest administrative unit in Ethiopia, similar to a ward, a neighborhood, or a localized and delimited group of households. *Kebele* including a forest (core zone) in its jurisdiction were considered to be 'near to forest', while others were 'far from forest'. Similarly, those *kebele* located near a road were assumed to have better access to markets, and were considered as 'near market', while others were 'far from market'. By using these levels of the two factors, four proximity categories were constructed to which all *kebele* of the reserve were assigned, and then two *kebele* from each category were selected randomly. Thus, a total of eight sample *kebele* were used for data collection (Table 1). Based on the local administration office data, about 4300 (N) households dwell in eight sample *kebele*, so 300 (*n*) households were assigned for the total sample [41]. Sampling intensity was proportionally allocated to each sample *kebele* based on population size. Finally, sample households were randomly selected for data collection.

Table 1. Sampled *kebele*, altitudes, and size.

Kebele	**Altitudinal Range (m.a.s.l)**	**Sample Size (*n*)**
Wabo	1570 to 1624	27
Wutete	1565 to 1672	45
Sololo	1624 to 1688	43
Wangene	1562 to 1890	44
Weyira	1789 to 1973	26
Werebo	1725 to 1892	45
Beteli Gebecha	1754 to 1819	45
Elemo	1906 to 1981	25

2.3. Data Collection and Methodology

Data were collected from December 2014 to February 2015. First, key informant interviews substantiated by field observation were applied to understand the rationale of local farming and to identify the predominant AFPs. Based on this result, the household survey was designed, including a separate dataset for each of the identified AFPs. The household head was questioned on the socioeconomic and biophysical attributes of the household and agroforestry plots. The characterization of the AFP included the location, size, and spatial arrangements of the components, the identification of tree species and obtained products, and the uses and marketability of these. Field observation complemented data gathering.

Plant species identification was supported by a local taxonomist and specialized literature [42–45]. Plant uses and services were obtained from each system and grouped in accordance with the functional groups set by Mendez [46] and Abebe [8]. The food edibility potential was evaluated in two steps. Species were identified as edible and non-edible by comparing first-hand observation with secondary resources [39,44,45,47]. The 'edible' category was further subdivided into 'potentially-edible' and 'active-food' species. The former refers to edible plant species not primarily used as food, while the latter refers to species primarily cultivated for food. 'Active-food' was further re-classified into 10 plant food groups according to the United Nations Food and Agriculture Organization (FAO) [48]: 'cereals', 'white root and tubers', 'vitamin-A-rich vegetables and tubers', 'dark green leafy vegetables', 'other vegetables', 'vitamin-A-rich fruits', 'other fruits', 'legumes, nuts, and seeds', 'sweets', and 'spices, condiments, stimulants, beverages, and additives'.

The potential of each agroforestry practice to generate income was assessed in two steps. First, we estimated the amount and major sources of annual income of the household, both on farm and off farm, establishing the relationship with the species and products of each AFP. Second, we listed all species and their actual marketability status, classified into three classes: (i) actively-marketed species, either cited as cash crops by at least one respondent and/or observed in local markets; (ii) passively-marketed species, mentioned by key informants or in the literature, but occasionally cited by farmers and rarely observed in markets; and (iii) non-marketed species, which included all species not belonging to the previous two classes [39,44,45,47].

The species richness of each category, i.e., food group, utility, edibility, and marketability, was calculated using Menhinick's index Equation (1) [49].

$$D = S/\sqrt{N} \tag{1}$$

where D = Menhinick's index, S = number of species of a given use type/food group of a given plot, and N = total species per plot.

Statistical analyses included the calculation of descriptive statistics for relevant biophysical and socioeconomic variables, and were conducted as a post-hoc analyses for subsequent testing, such as one-way ANOVA, Tukey's test, and Pearson correlation analysis. The variability in different categories of each practice was tested separately across households. The tests were implemented using Minitab 17.0 software [50].

3. Results

3.1. Household Socioeconomic Profile

Males headed the majority of Yayu households (84%), where 44% of the households were within the medium wealth class, 17% in the rich class, and 39% in the poor class. Average family size was about five individuals, ranging between four and six. The average age of the respondents was 44.3 years. Ethnically, 75.6% of the households belonged to the Oromo ethnic group, followed by Amhara (19.3%) and Tigreway (4.6%). The majority of the respondents were native to the area, and 23.7% (mostly the two latter ethnic groups) were settlers. The illiteracy rate was 41.4%; 36.8% had attended primary school, and 18.3% had attended school to grade six or beyond. The average landholding size was 4.1 ± 3.2 ha per household (Figure 3).

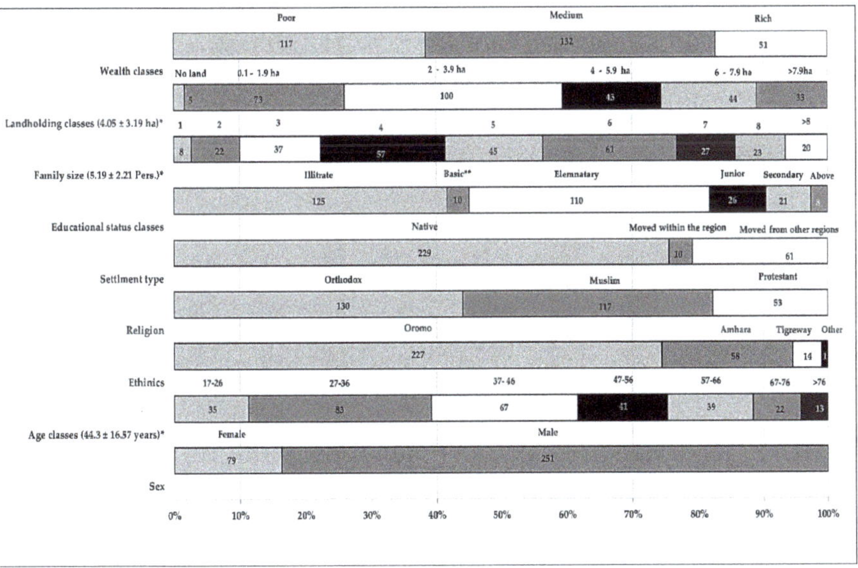

Figure 3. Demographic, socioeconomic, and cultural attributes of sample households in Yayu. * Average ± standard deviation; ** Basic reading and writing skill from non-formal school attendance, e.g., traditional/religious school, etc.

3.2. Agroforestry Practices and Purposes

The collected information revealed that almost all of the farming activities of the area involve deliberately perennial woody species, in three distinct types of niches, i.e., homestead, coffee plot, and farmland, which locally identified as *guwaro*, *laffa bunna*, and *laffa qonna*, respectively. By adopting the classification scheme of Nair [51], the locally recognized farming practices were identified as homegarden (HG), multistorey-coffee-system (MCS), and multipurpose-trees-on-farmland (MTF). An HG is a complex multispecies production system practiced around the homestead, locally named *guwaro*. MCS is locally named *laffa bunna*, literally translated as coffee land. Although coffee cultivation is present in most land-use systems, MCS is distinguished by involving naturally grown and/or planted coffee with mostly native shade trees, resembling a multi-strata forest. The third type, MTF, locally known as *laffa qonna*, literally farmland, refers to lands designated for the production of annual crops that deliberately integrate perennial woody species to increase or optimize plot output.

About 81% of the respondents involve in all three practices. MCS alone is practiced by 97% of the households, HG by 93%, and MTF by 85%. HG covers the smallest area (average 0.08 ha), and MCS the largest (2.6 ha) (Figure 4a). Concerning the primary purpose, MCS is used entirely for income generation, 66% of MTF is devoted to food production, in some cases also to wood and cash crop cultivation, and HG focuses on food (40%) and cash crop production (40%) (Figure 4b). Regarding the number of specific purposes/benefits per practice (annual crop production, fruit production, cash crop production, vegetable production, etc.), the highest was in HG (max. eight) per household, and in more than 90% of the households at least three specific outputs were generated. In contrast, the lowest value was found in MCS (max. three) per household (Figure 4c). The main users or decision makers of MCS and MTF are dominantly males (>83%). Females dominated in HG (62%) (Figure 4d).

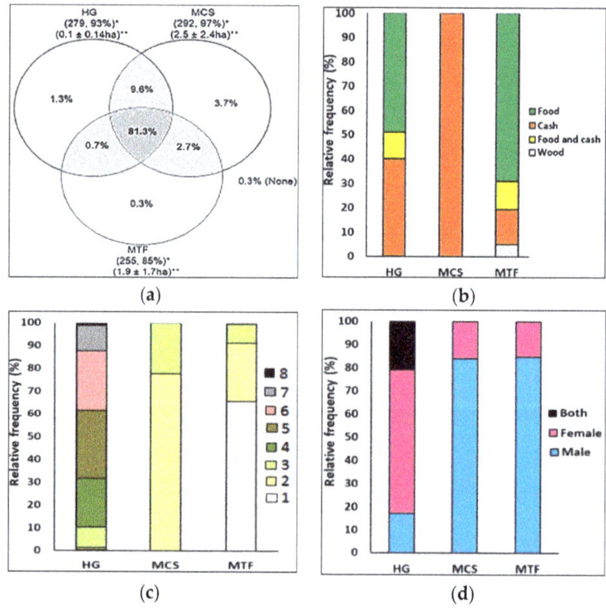

Figure 4. Characteristics of predominant agroforestry practices in Yayu. (**a**) Classification, number, and size of households * Household count, relative proportion; ** size (mean ± SD). (**b**) Relative frequency of main purpose of outputs per practice. (**c**) Relative frequency of number of benefits per practice. (**d**) Relative frequency of the main users per gender per practice. Homegardens (HG), multistorey-coffee-systems (MCS), and multipurpose-trees-on-farmlands (MTF).

3.3. Floristic Composition

One hundred and twenty-seven plant species from 47 families were identified in all three AFPs. The highest number was found in HG (88), followed by MCS (65) and MTF (55) (Figure 5a). About 68.5% were perennial (tree and shrubs); herbs were absent in MCS; 69% of the species were native to the Yayu area (Figure 5b); and most herbs and exotic species existed in HG. A full list of species identification and characterization is available on demand.

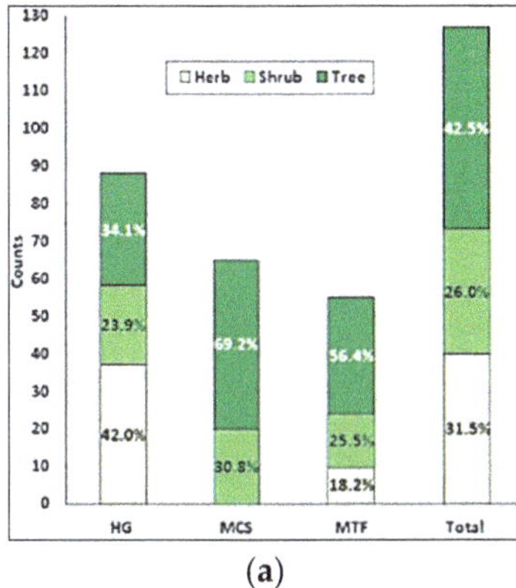

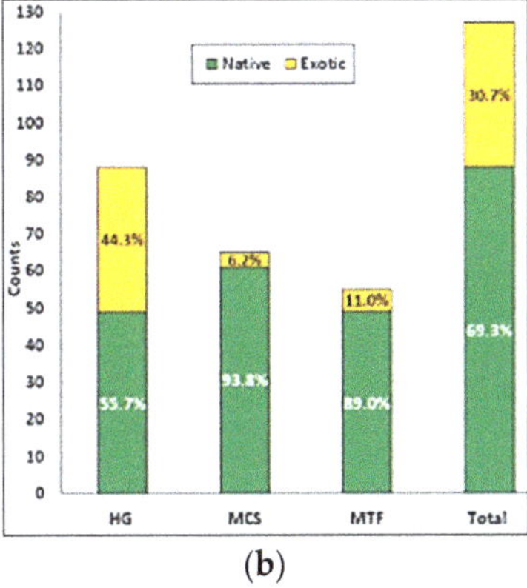

(a) (b)

Figure 5. Floristic composition of local agroforestry practices in Yayu, southwestern Ethiopia. (**a**) Relative proportion per growth habit. (**b**) Relative proportion per origin. Homegardens (HG), multistorey-coffee-systems (MCS), and multipurpose-trees-on-farmlands (MTF).

Concerning the frequency of occurrence of species, *Coffea arabica*, *Mangifera indica*, *Persea americana*, and *Brassica oleracea* were dominant in more than 70% HG. Besides coffee, present in all MCS, shade tree species like *Cordia africana* and *Albizia gummifera* were present in more than 70% of MCS. In contrast, MTF was dominated by *Zea mays* (more than 95%), followed by *Sesbania sesban* (33%), *Eragrostis tef* (31%), *Sorghum bicolor* (26.3%), and *Eucalyptus grandis* (20.8%). The multipurpose tree species *Vernonia amygdalina* was the only species found in all three practices (Figure 6).

3.4. Plant Uses and Services

The existing 10 different types of plant uses and services [46] were observed in the three AFPs, i.e., food; spices, condiments, and other food and beverage additives; stimulants; fodder; fuel; timber; non-timber tree products; shade trees for coffee; other services, e.g., live fences, windbreaks, demarcation, recreation, and ornamental; and medicine.

Almost all uses were observed in all three practices. Only 'food' in MCS, 'shade trees for coffee' in MTF, and 'spices, condiments, and other foods and beverage additives' were missing in both practices. Regarding the species count for each use, 'stimulants' (2) and 'fuel' (52) were the two extremes. Overall, 'food' scored significantly high ($p < 0.01$), with 7.8 species per plot in HG, whereas 'fuel' was the highest in MCS (7.1). Species richness per plot and count showed similar trends in all uses and service categories (Table 2).

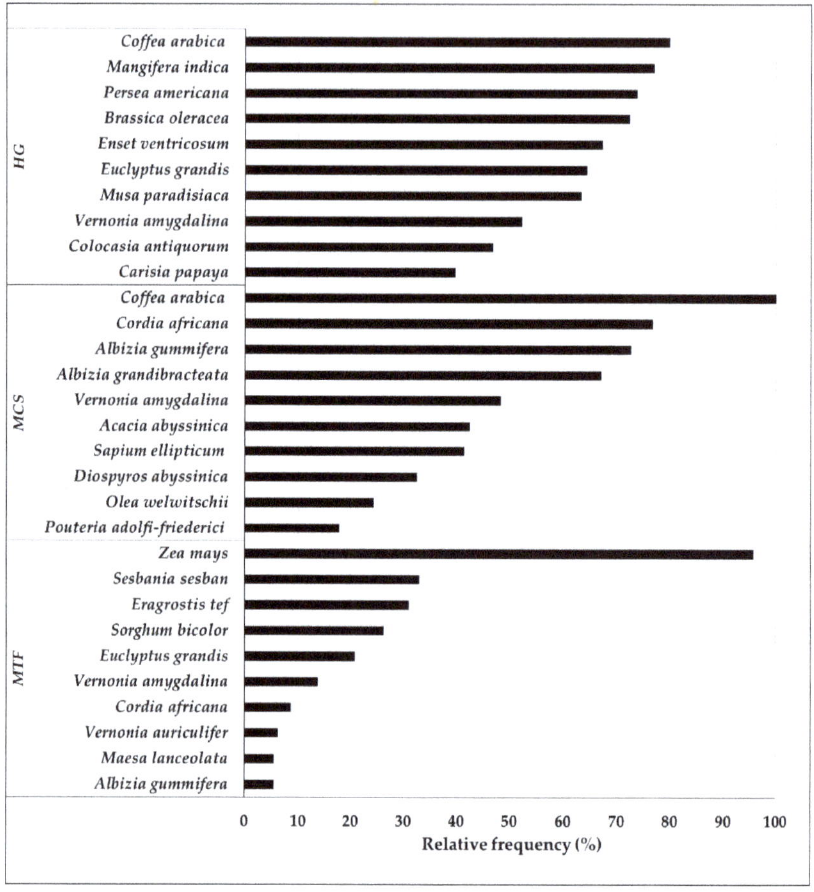

Figure 6. Relative frequency of the 10 most frequent species found in each of the three agroforestry practices in Yayu. Homegardens (HG), multistorey-coffee-systems (MCS), and multipurpose-trees-on-farmlands (MTF).

3.4.1. Food Production Potential

Out of the 127 species identified in all three practices, 80 were edible, while 55 were managed as 'active-food'. Except three of the 'active-food' species, the rest were observed in HG, with an average of nine species. The highest number of 'potentially-edible' species was found in MCS (21), and the highest number of 'active-food' species was found in HG (52) (Table 2). The ratio between 'potential' and 'active-food' reveals the untapped value of edible species. In MCS, both in total and at the plot level, the ratio reached 21:1 and 1.7:1, respectively (Table 3).

The 'active-food' category was subdivided into 10 different food groups. As expected, HG exhibited the largest variety of species of all food groups. The food group 'other fruits' scored the highest in all AFPs, whereas 'spices, condiments, and beverages' and 'cereals' showed higher species richness in MCS and MTF. The average species number and richness per household showed a significant variation in HG and MTF ($p < 0.01$) (Table 4).

Table 2. Count, percentage, average ($\pm$ standard deviation) number per household, and richness per household of species of 10 plant uses and service categories in three AFPs and p-value for one-way ANOVA test among categories.

Use and Service Category	Count of Total Species (%)				Average Number of Species/HH ($\pm$SD)			Average Richness/HH ($\pm$SD)		
	All	HG	MCS	MTF	HG	MCS	MTF	HG	MCS	MTF
Food	44 (34.6)	42 (46.6)	0 (0)	10 (18.2)	7.8 [A] ($\pm$2.9)	0.0 [F] ($\pm$0.0)	1.8 [BC] ($\pm$1.0)	1.6 [A] ($\pm$0.5)	0.0 [F] ($\pm$0.0)	0.9 [B] ($\pm$0.4)
Spices, condiments, and other food and beverage additives	9 (7.1)	10 (10.2)	0 (0)	0 (0)	0.6 [FG] ($\pm$0.9)	0.0 [F] ($\pm$0.0)	0.0 [E] ($\pm$0.0)	0.1 [FG] ($\pm$0.2)	0.0 [F] ($\pm$0.0)	0.0 [D] ($\pm$0.0)
Stimulants	2 (1.6)	2 (2.3)	2 (1.5)	1 (1.8)	1.2 [EFG] ($\pm$0.7)	1.0 [EF] ($\pm$0.0)	0.1 [DE] ($\pm$0.4)	0.2 [EF] ($\pm$0.1)	0.4 [E] ($\pm$0.1)	0.1 [D] ($\pm$0.2)
Fodder	41 (32.3)	27 (29.5)	31 (46.2)	29 (52.7)	2.7 [CD] ($\pm$2.7)	3.6 [BC] ($\pm$3.4)	2.0 [B] ($\pm$2.1)	0.6 [C] ($\pm$0.5)	1.3 [C] ($\pm$1.1)	1.0 [B] ($\pm$0.8)
Fuel	52 (40.9)	27 (29.5)	46 (69.2)	36 (65.5)	4.5 [B] ($\pm$2.1)	7.1 [A] ($\pm$ 3.6)	2.9 [A] ($\pm$1.9)	0.9 [B] ($\pm$0.4)	2.5 [A] ($\pm$0.5)	1.4 [A] ($\pm$0.5)
Timber	34 (26.8)	17 (18.2)	31 (46.2)	24 (43.6)	1.5 [EF] ($\pm$1.1)	3.4 [C] ($\pm$2.6)	0.9 [CD] ($\pm$1.1)	0.3 [DE] ($\pm$0.2)	1.1 [C] ($\pm$0.6)	0.4 [C] ($\pm$0.4)
Non-timber tree products	10 (7.9)	5 (4.5)	10 (13.8)	3 (5.5)	0.1 G ($\pm$0.4)	0.3 [F] ($\pm$0.7)	0.1 [DE] ($\pm$0.3)	0.0 G ($\pm$0.1)	0.1 [F] ($\pm$0.2)	0.01 [D] ($\pm$0.1)
Shade trees for coffee	26 (20.5)	20 (21.6)	25 (36.9)	0 (0)	2.0 [CDE] ($\pm$1.7)	4.7 [B] ($\pm$2.2)	0.0 [E] ($\pm$0.0)	0.4 [D] ($\pm$0.3)	1.6 [B] ($\pm$0.5)	0.0 [D] ($\pm$0.0)
Other services, e.g., live fences, windbreaks, etc.	18 (14.2)	16 (17)	15 (21.5)	12 (21.8)	1.9 [DE] ($\pm$1.2)	2.0 [DE] ($\pm$1.4)	0.9 [CD] ($\pm$0.9)	0.4 [D] ($\pm$0.3)	0.7 [D] ($\pm$0.5)	0.5 [C] ($\pm$0.4)
Medicine	21 (16.5)	21 (22.7)	9 (12.3)	7 (12.7)	3.1 [C] ($\pm$1.67)	2.1 [D] ($\pm$1.1)	0.4 [DE] ($\pm$0.8)	0.6 [C] ($\pm$0.3)	0.8 [D] ($\pm$0.3)	0.2 [CD] ($\pm$0.3)
p-value					<0.01	<0.01	<0.01	<0.01	<0.01	<0.01

Categories with at least one similar superscript do not significantly differ at least at α = 0.05; HH = household; HG = homegardens; MCS = multistorey-coffee-systems; MTF = multipurpose-trees-on-farmlands.

Table 3. Count, percentage, average ($\pm$ standard deviation) number per household, and richness per household of species of three edibility categories in three agroforestry practices (AFPs), and a 'potential' to 'active' ratio in each practice.

Edibility Category/Ratio	Count of Total Food Species (%)				Average Number of Species/HH ($\pm$SD)			Average Richness/HH ($\pm$SD)		
	All	HG	MCS	MTF	HG	MCS	MTF	HG	MCS	MTF
Total edible	80 (62.9)	63 (71.6)	22 (33.8)	25 (45.5)	9.3 ($\pm$ 3.6)	2.7 ($\pm$1.4)	2.3 ($\pm$1.3)	2.6 ($\pm$0.6)	0.9 ($\pm$0.3)	1.2 ($\pm$0.4)
Potentially edible	25 (19.7)	11 (12.5)	21 (32.3)	14 (25.6)	0.3 ($\pm$0.6)	1.7 ($\pm$1.4)	0.4 ($\pm$0.7)	0.1 ($\pm$0.2)	0.6 ($\pm$0.4)	0.2 ($\pm$0.3)
Active food	55 (44)	52 (59.1)	1 (1.5)	11 (20.0)	9.0 ($\pm$3.4)	1.0 ($\pm$0.0)	1.9 ($\pm$1.1)	2.5 ($\pm$0.6)	0.4 ($\pm$0.1)	0.96 ($\pm$0.4)
Ratio potential: active food	0.5	0.2	21.0	1.3	0.0 ($\pm$0.1)	1.7 ($\pm$1.4)	0.3 ($\pm$0.5)	0.0 ($\pm$0.1)	1.7 ($\pm$1.4)	0.3 ($\pm$0.5)

HH = household; HG = homegardens; MCS = multistorey-coffee-systems; MTF = multipurpose-trees-on-farmlands.

Table 4. Count, percentage, average (± standard deviation) number per household, and richness per household of species of 10 food groups in three AFPs and *p*-value for one-way ANOVA test among food groups.

Food Group	Count of Active Food Species (%)				Average Number of Species/HH (± SD)			Average Richness/HH (± SD)		
	All	HG	MCS	MTF	HG	MCS	MTF	HG	MCS	MTF
Cereals	6 (13.6)	3 (7.1)	0.0 (0.0)	6 (54.5)	0.1 [E] (±0.4)	0.0 (±0.0)	1.6 [A] (±0.8)	0.1 [D] (±0.1)	0.0 (±0.0)	1.2 [A] (±0.3)
White roots and tubers	6 (13.6)	6 (14.3)	0.0 (0.0)	0 (0.0)	1.5 [B] (±1.2)	0.0 (±0.0)	0.0 (±0.0)	0.4 [BC] (±0.3)	0.0 (±0.0)	0.0 [B] (±0.0)
Vitamin A-rich vegetables and tubers	4 (9.1)	4 (9.5)	0.0 (0.0)	0 (0.0)	0.5 [DE] (±0.6)	0.0 (±0.0)	0.0 (±0.0)	0.2 [D] (±0.2)	0.0 (±0.0)	0.0 [B] (±0.0)
Dark green leafy vegetables	4 (9.1)	4 (9.5)	0.0 (0.0)	0 (0.0)	1.0 [C] (±0.7)	0.0 (±0.0)	0.0 (±0.0)	0.3 [C] (±0.2)	0.0 (±0.0)	0.0 [B] (±0.0)
Other vegetables	5 (11.4)	5 (11.9)	0.0 (0.0)	0 (0.0)	0.5 [DE] (±0.7)	0.0 (±0.0)	0.0 (±0.0)	0.1 [D] (±0.2)	0.0 (±0.0)	0.0 [B] (±0.0)
Vitamin A-rich fruits	3 (6.8)	3 (7.1)	0.0 (0.0)	0 (0)	1.2 [BC] (±0.7)	0.0 (±0.0)	0.0 (±0.0)	0.4 [C] (±0.2)	0.0 (±0.0)	0.0 [B] (±0.0)
Other fruits	12 (27.3)	12 (28.6)	0.0 (0.0)	0 (0)	2.4 [A] (±1.4)	0.0 (±0.0)	0.0 (±0.0)	0.8 [A] (±0.4)	0.0 (±0.0)	0.0 [B] (±0.0)
Legumes, nuts, seeds	5 (11.4)	5 (11.9)	0.0 (0.0)	4 (36.4)	0.2 [DE] (±0.4)	0.0 (±0.0)	0.1 [B] (±0.4)	0.1 [D] (0.1)	0.0 (±0.0)	0.1 [B] (±0.2)
Sweets	1 (2.3)	1 (2.4)	0.0 (0.0)	0 (0)	0.1 [E] (±0.3)	0.0 (±0.0)	0.0 (±0.0)	0.0 [D] (±0.1)	0.0 (±0.0)	0.0 [B] (±0.0)
Spices, condiments, and beverages	9 (20.5)	9 (21.4)	1.0 (1.5)	0 (0)	0.7 [CD] (±0.9)	1.0 (±0.0)	0.0 (±0.0)	0.6 [B] (±0.3)	1.0 (±0.0)	0.1 [B] (±0.2)
p-value					<0.01	N.A	<0.01	<0.01	N.A	<0.01

Food groups with at least one similar superscript do not significantly differ at least at $\alpha = 0.05$; N.A = test not applicable; HH = household; HG = homegardens; MCS = multistorey-coffee-systems; MTF = multipurpose-trees-on-farmlands.

3.4.2. Income Generation Potential

Farming in all three AFPs accounts for almost 90% of the households' income; MCS has the largest share (60%), mainly from selling *Coffea arabica* (Figure 7a,c). In HG, *Coffea arabica* generates 45% of the cash, the rest is provided mainly by fruits and livestock-related activities (Figure 7b). As MTF is mostly devoted to food production for self-consumption, the contribution to the households' income was slightly lower than that in MCS and HG (Figure 7a). Cash was rather generated through the sale of cash crops such as *Catha edulis* (52%), and annual crops such as *Eragrostis tef* and *Zea mays* (38%) (Figure 7d).

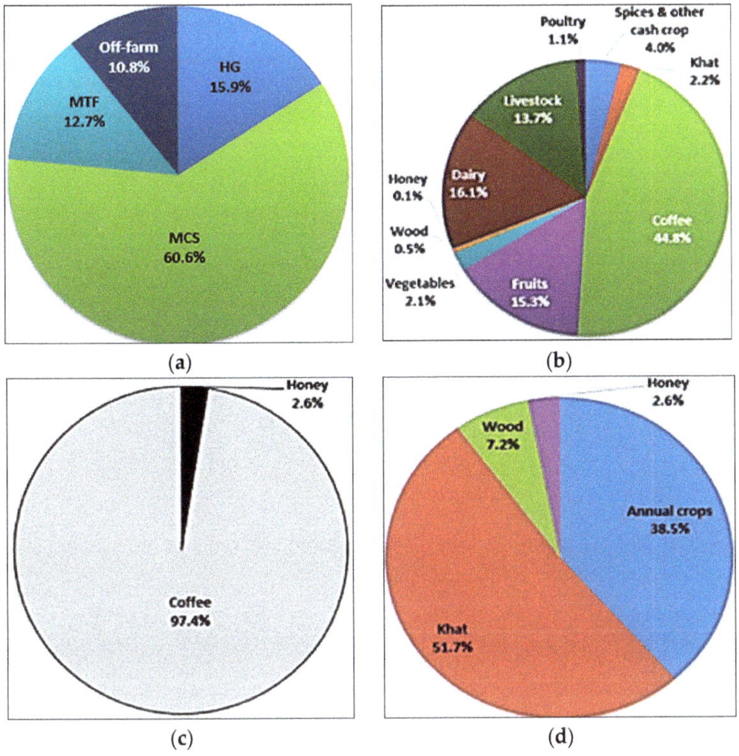

Figure 7. Income generation in households by (**a**) main agricultural and non-agricultural activities; (**b**) homegardens; (**c**) multistorey-coffee-systems; and (**d**) multipurpose-trees-on-farmlands.

Among the 127 species identified, 50 (39.1%) were reported as 'actively-marketed species'. In terms of species composition, HG showed the highest percentage of both active and passively (occasionally) marketed species (47), followed by MTF scoring a much lower species count (nine). On the other hand, only one actively-marketed species was reported in MCS (*Coffea arabica*). On average, the largest number of cash crops per households was recorded in HG (7.6). Regardless of the mode of utilization, whether for self-consumption or selling, almost all households had at least one actively-marketed species. The species counts and richness of all marketed species categories in MTF were found to differ significantly ($p > 0.01$). The actively-marketed species category was significantly different ($p < 0.01$) from the passively-marketed and non-marketed category in all practices except in MCS (Table 5).

Table 5. Count, percentage, average ($\pm$ standard deviation) number per household, and richness per household species of three marketability categories in three AFPs and p-value for one-way ANOVA test among categories.

Marketability Category	Species Count (%)				Average Number of Species/HH ($\pm$SD)			Average Species Richness/HH ($\pm$SD)		
	All	HG	MCS	MTF	HG	MCS	MTF	HG	MSC	MTF
Actively-marketed species	50 (39.4)	47 (53.4)	1 (1.5)	9 (16.4)	7.6 [A] ($\pm$2.8)	1 [B] ($\pm$0.0)	2.1 [A] ($\pm$1.1)	2.3 [A] ($\pm$0.6)	0.4 [B] ($\pm$0.1)	1.1 [A] ($\pm$0.4)
Passively-marketed species	18 (14.2)	13 (14.7)	11 (16.9)	10 (18.1)	2.7 [B] ($\pm$1.6)	1.3 [B] ($\pm$0.9)	0.4 [B] ($\pm$0.6)	0.5 [B] ($\pm$0.3)	0.5 [B] ($\pm$0.3)	0.2 [C] ($\pm$0.3)
Non-marketed species	59 (46.5)	28 (31.8)	53 (81.5)	36 (65.5)	2.2 [B] ($\pm$1.6)	5.7 [A] ($\pm$3.6)	1.5 [C] ($\pm$1.5)	0.6 [B] ($\pm$0.4)	1.9 [A] ($\pm$0.7)	0.7 [B] ($\pm$0.5)
p-value					<0.01	<0.01	<0.01	<0.01	<0.01	<0.01

Categories with similar superscript do not significantly differ at least at α = 0.05; HH = household; HG = homegardens; MCS = multistorey-coffee-systems; MTF = multipurpose-trees-on-farmlands.

4. Discussion

4.1. Agroforestry Practices and Purposes

The three major AFPs, i.e., HG, MCS, and MTF, have their own primary production purposes and specific management, which enable smallholder farm households to diversify their production across the year. Hence, all three are important for sustaining the livelihoods of farmers in Yayu. This was confirmed by the fact that more than 80% of the households in Yayu practice all three agroforestry systems.

However, each practice plays a predominant role. MCS is mainly used to generate money and, for the majority of households, is the main, if not the only, source of cash. Most farmers use MTF to produce food, and HG is used for both a source of food and cash to supplement the other two practices. In the absence of either one or two practices, the importance of HG increases, becoming the main source of food and/or income. Similar findings were obtained by Kebebew and Urgessa [52] in the Jimma area, southwestern Ethiopia.

Besides these, each AFP provides other benefits. This is especially true in the case of HG, as management practices tend to encourage the production of useful by-products such as fuel, fodder, and timber. Contrarily, MCS and MTF have rather specific purposes, and activities not directly contributing to yield maximization, as in the case of MCS, are discouraged [38].

In the three AFPs, gender roles differ. Women are mostly in charge of the management and utilization of HG, securing the annual food provision and supplementary cash for the household [40,53,54]. In contrast, the management and benefits of MCS and MTF are mostly controlled by males—generally heads of the households [53,55].

4.2. Predominant Species Composition

The highest number of species was found in HG, followed by MCS and MTF. Only 19% of the identified species were found in all practices, and 52% occurred in only one of the three AFP. So, species distribution is practice/system-specific. The number of species in HG is similar to values reported in other studies [15,16,18,56].

Regarding growth habits, as expected, perennial species were dominant but also shrubs were common, although fewer in MCS and MTF due to the regular clearings carried out to prevent competition.

Concerning the species diversity, based on their origin, about 70% of the species identified were native, confirming that the Yayu area is naturally endowed with high biodiversity [39]. A higher number of exotic species were observed in HG, followed by MTF and MCS. This confirms that coffee forest production (MCS) is more environmentally friendly than the other two practices. This finding validates those of Muleta et al. [38] and Gole et al. [22], who observed that local communities exert their experience in managing the naturally grown coffee for commercial purposes. Nevertheless, there is a remarkable dominance of individual species, especially in MCS and MTF, by *Coffea arabica* and *Zea mays*, respectively. The dominance of *Zea mays* in MTF is consistent with similar reports from the Koga watershed in northwestern Ethiopia [57]. In contrast, the species frequency distribution in HG is rather flat, as the HG approach has multiple production objectives, i.e., stimulants, fruits, vegetables, roots and tubers, and timber. However, Kebebew and Urgessa [52] reported that fruit trees were a dominant group of species in HG of Jimma.

4.3. Species Uses and Services

All three practices provide additional uses and services besides food. HG was found to be the most versatile, as it delivers 10 different groups of uses and services. Similar values were observed in Abebe [8], Senbeta et al. [39], and Méndez [46]. The types of uses and services by practice were cross-analyzed, e.g., food uses in HG were inversely correlated with the MTF size, i.e., as the size of MTF increases, the number of food crops in HG plots decreases (Table A1, Appendix A). According to the local farmers, this relation is common in the area because those households who produce a sufficiently large amount of cereals from their MTF are less interested in using their HG for growing food. Similarly, the production of 'stimulants', specifically *Coffea arabica*, in HG, decreases as the households have larger MCS plots (Table A1, Appendix A). Both cases show the complementarity that exists among the AFPs of Yayu.

Most uses and services except 'food', 'spices, condiments, and other food and beverage additives', and 'shade trees for coffee' were provided by all AFPs. Local farmers confirmed that 'fuel' is mainly a by-product of MCS pruning, weeding, thinning, and clearing. Meanwhile, in HG-specific species, such as *Vernonia amygdalina,* are cultivated solely for the production of fuel. The situation is similar with regard to 'fodder', where multipurpose trees supplement the hay obtained from the communal grazing lands [40,55].

On the other hand, some uses and services are limited to specific practices. For instance, 'shade trees for coffee' was observed only in MCS and HG, as *Coffea arabica* is hardly present in MTF. Similarly, 'other services' such as live fences and hedges are more important for MTF and HG than for MCS, as was reported for the area by Etissa et al. [56].

4.4. Food Production Potential

MTF and HG were found to be the main food-supplying practices. The number of all edible species identified (80) was considerably higher than in other areas with a similar ecological profile in Ethiopia, e.g., 23 by Senbeta et al. [39] in Yayu, 59 by Abebe [58] in HG of Sidama. However, differently from these studies, which considered only native species, the present study included both native and exotic species.

HG has the larger share of active foods species (82.5%) compared to 1.5% in MCS and 20% in MTF. The dominant active food species in MTF are *Zea mays, Sorghum bicolor, Eleusine coracana,* and *Eragrostis tef,* which are ingredients of the traditional food *Enjera* (main source of carbohydrates). Other second-order species are *Vicia faba* and *Pisum sativum,* which are leguminous providers of protein. Similar values were observed in the Jimma zone [52] and the upper Blue Nile basin [57].

Noticeable is the low supply of staple foods by MTF during the 'food gap', the window of time between seasons. This is generally filled with food cultivated in HG [40,53], including species such as *Ensete ventricosum,* which is available throughout the year [6–8,59], and *Colocasia antiquorum, Dioscoreaal alta,* and *Solanum tuberosum,* which are available during the 'food gap' specifically, and are complemented by leafy vegetables such as *Brassica oleraceae* spp. and *Brassica carinata* [40,55,60,61].

Concerning the presence of 'active-foods', MTF is dominated by cereals but lacks vegetables and fruits, which are mostly provided by HG, endowed by a broader diversity of active food species from different food groups, e.g., *Carica papaya, Prunus persica, Daucus carot, Cucurbita pepo, Capsicum frutescens, Brassica oleracea, Brassica carinata,* etc., which are also key sources of micronutrients.

4.5. Income Generation Potential

Household economic capability to acquire food in the market is key for the food access pillar [62,63]. As shown, 95% of the monetary income in Yayu comes from selling *Coffea arabica* harvested from MCS. This cash is used not only to buy food but also to cover other expenditures [40,54].

However, this high dependency on *Coffea arabica* creates a concomitant cash shortage. Alternatively, HG provides a diversity of merchantable products, e.g., dairy products, fruit, livestock, spices, and even other cash crops that are sold throughout the year, but especially during that 'cash gap'. The findings of Etissa et al. [56] confirmed this. Similarly, in MTF, with *Catha edulis* being harvested several times throughout the year, can generate a continuous flow of cash, and in the case of annual crops, whenever surplus is achieved it may also be sold. An interesting paradox occurs with *Eragrostis tef,* the most valued staple, which sometimes is sold to buy cheaper staples like *Zea mays* and *Sorghum bicolor.*

Regarding the species richness of each marketability category, most species cultivated in HG, such as *Catha edulis*, *Musa paradisiaca*, *Mangifera indica*, and *Rhamnus prinoides*, were actively sold in the area. Contrastingly, MCS and MTF were dominated by non-marketable tree species in number but not in area. This is likely due to the priority given to *Coffea arabica* in MCS, and to annual crops such as *Zea mays* and *Sorghum bicolor* in MTF. In general, the non-marketed species exceeded the marketable categories, suggesting a potential of improvement via conservation, transformation, and/or marketing.

4.6. Relationships among Household and Agroforestry Attributes

The correlation analysis conducted between attributes of the household and agroforestry practices (see Table A1, Appendix A) disclosed that altitude and proximity to market are the two most important household attributes, which influencing the species composition and richness of a given AFP in Yayu. As the traditional coffee production system causes minimum damage on the existing vegetation and as it exists in a biosphere reserve, the residual influence of the original vegetation on the current species composition of AFPs was expected. The original vegetation is mainly a result of the local topography and ecological factors. These results agree with those of Addi et al. [64], who focused on the correlation of natural vegetation composition and altitude in southwestern Ethiopia, and those of Bajigo et al. [17], who observed a similar association of altitude with woody species diversity in HG of the Wolayita zone.

Similarly, the market has a considerable influence on the type of species grown and their purpose of management under each AFP of Yayu. For instance, in Yayu, households located near to the market had better species richness than those farther away. In facts, both negative and positive influences of the market on different traits of AFPs have been reported. For instance, in line with the present study, the authors of Reference [8] reported a positive correlation between the species richness of agroforestry plots and proximity to markets in the Sidama zone in Ethiopia. In contrast, Reference [65], studying HG diversity in Indonesia, revealed that HG near to markets tended to be dominated by a few commercial crops. In Yayu, the major reason for the higher species richness in HG near to markets may be the dominance that *Coffea arabica* already has on MCS, while the others are purposefully managed to meet household and market demands.

Gender also correlates with the richness of some species groups, depending on the purpose of management, plot location, and labor demand of the species. For example, cultivating legumes and cereals in MTF plots requires higher cropping and guarding labor, which the female-headed households often lack. According to the local people, the females in those households are often widows or divorced mothers. These avoid labor-demanding crops in their MTF plots and convert a share to cash-generating tree species such as *Eucalyptus grandis* and give the rest to sharecroppers [53]. On the other hand, the HG plots of female-headed households were rich in food groups such as vitamin A-rich dark green vegetables and tuber and root crops, which are of great importance for the food security of these households during shortage times. A study conducted on the driving forces of changes in the structure of traditional HG agroforestry of southern Ethiopia reported a more significant relation between women and food crops than between women and cash crops grown in HG [66].

In general, the migrant and resettled households had AFP plots that were less rich in native useful species, including edible ones, than the native households, except for the actively-marketed species. This because they have relatively less knowledge about the type and uses of native plant species. According to the local people, the resettled households change the species of their AFP into merchantable exotic species more frequently than the natives. This implies that the impact of migration has a considerable impact on the environment, especially on non-marketable species. Lemenih et al. [67] confirmed the negative relationship between migration and environmental management as a lack of formal or informal structure and poor social capital with respect to the native environment.

4.7. The Five Pillars of Food Security

This study demonstrated that each AFP has considerable potential, but differentiated contribution toward the five pillars of FNS of small farming households of Yayu. MTF primarily contributes to the 'availability' pillar by serving as the main supply of the annual food. MCS is solely to contributor to the 'access' pillar by generating the annual cash the household uses to purchase food and other food-related inputs from the market. The HG plots of smallholders usually are used to fill the seasonal and unexpected food and cash supply gaps by producing crops with different harvesting seasons, which ultimately enhance the stability of the above two pillars. Furthermore, the diversity in all three AFPs plays an important role for the 'utilization' pillar by providing fruits, vegetables, fuel, and fodder, which ultimately increases the access of the household to diverse and healthy foods. The final pillar, 'sovereignty', is mainly assured by the HG as its management and selling of products are mostly controlled by the female family members, who manage the type of food prepared for the household.

5. Conclusions

Yayu agroforestry practices (AFPs) constitute a remarkable case study concerning potential local-based efforts to improve food and nutrition security (FNS). The local farmers use three different AFPs, namely homegarden (HG), multistorey-coffee-system (MCS), and multipurpose-trees on-farmland (MTF), with different spatial and temporal arrangements, structures, and compositions. Each has a main purpose, i.e., MTF for food production, MCS for cash generation, and HG for both. Inter- and intra-practice variations exist with respect to species composition and utilization. Making the best use of their differences, farmers manage and utilize these practices in a synchronized way to sustain their livelihoods.

Based on the correlation analysis, it is concluded that the species composition, structure, and mode of utilization of a given AFP in Yayu is a function of household attributes such as elevation, proximity to market, gender, and settlement history.

This study also acknowledges the existence of a knowledge gap regarding the detailed contributions of AFPs to the current FNS of smallholder farm households in Yayu. Thus, empirical research should assess the FNS status of smallholder farm households to relate this with the observed attributes of each AFP. As shown, the Yayu area is endowed with untapped resources of edible and marketable plants, whose contributions should be explored in depth, particularly of those within local AFPs, toward the enhancement of FNS as well as the living standards of smallholder farm households of Yayu and other similar areas.

Author Contributions: Conceptualization, O.J., D.C.-C., and M.v.N.; field work, analysis, drafting, O.J.; review and editing, O.J., D.C.C., and M.v.N.

Funding: This research was funded by the German Federal Ministries of Education and Research (BMBF) and of Cooperation and Development (BMZ), grant number FKZ 031 A258 A.

Acknowledgments: We are greatly indebted to Ato. Abdata W.; Ato. Alebacew T.; Ato. Bekele D.; Ato. Bekele M.; W/ro Bekabil Z.; Ato. Bekele Z; Ato. Fiqadu B.; W/ro Hora A.; Ato. Korsa T.; W/ro Nuri F.; Ato. Olana D for their cooperation during data collection. We thank also the ECFF (Environment and Coffee Forest Forum) and the Agricultural offices of Yayu and the surrounding Woreda—their support and facilitation were indispensable for realization of the field work.

Conflicts of Interest: The authors declare no conflict of interest.

Appendix

Table A1. Pearson correlation coefficients between attributes of three agroforestry practices and household characteristics in Yayu.

Attributes		Appendix A	Appendix A	Appendix A	Appendix A	Appendix A	Appendix A	Appendix A	Appendix A	Appendix A	Appendix A	Appendix A	Appendix A	Appendix A	Appendix A	Appendix A
Type of main purpose	HG	−0.16 **	−0.18 **	−0.03	−0.17 **	0.11	−0.17	0.13 *	−0.01	0.11	0.15 *	0.14 *	0.16 **	0.28 **	0.14 *	0.11
	MCS	−0.04	−0.05	0.01	−0.04	0.04	0.05	−0.01	−0.04	−0.03	0.04	0.14 *	0.18 **	0.07	0.17 **	0.09
	MTF	−0.25 **	−0.22 **	0.06	0.03	0.14 *	0.01	−0.01	−0.08	0.03	0.03	0.14 *	0.16 **	0.15 **	0.16 **	0.07
Number of benefits	HG	−0.24 **	−0.19 **	−0.06	−0.12 *	0.15 *	−0.2 **	0.17 **	−0.08	0.04	0.22 **	0.23 **	0.22 **	0.51 **	0.23 **	0.10
	MCS	−0.15 **	−0.15	0.09	−0.17 **	0.05	−0.02	0.04	0.01	0.06	0.20 **	0.3 **	0.34 **	0.25 **	0.34 **	0.15 **
	MTF	−0.31 **	−0.30 **	−0.02	−0.08	0.17 **	−0.14	0.11	−0.03	0.06	0.15 *	0.23 **	0.24 **	0.28 **	0.22 **	0.14 *
Total no. of species	HG	−0.33 **	−0.30 **	−0.12	0.01	0.08	−0.15 *	0.16 **	0.01	0.15 *	0.12 *	0.09	0.13 *	0.17 **	0.25 **	−0.12
	MCS	−0.27 **	−0.33 **	0.08	−0.08	0.06	−0.16 **	0.18 **	0.10	0.06	0.17 **	0.14 *	0.25 **	0.11	0.33 **	−0.01
	MTF	−0.13 *	−0.12	0.09	−0.10	0.05	−0.14 *	0.13 *	0.08	0.10	0.22 **	0.09	0.17 **	0.15 *	0.20 **	0.03
Number of native species	HG	−0.17 **	−0.09	−0.18 **	−0.02	0.04	−0.30 **	0.27 **	−0.01	0.18 **	0.07	0.09	0.10	0.17 **	0.17 **	−0.05
	MCS	−0.27 **	−0.33 **	0.08	−0.08	0.05	−0.16 **	0.18 **	0.10	0.06	0.17 **	0.14 *	0.24 **	0.11 *	0.32 **	−0.01
	MTF	−0.07	−0.07	0.11	−0.08	0.06	−0.14 *	0.12	0.08	0.13 *	0.21 **	0.07	0.15 *	0.12	0.18 **	0.02
Number of exotic species	HG	−0.36 **	−0.38 **	−0.02	0.03	0.08	0.05	0.00	0.02	0.06	0.12 *	0.06	0.11	0.11	0.23 **	−0.13 *
	MCS	−0.08	−0.09	0.07	0.02	0.07	0.00	0.03	0.05	0.02	0.07	0.08	0.14 *	0.05	0.23 **	−0.06
	MTF	−0.21 **	−0.19 **	−0.04	−0.10	−0.03	−0.04	0.04	0.01	−0.05	0.08	0.10	0.10	0.12	0.11	0.04
Number of species cultivated as food	HG	−0.26 **	−0.3 **	−0.02	0.08	0.13 *	−0.12 *	0.15 *	0.05	0.09	0.04	−0.04	0.04	−0.05	0.13 *	−0.17 **
	MTF	0.32 **	0.38 **	−0.16 *	−0.19 **	−0.02	−0.08	0.04	0.04	0.00	0.07	0.14 *	0.17 **	0.11	0.03	0.33 **
Number of species cultivated as stimulant	HG	−0.29 **	−0.26 **	0.09	−0.10	−0.02	−0.2 **	0.21 **	0.06	0.11	0.04	0.15 *	0.06	0.14 *	0.11	−0.06
	MCS	0.29 **	0.35 **	−0.09	0.13 *	−0.05	0.15 **	−0.18 **	−0.12 *	−0.06	−0.18 **	−0.11	−0.15 **	−0.11	−0.21 **	0.02
	MTF	−0.34 **	−0.34 **	0.12	−0.02	0.10	−0.08	0.11	−0.05	0.13 *	0.13 *	0.12	0.22 **	0.18 **	0.29 **	−0.10
Richness of edible species total	HG	−0.35 **	−0.36 **	−0.01	0.09	0.11	−0.10	0.11	0.00	0.09	0.05	−0.02	0.06	−0.01	0.17 **	−0.20 **
	MCS	−0.04	−0.08	0.02	0.06	−0.07	−0.03	0.05	−0.07	0.02	0.07	0.00	0.06	−0.06	0.13 *	−0.01
	MTF	0.15 *	0.13 *	−0.01	−0.13 *	0.02	−0.15 *	0.14 *	0.12	0.11	0.11	0.02	0.11	0.02	0.07	0.13 *
Richness of active-food species	HG	−0.34 **	−0.37 **	0.03	0.11	0.11	−0.11	0.11	−0.01	0.10	0.02	−0.06	0.02	−0.05	0.13 *	−0.21 **
	MTF	0.19 **	0.25 **	−0.12	−0.2 **	0.03	−0.12	0.09	0.02	0.07	0.12	0.19 **	0.26 **	0.18 **	0.15 *	0.30 **
Richness of actively-marketed species	HG	−0.35 **	−0.38 **	0.07	0.02	0.08	0.03	−0.03	−0.02	0.04	0.04	−0.12 *	0.05	0.00	0.16 *	−0.20 **
	MCS	0.25 **	0.29 **	−0.10	0.11	−0.03	0.11	−0.14 *	−0.13 *	−0.05	−0.18 **	0.02	−0.09	−0.10	−0.13 *	0.00
	MTF	−0.08	−0.05	−0.10	−0.14 *	0.13 *	−0.13 *	0.12 *	−0.04	0.04	0.17 **	0.22 **	0.32 **	0.27 **	0.27 **	0.26 **
Richness of cereals	HG	0.00	−0.04	−0.09	0.08	−0.12	−0.08	0.11	0.14 *	0.07	−0.05	−0.18 **	−0.25 **	−0.13 *	−0.13 *	−0.47 **
	MTF	0.28 **	0.27 **	−0.12	−0.14 *	0.02	−0.07	0.04	0.07	0.052	0.14 *	0.16 **	0.15 *	0.09	0.05	0.27 **
Richness of roots and tubers	HG	−0.08	−0.14 *	−0.11	0.15 *	0.19 **	−0.34 **	0.40 **	−0.05	0.16 **	0.02	−0.07	−0.01	−0.02	0.06	−0.03
Richness of legumes and nuts	HG	−0.01	−0.05	0.07	0.06	−0.03	−0.08	0.01	0.08	0.08	−0.09	−0.21 **	−0.21 **	−0.13 *	−0.12	−0.27 **
	MTF	0.14 *	0.25 **	−0.11	−0.15 *	−0.09	−0.07	0.04	0.03	−0.02	−0.05	0.10	0.06	0.04	−0.03	0.18 **
Richness of vitamin A-rich vegetables	HG	−0.01	0.10	0.08	−0.02	0.05	0.20 **	−0.19 **	−0.05	−0.10	0.03	0.03	0.05	0.01	−0.03	0.07
Richness of dark green leafy vegetables	HG	0.13 *	0.19 **	−0.05	0.02	−0.09	−0.12 *	0.13 *	0.01	0.04	−0.03	−0.13 *	−0.09	−0.14 *	−0.11	−0.05
Richness of vitamin A-rich fruits	HG	−0.26 **	−0.23 **	−0.10	−0.10	−0.03	0.01	−0.02	−0.05	0.05	0.05	0.06	0.09	0.11	0.13 *	−0.07
Richness of other vegetables	HG	0.07	0.06	0.18 **	0.02	0.00	0.39 **	−0.39 *	−0.14 *	−0.17 **	0.00	0.02	0.08	−0.01	0.08	0.06
Richness of other fruits	HG	−0.26 **	−0.38 **	−0.01	0.01	0.15 *	−0.10	0.11	0.07	0.06	0.09	0.03	0.18 **	0.09	0.26 **	0.02

* significant correlation at $p < 0.05$; ** significant correlation at $p < 0.01$; β nominal variables.

References

1. Garrity, D.P. Agroforestry and the achievement of the Millennium Development Goals. *Agrofor. Syst.* **2004**, *61*, 5–17. [CrossRef]
2. Hasan, M.K.; Alam, A.A. Land degradation situation in Bangladesh and role of agroforestry. *J. Agric. Rural Dev.* **2006**, *4*, 19–25. [CrossRef]
3. Kalaba, K.F.; Chirwa, P.; Syampungani, S.; Ajayi, C.O. Contribution of agroforestry to biodiversity and livelihoods improvement in rural communities of Southern African regions. In *Tropical Rainforests and Agroforests under Global Change*; Tscharntke, T., Leuschner, C., Veldkamp, E., Eds.; Springer: Berlin, Germany, 2010; pp. 461–476. ISBN 978-3-642-00493-3.
4. Bishaw, B.; Neufeldt, H.; Mowo, J.; Abdelkadir, A.; Muriuki, J.; Dalle, G.; Assefa, T.; Guillozet, K.; Kassa, H.; Dawson, K.; et al. *Farmers' Strategies for Adapting to and Mitigating Climate Variability and Change through Agroforestry in Ethiopia and Kenya*; Oregon State University: Corvallis, OR, USA, 2013.
5. Mbow, C.; Van Noordwijk, M.; Luedeling, E.; Neufeldt, H.; Minang, A.P.; Kowero, G. Agroforestry solutions to address food security and climate change challenges in Africa. *Curr. Opin. Environ. Sustain.* **2014**, *6*, 61–67. [CrossRef]
6. Abate, T.; Hiebsch, C.; Brandt, S.A.; Gebremariam, S. Enset-based sustainable agriculture in Ethiopia. In *Proceedings from the International Workshop on Enset*; Abate, T., Hiebsch, C., Brandt, S.A., Gebremariam, S., Eds.; IAR: Addis Abeba, Ethiopia, 1996.
7. Brandt, S.; Spring, A.; Clitton, H.; McCabe, J.T.; Tabogie, E.; Diro, M.; Woldemichael, G.; Yintiso, G.; Shigeta, M.; Tesfaye, S. *The "Tree against Hunger": Enset-Based Agricultural Systems in Ethiopia*; American Association for the Advancement of Science: Washington, DC, USA, 1997.
8. Abebe, T. Determinants of crop diversity and composition in Enset-coffee agroforestry homegardens of Southern Ethiopia. *J. Agric. Rural Dev. Trop. Subtrop.* **2013**, *114*, 29–38.
9. Van Noordwijk, M.; Bizard, V.; Wangpakapattanawong, P.; Tata, H.L.; Villamor, G.B.; Leimona, B. Tree cover transitions and food security in Southeast Asia. *Glob. Food Secur.* **2014**, *3*, 200–208. [CrossRef]
10. Ickowitz, A.; Powell, B.; Salim, M.A.; Sunderland, T.C.H. Dietary quality and tree cover in Africa. *Glob. Environ. Chang.* **2013**, *24*, 287–294. [CrossRef]
11. Jamnadass, R.; Place, F.; Torquebiau, E.; Malézieux, E.; Liyama, M.; Sileshi, G.; Kehlenbeck, K.; Masters, E.; McMullin, S.; Dawson, I. *Agroforestry, Food and Nutritional Security*; Working Paper No. 170; ICRAF: Nairobi, Kenya, 2013. [CrossRef]
12. Food and Agriculture Organization (FAO). *The State of Food and Agriculture. Biofuels: Prospects, Risks and Opportunities*; Food and Agriculture Organization: Rome, Italy, 2008.
13. Altieri, M.A.; Funes-Monzote, F.R.; Petersen, P. Agroecologically efficient agricultural systems for smallholder farmers: Contributions to food sovereignty. *Agron. Sustain. Dev.* **2012**, *32*, 1–13. [CrossRef]
14. Wilson, A. Household Food Security and Food Sovereignty: Framing the Future of Hunger and Agriculture. University Honor's Thesis, Portland State University, Portland, OR, USA, 2015.
15. Abebe, T. Diversity in Homegarden Agroforestry Systems of Southern Ethiopia. Ph.D. Thesis, Wageningen University, Wageningen, The Netherlands, 2005.
16. Mekonnen, E.L.; Asfaw, Z.; Zewudie, S. Plant species diversity of homegarden agroforestry in Jabithenan district, North-Western Ethiopia. *Int. J. Biodivers. Conserv.* **2014**, *6*, 301–307. [CrossRef]
17. Bajigo, A.; Tadesse, M. Woody species diversity of traditional agroforestry practices in Gununo watershed in Wolayitta zone, Ethiopia. *For. Res.* **2015**, *4*, 155. [CrossRef]
18. Mengitu, M.; Fitamo, D. Plant species diversity and composition of the homegardens in Dilla Zuriya Woreda, Gedeo Zone, SNNPRS, Ethiopia. *Plant* **2015**, *3*, 80–86. [CrossRef]
19. Poschen, P. An evaluation of the *Acacia albida*-based agroforestry practices in the Hararghe highlands of Eastern Ethiopia. *Agrofor. Syst.* **1986**, *4*, 129–143. [CrossRef]
20. Tesemma, M.N. The Indigenous Agroforestry Systems of the South-Eastern Rift Valley Escarpment, Ethiopia: Their Biodiversity, Carbon Stocks, and Litterfall. Ph.D. Thesis, University of Helsinki, Helsinki, Finland, 2013.
21. Asfaw, Z.; Ågren, G.I. Farmers' local knowledge and topsoil properties of agroforestry practices in Sidama, Southern Ethiopia. *Agrofor. Syst.* **2007**, *71*, 35–48. [CrossRef]

22. Gole, W.T.; Senbeta, F.; Tesfaye, K.; Fite, G. *Yayu Coffee Forest Biosphere Reserve Nomination Form*; Ethiopian MAB National Committee: Addis Ababa, Ethiopia, 2009.
23. Assefa, A.D. Local Institutions and Their Influence on Forest Resource Management in Southwest of Ethiopia: The Case of Yayu Forest. Master's Thesis, Addis Ababa University, Addis Ababa, Ethiopia, 2010.
24. Reliefweb. The Ethiopia Network on Food Security. 2002. Available online: http://reliefweb.int/sites/reliefweb.int/files/resources/9CF3EE6079D3BD7785256B61004C77EB-usaid_eth_14feb.pdf (accessed on 28 March 2017).
25. Famine Early Warning Systems Network (FEWS NET). Ethiopia: Food Security Update. 2005. Available online: http://www.fews.net/sites/default/files/documents/reports/Ethiopia_200503en.pdf (accessed on 28 August 2017).
26. Famine Early Warning Systems Network (FEWS NET). Ethiopia: Food Security Outlook. 2009. Available online: http://www.fews.net/sites/default/files/documents/reports/ethiopia_OL_04_2009_final.pdf (accessed on 28 August 2017).
27. Famine Early Warning Systems Network (FEWS NET). Ethiopia: Food Security Outlook. 2013. Available online: http://www.fews.net/sites/default/files/documents/reports/Ethiopia_OL_10_2013_0.pdf (accessed on 28 August 2017).
28. Famine Early Warning Systems Network (FEWS NET). Ethiopia: Food Security Outlook. 2015. Available online: http://www.fews.net/sites/default/files/documents/reports/Ethiopia_OL_2015_10_0.pdf (accessed on 28 August 2017).
29. Famine Early Warning Systems Network (FEWS NET). Ethiopia: Famine Early Warning Systems Network. 2017. Available online: http://www.fews.net/east-africa/ethiopia (accessed on 28 August 2017).
30. Taye, E. *Report on Woody Plant Inventory of Yayu National Forestry Priority Area*; IBCR/GTZ: Addis Ababa, Ethiopia, 2001; pp. 2–21.
31. Yeshitela, K.; Bekele, T. Plant community analysis and ecology of Afromontane and transitional rainforest vegetation of southwestern Ethiopia. *SINET Ethiop. J. Sci.* **2002**, *25*, 155–175. [CrossRef]
32. Gole, W.T. Vegetation of the Yayu Forest in SW Ethiopia: Impacts of Human Use and Implications for In Situ Conservation of Wild Coffea arabica, L. Populations. Ph.D. Thesis, University of Bonn, Bonn, Germany, 2003.
33. Senbeta, F.; Schmitt, C.; Denich, M.; Demissew, S.; Velk, P.L.; Preisinger, H.; Woldemariam, T.; Teketay, D. The diversity and distribution of lianas in the Afromontane rain forests of Ethiopia. *Divers. Distrib.* **2005**, *11*, 443–452. [CrossRef]
34. Senbeta, F.; Denich, M. Effects of wild coffee management on species diversity in the Afromontane rainforests of Ethiopia. *For. Ecol. Manag.* **2006**, *232*, 68–74. [CrossRef]
35. Gole, W.T.; Borsch, T.; Denich, M.; Teketay, D. Floristic composition and environmental factors characterizing coffee forests in southwest Ethiopia. *For. Ecol. Manag.* **2008**, *255*, 2138–2150. [CrossRef]
36. Ettisa, T.A. Diversity of Vascular Epiphytes along Disturbance Gradient in Yayu Forest, Southwest Oromia, Ethiopia. Master's Thesis, Addis Ababa University, Addis Ababa, Ethiopia, 2010.
37. Tafesse, A. Agroecological Zones of South Western Ethiopia. Ph.D. Thesis, University of Trier, Trier, Germany, 1996.
38. Muleta, D.; Assefa, F.; Nemomissa, S.; Granhall, U. Socioeconomic benefits of shade trees in coffee production systems in Bonga and Yayu Hurumu districts, southwestern Ethiopia: Farmers' perceptions. *Ethiop. J. Educ. Sci.* **2011**, *7*, 39–55.
39. Senbeta, F.; Gole, T.W.; Denich, M.; Kellbessa, E. Diversity of useful plants in the coffee forests of Ethiopia. *Ethnobot. Res. Appl.* **2013**, *11*, 49–69. [CrossRef]
40. Jemal, O.M.; Callo-Concha, D. *Potential of Agroforestry for Food and Nutrition Security of Small-Scale Farming Households: A Case Study from Yayu, Southwestern Ethiopia*; Working Paper 161; ZEF, University of Bonn: Bonn, Germany, 2017.
41. Israel, G.D. *Determining Sample Size*; Fact Sheet PEOD-6; Florida Cooperative Extension Service, Institute of Food and Agricultural Sciences, University of Florida: Gainesville, FL, USA, 1992.
42. Mooney, H.F. *Glossary of Ethiopian Plant Names*; Dublin University Press: Dublin, Germany, 1963.
43. Mesfin, T.; Hedberg, I. *Flora of Ethiopia and Eritrea*; National Herbarium, Addis Abeba University: Uppsala, Sweden, 1995; Volume 2, pp. 71–106.

44. Bekele, A. *Useful Trees and Shrubs for Ethiopia: Identification, Propagation and Management for 17 Agroclimatic Zones*; World Agroforestry Center: Nairobi, Kenya, 2007.

45. Teketay, D.; Senbeta, F.; Barklund, P. *Edible Wild Plants in Ethiopia*; Addis Ababa University Press: Addis Ababa, Ethiopia, 2010.

46. Mendez, E. An assessment of tropical homegarden as example sustainable local agroforestry systems. In *Agroecosystem Sustainability: Developing Practical Strategy*; Gliessman, S.R., Ed.; CRC Press: Boca Raton, FL, USA, 2000; pp. 51–66.

47. Molla, E.L.; Asfaw, Z.; Kelbessa, E.; Van Damme, P. Wild edible plants in Ethiopia: A review on their potential to combat food insecurity. *Afr. Focus* **2011**, *24*, 71–121.

48. Kennedy, G.; Ballard, T.; Dop, M.C. *Guidelines for Measuring Household and Individual Dietary Diversity*; Food and Agriculture Organization of the United Nations: Rome, Italy, 2011.

49. Marrugan, A.E. *Ecological Diversity and Its Measurement*; Chapman & Hall: London, UK, 1988; pp. 154–196. ISBN 978-94-015-7360-3.

50. Minitab Inc. Minitab 17 Statistical Software. 2013. Available online: www.minitab.com (accessed on 28 August 2015).

51. Nair, P.K.R. Classification of agroforestry systems. *Agrofor. Syst.* **1985**, *3*, 97–128. [CrossRef]

52. Kebebew, Z.; Urgessa, K. Agroforestry perspective in land use pattern and farmers coping strategy: Experience from southwestern Ethiopia. *World J. Agric. Sci.* **2011**, *7*, 73–77.

53. Bekele, D.; (Weyira *kebele*, Elemo, Ethiopia); Bekele, M.; (Environment and Coffee Forest Forum, Yayu, Ethiopia); Bekabil, Z.; (Werebo *kebele*, Elemo, Ethiopia); Korsa, T.; (Oromiya forest and wild life enterprise, Wutete, Ethiopia). Personal communication, 2014.

54. Hora, A.; (Elemo *kebele*, Elemo, Ethiopia, Ethiopia); Fiqadu, B.; (Beteli Gebecha *kebele*, Elemo, Ethiopia); Nuri, F.; (Sololo *kebele*, Qumbabe, Ethiopia); Alebacew, T.; (Environment and Coffee Forest Forum, Yayu, Ethiopia); Bekele, Z.; (Wangene *kebele*, Hurumu, Ethiopia). Personal communication, 2014.

55. Abdata, W.; (Weyira *kebele*, Elemo, Ethiopia); Alebacew, T.; (Environment and Coffee Forest Forum, Yayu, Ethiopia); Bekele, M.; (local expert, Wutete, Illubabor, Ethiopia); Olana, D.; (Environment and Coffee Forest Forum, Wabo, Ethiopia). Personal communication, 2014.

56. Etissa, E.; Gole, T.; Teshome, A.; Abebie, T.G. Horticultural crops diversity and cropping in the smallholders home gardens in the transitional area of Yayu Coffee Forest Biosphere Reserve, Ethiopia. In Proceedings of the Conference on International Research on Food Security, Vienna, Austria, 18–21 September 2016.

57. Agidie, A.; Ayele, B.; Wassie, A.; Hadgu, K.M.; Aynekulu, E.; Mowo, J. Agroforestry practices and farmers' perception in Koga watershed, upper blue Nile basin, Ethiopia. *Agric. For.* **2013**, *59*, 75–89.

58. Abebe, T.; Wiersum, K.F.; Bongers, F. Spatial and temporal variation in crop diversity in agroforestry HGs of southern Ethiopia. *Agrofor. Syst.* **2010**, *78*, 309–322. [CrossRef]

59. Negash, A.; Niehof, A. The significance of enset culture and biodiversity for rural household food and livelihood security in southwestern Ethiopia. *Agric. Hum. Values* **2004**, *21*, 61–71. [CrossRef]

60. Asfaw, Z. Conservation and use of traditional vegetables in Ethiopia. In *Traditional Vegetables in Africa, Proceedings of the IPGRI International Workshop on Genetic Resources of: Conservation and Use, Nairobi, Kenya, 29–31 August 1995*; Guarino, L., Ed.; ICRAF: Nairobi, Kenya, 1997.

61. Ebert, A.W. Potential of underutilized traditional vegetables and legume crops to contribute to food and nutritional security, income and more sustainable production systems. *Sustainability* **2014**, *6*, 319–335. [CrossRef]

62. Riely, F.; Mock, N.; Cogill, B.; Bailey, L.; Kenefick, E. *Food Security Indicators and Framework for Use in the Monitoring and Evaluation of Food Aid Programs*; Nutrition Technical Assistance Project (FANTA)/U.S. Agency for International Development (USAID): Washington, DC, USA, 1999.

63. Food and Agriculture Organization (FAO). *Integrated Food Security and Humanitarian Phase Classification (IPC) Framework*; Policy Brief Issue 3; Food and Agriculture Organization: Rome, Italy, 2006.

64. Addi, A.; Soromessa, T.; Kelbessa, E.; Dibaba, A.; Kefalew, A. Floristic composition and plant community types of Agama Forest, an Afromontane Forest in Southwest Ethiopia. *J. Ecol. Nat. Environ.* **2016**, *8*, 55–69. [CrossRef]

65. Wiersum, K.F. Diversity and change in homegarden cultivation in Indonesia. In *Tropical Homegardens*; Kumar, B.M., Nair, P.K.R., Eds.; Springer: The Netherlands, 2006; pp. 13–24.

66. Gebrehiwot, M. Recent Transitions in Ethiopian Homegarden Agroforestry: Driving Forces and Changing Gender Relations. Licentiate Thesis, Swedish University of Agricultural Sciences, Umeå, Swedish, 2013.

67. Lemenih, M.; Kassa, H.; Kassie, G.T.; Abebaw, D.; Teka, W. Resettlement and woodland management problems and options: A case study from north-western Ethiopia. *Land Degrad. Dev.* **2014**, *25*, 305–318. [CrossRef]

 © 2018 by the authors. Licensee MDPI, Basel, Switzerland. This article is an open access article distributed under the terms and conditions of the Creative Commons Attribution (CC BY) license (http://creativecommons.org/licenses/by/4.0/).

 *sustainability*

Article

Family Farms, Agricultural Productivity, and the Terrain of Food (In)security in Ethiopia

Till Stellmacher and Girma Kelboro *

Center for Development Research (ZEF), University of Bonn, Genscherallee 3, 53113 Bonn, Germany;
t.stellmacher@uni-bonn.de
* Correspondence: gmensuro@uni-bonn.de; Tel.: +49-228-73-4917

Received: 30 September 2018; Accepted: 7 September 2019; Published: 12 September 2019

Abstract: Despite economic development and social improvements, millions of family farmers in Ethiopia are still struggling with food insecurity. Lack of technology adoption by family farmers is often considered as the root cause for low agricultural productivity and persistence of food insecurity. Based on a study of family farms in southwestern Ethiopia, we show the complex nexus between family farming, food insecurity, and agricultural productivity. We collected qualitative and quantitative data through 300 sample household interviews; expert interviews with elders and village chairmen, agricultural extension agents, farmers' cooperative heads, as well as experts in NGOs, research institutes, and state agencies; and on-farm observations with in-depth interviews and discussions with individual farmers. Our findings illustrate that everyday experiences, culture, knowledge, and priorities of farmers coupled with ecological and political factors play crucial roles—and need more consideration than the classic 'lack of technology' theorem.

Keywords: family farming; farmland; food security; rural development

---

1. Introduction

Global demand for food is expected to massively increase until 2050 [1]. The business-as-usual scenarios will even be worse for countries in Africa, the continent with the strongest population increase where a quarter of its people suffer from severe food insecurity. For 'food security', we adopt the definition developed at the World Food Summit in 1996 [2]: "food security exists when all people, at all times, have physical and economic access to sufficient, safe and nutritious food to meet their dietary needs and food preferences for a healthy and active life." Strengthening the local productivity of the agricultural sector is envisaged a way to decrease food insecurity [3]. At the same time, agriculture is under pressure as it increasingly competes with other sectors for land, water, energy, and labor. Consideration of environmental concerns in agriculture is becoming more and more important. Climate change has become another dimension to consider [4]. While the agronomic and technical questions of how to produce more of what kind of food in which farming systems are well considered in science and extension work, the socio-cultural frames and local capacities of family farms to manage productivity have long been undervalued and are not yet fully understood [5]. According to FAO [6], "Family farming (which includes all family-based agricultural activities) is a means of organizing agricultural, forestry, fisheries, pastoral and aquaculture production which is managed and operated by a family and predominantly reliant on family labor, including both women's and men's." Family farms constitute 98% of all farms worldwide [7]. In the last years, the approach to and discourse on food security shifted from a national "food first" perspective to a more integrated local "livelihood" perspective and from "objective indicators" to "subjective perception".

Ethiopia has been struggling to feed its growing population in the last decades. In the 1970s/80s, the country became internationally synonymous for drought and famine. Despite generally positive economic achievements in the last two decades, food insecurity still affects about 40–50% of Ethiopia's population [8,9]. Most food insecure people in Ethiopia are those who produce food themselves, namely the millions of family farmers in the countryside. Family farms employ 81% of the total population and produce up to 96% of the total agricultural GDP of the country [10]. More than 40% of Ethiopian family farmers are food deficient [11]. Relevant policies and practices to address the problem in an integrated and locally applicable manner require a solid understanding of the problem context, potentials, and challenges.

Ethiopia is on the eve of a deep agrarian transformation. Large-scale investments into agriculture are being made, and technical innovations such as the adoption of 'modern' fertilizer and 'improved' seed helped to increase agricultural productivity. However, factors such as shortage of land, land degradation, and effects of climate change hold family farmers into food insecurity circles. When it comes to reasons for a low technology adoption, factors such as a lack of purchasing power, lack of a formal credit market, and a lack of knowledge and (formal) education are often mentioned [12–16]. Underlying cultural and socio-psychological concerns that feature the everyday lives of farmers are, however, only rarely brought into the picture.

The Ethiopian state regards agriculture as a major strategic sector in which significant investments have been made. Efforts to improve agricultural productivity focuses on state-run extension work with rather technical and top-down approaches. The country has trained and employed around 50,000 local agricultural extension workers, known as Development Agents (DAs), the highest number in Africa and the fourth largest in the world next to China, India, and Indonesia [17]. The effectivity and efficiency of the extension work, however, proved to remain very low over decades [18].

In this paper, we argue that agricultural productivity and food (in)security should be understood in the context of complex family farming environments. The importance of contextualizing the support towards food security in Ethiopia is underlined [19]. We illustrate and support consideration of farmers' realities on the ground which include cultural and socio-psychological constraints and tradeoffs in research, policy, and practices to realize food security in the country. This paper is an outcome of a study which illustrates agricultural productivity and food (in)security from a bottom-up perspective of family farming households in the Yayu area, Southwest Ethiopia.

2. Agriculture in Ethiopia, Productivity, and the Drive for Food Security

2.1. Agricultural Policies and Strategies

Family farms and food (in)security in Ethiopia are situated in a web of interactions in which locally embedded daily practices and perceptions interact with top-down rural and agricultural policies and strategies which define and interpret agriculture in the context of national economic development. In addition to policies addressing rural poverty, most notably the Sustainable Development and Poverty Reduction Program (SDPRP) and the Plan for Accelerated and Sustained Development to End Poverty (PASDEP), Ethiopia has adopted a specific policy response to agricultural productivity and food insecurity, the Agricultural Development-Led Industrialization (ADLI) strategy [20]. The key assumption is that agriculture can develop rural areas by enhancing purchasing power of the rural population, which are mainly family farming households. Through commercialization, it is envisaged that the growth in agriculture will then spur growth in the industrial sector [21]. Subsequently, its Food Security Program (FSP) was incorporated in the national poverty reduction strategy [22]. The FSP contains the Productive Safety Net Program (PSNP) which aimed at enabling family farming households chronically vulnerable to food insecurity to resist shocks, create assets, and become permanently food secure. PSNP uses a combination of income generating measures and capacity building programs together with cash and food transfers based on individual needs, primarily in the most food insecure seasons [23].

The general state approach to increase food security of family farming households in Ethiopia is to 'modernize' farming and increase productivity. To archive this aim, two main paths are followed, namely technologization and specialization. Technologization largely brings about new inputs such as farming equipment, 'improved' seeds and chemical inputs (fertilizer, herbicides, fungicides, insecticides). Specialization focuses on few crops and practices considered best suited to a certain agro-ecological zone. In the national strategy, the Yayu area belongs to the southwestern region of the country which is defined to be best suited for the production of coffee, spices, and maize.

2.2. Analytical Framework

The classic path to achieve food security in Africa is by adoption of agricultural technologies and enhancing agricultural productivity [24]. The types of technologies selected and introduced are based on the 'success' of the green revolution in Asian and Latin American countries. These countries intensified agricultural production by using 'improved' seeds, fertilizer, and agro-chemicals and increased the amount of yield per hectare. Farmers were sensitized primarily through centrally-planned interventions with the state being the main player in the process. The programs were supported by big international donors such as the World Bank and IMF. Emphasis was given to training and capacity building of farmers to enable technology use. Generally, this technology-driven approach enabled Asian and Latin American countries to massively increase food security.

In the Ethiopian context, family farming comprises mixed farming practices which involve crop and livestock production. Agricultural Development-Led Industrialization (ADLI) is the policy direction which guides the national development in which family farming is situated. Ethiopia considers land and cheap labor as its comparative advantage to build its 'rapid' economic growth [20]. This is the main reason for the country to justify its ADLI. Rural and agricultural development policies are framed around ADLI (see Figure 1 for illustration of how the policies are linked to family farming and then to food security). ADLI premises on increasing agricultural productivity to achieve household food security on the one hand and production of industrial and export crops on the other. With this, agriculture is expected to play a leading role in household food security and national economic growth. Intensification of family farms through extension programs aims at household food security while large-scale investments are promoted to produce export and industrial products [25].

In this framework, Ethiopia puts emphasis on the classical agricultural innovation approach with a massive agricultural extension program aiming to transfer technological packages to family farms in order to promote their technologization and specialization. All across the country, Development Agents (DAs) provide farmers with a package of training, 'improved' seeds, and chemical inputs based on the specialization recommended for the region [26].

Family farming households, on their part, have locally-based and culturally-embedded ways of farming and social life [27]. Family farming in Ethiopia is far from mechanization and is often carried out through labor-intensive and locally developed farm tools such as oxen-drawn ploughs. Family farming is much more than an economic activity. It is also the social and cultural basis of livelihoods. It provides identity, living spaces, cultures of solidarity and collaboration, social resilience, and insurance systems [28]. Figure 1 below provides the analytical framework. The arrows in the figure show the direction of influence or feedback loops.

In reality, the state-supported extension services interact—and often clash—with the social and cultural make-up of local family farming systems.

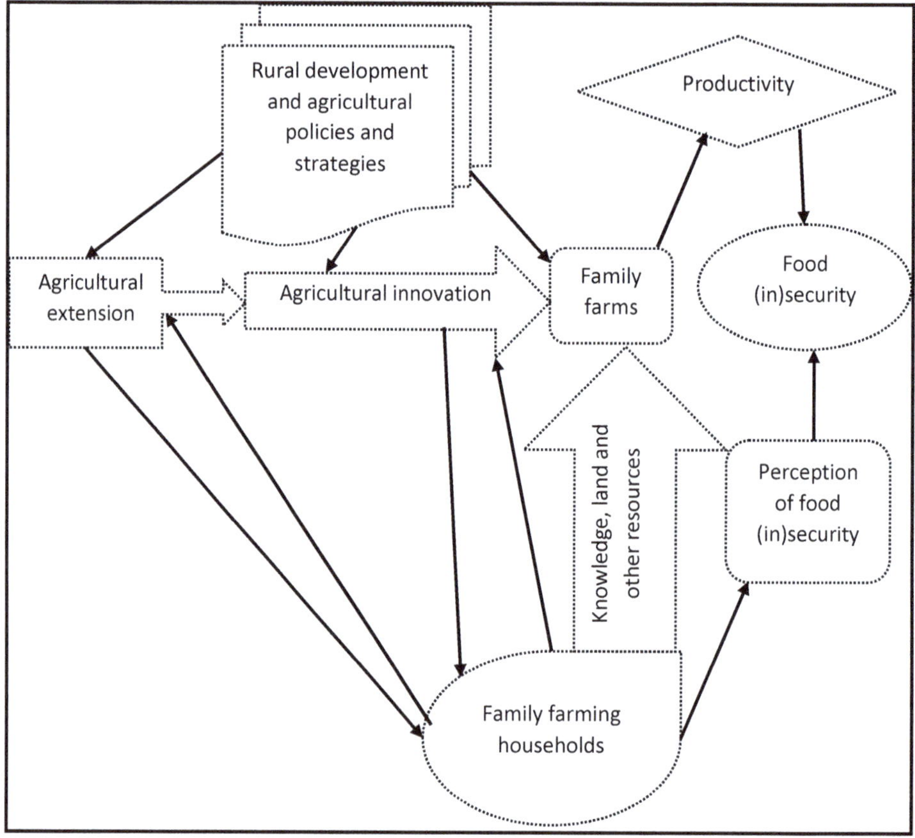

Figure 1. Analytical framework for food (in)security in Ethiopia.

3. Materials and Methods

A bottom-up perspective on family farming, agricultural productivity and food (in)security is shown by using the Yayu area, Oromia Regional State, in southwestern Ethiopia as a case. The empirical research was conducted as part of the BiomassWeb WP 2.2 together with WP 4.3. The Yayu area was chosen as a BiomassWeb case study jointly with all partners in a workshop in 2014. The study was conducted in eight *kebeles* (the smallest administrative unit in Ethiopia) (Wabo, Wutete, Sololo, Wangene, Weyira, Werebo, Beteli, Gebcha and Elemo) located in six *woredas* (districts which are divided in *kebeles*) (Yayu, Dorani, Hurumu, Bilo-Nopha, Alge-Sachi, and Chorra) in 2014/15. Selection of these *kebeles* and sample households followed stratification based on access to market in agreement with Jemal and colleagues [29]. The purposive selection of interview partners was combined with random sampling for qualitative and quantitative data collection. Three hundred family farmers were interviewed with semi-structured questionnaires; on-farm observation and discussions were carried out (Figure 2); focus group discussions were held, and expert interviews conducted with elders and village chairmen, agricultural extension agents, farmers' cooperative heads, as well as experts in NGOs, research institutes, and state agencies. Combination of the different data collection tools provided an in-depth and a broader understanding of the family farming practices, opportunities, and challenges on the local level.

Figure 2. A farmer showing his farm during on-farm discussion.

4. Results

4.1. Characteristics of the Study Site and Family Farming Households in Yayu

The study area is situated in southwestern Ethiopia at an altitude between 1140 and 2562 m.a.s.l. The agro-ecological zone is mid-altitude locally known as Woina Dega. The area receives high rainfalls (on average about 2100 mm/year) and is covered by some tracts of Afromontane rainforest. The majority of the households are ethnically Oromo. The area is relatively well accessible via a paved road from Belele to Metu and can be reached from Addis Ababa by car in one day.

Most family farms in the study area are headed by males (84%) who are on average 44.3 years old. The average family farming household size is 5.2 persons. Our interviewees perceive that the main problems of their farm are lack of land, high costs of agricultural inputs, livestock diseases, and wildlife impact on crops and livestock. Farmlands are primarily obtained through state allocation through the local state (kebele) administration (mentioned by 45% on the question how they got the largest part of their land); 43% mention inheritance, 6% sharecropping, and 2% purchase. The current average farmland holding size is around 1 ha., and 17% of the farmers are landless. The main ways to increase or obtain farmland are sharecropping, land renting, purchase, and entering into the forest.

Generally, family farms in the Yayu area use complex combinations of a large number of different annual and perennial crops, horticulture, agro-forestry, and livestock. Most of the farmers produce both cash and food crops. Coffee is unequivocally the main cash crop in the area. The average income from coffee per year is 11,715 Ethiopian Birr per household. Vegetables and fruits are intercropped around the homesteads for easy management and control against wild animals.

Seasonal food insecurity is a major concern for family farms in the Yayu area. 37% of all sample family farming households face food shortages each year. Most critical is June, July, and August. These months are shortly before the harvesting time, when the stocked food runs low. During times of food insecurity, families eat mostly only once a day, instead of three times a day.

Family farming in the Yayu area is not only about food production, but also provides identity. Those who have lived there for long belong to the area and the social and ecological system there. People know each other well. The young strongly believe that they have a legitimate right to inherit the property from their parents.

4.2. Defining Food (In)security: Global Standards and Local Culture

Food (in)security differs in its external (global) and internal (local) understanding and connotation. The global standards to measure food (in)security are related to the extent of access to safe food of the required quality and quantity at all times in line with the definition formulated after the 1996 World Food Summit in Rome. Food quality implies nutrient content of the food consumed, whereas access implies capacity to have the required food through production and/or purchase.

In the Yayu area, food (in)security is culturally inculcated in a different way. Food security is defined by having the main traditional staple food to eat three times a day. The types of food served also vary depending on the time of the day and age of a family member. In the mornings, adults usually drink coffee and eat roasted maize, barley, or pieces of bread. Lunch is usually eaten at home or on the farmland in a short break. Nevertheless, lunch time is commonly considered as the time for a main dish which constitutes mainly maize bread and vegetables. Similar staples are eaten for dinner. Milk is usually part of the food provided to children in families who are said to be relatively well off. Generally, those who depend on cabbage, vegetables, and root crops are traditionally considered poorer families, whereas those who consume animal products are considered richer families.

4.3. Local Resources: Farming Knowledge and Land Acquisition

Family farming in the Yayu area is an age-old practice inherited through generations. A typical family farm in Yayu consists of a coffee field, a plot for cereal production, a homestead garden with vegetables, forest products such as honey and fruits, and some livestock. This high diversity is crucial for the resilience of family farming households in times of food insecurity. A farmer (interviewed on 1 November 2014), for example, produces the following crops: coffee, maize, sweet potato, potato, chat, apple, avocado, banana, papaya, lemon, mango, orange, tomato, and cassava. He emphasized the importance of having diverse crops: "I collect about 8000–10000 Kgs and 6000–7000 Kgs of sweet potato and potato respectively. The problem with tomato is that, sometimes, leaf disease damages it. But I always have another option. Income from sweet potato and coffee complements each other. I harvest about 35–100 quintals of coffee every year."

However, diversity is not accommodated in state extension strategies which focus on specialization. In all national and regional rural and agricultural development programs the Yayu area is considered to be specialized in the production of coffee, spices, and maize. In an interview, an agricultural extension officer in Yayu (in October 2014) mentioned that the Yayu area is not an area of attention for national specialization in maize. On the one hand, the officer further noted that the specialization program is not fully operationalized though there are some efforts to support farmers in their production of coffee and spices. On the other hand, there is latent hesitance among the district officers in promoting maize in the area. In practice, however, the national specialization program shows little flexibility and does not pay tribute to particular local practices and needs.

Experiences and intragenerational and intergenerational knowledge exchange is vital for family farming. Farmers evaluate new technologies and compare between different introduced and local options (Figure 3). Based on their assessments, farmers can for example show preference to one 'improved' seed variety over another. The statement of one farmer (interviewed on 18 September 2014) is exemplary for such comparisons:

"The maize varieties known as 660, 661 and Shone are introduced to us by the extension service. We prefer Shone and 661. 660 variety does not give a good yield. However, we want to continue producing the indigenous variety as well, though not promoted or supported by the extension. We can sow it earlier with compost, without artificial fertilizer brought to us through the extension service. There is no need for us to wait until the seed comes as it is in our hands. Early seeding means early availability for consumption of the new (pre-mature) maize. It also produces a bigger size maize. One head of maize produces more than a glass of maize cereals. Further, it withstands dry weather conditions and produces good quality flour."

Figure 3. A farmer in Yayu showing his preferred maize variety.

Farmers also have different perceptions regarding the effect of fertilizer and 'improved' seeds on yield. Although they have been trained that the technologies increase crop yield, there are variations among farmers in associating higher yield with technology use. For 40% of the farmers in the Yayu area, the maize yield did not increase between the years 2008 and 2013.

The main problem of family farms in the Yayu area is a lack of farmland, mainly for cereal and coffee production. Over generations, the plots inherited from fathers to their children had become smaller and smaller. At the time of our study, however, the *kebele* administration told us that they had no more agricultural land available to be allocated to a new generation of farmers.

Under these conditions, family farmers see farming in an ambivalent way. Although it is likely to remain a key employer and 'way of life' also for the next generation, most farmers do not perceive family farming as a recommendable and sustainable activity that is worth investing in. Our respondents were largely unhappy about depending on farming for their livelihood. Therefore, they do not recommend it as a future economic option for their children. They associate farming with 'rural life' characterized by poverty, as well as poor transport facilities, markets, health services, and schools. Living in urban centers is preferred for the diversity of income opportunities and the possibility to generate income in a permanent position throughout the year. Urban life is considered 'clean' and 'modern'. Family farming households largely perceive that achieving 'better living conditions' is not possible through farming. Here is an illustrative statement from a farmer during our interview: "I have never advised my children even for one day to become farmers; I advise them to go to school and lead a better 'urban' life." (interviewed on 6 October 2014). Mostly those who have not attended school have been considered to have a very limited economic opportunity and have farming as the only possibility for their livelihood. Another farmer regrets that his life ended up as a farmer: "I have become a farmer since I did not go to school" (interviewed on 7 December 2014). Over a long time in Ethiopia, urban dwellers used to associate farming with backwardness and use the term *gebere* (farmer in Amharic) to insult people from rural communities. Most farmers in the Yayu area do not see the future of their children in family farming, given the experiences from their ancestors, the current conditions and own future outlook. 77% of the interviewed farmers said that they do not recommend family farming as a main source of income for their children in the future. This means that motivations for long term and substantial investments in farm productivity are often weak. In this line, being asked what family farmers would do if they would suddenly win 10,000 Birr in a lottery, direct investments in farm productivity were rarely mentioned. Instead, many farmers wanted to use the money to take a step out of farming. The two most often mentioned answers were that farmers wanted to use the 10,000 Birr prize money to 'build a house in town' or 'build a business center in town'. There are certainly many family farmers who would eagerly like to invest more in their farms' productivity and many already successfully do it, however, the socio-psychological aspects that limit the intrinsic motivations of family farmers to invest long term and substantially into their own farm have to be recognized. This can also be used to increase the effectivity and efficiency of the agricultural extension service.

5. Discussion, Conclusions, and Recommendations

Differences in understanding food security between Ethiopian family farmers and the global development actors show different underlying perceptions and language. In the Yayu area and elsewhere in Ethiopia, farmers perceive food insecurity as the inability to have the usual food during common dining times. The food composition also differs depending on the area, culture, season, and time of the day. The global food security discourse should, therefore, be broken down to the local level and realities. The work on the understanding of the concept should then be supported by approaches that engage the family farmers in using technologies and knowledge developed through joint efforts.

In the classic technology transfer approaches, farmers are rather considered as passive recipients. They are expected to learn from DAs and implement the agricultural technological package as prescribed and trained by the DAs. To increase agricultural productivity, extension works attempt to replace age-old practices of diversification and intercropping by specialization and intensification. Farm intensification is a characteristic feature of technologization. This emanates from the assumption that the problem of lower productivity is a consequence of lack of agricultural knowledge and technology especially among family farmers. Our results showed that such a linear cause–effect relationship is not possible given the complexity of the local realities. Low agricultural productivity is an outcome of an interplay of multiple factors, which include lack of understanding of the culture, knowledge and needs of farmers, and the underlying problems of land shortage, market access, and ineffective extension approaches.

Generally, the path of top-down technologization and specialization remains questionable as it often brings negative consequences to both farmers and the environment. After the introduction of 'improved' seed varieties in the Yayu area, for example, farmers have lost the culture of keeping the land races. As a result, they have become dependent on 'improved' varieties, although their seeds are often not available at the appropriate time through the extension system nor through private seed supplying companies.

Specialization is also problematic. Farmers in the Yayu area produce multiple crops, vegetables, and fruits throughout the year. Such a diverse production system is a security mechanism for food and nutrition security. Specialization on 1 ha of land, therefore, does not allow for integrated intra- and inter-family farming safety nets. The environmental consequences of agricultural specialization, from land erosion to water shortage, are widely recognized through the world.

'Farming' and 'rural life' have not been sufficiently supported in Ethiopia over generations. New concepts of rural life—and rural economy beyond family farming—are needed. Our finding goes in line with a study which shows that smallholder farming in developing countries has a structural image problem and a lack of societal recognition [30]. Especially the young and better educated farm household members often view agriculture as a risky and laborious venture that does not pay well—and is not prestigious.

Family farms in the Yayu area illustrate an interesting bottom-up perspective that contributes to the debate on food insecurity and agricultural productivity. Family farmers traditionally follow a multifunctional agriculture for subsistence and income. The high diversity of products produced strengthens the resilience of family farmers towards food insecurity. However, national and regional agricultural strategies and extension services are rather directed to product specialization and technical approaches.

Our findings further show that the nature of family farming is not static. Cultures and identities in rural Ethiopia are rapidly changing. Many farmers and their younger relatives do not see their (and their children's) future and career in agriculture. Some have little intrinsic motivation to invest long term and substantially into the productivity of their farm. Upon this backdrop, prevailing assumptions that one has to help family farmers to specialize and increase their productivity to attain food security, which will then spur sustainable rural development, need to be challenged. A new dimension on the future of family farming is needed. Specific focus needs to be given to the question of what will happen inter-generationally in few years' time when hundreds of millions of family farmers—not only in Africa but across the globe—will pass their farms over to their children in a few years.

Author Contributions: Conceptualization and writing—original draft preparation, G.K. and T.S.; Writing—review and editing, G.K. and T.S.

Funding: This research was funded by the German Federal Ministry of Education and Research (BMBF) and Federal Ministry of Economic Cooperation and Development (BMZ) through the BiomassWeb project (www.biomassweb.org), grant number FKZ 031 A258 A.

Acknowledgments: We would like to acknowledge the support for the field research we received from the Environment and Coffee Forest Forum (ECFF). We are also grateful to the farmers and local government officers for their willingness to participate in the study.

Conflicts of Interest: The authors declare no conflict of interest. The funders had no role in the design of the study; in the collection, analyses, or interpretation of data; in the writing of the manuscript, or in the decision to publish the results.

References

1. Valin, H.; Sands, R.D.; van der Mensbrugghe, D.; Nelson, G.C.; Ahammad, H.; Blanc, E.; Bodirsky, B.; Fujimori, S.; Hasegawa, T.; Havlik, P.; et al. The future of food demand: Understanding differences in global economic models. *Agric. Econ.* **2014**, *45*, 51–67. [CrossRef]

2. Pinstrup-Andersen, P. Food security: Definition and measurement. *Food Secur.* **2009**, *1*, 5–7. [CrossRef]

3. Latek, M. *Towards Food Security in Africa: Are International Private-Public Initiatives Paving the Way?* European Parliament: Strasbourg, France, 2017.

4. Godfray, H.C.J.; Beddington, J.R.; Crute, I.R.; Haddad, L.; Lawrence, D.; Muir, J.F.; Pretty, J.; Robinson, S.; Thomas, S.M.; Toulmin, C. Food security: The challenge of feeding 9 billion people. *Science* **2010**, *327*, 812. [CrossRef]

5. Pangaribowo, E.; Gerber, N.; Torero, M. *Food and Nutrition Security Indicators: A Review*; LEI Wageningen UR: Wageningen, The Netherlands, 2013.

6. International Year of Family Farming 2014. Available online: http://www.fao.org/fileadmin/user_upload/iyff/docs/Final_Master_Plan_IYFF_2014_30-05.pdf (accessed on 21 June 2019).

7. Graeub, B.E.; Chappell, M.J.; Wittman, H.; Ledermann, S.; Kerr, R.B.; Gemmill-Herren, B. The state of family farms in the world. *World Dev.* **2016**, *87*, 1–15. [CrossRef]

8. Regassa, N.; Stoecker, B.J. Household food insecurity and hunger among households in Sidama district, southern Ethiopia. *Public Health Nutr.* **2011**, *15*, 1276–1283. [CrossRef]

9. Shone, M.; Demissie, T.; Yohannes, B.; Yohannis, M. Household food insecurity and associated factors in West Abaya district, Southern Ethiopia, 2015. *Agric. Food Secur.* **2017**, *6*, 1–9. [CrossRef]

10. CSA. *Report on Area and Production and Farm Management Practices of Belg Season for Private Peasant Households*; Statistical Bulletin 578; Central Statistical Agency of the Federal Democratic Republic of Ethiopia: Addis Ababa, Ethiopia, 2016.

11. CSA; WFP. *Comprehensive Food Security and Vulnerability Analysis (CFSVA): Ethiopia*; Central Statistical Agency of Ethiopia: Addis Ababa, Ethiopia, 2014.

12. Asfaw, A.; Admassie, A. The role of education on the adoption of chemical fertilizer under different socioeconomic environments in Ethiopia. *Agric. Econ.* **2004**, *30*, 215–228. [CrossRef]

13. Fufa, B.; Hassan, R.M. Determinants of fertilizer use on maize in Eastern Ethiopia: A weighted endogenous sampling analysis of the extent and intensity of adoption. *Agrekon* **2006**, *45*, 38–49. [CrossRef]

14. Dercon, S.; Christiaensen, L. Consumption risk, technology adoption and poverty traps: Evidence from Ethiopia. *J. Dev. Econ.* **2011**, *96*, 159–173. [CrossRef]

15. Abebe, G.K.; Bijman, J.; Pascucci, S.; Omta, O. Adoption of improved potato varieties in Ethiopia: The role of agricultural knowledge and innovation system and smallholder farmers' quality assessment. *Agric. Syst.* **2013**, *122*, 22–32. [CrossRef]

16. Abate, G.T.; Rashid, S.; Borzaga, C.; Getnet, K. Rural finance and agricultural technology adoption in Ethiopia: Does the institutional design of lending organizations matter? *World Dev.* **2016**, *84*, 235–253. [CrossRef]

17. Swanson, B.E.; Davis, K. *Status of Agricultural Extension and Rural Advisory Services Worldwide*; Global Forum for Rural Advisory Services (GFRAS): Lausanne, Switzerland, 2014.

18. Leta, G.; Kelboro, G.; Stellmacher, T.; Hornidge, A.-K. *The Agricultural Extension System in Ethiopia: Operational Setup, Challenges and Opportunities*; ZEF Working Paper 158; ZEF Center for Development Research University of Bonn: Bonn, Germany, 2017.

19. Von Braun, J.; Olofinbiyi, T. Famine and food insecurity in Ethiopia. In *Food Policy for Developing Countries: Case Studies*; Pinstrup-Andersen, P., Cheng, F., Eds.; Cornell University: Ithaca, NY, USA, 2007.

20. MOFED. *Government of the Federal Democratic Republic of Ethiopia Rural Development Policy and Strategies*; Ministry of Economic and Economic Development: Addis Ababa, Ethiopia, 2003.

21. FAO Ethiopia. *Food and Agriculture Policy Trends*; Food and Agriculture Policy Decision Analysis Team, FAO: Rome, Italy, 2014.

22. Pankhurst, A.; Rahmato, D. Food security, safety nets and social protection. In *Food Security, Safety Nets and Social Protection in Ethiopia*; Rahmato, D., Pankhurst, A., van Uffelen, J.-G., Eds.; Forum for Social Studies: Addis Ababa, Ethiopia, 2013.

23. FAO. *Productive Safety Net Programme Fact. Sheet*; Food and Agricultural Organization: Rome, Italy, 2012.

24. Juma, C. *The New Harvest: Agricultural Innovation in Africa*; Oxford University Press: Oxford, UK, 2011.

25. Stellmacher, T. *Socio-Ecological Change in Rural Ethiopia: Understanding Local Dynamics in Environmental Planning and Natural Resource Management*; Peter Lang Publishing: Frankfurt, Germany, 2015.

26. FAO. *The State of Food and Agriculture: Innovation in Family Farming*; Food and Agriculture Organization of the United Nations: Rome, Italy, 2014.

27. Nyang'au, I.M.; Kelboro, G.; Hornidge, A.-K.; Midega, C.A.O.; Borgemeister, C. Transdisciplinary research: Collaborative leadership and empowerment towards sustainability of Push–Pull Technology. *Sustainability* **2018**, *10*, 2378. [CrossRef]

28. Jayne, T.S.; Mather, D.; Mghenyi, E. Principal challenges confronting smallholder agriculture in sub-Saharan Africa. *World Dev.* **2010**, *38*, 1384–1398. [CrossRef]

29. Jemal, O.; Callo-Concha, D.; van Noordwijk, M. Local agroforestry practices for food and nutrition security of smallholder farm households in southwestern Ethiopia. *Sustainability* **2018**, *10*, 2722. [CrossRef]

30. Proctor, F.J.; Lucchesi, V. *Small-Scale Farming and Youth in an Era of Rapid Rural Change*; IIED: London, UK; HIVOS: The Hague, The Netherlands, 2012.

© 2019 by the authors. Licensee MDPI, Basel, Switzerland. This article is an open access article distributed under the terms and conditions of the Creative Commons Attribution (CC BY) license (http://creativecommons.org/licenses/by/4.0/).

sustainability

Article

Socioeconomic Indicators of Bamboo Use for Agroforestry Development in the Dry Semi-Deciduous Forest Zone of Ghana

Daniel S. Akoto [1,2,*], Manfred Denich [1], Samuel T. Partey [2,3], Oliver Frith [2], Michael Kwaku [2], Alex A. Mensah [4] and Christian Borgemeister [1]

[1] Center for Development Research (ZEF), University of Bonn, 53113 Bonn, Germany; m.denich@uni-bonn.de (M.D.); cborgeme@uni-bonn.de (C.B.)

[2] International Bamboo and Rattan Organization (INBAR), PMB, Fumesua, 00233 Kumasi, Ghana; s.partey@cgiar.org (S.T.P.); o.frith@irri.org (O.F.); mkwaku@inbar.int (M.K.)

[3] International Crops Research Institute for the Semi-Arid Tropics (ICRISAT), BP 320 Bamako, Mali

[4] Department of Southern Swedish Forest Research Center, Swedish University of Agricultural Sciences, 23053 Alnarp, Sweden; alap0001@stud.slu.se

*   Correspondence: akoto.sarfo@uenr.edu.gh; Tel.: +233-20-7772-981

Received: 30 May 2018; Accepted: 26 June 2018; Published: 5 July 2018

Abstract: Bamboo agroforestry is currently being promoted in Ghana as a viable land use option to reduce dependence on natural forest for wood fuels. To align the design and introduction of bamboo agroforestry to the needs of farmers, information on the determinants of bamboo acceptability and adoption is necessary. It is, therefore, the aim of this study to determine how socioeconomic factors, local farming practices and local knowledge on bamboo may influence its acceptability and adoption as a component of local farming systems. Data were collected from 200 farmers in the dry semi-deciduous forest zone of Ghana using semi-structured questionnaire interviews. The results show that farmers' traditional knowledge on bamboo including its use for charcoal production and leaves for fodder are influential determinants of bamboo adoption. Among the demographic characteristics of farmers, age and gender are the most significant predictors. It is also evident that the regular practice of leaving trees on farmlands and type of cropping system may influence bamboo integration into traditional farming systems.

Keywords: adoption; land-use; deforestation; food security; renewable energy

---

1. Introduction

Deforestation emanating from excessive wood extraction for wood fuels continues to be a major agent of land degradation. The rate of deforestation in Ghana stands at 112.54 km^2 per annum, largely attributed to expansion of agriculture [1,2]. According to Afrane [3], about 73% of rural and 48% of urban households in Ghana depend on firewood and charcoal, respectively, for domestic and industrial use. Charcoal supply predominantly comes from savanna zones, and with lesser amount from deciduous and rainforests. Charcoal production is the next most dependent livelihood of the dry semi-deciduous forest zone (DSFZ) after farming and used as a secondary activity to support income from farming activities [4].

Owing to the increased sourcing of wood biomass from primary forests for charcoal production, government and scientists are advocating for the production and use of bamboo to reduce pressure on the major commercial timber species sourced as fuelwood. Due to development initiatives, such as the Bamboo and Rattan Development Programme (BARADEP), bamboo plantation establishment has increased in Ghana. This notwithstanding, monoculture bamboo plantations on agricultural lands may impact adversely on food security unless integrated systems with arable crops and/or livestock are given due consideration. In many parts of Asia, the integration of bamboo on croplands is confirmed a suitable approach for increased productivity of food crops and non-food biomass [5]. In Ghana, science-based bamboo agroforestry systems are limited and data to prove their suitability are lacking.

Currently, the International Bamboo and Rattan Organization (INBAR) is piloting a bamboo agroforestry system as a land use option for food security and renewable energy production in the DSFZ of Ghana. As an innovative development-oriented project, filling knowledge gaps on the determinants of bamboo use and adoption as a component of traditional farming systems is imperative to achieve large scale landscape adoption. In addition, knowledge on adoption determinants of bamboo will help in designing a bamboo-based agroforestry system that is tailored to the needs of farming communities. It was therefore the aim of this study to determine how socioeconomic factors, local farming practices and local knowledge on bamboo may influence its acceptability and adoption as a component of local farming systems.

Conceptual Framework of the Study

Several analytical frameworks have been used for the analysis of adoption of agroforestry technologies. Biot et al. [6] grouped these approaches into three major types: top-down interventions, populist or farmer-first, and neoliberal approaches. Stemming from the concept of farmer-first and sustainable livelihood principles, Shiferaw et al. [7] developed a wider conceptual framework for analyzing factors stimulating the successful adoption and adaptation of smallholder technologies. Given the focus of this study, the conceptual framework developed by Mercer [8] and modified by Zerihun et al. [9] is considered appropriate. The framework focuses on the adoption of already existing agroforestry technologies. However, this framework could be seen as very broad and complex to analyze the adoption rate and institutional setup of agroforestry technologies synchronously because institutional arrangements other than farmers were not directly evaluated to see their impact on adoption. Again, this study explored the willingness of farmers to accept bamboo agroforestry in the face of current wood energy needs and diversified income expectations of farmers in the DSFZ. Specifically, we modeled the interaction of explanatory variables, such as farmer characteristics, cropping systems, farming practices, bamboo ethnobotany, to predict the potential adoption of bamboo agroforestry in the DSFZ (Figure 1). These interactions facilitate farmer decision making processes and culminate in either adoption or non-adoption of technologies.

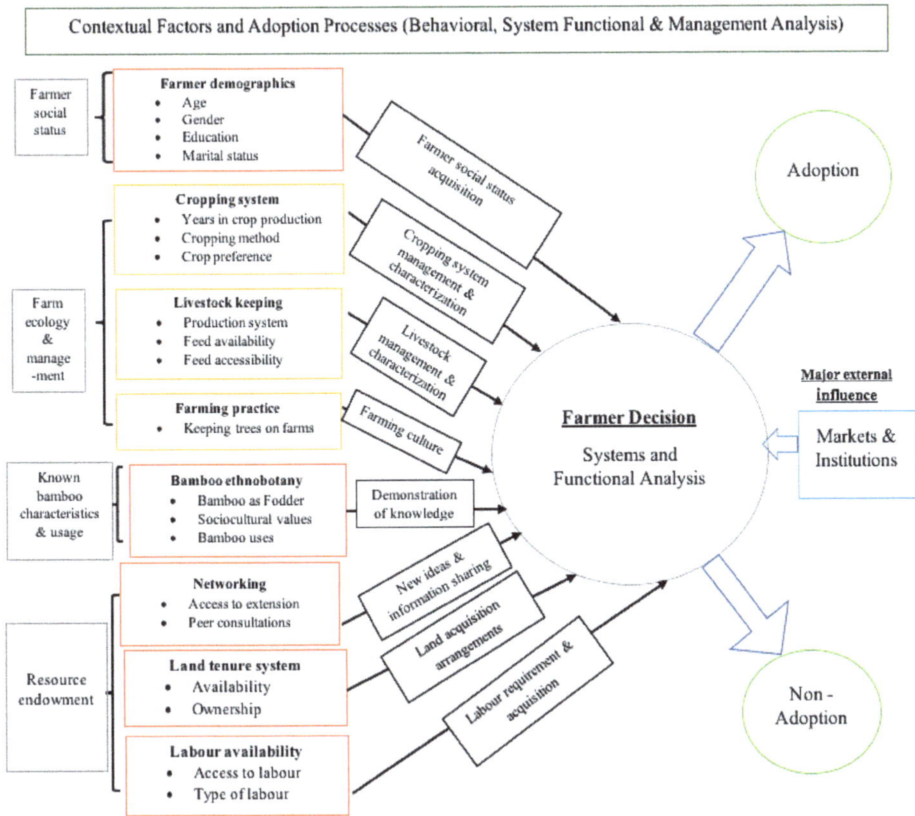

Figure 1. Analytical framework for modeling adoption potential of bamboo agroforestry. Source: Adapted from Zerihun et al. [9].

2. Materials and Methods

2.1. Study Area Description

The study was conducted in the Mampong, Ejura-Sekyedumase Municipals and Sekyere Central, Kumawu-Sekyere and Sekyere-Afram Plains Districts of Ghana (Figure 2). These areas are located within the dry semi-deciduous forest zone of Ghana (DSFZ), with a characterized bimodal rainfall pattern (with an average annual rainfall of 1270 mm). The major rainy season starts in March and peaks in May. There is a minor dint in July and a peak in August, ending in November. December to February is the drier season, which is warm and dusty (in the driest period). Mean annual temperature is 27 °C with variations in mean monthly temperature ranging between 22 °C and 30 °C throughout the year. The soil type of the study site is sandy loam (Ejura—Denteso Association). The study area falls within the zone, considered as the major food basket of Ghana and has the highest production of charcoal and fuelwood from natural sources.

Subsistence agriculture is the major economic activity employing about 65% of the population. The bulk of agricultural production is from manually cultivated rainfed crops. Major crops include: maize, cowpea, cassava, yam, and plantain. The DSFZ was chosen because of its unique characteristic features which combine those of the forest and savanna zones.

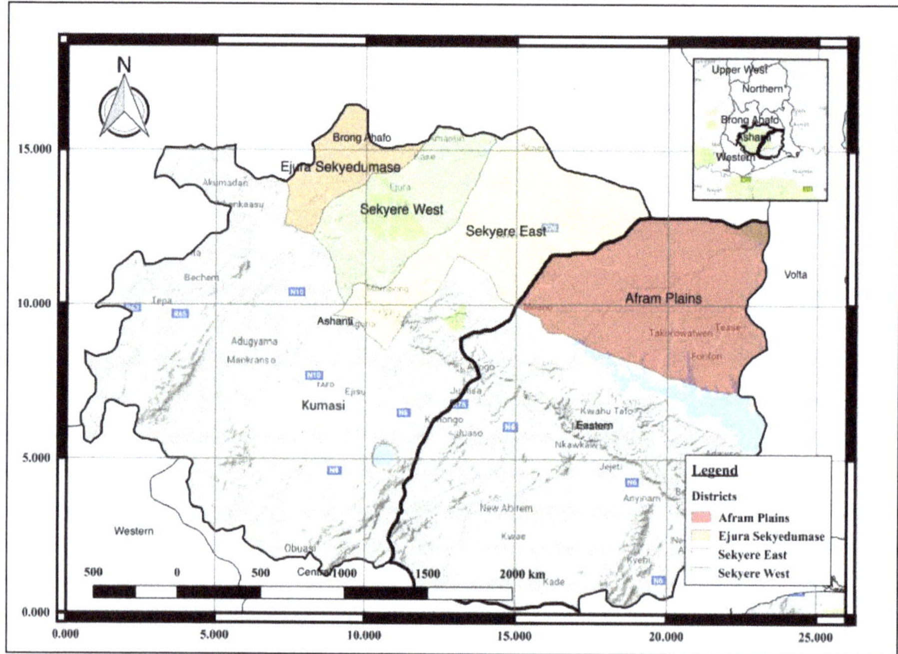

Figure 2. District map of Ghana showing the study sites in the DSFZ (map overlaid on Google satellite scene (2018) of Ghana).

2.2. Data Collection, Sampling Procedure and Analysis

A systematic purposive sampling method was adopted to select 200 household heads with farming as their primary occupation. Farmers (specifically, vegetable, yam, beans, and maize and cassava farmers) from 20 communities of five districts (four from each district) were selected for a household survey. The number of households interviewed in each community was estimated according to the recommendations by Edriss [10]:

$$n = \frac{N}{1 + N\left(e\right)^2} \tag{1}$$

where 'n' is the sample size, 'N' is the population size, and 'e' is the level of precision equal to 0.05 at 95% confidence level.

A semi-structured questionnaire was administered during the household survey to obtain information on the socio-economic variables that are likely to influence adoption of bamboo agroforestry. We modeled the potential adoption of bamboo agroforestry by designating a value of 1 if the farmer is willing to plant bamboo on his/her farmland (potential adopter) and 0 if the farmer is unwilling to plant bamboo on his/her farmland (potential non-adopter).

Primary data collected were analyzed using binary logistic regression as recommended by Masangano [11] to model adoption of bamboo agroforestry on a set of predictive variables from farmer characteristics (age, education level, gender, marital status), bamboo ethnobotany (knowledge and use), and agronomic practices (cropping system patterns, farming practices) at a 5% probability level using Statistical Package for Social Sciences (SPSS ver. 20.0). The best set of model predictors were evaluated using the log-likelihood criterion, significance of the Chi-square test statistic and the overall model performance adjudged by the stronger power of coefficient of determination (R^2).

Therefore, the interpretation of results focused on statistical significance, the direction of the regression coefficients (either positive or negative), and the odds-ratio [Exp (β)]. Cross-tab analyses were also conducted to estimate the proportion of potential adopters and non-adopters of bamboo agroforestry.

The main limitation of the survey was that it could not acquire all information needed for the pragmatic diagnosis of bamboo integrated farming system problems, because bamboo agroforestry is yet to be practiced with no field demonstrations existing in all sampled communities. As a result, detailed information on traditional farming practices adopted by farmers and their bamboo ethnobotany, energy (fuelwood) needs and crisis, soil fertility and management, and crop yield trends were collected through focus group discussions with farmers to validate the answers in the questionnaires.

3. Results and Discussion

3.1. Farmers' Characteristics/Demographics as Indicator for Adoption

Table 1 summarizes model results on using farmer characteristics as predictive variables for agroforestry adoption. The best set of model predictors are statistically significant ($p < 0.001$) with 87% correct prediction at the 5% level. The results show that gender and age of farmers can significantly predict the potential adoption of bamboo agroforestry. The odds ratio for age is 1.092 with a positive coefficient of 0.088 signifying that adults have larger potential to adopt bamboo agroforestry than the youth. The study found the majority of respondents within the ages of 31–45 years (40%). Within the ages of 31–45 (27%) are potential adopters whilst 24 (12%) are potential non-adopters. This age-influenced adoption trend could be attributed to the perception of young farmers; most of them see farming as a secondary occupation and use it to supplement their monetary income relative to older farmers whose major source of livelihood is farming and thus have a stronger likelihood to accept new farming technologies. This finding is inconsistent with reports by Rogers [12] and Ajayi [13], which highlight a decreasing potential of agroforestry technology adoption with increasing age. The effect of age on technology adoption has been well studied and reported to be context-specific. For instance, while Rogers [12] and Ajayi [13] found age could influence agricultural technology adoption, Waswa [14], Ndiema [15] and Njuguna et al. [16] found no relationship between age and technology adoption On the other hand, other researchers have found age to be positively correlated with technology adoption [17–21] citing people between the ages of 18–43 as more active and ready to take risks by adopting new technologies.

Table 1. Parameter estimates of modeling farmers' characteristics for predicting bamboo agroforestry adoption.

Variables [1]	Coefficients	Std. Error	*p*-Value	Exp (β)
Age	0.088	0.065	0.000	1.092
Gender	0.002	0.028	0.030	1.002
Education level	−0.853	0.090	0.059	0.426
Marital status	0.006	0.041	0.102	1.006
Constant	−1.889	0.210	0.000	0.151

[1] The best set of model predictors (−2 Log-likelihood = 95.9) was significant (Chi-square test value = 58.04, $p < 0.001$) and overall model predictions on adoption of 87.4%. Std. error is the standard error and Exp (β) is the odds-ratio representing the likelihood of adoption.

Gender analysis is also significant to the adoption model with men found to be more likely to adopt bamboo agroforestry than women. Majority of the farmers are males (80%) of which potential adopters are 136 (68%) and 24 (12%) estimated as potential non-adopters. Female farmers numbered 40 with 38 (19%) characterized as potential adopters whilst only 2 (1%) were non adopters. Although the female respondents constitute a smaller percentage of respondents, majority of them shows keen interest in adopting bamboo agroforestry. Nevertheless, the decisions on the choice of new technologies by women farmers are strongly dependent on their husbands as male household heads tend to have more access to land. This is in agreement with Scherr [22] who found in her studies on economic factors influencing farmer adoption of agroforestry that females are not permitted to make decisions to adopt agroforestry technologies without consulting the family head (mostly males). This adds to the growing concern of gender inequalities on household decisions and access to farm resources. The lower proportion of sampled women-farmers in the study area could be linked to situations where women do not have headship to land and tree tenure due to the largely patrilineal inheritance systems [23]. Issues of gender in agricultural technology adoption have been explored over the years, and studies report mixed evidence concerning the different roles females and males play in the adoption of technology [24]. In comparing these facts, Morris and Doss [25] report no significant relationship between gender and probability to adopt improved maize in Ghana. Conversely, other studies have shown differences in gender norms and culture play significant roles in technology adoption [26–28]. In Nigeria, Obisesan [29] found gender to be a significant determinant of the adoption of improved cassava production approaches. Similarly, Lavison [30] also reports male farmers are more likely to adopt organic fertilizer as compared to their female colleagues.

Unlike age and gender, the study shows that the level of education and marital status are not significant determinants of bamboo agroforestry adoption. Although this contradicts the assertion that education (formal and informal) or training increase the rate of technology adoption [31], it may be context-specific. Other studies, such as Amudavi [32] and Ndiema [15], also found that the educational status of a farmer was not a significant indicator of technology adoption. The results of the current study suggest that, in the study region, the kind of education or training that can facilitate the adoption of an innovation might be the education on the innovation itself and how well farmers are exposed to the innovation; and not necessarily academic education [33–38].

Many new practices stemming from a top-down approach and overlooking socio-economic realities often produce disappointing results for implementing agencies [39]. However, understanding the prevailing social values can positively influence the adaptation and commitment to both existing and introduced technologies [40]. In addition, studies on agroforestry adoption are becoming increasingly important to researchers. It is therefore essential to monitor socio-economic concepts in agroforestry to delineate strengths and weaknesses in the current state of knowledge and to foster guidance for further investigation and optimal decision making through productive feedback loops between researchers and farmers [41].

3.2. Farming Practices as Indicator for Adoption of Bamboo Agroforestry

Farming practices such as keeping trees on farms and the type of tree species left on farms had significant effect and explained 79.2% of the adoption potential of bamboo agroforestry (Table 2).

Table 2. Farming practices to predict bamboo agroforestry adoption in the DSFZ.

Variables [1]	Coefficients	Std. Error	*p*-Value	Exp (β)
Keeping trees on farms	1.866	0.010	0.001	0.155
Type/preferred tree species	−1.021	0.020	0.04	1.200
Constant	−1.889	0.210	0.000	0.151

[1] The best set of model predictors (−2 Log-likelihood = 158.4) was significant (Chi-square test value = 116.09, $p < 0.05$) and overall model predictions on adoption of 79.2%. Std. error is the standard error and Exp (β) is the odds-ratio representing the likelihood of adoption.

For the 194 farmers who kept trees on their farms, 168 (85%) were potential adopters and 26 (13%) were potential non-adopters. However, all the farmers (4) who do not leave trees on their farms were potential adopters (2%). Figure 3 shows preferred trees left on farmlands for economic and environmental benefits. The farmers report several reasons for leaving trees on farms such as economic gains, shade, soil and water conservation, fodder, and fuelwood provision. This implies that farmers would most assuredly accept to plant any woody perennial on their farm if they knew the ecological and economic functions of such woody perennial. It could be further deduced that farmers' decision for keeping trees on their farmlands have ecological justifications and importance since trees maintain and improve soil fertility through processes of nitrogen fixation and nutrient uptake from deep soil horizons [42]. Furthermore, trees improve the structural properties of the soil with their rooting systems by reducing soil erosion and increasing soil water infiltration [43]. Alavalapati and Nair [44] recounted that farmers mostly implement agroforestry systems to provide household needs such as food, fodder, and fuelwood. This system may not be imperative to the conventional 'agroforester' such as social benefits or community acceptability of the system [39,45]. Technology adoption has many policy implications on agricultural and agroforestry development and technology-specific attributes have been shown in the past to significantly determine farmers' decision to adopt a technology [46]. The economic value of woody perennials is a key factor in farmers' adoption of a technology [22]. According to Glover et al. [47], environmental or ecological potential of a woody perennial is critical in influencing an adoption decision.

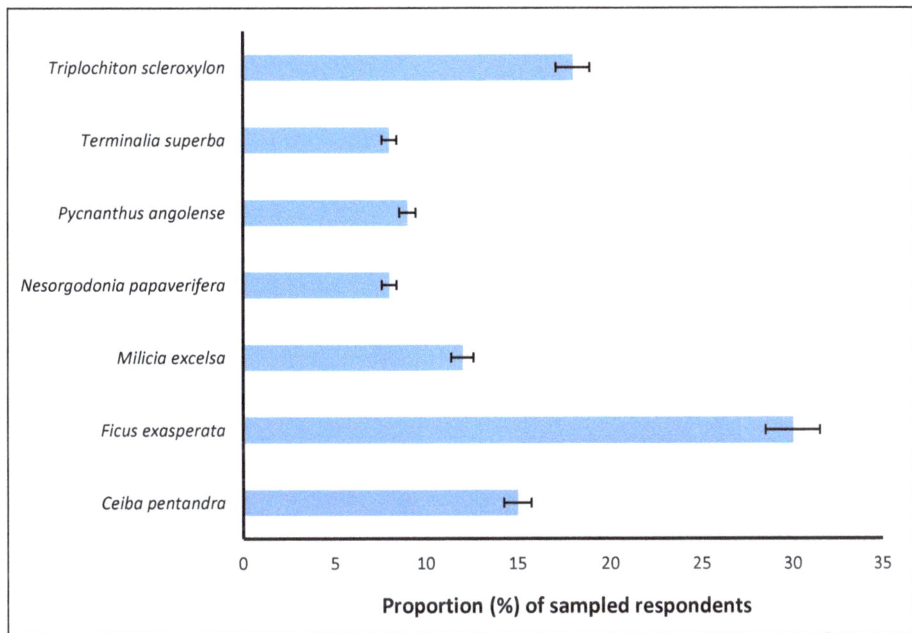

Figure 3. Preferred tree species left on farms by respondents. Error bars show percentage (5%) deviation of sampled respondents.

3.3. Characterizing Farmers' Cropping Systems as a Predictor for Adoption

The set of predictors such as crop production objective, crop preference and cropping method which characterize cropping systems in the study area were found to be significant ($p < 0.001$) determinants of bamboo agroforestry adoption (Table 3). It was observed that most (67%) of the sampled respondents were farmers who grow crops mainly for commercial purposes. Out of this, 61% were potential adopters and 6% were potential non-adopters. With recent climate change and land degradation impacts on crop yields and poverty trends especially in sub-Saharan Africa (SSA), most farmers are shifting from the traditional subsistence farming to become more commercially-oriented. Such farmers are keener to explore innovations that seek to increase production and farm incomes [48–50]. Other findings have also supported the assertion that capital or economic situation of a farmer influences technology adoption [31,51,52]. In SSA, bamboo-based agroforestry systems are now developing, but literature on their economic feasibilities is limited. However, with livelihoods in SSA mostly tied to agriculture and forestry, investments in bamboo-based agroforestry systems may contribute to rural poverty reduction and improved livelihoods in the region [53]. Like most agroforestry systems, bamboo-based agroforestry systems are expected to open new income streams by diversifying agroecosystems and offering multiple economic benefits from the sale of grains and vegetables from short-duration crops (integrated with bamboo), supply of fodder for livestock, and the sale of processed bamboo culms as wood fuels, charcoal, timber or industrial raw materials [54]. In most of SSA, it is common for rural households to diversify income streams as a pathway to reduce vulnerability to the failure of their primary income generating activities [48,55].

Table 3. Parameter estimation of farmers' cropping system in predicting bamboo agroforestry adoption.

Variables [1]	Coefficients	Std. Error	p-Value	Exp (β)
Number of years in crop production	−0.273	0.047	0.961	0.761
Primary objective for growing crops	17.368	0.049	0.02	0.031
Crop preference	1.357	0.110	0.01	3.886
Regular cropping method	−1.537	0.106	0.03	0.754
Meeting crop production target	1.637	0.031	0.02	5.142
Challenges with soil fertility	1.959	0.031	0.084	7.091
Access to fertilizer	−0.708	0.033	0.490	0.493
Constant	−25.382	0.024	0.998	0.000

[1] The best set of model predictors (−2 Log-likelihood = 118.8) was significant (Chi-square test value = 35.22, $p < 0.001$) and overall model predictions on adoption of 86.9%. Std. error is the standard error and Exp (β) is the odds-ratio representing the likelihood of adoption.

Although bamboo raw materials have no guaranteed prices in Ghana, potential bamboo farmers have the advantage of benefiting from the emergent and growing bamboo economy and industry in the country. Ghana seems to be on a pathway to advancing bamboo resource development following the increasing market for bamboo products and the presence of the West Africa sub-regional office INBAR in Kumasi. The Government of Ghana and INBAR have collaborated in establishing the Bamboo and Rattan Development Programme (BARADEP). Aside the small-scale traditional bamboo basketry/craft shops and the famous Bamboo Bicycle Producing Company at Toase-Nkawie in Kumasi, there are many other private bamboo initiatives ranging from small cottage to large-scale enterprises including Global Bamboo Products in Anyinam, KWAMOKWA bamboo plantations all in Kumasi, Greater Accra Bamboo and Rattan Handicrafts Association, Brotherhood Cane/Rattan Weavers Association., Links Handicrafts Association., New Vision Handicrafts Association, Pioneer Bamboo Manufacturing Co., Ltd. (Accra, Ghana), Assin Foss and TTom Bamboo Toothpick Processing Company, Tandan. Also, many foreign investors have taken advantage of this initiative, sound political environment and the favorable climatic conditions for bamboo development and have a dozen of large-scale plantations and processing centers in Ghana. One of the most successful of such initiatives is the EcoPlanet Bamboo project, with its vision as a bamboo plantation and processing company that focuses on the provision of a secure and certified source of fiber for timber manufacturing industries and markets globally. Darlow Enterprises is another bamboo company of importance which is also resident in Ghana. With headquarters in Belize and the Philippines, Darlow Enterprises focuses on bamboo charcoal production. All these enterprises present a great market channel for bamboo trade from which potential bamboo farmers could link up and benefit financially through the trading of bamboo products.

3.4. Bamboo Ethnobotany as a Predictive Variable for Adoption of Bamboo Agroforestry

Ethnobotany focuses on how plants have been or are used, managed and perceived in human societies, and it expresses how plants are used for various needs (clothing, conservation techniques, shelter, food, medicine, hunting, magico-religious concepts) as well as their general economic, and sociological importance in societies [56]. From the study, local knowledge on bamboo characteristics and usage record a strong prediction (88.9%) to the model of potential adoption. Farmers' readiness to try bamboo fodder on their livestock, readiness to incorporate bamboo cultivation on farms for fodder, the visibility of bamboo by farmers, personal planting of bamboo, bamboo use, and farmers' readiness to produce bamboo charcoal are statistically significant to the model (Table 4).

Table 4. Model estimates of farmers' bamboo ethnobotany to predict bamboo agroforestry adoption.

Variables [1]	Coefficients	Std. Error	*p*-Value	Exp (β)
Knowledge on bamboo leaves used as fodder	−0.769	0.026	0.067	0.463
Livestock fed with bamboo leaves before	−20.505	0.018	0.098	0.000
Readiness to try bamboo fodder	−1.840	0.033	0.000	0.159
Readiness to incorporate bamboo cultivation on farm as fodder	−1.040	0.035	0.005	0.219
Seen/heard bamboo	3.727	0.017	0.033	1.316
Personally planted bamboo before	2.321	0.011	0.040	8.364
Taboos/beliefs associated with the use or planting of bamboo	−0.603	0.017	0.471	0.547
Knowledge on bamboo charcoal	−0.006	0.023	0.836	0.994
Production of bamboo charcoal before	1.243	0.000	0.060	1.222
Readiness to produce bamboo charcoal	1.456	0.011	0.001	4.562
Personally used/seen someone using bamboo	2.343	0.028	0.004	3.561
Constant	−12.382	0.024	0.998	0.000

[1] The best set of model predictors (−2 Log-likelihood = 11.9) was significant (Chi-square test value = 12.93, $p < 0.001$) and overall model predictions on adoption of 88.9%. Std. error is the standard error and Exp (β) is the odds-ratio representing the likelihood of adoption.

Out of the 186 farmers who appeared to possess some level of knowledge on bamboos, 164 (88%) were potential adopters and 22 (12%) are potential non-adopters. Bamboo use in Ghana is tied to fencing, use as television poles, roofing etc. (Figure 4). Cross-tab analysis also shows that 124 (78%) farmers are ready to try bamboo fodder, whilst 36 (22%) preferred otherwise. It is argued that new or existing interventions to encourage tree planting on farmlands need to be centered or realigned on farmers' comprehension of tree management in the domains of household livelihood schemes, such as fodder needs, energy needs and supplement of income, stressing that information about farmers' perceptions of the significance of trees and the constrictions they face in increasing tree resources are rare [56–60]. In the DSFZ, livestock is mostly kept on free-range systems where prolonged drought limit access to nutritive food materials [61]. As bamboos are drought-tolerant and produce relatively high nutritive fodder, their integration into farming systems can supply supplementary feed materials for livestock [62]. While the increased knowledge and use of bamboo have the potential to support its integration into traditional farming systems, issues of land tenure, labour availability and economic importance have to be critically assessed.

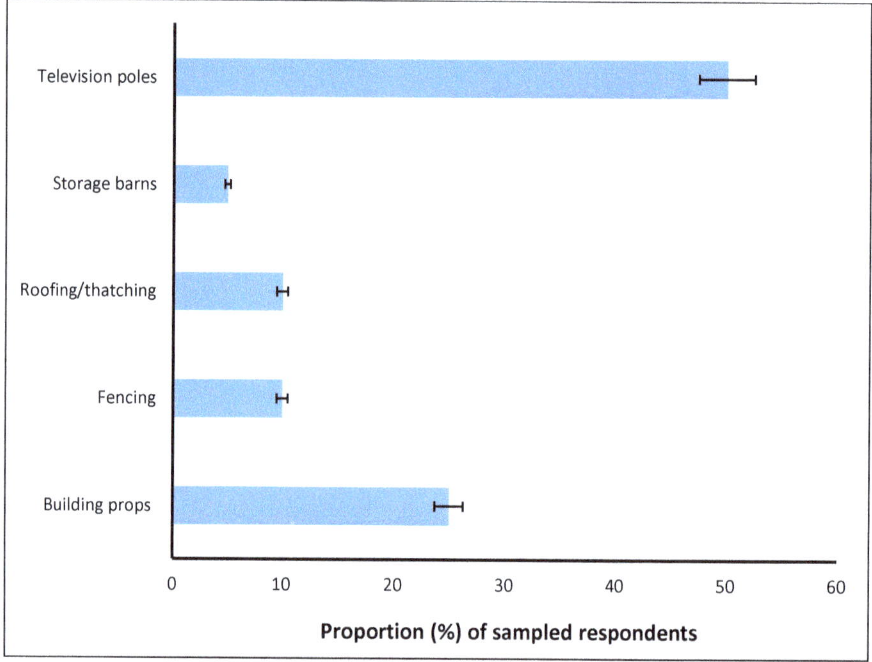

Figure 4. Predominant uses of bamboo known to sampled respondents in the DSFZ. Error bars show percentage (5%) deviation of sampled respondents.

4. Conclusions

The study identifies the socioeconomic factors and farming practices that influence the adoption of bamboo agroforestry in the dry semi-deciduous forest zone of Ghana. Factors found to be statistically significant were: age, gender, cropping method, crop preferences, primary objective for growing crops such as market availability and early maturity, role of bamboo as fodder plant, uses and benefits of bamboo, cropping system and farming practice. The present study provides key contributions for future bamboo-based agroforestry design and some directions for agricultural development policies. The key influential socioeconomic indicators identified in the study may provide a guide for rolling local farmers into the 2016–2040 National Forest Plantation Development Strategy of Ghana which seeks to establish 1000 hectares of bamboo plantations annually; culminating into the agenda to plant 50,000 hectares of bamboo plantations over the next 25 years by employing agroforestry principles. In the quest to develop Ghana's bioenergy sector, bamboo biomass has been included to ensure sustainability of the Ghana Bioenergy Policy initiative. The Government of Ghana may benefit essentially by using this study as a basis for further studies in developing a comprehensive national database to offer more insight into bamboo-socioeconomic implications to enhance the achievement of this goal. There is also the opportunity for Ghana to revitalize the almost 'defunct' national agroforestry policy by drawing lessons from this study by critically considering those socioeconomic indicators identified.

Author Contributions: "Conceptualization, D.S.A., M.D., and S.T.P.; Methodology, D.S.A., S.T.P., M.D., and A.A.M.; Software, D.S.A., S.T.P., and A.A.M.; Validation, D.S.A., M.K., S.T.P., M.D., and C.B.; Formal Analysis, D.S.A., S.T.P., and A.A.M.; Investigation, M.D., C.B., and O.F..; Resources, M.D., C.B., O.F., and M.K.,; Data Curation, D.S.A., S.T.P., and M.D.; Writing-Original Draft Preparation, D.S.A.; Writing-Review & Editing, D.S.A., S.T.P., M.D., and C.B.; Visualization, S.T.P., M.D., and C.B.; Supervision, M.D., S.T.P., and C.B.; Project Administration, M.D., O.F., and M.K.; Funding Acquisition, M.D., M.K. and O.F."

Funding: This paper was produced as an activity of BiomassWeb (grant no. 031A258A), a project funded by the German Federal Ministry of Education and Research (BMBF) in the context of the initiative GlobE "Securing the Global Food Supply". BiomassWeb is managed by the Center for Development Research (ZEF)-Universität of Bonn, Germany.

Conflicts of Interest: The authors declare no conflicts of interest.

References

1. FAO. *Agriculture Organization: Global Forest Resources Assessment*; FAO: Rome, Italy, 2010.
2. Olderman, L.R.; Eng, W.; Hakkeling, R.T.A.; Sombroek, W.G. *World Map of the Status of Human-Induced Soil Degradation: An Explanatory Note*; Global Assessment of Soil Degradation (GLASOD): Nairobi, Kenya, 1991. Available online: http://www.the-eis.com/data/literature/World%20map%20of%20the%20status%20of%20human-induced%20soil%20degradation_1991.pdf (accessed on 10 January 2018).
3. Afrane, G. Examining the potential for liquid biofuels production and usage in Ghana. *Energy Policy* **2012**, *40*, 444–451. [CrossRef]
4. Benhin, J.K.A.; Barbier, E.B. The Effects of the Structural Adjustment Program on Deforestation in Ghana. *Agric. Resour. Econ. Rev.* **2001**, *30*, 66–80. [CrossRef]
5. Mailly, D.; Christanty, L.; Kimmins, J.P. Without bamboo, the land dies: Nutrient cycling and biogeochemistry of a Javanese bamboo talun-kebun system. *For. Ecol. Manag.* **1997**, *91*, 155–173. [CrossRef]
6. Biot, Y. *Rethinking Research on Land Degradation in Developing Countries*; World Bank Discussion Papers: Washington, DC, USA, 1995; pp. 63–289.
7. Shiferaw, B.A.; Okello, J.; Reddy, R.V. Adoption and adaptation of natural resource management innovations in smallholder agriculture: Reflections on key lessons and best practices. *Environ. Dev. Sustain.* **2009**, *11*, 601–619. [CrossRef]
8. Mercer, D.E. Adoption of agroforestry innovations in the tropics: A review. *Agrofor. Syst.* **2004**, 61–62. [CrossRef]
9. Zerihun, M.F.; Muchie, M.; Worku, Z. Determinants of agroforestry technology adoption in Eastern Cape Province, South Africa. *Dev. Stud. Res.* **2014**, *1*, 382–394. [CrossRef]
10. Edriss, A. *Introduction to Statistics*; Bunda College of Agriculture, University of Malawi: Lilongwe, Malawi, 2006.
11. Masangano, C. Diffusion of Agroforestry Technologies 1996. Available online: http://www.msu.edu/user/masangn/agrof.html (accessed on 10 January 2018).
12. Rogers, E.M. Diffusion of Innovations (4th Eds.). Available online: https://scholar.google.com/scholar?hl=en&as_sdt=0%2C5&q=Rogers+E.M.+Diffusion+of+Innovations%2C+4th+%28edn%29.+The+Free+Press.+1995.+New+York%2C+USA.&btnG (accessed on 10 January 2018).
13. Ajayi, O.C.; Akinnifesi, F.K.; Sileshi, G.; Chakeredza, S. Adoption of renewable soil fertility replenishment technologies in the southern African region: Lessons learnt and the way forward. *Nat. Resour. Forum* **2007**, *31*, 306–317. [CrossRef]
14. Waswa, S.I. Factors Related to Adoption of selected Agroforestry Technologies by Small Scale Farmers as a Response to Environmental Degradation: The Case of Njoro Kenya Rongai and Njoro Divisions of Nakuru District. Master's Thesis, Egerton University, Njoro, Kenya, 2000.
15. Ndiema, A.C. Factors Affecting the Adoption of Wheat Production Technologies by Farmers in Njoro and Rongai Divisions of Nakuru Districts, Kenya. Master's Thesis, Egerton University, Njoro, Kenya, 2002.
16. Njuguna, I.M.; Munyua, C.N.; Makal, S.K. Influence of demographic characteristics on adoption of improved potato varieties by smallholder farmers in Mumberes Division, Baringo County, Kenya. *J. Agric. Ext. Rural Dev.* **2015**, *7*, 114–121.
17. Okuthe, I.K.; Ngesa, F.U.; Ochola, W.W. The socio-economic determinants of the adoption of improved sorghum varieties and technologies by smallholder farmers: Evidence from South Western Kenya. *Int. J. Humanit. Soc. Sci.* **2013**, *3*, 18.
18. Ragland, J.; Lal, R. *Technologies for Sustainable Agriculture in the Tropics*; American Society of Agronomy: Madison, WI, USA, 1993.
19. Aboud, A.A. *Relevance of the Novel Development Paradigms in Environmental Conservation: Evidence of Njoro Farmers, Kenya*; Egerton University: Njoro, Kenya, 1997.

20. Wasula, S.L. Influence of Socio-Economic Factors on Adoption of Agroforestry Related Technology. The Case of Njoro and Rongai Districts, Kenya. Master's Thesis, Egerton University, Njoro, Kenya, 2000.
21. Atibioke, O.A.; Ogunlade, I.; Abiodun, A.A.; Ogundele, B.A.; Omodara, M.A.; Ade, A.R. Effects of farmers' demographic factors on the adoption of grain storage technologies. *J. Res. Humanit. Soc. Sci.* **2012**, *2*, 56–63.
22. Scherr, S.J. Economic factors in farmer adoption of agroforestry: Patterns observed in Western Kenya. *World Dev.* **1995**, *23*, 787–804. [CrossRef]
23. Nyaga, J.; Barrios, E.; Muthuri, C.W.; Öborn, I.; Matiru, V.; Sinclair, F. Evaluating factors influencing heterogeneity in agroforestry adoption and practices within smallholder farms in Rift Valley, Kenya. *Agric. Ecosyst. Environ.* **2015**, *212*, 106–118. [CrossRef]
24. Bonabana-Wabbi, J. Assessing Factors Affecting Adoption of Agricultural Technology: The Case of Integrated Pest Management (IPM). Master's Thesis, Kumi, Uganda, 2002.
25. Morris, M.; Doss, C. How does gender affect the adoption of agricultural innovations? The case of improved maize technology in Ghana. In Proceedings of the American Agricultural Economics Association (AAEA) Annual Meeting, Nashville, TN, USA, 8–11 August 1999.
26. Mesfin, A. Analysis of factors Influencing Adoption of Triticale and Its Impact. The Case Farta Wereda. Master's Thesis, School of Graduate Studies of Alemaya University, Dire Dawa, Ethiopia, 2005.
27. Omonona, B.; Oni, O.; Uwagboe, O. Adoption of improved Cassava varieties and its impact on Rural Farming Households in Edo State, Nigeria. *J. Agric. Food Inf.* **2005**, *7*, 40–45. [CrossRef]
28. Mignouna, B.; Manyong, M.; Rusike, J.; Mutabazi, S.; Senkondo, M. Determinants of Adopting Imazapyr-Resistant Maize Technology and its Impact on Household Income in Western Kenya. *AgBioForum* **2011**, *14*, 158–163.
29. Obisesan, A. *Gender Differences in Technology Adoption and Welfare Impact among Nigerian Farming Households*; Munich Personal RePEc Archive Paper No. 58920, Posted 1; University of Ibadan: Ibadan, Nigeria, 2014.
30. Lavison, R. Factors Influencing the Adoption of Organic Fertilizers in Vegetable Production in Accra. Master's Thesis, University of Ghana, Legon-Accra, Ghana, 2013.
31. Venkatesh, V.; Michael, G.M.; Phillip, L.A. A longitudinal field investigation of gender differences in individual technology adoption decision-making processes. *Org. Behav. Hum. Decis. Process.* **2000**, *83*, 33–60. [CrossRef] [PubMed]
32. Amudavi, M.A. Influence of Socio-Economic Factors on Adoption of Maize Related Technology. The Case of Smallholder Farmers in Hamisi Division of Vihiga District, Kenya. Master's Thesis, Egerton University, Njoro, Kenya, 1993.
33. Kafle, B.; Shah, P. Adoption of improved potato varieties in Nepal: A case of Bara District. *J. Agric. Sci.* **2012**, *7*, 14–22. [CrossRef]
34. Wafuke, S. Adoption of Agroforestry Technology among Small Scale Farmers in Nzoia Location, Lugari District, Kenya. Master's Thesis, Egerton University, Njoro, Kenya, 2012.
35. Thangata, P.H.; Alavalapati, J.R.R. Agroforestry adoption in southern Malawi: The case of mixed intercropping of Gliricidia sepium and maize. *Agric. Syst.* **2003**, *78*, 57–71. [CrossRef]
36. Ayalew, A.M. Factors Affecting Adoption of Improved Haricot Bean Varieties and Associated Agronomic Practices in Dale Woreda, SNNPRS. Master's Thesis, Hawassa University, Hawassa, Ethiopia, 2011; p. 109.
37. Tegegne, Y. Factors Affecting Adoption of Legume Technologt and Its Impact on Income of Farmers: The Case of Sinana and Ginir Woredas of Bale Zone. Master's Thesis, Haramaya University, Hawassa, Ethiopia, 2017.
38. Ntshangase, N.L.; Muroyiwa, B.; Melusi, S. Farmers' perception and factors influencing the adoption of no-till conservation agriculture by small-scale farmers in Zashuke, Kwazulu-Natal Province. *Sustainability* **2018**, *10*, 555. [CrossRef]
39. Buck, L.E.; Lassoie, J.P.; Fernandes, E.C.M. *Agroforestry in Sustainable Agricultural Systems*; CRC Press: Boca Raton, FL, USA, 1998.
40. Pattanayak, S.K.; Mercer, D.E.; Sills, E.; Yang, J.C. Taking stock of agroforestry adoption studies. *Agrofor. Syst.* **2003**, *57*, 173–186. [CrossRef]
41. Rule, L.C.; Flora, C.B.; Hodge, S.S. Social dimensions of agroforestry. In *North American Agroforestry: An Integrated Science and Practice*; American Society of Agronomy: Madison, WI, USA, 2000; pp. 361–386.
42. Nair, V.; Nair, P.K.; Kalmbacher, S.; Ezenwa, I. Reducing nutrient loss from farms through silvopastoral practices in coarse-textured soils of Florida, USA. *Ecol. Eng.* **2007**, *29*, 192–199. [CrossRef]

43. Ayres, E.; Steltzer, H.; Berg, S.; Wallenstein, M.D.; Simmons, B.L.; Wall, D.H. Tree species traits influence soil physical, chemical, and biological properties in high elevation Forests. *PLoS ONE* **2009**, *4*, e5964. [CrossRef] [PubMed]

44. Alavalapati, J.; Nair, P.K.R.; Barkin, D. Socioeconomic and Institutional Perspectives of Agroforestry. In *World Forests, Markets and Policies, World Forests*; Springer: Dordrecht, The Netherlands, 2001; pp. 71–83.

45. Kurtz, W.B. Economics and policy of agroforestry. In *North American Agroforestry: An Integrated Science and Practice*; American Society of Agronomy: Madison, WI, USA, 2000; pp. 321–360.

46. Idrisa, Y.L.; Ogunbamenu, B.O.; Amaza, P.S. 'Influence of farmers' socio-economic and technology characteristics on soybeans seeds technology adoption in Borno State, Nigeria. *Afr. J. Agric. Res.* **2010**, *5*, 1394–1398. [CrossRef]

47. Glover, E.K.; Hassan, B.A.; Mawutor, K.G. Analysis of socio-economic conditions influencing adoption of agroforestry practices. *Int. J. Agric. For.* **2013**, *3*, 178–184.

48. Castle, M.H.; Lubben, B.D.; Luck, J.D. *Factors Influencing the Adoption of Precision Agriculture Technologies by Nebraska Producers*; Agricultural Economics: Lincoln, NE, USA, 2016; p. 49.

49. Akudugu, M.; Guo, E.; Dadzie, S. Adoption of Modern Agricultural Production Technologies by Farm Households in Ghana: What Factors Influence their Decisions? *J. Biol. Agric. Healthcare* **2012**, *2*, 3.

50. Factors Influencing the Adoption of Organic Fertilizers in Vegetable Production in Accra. Available online: http://ugspace.ug.edu.gh/handle/123456789/5410 (accessed on 10 January 2018).

51. Caswell, M.K.; Fuglie, C.; Ingram, S.J.; Kascak, C. *Adoption of Agricultural Production Practices: Lessons Learned from the US*; Agriculture Economic Report; Department of Agriculture Area Studies Project, US Department of Agriculture, Resource Economics Division, Economic Research Service: Washington, DC, USA, 2001; p. 792.

52. Girmay, G.D.; Abadi, N.; Hizikias, E.B. Contribution of *Oxytenanthera abyssinica* (A. Rich) Muro and determinants of growing in homestead agroforestry system in Northern Ethiopia. *Ethnobat. Res. Appl.* **2016**, *14*, 479–490.

53. FAO. *Climate Smart Agriculture: Policies, Practices and Financing for Food Security, Adaptation and Mitigation*; Food and Agriculture Organization: Rome, Italy, 2010.

54. Babulo, B.; Muys, B.; Nega, F.; Tollens, E.; Nyssen, J.; Deckers, J.; Mathijs, E. Household livelihood strategies and forest dependence in the highlands of Tigray, Northern Ethiopia. *Agric. Syst.* **2008**, *98*, 147–155. [CrossRef]

55. Kusimi, J.M. Characterizing land disturbance in Atewa range forest reserve and buffer zone. *Land Use Policy* **2015**, *49*, 471–482. [CrossRef]

56. Evans, R.S. The importance of ethnobotany in environmental conservation. *Am. J. Econ. Soc.* **1994**, *53*, 202–206.

57. Atangana, A.; Khasa, D.; Chang, S.; Degrande, A. Socio-cultural aspects of agroforestry and adoption. *Trop. Agrofor.* **2014**, 323–332. [CrossRef]

58. Belay, T.B.; Linder, A.; Pretzsch, J. Indicators and determinants of small-scale bamboo commercialization in Ethiopia. *Forests* **2013**, *4*, 710–729.

59. Hogarth, N.J.B.; Belcher, H. The contribution of bamboo to household income and rural livelihoods in a poor and mountainous county in Guangxi, China. *Int. For. Rev.* **2013**, *15*, 71–81. [CrossRef]

60. Wang, G.L.; Innes, S.; Dai, G. Achieving sustainable rural development in Southern China: The contribution of bamboo forestry. *Int. J. Sustain. Dev. World Ecol.* **2008**, *15*, 484–495. [CrossRef]

61. Satya, S.; Lalit, M.; Bal, P.S.; Naik, S.N. Bamboo shoot processing: Food quality and safety aspect (a review). *Trends Food Sci. Technol.* **2010**, *21*, 181–189. [CrossRef]

62. Partey, S.T.; Sarfo, D.A.; Frith, O.; Kwaku, M.; Thevathasan, N.V. Potentials of Bamboo-Based Agroforestry for Sustainable Development in Sub-Saharan Africa: A Review. *Agri. Res.* **2017**, *6*, 22–32. [CrossRef]

© 2018 by the authors. Licensee MDPI, Basel, Switzerland. This article is an open access article distributed under the terms and conditions of the Creative Commons Attribution (CC BY) license (http://creativecommons.org/licenses/by/4.0/).

sustainability

Article

Tree Species Diversity and Socioeconomic Perspectives of the Urban (Food) Forest of Accra, Ghana

Bertrand F. Nero [1,2,*], Nana Afranaa Kwapong [1,3], Raymond Jatta [1] and Oluwole Fatunbi [1]

1   Forum for Agricultural Research in Africa (FARA), 12 Anmeda Street, PMB CT 173, Accra 233, Ghana; nkwapong@faraafrica.org (N.A.K.); rjatta@faraafrica.org (R.J.); ofatunbi@faraafrica.org (O.F.)
2   Department of Forest Resources Technology, Kwame Nkrumah University of Science and Technology, PMB, University Post Office, Kumasi 233, Ghana
3   Department of Agricultural Extension, University of Ghana, P.O. Box LG 68, Legon, GA-492-3175 Accra, Ghana
*   Correspondence: bfnero@knust.edu.gh or bnero@faraafrica.org

Received: 30 June 2018; Accepted: 21 September 2018; Published: 25 September 2018

Abstract: Urban and peri-urban forestry has emerged as a complementary measure to contribute towards eliminating urban hunger and improved nutritional security. However, there is scanty knowledge about the composition, diversity, and socioeconomic contributions of urban food trees in African cities. This paper examines the diversity and composition of the urban forest and food trees of Accra and sheds light on perceptions of urbanites regarding food tree cultivation and availability in the city. Using a mixed methods approach, 105 respondents in six neighborhoods of Accra were interviewed while over 200 plots (100-m^2 each) were surveyed across five land use types. Twenty-two out of the 70 woody species in Accra have edible parts (leaves, fruits, flowers, etc.). The food-tree abundance in the city is about half of the total number of trees enumerated. The species richness and abundance of the food trees and all trees in the city were significantly different among land use types ($p < 0.0001$) and neighborhood types ($p < 0.0001$). The diversity of food-bearing tree species was much higher in the poorer neighborhoods than in the wealthier neighborhoods. Respondents in wealthier neighborhoods indicated that tree and food-tree cover of the city was generally low and showed greater interest in cultivating food (fruit) trees and expanding urban forest cover than poorer neighborhoods. These findings demonstrate the need for urban food policy reforms that integrate urban-grown tree foods in the urban food system/culture.

Keywords: mixed methods; richness; edible; food bearing; neighborhoods

---

1. Introduction

Malnutrition, poverty and food insecurity, once thought to be characteristic of rural areas, have become prevalent in cities of rapidly urbanizing regions such as Africa [1]. About 40% of Africa's urban population lives below the poverty line (US$ 2.0 per day), while an estimated 72% lives in slums [2,3]. Furthermore, four out of every five urban households in Africa are food insecure with the urban poor being the most severely affected [1] and about one-fifth of them spend up to 70% of their income on food [4]. Incorporating tree-food cultivation and conservation in the urban landscape and within human settlements could enhance food security and complement traditional food security strategies including agriculture [5,6]. However, data about the contribution of trees/forests to food and nutritional security especially in urban areas are lacking or at best scanty and fragmentary.

Urbanization has massive implications for food security in cities of many developing countries. It is projected to cause about 3% and 9% loss of prime agricultural lands and food production, respectively, in Africa by 2030 [7]. It will also trigger declines in the agricultural labor force and further depletion of urban green spaces [8,9]. These transformations have inspired donor-aided importation of cereals to sustain the burgeoning urban populations in many West African countries [10]. Consequently, there is an evolution of the urban food culture towards increased consumption of cheap fast foods [11,12], which together with the traditional staples of cereals, roots, and tubers is gradually resulting in increased cases of malnutrition, obesity, and stunting [12,13]. Food and nutritional insecurity is more severe in some urban areas than in rural areas [13]. Addressing these challenges requires locally relevant knowledge, supportive policies, and optimal physical and social environments [14].

One target of the sustainable development goals is to eliminate hunger and all its forms by 2030. Similarly, the African Union Summit at Malabo, Equatorial Guinea, in 2014 adopted the declaration to end hunger by doubling agricultural productivity levels and eliminating child malnutrition by 2025 [12,15]. However, there is growing evidence that agricultural strategies alone fall short of eliminating hunger, result in unbalanced diets that lack nutritional diversity, enhance exposure of the most vulnerable groups to volatile food prices, and fail to recognize the long-term consequences of intensified agricultural systems [16,17]. In parallel, there is growing evidence that forest- and tree-based systems can play significant roles in complementing agricultural production in the food security discourse. Food security also includes nutrition security and access to food, which ensures adequate macro- and micronutrient intake without excessive intake of calories [17].

Forest directly provides a diversity of healthy foods rich in micronutrients and fiber but low in sodium and refined fats and sugar, i.e., foods that are culturally valued and integral to the local food systems and food sovereignty, and that fill seasonal or cyclical food gaps and serve as safety nets or buffers in times of critical food shortage [16–18]. Forests within cities directly supply fruits, vegetables, seeds, oils, and nuts, which provide essential nutrients including vitamins and minerals [18]. Urban forests indirectly supply food through urban apiculture, silvopastoral practices (e.g., *Piper nigrum*), taungya systems, and snail and grasscutter farming [18]. Urban agroforestry is a promising option that could be integrated into a broader concept of urban food forestry [18–20] to sustain food security and environmental wellbeing. The inclusion of trees in food and nutrition security is fairly recent [5]. However, few studies have characterized urban food tree species diversity and composition within cities in West Africa [21,22] despite evidence that children who live in areas with high forest/tree cover enjoy greater dietary diversity and more nutritious foods [23].

Ghana, like most countries in sub-Saharan Africa (SSA), is undergoing rapid urbanization with an urban population (which now stands at 54%) projected to increase to 75% by 2050 [24,25]. Increases in human population lead to urban sprawl, which results in loss of agricultural and natural vegetation cover. Besides, food insecurity in Ghana is most prevalent in urban areas [26] where urban diets consist more of animal, sugary, and fatty foods and less of fruits and vegetables [27,28]. However, to date there is no known study on perceptions of the urbanites or composition and diversity of urban food-tree species across the variety of land uses and neighborhood types in this country. Additionally, about two out of every three school-age children in urban Ghana experience at least one nutritional deficit and are either stunted, underweight or anaemic with calcium intake being very low [28].

2. Materials and Methods

2.1. Study Area

The study was conducted in Accra (5°35″ N, 0°60″ W), Ghana. Accra is the capital of Ghana and is located on the Gulf of Guinea. It has a total land area of more than 250 km^2 and a population of about 2.27 million [25]. The greater Accra Metropolitan Area (AMA) is estimated to have a population of 4 million. It is a cosmopolitan city that is socioeconomically stratified based on environmental quality [29]. Neighborhood socioeconomic conditions have also been shown to correlate strongly with vegetation cover with poorer neighborhoods being less green and wealthier neighborhoods being more green [30]. The six study neighborhoods lie on a line from the coast towards the northern boundaries of the AMA, i.e., Jamestown/Ushertown (JT), Asylum Down (AD), Kanda (KN), Nima (NM), Roman Ridge (RR), and East Legon (EL). JT is one of the oldest neighborhoods of Accra and lies between the coast and Accra business center; AD and NM are located approximately 3 km inland separated by the Ring Road central. The Kanda highway separates NM and KN. RR and EL are approximately 5 and 10 km inland and west and north of the Accra International Airport, respectively.

Accra is in the equatorial climatic region, one of the driest regions in Ghana. Its mean annual precipitation and temperature are about 809 mm and 26.6 °C, respectively [31]. Humidity is generally high (65%–95%).

Savannah grasslands and thickets were the dominant vegetation type in the Accra plains prior to recent urbanization. Originally, a dry semi-deciduous forest of the "south-east outlier type" was the natural vegetation cover of the Accra plains. Over several millennia, human activities have successfully transformed this dry forest into the present savannah type [32]. The dry forest was a multistory closed canopy forest composed of several woody tree and liana species such as *Millettia thonningii*, *Diospyros* spp., *Ceiba* spp., etc. With such high tree diversity, the dry forest more likely supported more food-bearing trees than the savannah. On the other hand, the open savannah is easier to exploit for farming and grazing.

The study employed a mixed-methods approach in gathering the data. Mixed methods is a research approach where researchers collect, analyze, and integrate quantitative and qualitative data in a single study or a sustained long-term project [33]. The approach in this study is based on the fact that the research questions are best addressed by complementing both quantitative and qualitative data. The quantitative data were collected through vegetation inventory, while semi-structured questionnaires were used to collect qualitative data. Gathering both qualitative and quantitative data aided the generation of multiple perspectives about the contribution of urban forests to urban food security in Accra.

2.2. Vegetation Sampling Design

The plots were located in Accra in a stratified random sampling design in 2017 as part of the urban biomass study in Ghana [9,10]. Stratification was based on the relative homogeneity of the physiognomic units of land uses. Five land uses, i.e., home gardens (HG), cemeteries/sacred grooves (C), public parks (PP), institutional compounds (IC), and streets (S) were identified with the aid of land use maps and Google Earth images.

A total of 200 relevés and 5 streets were surveyed. Sample areas were 100 m^2 in the parks and cemeteries, while entire areas of the institutional compounds and home gardens were surveyed. The trees and shrubs within each relevé were counted by species and the number recorded. All trees were identified to the species level with the aid of tree identification experts and published tree identification guides such as those by Hawthorne and Gyakari [34] and Oteng-Amoako [35]. Samples of unidentified tree species were collected and processed following standard taxonomic procedures and subsequently forwarded to the University of Ghana and Forestry Research Institute of Ghana Herbaria. The selection of home gardens followed the systematic approach described below under the socioeconomic survey section.

2.3. Socioeconomic Survey

Data on the urbanites' perceptions and use of urban forest trees for food and their interest in cultivating and promoting urban food forestry were collected by means of semi-structured and structured interviews and observations. Respondents were drawn from three classes of neighborhoods, i.e., high income, middle income, and low income, and interviewed between September and November 2017. JT and NM were selected in the low-income category, KN and AD in the middle-income neighborhood and EL and RR in the high-income neighborhood [29,36,37]. Within each neighborhood, a systematic sampling approach was adopted in selecting the households for the socioeconomic survey. Transects were randomly laid across the neighborhood. For each transect, a house was randomly selected within a 50-m radius at the starting point and subsequently every 200 m along the transect. One adult individual present in the selected house was interviewed.

Overall, 96 representatives of urban households were interviewed using semi-structured questionnaires. The demographic characteristics of the respondents are presented in Table 1. About 55% of the respondents were females, 44% lived in low-income neighborhoods, 30% in high-income neighborhoods, and the rest in middle-income neighborhoods. Lack of interest and high bureaucratic hurdles in the middle- and high-income neighborhoods accounted for the low participation rates. About 17% of the respondents lived in government-owned houses, 23% in rented private houses, while 60% lived in family/self-owned houses. Roughly 62% were in informal employment, 33% were formerly employed, while 5% were unemployed. The educational background of the respondents was 37% secondary, 30% tertiary, and 30% primary or basic education, while the remaining 3% had never been to school.

Table 1. Socioeconomic characteristics of the study sample, $N = 96$.

Socioeconomic Variables	Number	Percent (%)
Neighborhood characteristics		
High income	29	30.21
Middle income	25	26.04
Low income	42	43.75
Gender		
Female	53	55.21
Male	43	44.79
House ownership		
Family	57	59.38
Government	17	17.17
Rented	22	22.92
Age range (years)		
20–40	54	56.25
40–60	32	33.33
>60	10	10.42
Employment		
Formal	32	33.33
Informal	59	61.46
Informal	5	5.21
Education		
Primary	29	30.21
Secondary	35	36.46
Tertiary	29	36.46
Uneducated	3	3.13

Verbal informal consent of the respondents was obtained from each individual who participated in the study, and the researchers adhered to appropriate international ethical guidelines. The aim and purpose of the study was explained to selected participants. The questionnaire used during the interviews was designed to gather data about the socioeconomic characteristics of the respondents, their interest in cultivating fruit trees and promoting urban forestry, as well as on their views about the urban forest and fruit tree cover in the city. The common tree fruits consumed among urbanites were also investigated. In all, about 15 questions were included in each questionnaire. In addition, the trees and shrubs present in the compounds of the selected houses were inventoried.

2.4. Data Analysis

Patterns of species diversity were analyzed using two types of diversity, i.e., α-diversity and β-diversity, and also evenness. Two aspects of α-diversity were analyzed, the first being species richness (S), which is defined as the number of species per unit sample plot or land use. Because S can be exaggerated by the presence of rare species, α-diversity was also measured with the Simpson's diversity index (D) (Equation (1)) [38]. It is a weighted expression of species richness and abundance of each species in the population and is calculated as

$$D = 1 - \frac{\sum n(n-1)}{N(N-1)} \tag{1}$$

where Σ = summation (total), n = number of individuals of each species, and N = the total number of individuals of all species in the population.

Evenness, defined as the relative abundance of species per unit area, was used to measure the similarity of relative abundance of species within sample plots (Equation (2)) [38]. It was estimated with the Pielou's evenness index (J) as

$$J = \frac{H}{S} \tag{2}$$

where H = Shannon diversity index and S is species richness as previously defined.

Chi-square test was conducted on abundance, S, D and J among land use and neighborhood types using SAS.

Beta (β) diversity was calculated to determine species turnover or the extent to which species diversity differs among land uses and neighborhood types within Accra. Various measures of β-diversity have been proposed. However, in the present study, Whittaker's diversity index (β_w) was used because it is widely regarded as a simple but highly effective measure of β-diversity [39]. It was calculated as:

$$B_w = \frac{S_{total}}{S_{ave}} \tag{3}$$

where S_{total} = overall species richness of the city and S_{ave} = α-diversity for each land use or neighborhood type.

In addition, correspondence analysis was applied to the data to illustrate floristic relationships between plant communities, land uses, and neighborhood types. Neighborhood type and land use type were considered nominal variables, while species abundance was a continuous variable. Analysis was conducted for all species and food-bearing tree species in Accra. A desk study was used to extract concentrations of some of the essential micronutrients supplied by these trees within Accra.

Descriptive statistics, such as percentages and frequencies, were used to analyze the data obtained from the questionnaires. Bar charts were generated using Microsoft Excel. In addition, we used a binomial generalized linear model (GLM) to determine whether respondent socioeconomic characteristics influence their views about urban fruit tree cover and their interest in promoting urban food forestry in Accra. More specifically, logistic regression was used to determine whether neighborhood type and other socioeconomic variables had an effect on the interest in cultivating fruit trees and promoting urban forestry. The dependent variable is binary, i.e., 1 if the respondent is interested in cultivating trees and 0 otherwise. Binary response data can be accommodated using a fixed-effects or random-effects logit model:

$$\text{Logit } (Y_i) = LN\ (p/1-p) = B_0 + B_1 X_1 + B_2 X_2 + \ldots \ldots + X_n B_n \tag{4}$$

where Y_{ij} denotes the ith survey response in the jth neighborhood type, B_0 and B_1 are coefficients to e estimated in the regression step, $X_{i,j}$ is a vector of independent variables (e.g., neighborhood type, household size, sex of respondents etc.).

3. Results

3.1. Overall Woody Species Diversity

A total of 798 trees and shrubs belonging to 70 species and 30 families were recorded during the period of survey. There was a significant difference in species richness (χ^2_4 = 33.4, p < 0.0001) and abundance (χ^2_4 = 362.4, p < 0.0001) for all trees among land use types (Table 2), although only the latter was significantly different (χ^2_4 = 267.8, p < 0.0001) with respect to neighborhood types (Table 3). Home gardens and institutional compounds had the highest Simpson's diversity index and Pieluo's evenness values among other land uses. Similarly, low-income neighborhoods had the highest Simpson's index and Pieluo's evenness of 0.96 and 0.93, respectively, among neighborhood types (Table 3). From i-tree canopy analysis, the tree cover increased with the wealth status of the neighborhood. JT and NM had a tree cover of 10.2 ± 4.15% and 14.0 ± 5.29%, AD and KN had 30.0 ± 10.0% and 37.5 ± 7.65%, while EL and RR had 46.7 ± 6.44% and 60.9 ± 5.88% of the total neighborhood land area, respectively.

Table 2. Woody tree species richness, abundance, Simpson's diversity index, and Pielou's evenness among land uses in Accra.

Land Use	Richness	Abundance	Simpson's D	Pielou's Evenness, J
Cemetery	12.0	68.0	0.496	1.247
Institutional compound	33.0	144.0	0.957	2.945
Home garden	47.0	368.0	0.945	3.149
Park	29.0	91.0	0.910	2.800
Street	12.0	127.0	0.829	1.870
Accra total	70.0	798.0	0.950	3.337
Chisq (χ^2)	33.4	362.36	0.163	0.406
p-value	**<0.0001**	**<0.0001**	0.997	0.982

Note: Bold numbers: highlighting areas where comparisons were statistically significant.

Table 3. Woody tree species richness, abundance, Simpson's diversity index, and Pielou's evenness among neighborhood types in Accra.

Neighborhood	Abundance	Richness	Simpson's D	Pielou's Evenness, J
High income	502	42	0.93	0.80
Middle income	215	33	0.92	0.86
Low income	131	35	0.96	0.93
Chisq (χ^2)	267.8	1.22	0.001	0.01
p-value	**<0.0001**	0.544	0.999	0.995

Note: Bold number: highlighting areas where comparisons were statistically significant.

A total of 3.4 ha was sampled in home gardens and institutional compounds in the six neighborhoods. The sampled area coverage among neighborhoods was 29% in EL, 26% in RR, 20% in KN, 8% in AD, 9% in NM, and 7% in JT. In addition, about 0.30 and 0.34 ha of land were sampled in cemeteries and parks, respectively. No park was located in either of the study neighborhoods.

3.2. Food-Tree Species Diversity

About half of the total number of trees enumerated (798) in this study had edible components (i.e., leaves or fruits) and belonged to 22 species and 17 families. Both species richness and abundance were significantly different among land uses (Table 4), while the former was not significantly different among neighborhood types (Table 5). Cemetery had the highest Simpson's index and Pielou's evenness values of 0.905 and 0.963, respectively, owing to its relatively low species richness and abundance. Meanwhile, food trees constituted only 10% of all the trees in cemeteries while the highest proportion among land uses was 66% in home gardens. The Simpson's diversity index and Pielou's evenness were highest in low income neighborhoods compared to the other neighborhood types (Table 5).

Table 4. Food-bearing tree diversity of different land use types in Accra, Ghana.

Land Use	Richness	Abundance	Simpson's D	Pielou's J	% Fruit Trees	% Fruit Tree Species
Cemetery	5.0	7.0	0.905	0.963	10.3	41.7
IC	12.0	76.0	0.877	0.873	52.8	36.4
HG	19.0	243.0	0.889	0.821	66.0	40.4
Park	7.0	19.0	0.819	0.868	20.9	24.1
Street	3.0	43.0	0.135	0.271	33.9	25.0
Total	22.0	388.0	0.867	0.761	48.6	31.4
Chisq (χ^2)	17.9	476.6	0.606	0.4063	56.3	08.5
p-value	**0.0013**	**<0.0001**	0.962	0.982	**<0.0001**	0.0758

Note: Bold numbers: highlighting areas where comparisons were statistically significant. IC = Institutional compound, HG = Home garden.

Table 5. Food-tree species diversity and abundance among neighborhood types in Accra.

Neighborhood	Abundance	Richness	Simpson's D	Pielou's J
High income	198	13	0.93	0.80
Middle income	122	12	0.92	0.85
Low income	69	16	0.96	0.90
Chisq (χ^2)	64.8	0.634	0.00	0.01
p-value	**<0.0001**	0.728	0.999	0.997

Note: Bold number: highlighting areas where comparisons were statistically significant.

3.3. Species Composition and β-Diversity of Food-Tree Species

Woody species of various land uses in Accra were separated using correspondence analysis (Figure 1). Dimension 1 explained 34.1% of the variation and appears to be determined by species such as *Veitchia merrillii*, *Zanthoxylum zanthoxyloides*, and *Plumera alba* in the parks, and *Millingtonia hortensis* and *Khaya senegalensis* in the streets; dimension 2 explained 30.1% of the variation and is determined more by food-bearing tree species such as *Persea americana*, *Vernonia amygdalina*, *Moringa oleifera*, and *Elaeis guineensis*, etc., in the home gardens and by *Millingtonia hortensis* and *Albizia lebbeck* in the streets. Institutional compounds and home gardens are grouped together, as are parks and cemeteries, with streets completely separated from these groups.

Woody species separation among neighborhood types followed a different pattern (Figure 2). Dimension 1 explains 57.1% of the variation and was influenced more by low-income and high-income neighborhoods. Dimension 2 explains 42.9% of the variation and was determined more by the middle-income neighborhoods. Species such as *Artocarpus incisus*, *Cassia spectabilis*, *Citrus reticulata*, and *Citrus limon*, as well as *Musa* spp., *Vernonia amygdalina*, and *Gliricidia sepium*, etc., also strongly influenced dimension 2.

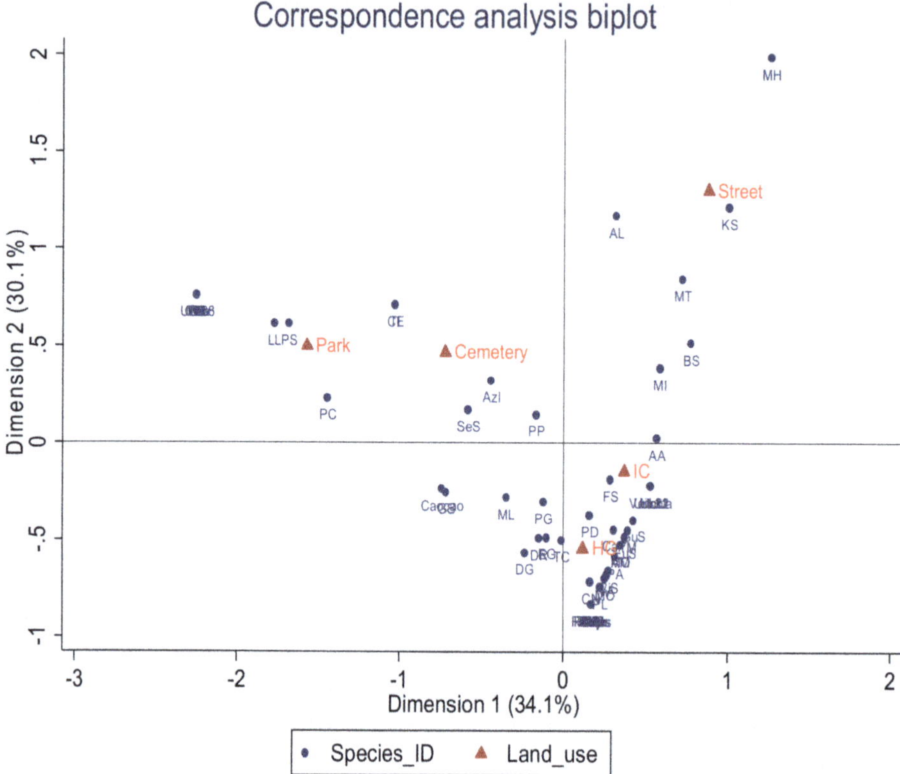

Figure 1. Correspondence analysis biplot of woody species among land uses in Accra. Land_use: IC = Institutional compound, HG = home garden. Species: AA = *Acacia auriculiformis*, AM = *Acacia mangium*, AL = *Albizia lebbeck*, AO = *Anarcadium occidentalis*, AnM = *Annona muricata*, AT = *Antiaris toxicaria*, AI = *Artocarpus incisus*, AS = *Araucaria* spp., Azl = *Azadirachta indica*, BS = *Blighia sapida*, Cap = *Calotropis procera*, CaP = *Carica papaya*, CaE = *Casuarina equisetifolia*, CS = *Cassia spectabilis*, CP = *Ceiba pentandra*, CeN = *Cestrum norturnum*, CE = *Chlorophora excelsa*, CL = *Citrus limon*, CR = *Citrus reticulata*, CiS = *Citrus sinensis*, CN = *Cocos nucifera*, CC = *Cola caricifolia*, DR = *Delonix regia*, DG = *Dialium guineense*, EG = *Elaeis guineensis*, EuS = *Eucalyptus* spp., FB = *Ficus bonsai*, FS = *Ficus* spp, GS = *Gliricidia sepium*, GA = *Gmelina arborea*, GuS = *Guaiacum sanctum*, HF = *Holarrhena floribunda*, KS = *Khaya senegalensis*, LL = *Leucaena leucocephela*, MI = *Mangifera indica*, MT = *Millettia thonningii*, MH = *Millingtonia hortensis*, MC = *Morinda citrifolia*, ML = *Morinda lucida*, MO = *Moringa oleifera*, MB = *Musa balbisiana*, NL = *Newbouldia laevis*, PP = *Peltophorum pterocarpum*, PA = *Persea americana*, PC = *Pinus caribaea*, PD = *Pithecellobium dulce*, PS = *Pithecellobium saman*, PlA = *Plumera alba*, PL = *Polyalthia longifolia*, PG = *Psidium guajava*, Ricinus = *Ricinus communis*, RR = *Roystonia regia*, SeS = *Senna siamea*, SR = *Sterculia rhinopetala*, TI = *Tamarindus indica*, TC = *Terminalia catappa*, TM = *Terminalia montalis*, Caocao = *Theobroma cacao*, Thuja = *Thuja orientalis*, Veitchia = *Veitchia merrillii*, VA = *Vernonia amygdalina*, ZZ = *Zanthoxylum zanthoxyloides*.

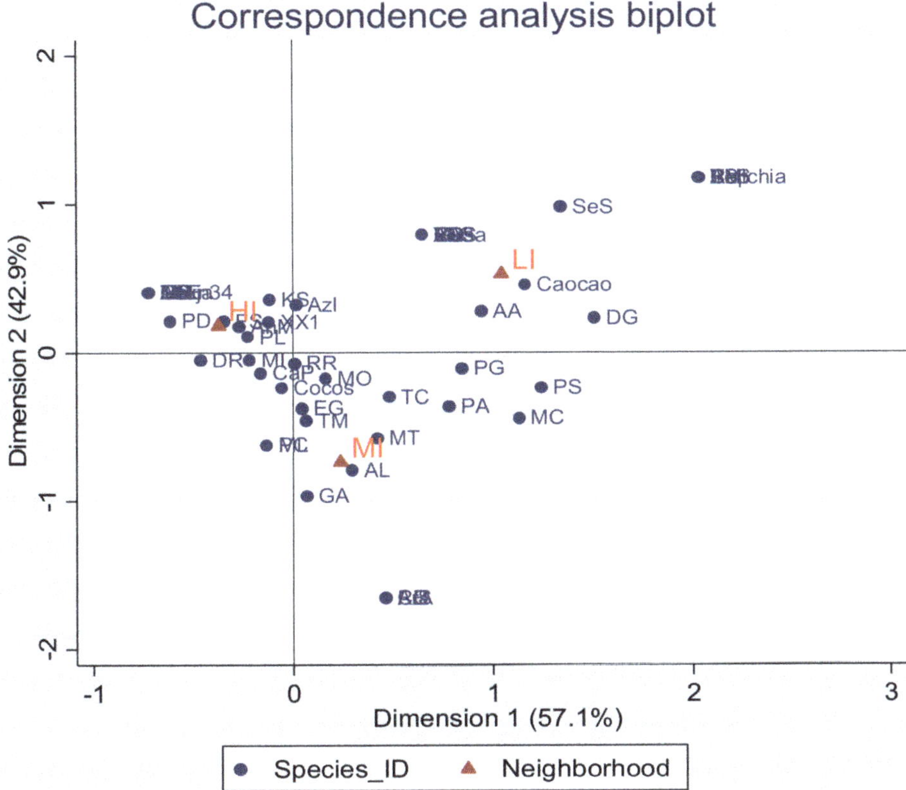

Figure 2. Correspondence analysis biplot of species in three neighborhood types in Accra. HI—high-income, MI—middle-income, and LI—low-income neighborhood types. Names of species are as in Figure 3.

Beta diversity analysis revealed that 75% of the species in the cemetery were similar to those of institutional compounds and home gardens while only 50% and 58% were similar to those of streets and parks, respectively. In contrast, 30% and 42% of species in IC were found in streets and parks, respectively, whereas 76% were in home gardens. About 75% and 55% of the species in streets and parks, respectively, were also in home gardens while 50% of the species in streets were also in home gardens. As a consequence of these similarities, Beta (β_w) diversity estimates between land uses were not significantly different, ranging between 1.4 (IC and HG) and 1.7 (HG and street, park and street etc.).

3.4. Nutrient Composition of Selected Common Urban Food-Tree Species in Accra

The most common food-bearing tree species in the survey based on abundance included *Mangifera indica* (mango), *Elaeis guineensis* (oil palm), *Cocos nucifera* (coconut), *Carica papaya* (pawpaw), *Terminalia catappa* (Indian almond), and *Moringa oleifera*. The five least abundant species were *Tamarindus indica*, *Artocarpus incisus*, *Citrus limon*, *C. reticulata*, and *Blighia sapida*.

The food-bearing tree species contain a range of major and micronutrients including vitamins essential for human wellbeing. Tree foods were consumed directly as fruit pulp, leaves, nuts, seeds, and juices like coconut water while others required further processing before being consumed. All the species enumerated are high in proteins, carbohydrates, crude fiber, crude fat, ash, and moisture content, although these vary among species. On average, protein, carbohydrate, fiber, fat, ash, and moisture content were 8.12, 39.5, 12.64, 10.3, 4.8 and 27.4 g/g, respectively. This indicates that tree foods could play a critical role in meeting the dietary needs of urbanites. In addition, the common minerals in the fruits, leaves and seeds of these tree species included calcium (Ca), potassium (K), sodium (Na), magnesium (Mg), zinc (Zn), iron (Fe), iodine (I), copper (Cu), manganese (Mn), and molybdenum (Mo). At least 50% of the food tree species listed in Accra supply 8% of the above nutrients (Table 6). At least 10 tree species can supply Cu and Mn while only two species (*Dialium guineensis* and *Elaeis guineensis*) supply I. Furthermore, a few of the tree species listed in Accra have the potential to supply niacin, riboflavin, thiamine, and the vitamins K, E, C, and A. Leaves and fruits of *Vernonia amygdalina*, *Tamarindus indica*, and *Annona muricata* can supply niacin. Mean concentrations of each mineral and the number of tree species bearing these nutrients are shown in Table 6. There are wide variations in the nutritional composition and concentrations of these nutrient elements among species. Of the sampled food-bearing trees, 6% supply niacin, 52% supply P, 63% supply Cu while over 80% supply all the other basic minerals (Table 6). About 98% of the trees provide energy when eaten as fruits, nuts, seeds, leaves, oils, and flowers. Overall 80% of the food bearing trees sampled is eaten as fruits, 31% as nuts, 28% as oils, 9% as leaves, and 29% as seeds.

3.5. Perceptions of Urban Forest Cover and Interest in Promoting Urban Forestry in Accra

In total, 58 of the 96 respondents (60%) expressed an interest in cultivating food trees and promoting urban forest expansion activities in Accra. The results show that neighborhood type and perceptions about the food-tree/forest cover significantly influenced the likelihood of one being interested in cultivating trees (Table 7). Results of the odd ratio analysis further reveal that people in high-income and middle-income neighborhoods were respectively 22 and seven times more likely to be interested in cultivating food/forest trees in Accra compared to people in low-income/slum neighborhoods. Overall interest in cultivating food-bearing trees in the high-income, middle-income, and low-income neighborhoods was respectively shown by 69%, 56% and 57% of the respondents interviewed. Similarly, people who perceived the city to be low in forest/tree cover were more likely than not to be interested in cultivating forest/food trees compared to people who perceived the tree cover of Accra to be high.

Level of awareness of urban forestry and urban food forestry in Accra was examined. The qualitative results reveal that level of awareness was only marginally different for both males and females (Figure 3). Nearly 100% of both male and female respondents were enthusiastic about promotion of urban forestry in Accra. The perception of people in high-income neighborhoods about the forest and fruit forest cover of Accra as well as the enthusiasm to promote urban forestry was higher than in the other neighborhood types (Figure 4). About 57% and 54% of the respondents in low-income neighborhoods had some level of awareness of urban forest and urban food-forest cover of Accra, respectively. Low-income neighborhoods were the least likely to promote urban forestry (95%) compared to other neighborhood types.

Table 6. Mean, minimum, and maximum micronutrient and selected vitamins in food-tree species in Accra, Ghana.

	Niacin	K	Ca	P	Zn	Na	Mg	Fe	Cu	Mn	Iodine (I)
Number of species	3.0	14.0	16.0	12.0	15.0	13.0	19.0	16.0	10.0	10.0	2.0
Proportion of sampled food tree species (%)	14.3	66.7	76.2	57.1	71.4	61.9	90.5	76.2	47.6	47.6	9.5
Number of trees	23	310	328	202	321	337	364	368	238	259	65
Proportion of sampled food trees (%)	5.9	79.9	84.5	52.1	82.7	86.9	93.8	94.8	61.3	66.8	16.8
Mean	1.2	275.3	62.8	232.5	21.2	40.2	54.8	4.1	8.9	5.3	18.8
Minimum	1.7	0.1	0.1	0.0	0.0	0.0	0.0	0.0	0.0	0.0	4.3
Maximum	1.9	673.5	250.1	1411.0	193.9	169.5	347.0	16.2	70.3	29.2	33.3

K—potassium, Ca—calcium, P—phosphorous, Zn—zinc, Na—sodium, Mg—magnesium, Fe—iron, Cu—copper, Mn—manganese.

Table 7. Logistic regression of factors influencing interest in cultivating food-bearing trees in Accra, Ghana.

Variable	DF	Wald Chi-Square	Pr > ChiSq
X1—Neighborhood type	2	6.8959	**0.0318 **
X2—House ownership	3	0.3936	0.9416
X3—Age range	2	0.8703	0.6472
X4—Sex of respondent	1	0.1161	0.7333
X5—Household size	16	6.7629	0.9776
X6—Employment of household head	2	2.4212	0.2980
X7—Level of education of respondent	3	3.5376	0.3159
X9—Satisfaction with quality and quantity of food	1	1.4275	0.2322
X10—Vegetation cover of Accra	1	1.4547	0.2278
X11—Food/Fruit tree cover of Accra	1	7.0229	**0.0080 ***
X12—Vegetation cover of neighborhood	3	5.5055	0.1383

**(*) Significant at $p = 0.05$.

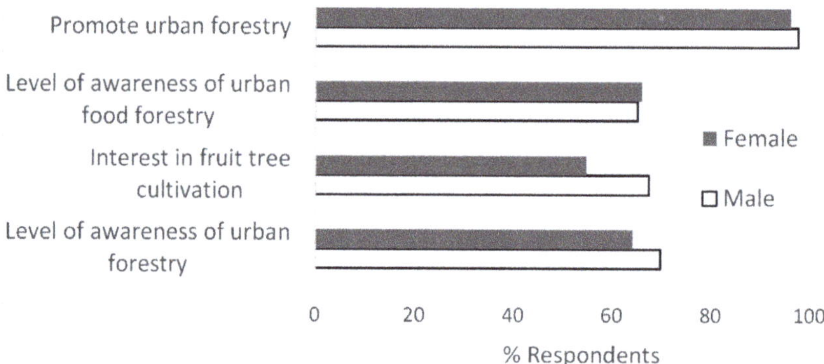

Figure 3. Views of male and female respondents on attributes of urban forestry in Accra, Ghana.

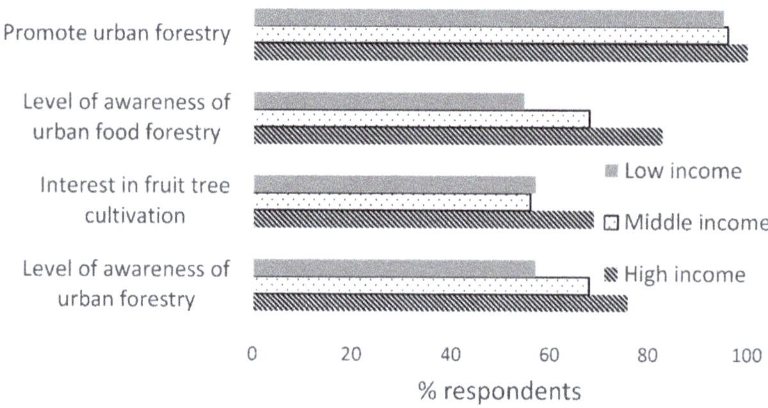

Figure 4. Views of respondents on level of awareness of the urban forest and promotion of urban forestry among neighborhood types in Accra, Ghana.

4. Discussion

4.1. Forest and Food-Tree Species Composition and Diversity

In the current study, we found that the city of Accra supports a fairly high woody species diversity, and that this diversity was significantly stratified among the vegetated land use types and neighborhood types (Tables 2 and 3). We further found that species such as *Azadirachta indica* (neem), *Mangifera indica* (mango), *Elaeis guineensis* (palm), *Cocos nucifera* (coconut), and *Polyalthia* spp. (weeping willow) were the five most common species in the city. These findings are consistent with findings of floral diversity studies in Abuja, Nigeria, and the Valley View University campus near Accra where woody species richness of 69 and 53, respectively, were observed [32,40]. However, species richness and diversity in the current study were lower than the 176 species and Shannon index of 3.70 reported in Kumasi [9,41] and the 297 tree species in Lome, Togo [42]. Yet the diversity and richness values in Accra conform with recent findings that cities in Africa support relatively high diversity even when compared to neighboring natural forest [40–43]. Although cities are typically located in biodiversity hotspots [44,45], their high diversity, intercity disparities in diversity, and species richness are inherent in the morphology and age of the city, and the natural ecological factors as well as building and population densities. Accra is located in the coastal savannah zone with many of the (common) species recorded in the current study classified as savannah species [32,46]. This explains its lower richness and diversity values compared to cities located in high forest zones, which possibly have lower population and building densities.

The within-city differences in woody species diversity and richness among land uses may be ascribed to several factors, including governance, ecological, and socioeconomic conditions as well as the culture of the people. Home gardens harbor the highest species diversity and compositional differences. Species richness and diversity of home gardens in Accra are similar to those of home gardens in cities such as Rio Claro, Brazil [47], and directly manifest their multifunctional and structural complexities [48] underpinned by various socio-ecological constraints [47,49]. Home gardens in Accra are varied in size and severely fragmented, and maintained for alimentary, fuelwood, aesthetics, and environmental conservation purposes. Consequently, their diversity reflects the diversity of the individuals residing in the city and their interests. On the other hand, parks, streets, and cemeteries are directly under the control and management of the local government. Diversity in these land uses is comparatively lower because these were planted with a few exotic tree species for specific limited functions. While *Azadirachta indica* predominates in the cemeteries, the most common species in parks and on streets were *Gliricidia sepium* and *Millingtonia hortensis*, respectively. Their primary functions are recreation (aesthetics), shade, and environmental conservation.

The current study also found that species abundance and richness were much higher in high-income neighborhoods compared to other neighborhood types while the reverse was true for Simpson's diversity index and Pielou's evenness. These findings corroborate previous studies that found that high-income neighborhoods had higher biodiversity than low-income neighborhoods [50,51]. This coupled with the low tree cover in low-income neighborhoods should be an issue of concern for policy makers, as residents in these low income neighborhoods are less likely to access and enjoy sufficient environmental benefits emanating from urban forest and tree diversity in these third world cities.

4.2. Urban Forest-Tree Diversity and Food/Nutritional Security

Residents of Accra, like their compatriots in other West African cities, have a low fruit and vegetable consumption culture [11]. While high fruit and vegetable prices deprive many of their preferred balanced diet (lack of access), unavailability of food resources can constrain food security in a given area. In the light of this, we explored the availability and distribution of food-bearing tree species in three neighborhood types within Accra. It was found that 31% of the tree species and nearly 50% of all trees enumerated were food bearing (Table 4). About 51% of these food trees were in the high-income neighborhoods and 18% in low-income neighborhoods. In contrast, food-tree diversity was highest in the low-income neighborhoods compared to the other neighborhood types. Similar to the findings of the present study, 18 food-bearing species were recorded on the campus of the University of Port Harcourt, Nigeria [22]. Our findings are also consistent with previous findings that plant species useful for food, medicines and other provisioning services were more frequently found in gardens of poorer residents than in those of wealthier residents [49]. Considering that increasing neighborhood wealth correlates with increasing vegetation cover [30], we surmise that the low tree abundance in low-income neighborhoods may be due to lack of space resulting from legacy effects of past development and lack of interest in cultivating/managing trees. Whereas poorer urbanites maintain small home gardens for provisioning of basic services such as alimentary, medicinal, income, and livelihood services which reflect their rural origins and cultural heritage, the relatively wealthier class invest in large home gardens for their aesthetic and recreational functions [47,49]. These disparities in green cover and food-tree populations among neighborhood types could have dire tree-food availability consequences especially in low-income neighborhoods where over 70% of the income is mostly spent on purchasing staple foods. The foregoing emphasizes the need for scientists and policy makers to consider judicious and scientific policies that ensure equity in the distribution of green spaces and biodiversity among neighborhoods in order to avoid issues of resource and environmental injustice [8,9]. In low-income neighborhoods, opportunities exist with respect to planting food-bearing trees along streets, within some family house compounds, and in the pockets of bare areas dotted around the neighborhoods.

Analysis of the nutrient contents and composition of food components of these tree species revealed that they can supply a very large amount of basic nutrients (proteins, carbohydrates, fats, fiber), vitamins, and mineral nutrients. Average nutrient concentrations of the selected minerals found were fairly high (Table 6). The edible parts of these trees included fruits, leaves, nuts, seeds, oils, and flowers. Our findings corroborate observations by the FAO that home gardens provide families with the non-staples they require all year round [52], and forest-tree products are usually higher in nutrient contents than domesticated crops and staples [18]. However, evidence of resource availability does not necessarily imply access and utilization. In urban Ghana, diets tend to be dominated by animal-based products of rice, pasta, meat, and fish mixed and less of fruits and vegetables [11,27]. Although such dietary diversification reduces the risk of communicable diseases such as Type 2 diabetes among adults, it is only partially true in most sub-Saharan African cities [53]. About 67% school-age children in parts of Accra are malnourished, suffering at least one nutritional deficit of being stunted, underweight or anemic [28]. Although energy intake of these school kids was adequate, there was a conspicuously low Ca intake [28]. Moreover, deficiencies in minerals such as Fe, Zn, and vitamin A in urban diets around the world are not uncommon [54]. In contrast, our current study reveals that 76% of the food-tree species contain high concentrations of Ca (0.05–250 mg/100 g), 76% contain Fe (0.0009–16.1 mg/100 g) and 71% contain Zn (0.0003–193 mg/100 g). Urban forest-tree species therefore are (1) a source of a diversity of healthy food, high in micronutrients and fiber, and low in fats and sugars; (2) high in food of cultural value and contribute significantly to the local food systems and sovereignty; and (3) help to fill seasonal and cyclical food gaps and act as a "safety net" when there is shortage [17]. The high incidences of nutrient deficiencies among urbanites in Ghana vis-à-vis the relatively high abundance of nutritious food-tree species in Accra suggests that urban tree foods may not be accessible to all.

Besides, urban food trees can provide several ecosystem services and goods. In Bobo Dioulasso, Burkina Faso, mango and cashew plantations have the potential to contribute 6% of the monthly food expenditures of the involved poor households, reduced the run-off co-efficient by 4% (reduction in flood risks and increased infiltration), sequestered over 1835.56 tons of CO_2 per ha, and significantly reduced ambient temperatures [55]. Considering this wide range of benefits, we suggest populating school compounds and open spaces in cities, particularly Accra, with food-bearing trees and encouraging the consumption of the fruits among the urbanites. Policies and strategies to expand the urban food forest cover, extend shelf-life of fruits, and enhance the consumption of fruits in African cities are critical.

4.3. Perceptions and Interest in Promoting Urban Food Forest

Restoring, protecting, and enhancing green infrastructure (forest and woodlands) in urban areas are not only ecologically and socially desirable, they are also economically viable [56]. As outlined above, the urban forest has the potential to provide a myriad of ecosystem goods and services, food security, climate regulation, aesthetics, regulation of storm run-off, reduction in erosion, and other supporting services etc. [57,58]. Within Africa, research on social and economic aspects of urban forestry are rather limited [59]. Our present study reveals that perceptions about the level of awareness of the urban forest in Accra and the willingness to promote urban forestry did not differ significantly among key socioeconomic variables, i.e., gender, sex of respondents, neighborhood type, type of house ownership, household size. At least 70% of the respondents in the high-income neighborhoods thought there is high awareness of urban forest(ry) in the city, with about 70% of the residents with high-school education and below asserting likewise. However, cultivating food-bearing trees in the city was probably significantly influenced by the neighborhood's wealth status and the view about the forest cover (Table 7). These findings agree with findings that 99% of respondents in Ibadan, Nigeria, were aware of the urban forest, its benefits and value for their wellbeing, and expressed a strong interest in preserving urban forest cover [60]. Proximity to trees/forests and the perceived (dis)incentives associated with forest are strong determinants of perceptions about the urban forest [60,61]. Within the Ayawaso East submetropolis of Accra, residents of relatively denser forested areas rank and prioritize urban forest higher than their compatriots in less dense forested areas [62]. Since high-income neighborhoods of most cities are considerably higher in forest/green cover than other neighborhood types [8,30,63], the perceptions of respondents in high-income neighborhoods in the present study are not out of place.

Why the less educated respondents believe there is high awareness level of the urban forest in the city is rather difficult to substantiate, since a lot of these people were located in the low-income and middle-income neighborhoods where tree cover is relatively low. We believe, however, that the issue is due to misinformation and lack of adequate knowledge about what constitutes an urban forest. These urbanites may have been deluded by the presence of the few trees in their neighborhoods coupled with past developmental legacies where old towns such as James/Usher Town were built devoid of green cover [64]. Since colonial times, neighborhoods such as Nima and James Town have been developed haphazardly either with limited planning or overstepping existing plans and laws. These historical patterns, unavailability of space, customary land use systems, and beliefs about a city without nature may have shaped the views of residents in neighborhoods with less forest cover who felt that the few trees in their neighborhoods were enough to meet their basic life quality needs. The situation creates a need for expanding formal educational programs as well as education about trees and forests in cities particularly among residents of low-income neighborhoods in Accra. Educational programs must be multicultural in character, conducted in the local languages using several media platforms, and above all connect people with their trees and green spaces both physically and psychologically [65].

Our findings that residents of high- and middle-income neighborhoods were more likely to plant trees than those in low-income neighborhoods corroborate previous findings. In one city each in Nigeria and Congo D.R., 90% and 86% of respondents, respectively, indicated their preference regarding cultivating trees in urban areas and were willing to participate in public tree planting programs [66,67]. People who are well educated, have relatively high income and may live in areas with sufficient land area around their houses are therefore more likely to cultivate trees or participate in public tree planting programs [63,66,68]. The high- and middle-income neighborhoods in Accra are occupied by the elites and the wealthy classes of society who have a good judgment of the value of trees and have the resources to plant and care for trees within their homesteads.

Some studies have concluded that people in neighborhoods that are relatively deprived (low income and education levels) are less likely to participate in public tree planting programs [63,69]. Reasons why people in low-income neighborhoods seem to be less interested in tree cultivation include (1) lack of space to plant trees; (2) benefits of trees take too long to be realized; (3) trees interfere with building foundations and roofs; (4) tree waste such as falling litter is expensive and time consuming to clean; and (5) some perceive trees or forest in human settlements to be associated with negative incentives such as serving as hideouts for criminals, habitats for dangerous pests, etc.

5. Conclusions

The study assessed the urbanites' perceptions and the diversity of urban forest of Accra and its potential to contribute to food security in cities in developing countries. About 31% of the woody species and nearly 50% of all tree individuals recorded have the potential to provide food either as fruits, seeds, nuts, leaves, oils, and many others. Home gardens harbor the highest species richness with high-income neighborhoods having high abundance of trees while the highest diversity values for food-tree species were found in low-income neighborhoods. In addition, it was observed that these woody species can serve as sources of several major and micronutrients as well as vitamins essential for addressing hidden hunger issues which frequently affect urbanites in developing countries owing to the changing food culture, high food prices, and overreliance on traditional staples. In particular, more than 70% of the food-tree species in the city could supply critical nutrients such as Fe, Zn, Mo, I and Ca. It is concluded that the urban forest is a vital source of food and nutrients and could play a critical role in complementing traditional agricultural practices in the fight against urban food insecurity and malnutrition in Africa.

However, urban food-tree diversity is spatially stratified among neighborhood types, a situation that could lead to environmental and food security injustices. Low-income neighborhoods are particularly disadvantaged in this regard. Therefore, widening socioeconomic inequalities in the city will exacerbate environmental inequality [29], and by default the inequality in the distribution of the urban forest cover. Policies and strategies to address food insecurity through urban forestry must be directed at reducing income and educational inequalities as well as at engaging in massive public and private tree planting exercises in all neighborhoods. Because tree-food production is seasonal, in order to ensure the availability of tree foods all year round it is important to consider the food production seasons of species when selecting species for urban food forest cultivation. Innovative strategies could be adopted to avoid seasonal food and nutritional insecurities within the urban landscape.

The present study restricted its scope to tree products and the perceptions of tree cover and interest in cultivating trees in the city of Accra. A complete perspective of the potential and contribution of urban trees to food security requires a thorough analysis of intra- and intercity urban forest and food security relations. Deliberate policies in favor of food-tree cultivation or urban food forestry as part of measures to address urban food insecurity should be a national and regional priority in Africa. Policies to reduce income and educational inequality in cities are critical to sustaining greener cities, which are ideals enshrined in the sustainable development goals.

Plans for follow-up research include studies to monitor tree-growth responses with respect to neighborhood characteristics and the factors that influence participation in tree planting programs within cities in these developing countries. Urban forest structure and the diversity relationship to urban health will be another primary research focus. It may also be worthwhile to increase sampling efforts and number of neighborhoods in a follow-up study in order to provide more robust data on the dietary pattern and nutritional habits of neighborhoods in relation to environmental quality and vegetation cover.

Author Contributions: Conceptualization, B.F.N. and O.F.; Methodology, B.F.N.; Software, B.F.N.; Validation, B.F.N., N.A.K., R.J., and O.F.; Formal Analysis, B.F.N.; Investigation, B.F.N.; Resources, B.F.N., N.A.K., R.J.; Data Curation, B.F.N.; Writing—Original Draft Preparation, B.F.N.; Writing—Review & Editing, B.F.N., N.A.K., R.J., and O.F.; Visualization, B.F.N.; Supervision, B.F.N., O.F.; Project Administration, B.F.N.; Funding Acquisition, B.F.N., R.J., and O.F.

Funding: This research was funded by the German Federal Ministry of Education and Research (BMBF) and the German Federal Ministry of Economic Cooperation and Development (BMZ) through the BiomassWeb project, which is coordinated by the Center for Development Research (ZEF), University of Bonn, Germany, and the Forum for Agriculture Research in Africa (FARA) Accra, Ghana.

Acknowledgments: We are grateful to several individuals and institutions for their assistance with data collection and processing. Many thanks to all institutional heads and landlords and landladies in the six neighborhoods of Accra who permitted us to collect vegetation and socioeconomic data on their premises. Our profound gratitude to Jonathan Dabo of CSIR-FORIG for assisting with tree identification, Nadia Danso and Emmanual K. Asare for assisting with data collection. The contributions of the FARA staff who facilitated the paperwork of our operations at FARA are duly acknowledged. We are grateful to the four anonymous reviewers whose comments critically improved the quality of the paper.

Conflicts of Interest: The authors declare no conflict of interest.

References

1. Frayne, B.; Pendleton, W.; Crush, J.; Acquah, B.; Battersby-lennard, J.; Bras, E.; Chiweza, A.; Fincham, R.; Kroll, F.; Leduka, C.; et al. *The State of Urban Food Insecurity in Southern Africa*; African Food Security Urban Network: Cape Town, South Africa, 2010; ISBN 9780986982019.

2. Baker, J.L. *Urban Poverty: A Global View*; The World Bank Group: Washington, DC, USA, 2008; Available online: http://www.worldbank.org/urban/ (accessed on 20 September 2017).

3. Chen, S.; Ravallion, M. Absolute poverty measures for the developing world, 1981–2004. *Proc. Natl. Acad. Sci. USA* **2007**, *104*, 16757–16762. [CrossRef] [PubMed]

4. Maxwell, D.; Levin, C.; Armar-klemesu, M.; Ruel, M.; Morris, S.; Ahiadeke, C. *Urban Livelihoods and Food and Nutrition Security in Greater Accra, Ghana*; IFPRI: Washington, DC, USA, 2000; ISBN 0896291154.

5. CFS/HLPE. *Sustainable Forestry for Food Security and Nutrition*; Food and Agriculture Organization (FAO): Rome, Italy, 2017.

6. Pimentel, D.; McNair, M.; Buck, L.; Pimentel, M.; Kamil, J. The value of forests to world food security. *Hum. Ecol.* **1997**, *25*, 91–120. [CrossRef]

7. Bren d'Amour, C.; Reitsma, F.; Baiocchi, G.; Barthel, S.; Güneralp, B.; Erb, K.-H.; Haberl, H.; Creutzig, F.; Seto, K.C. Future urban land expansion and implications for global croplands. *Proc. Natl. Acad. Sci. USA* **2017**, *114*, 8939–8944. [CrossRef] [PubMed]

8. Nero, B.F. Urban Green Spaces Enhance Carbon Sequestration and Conserve Biodiversity in Cities of the Global South: Case of Kumasi, Ghana. Ph.D. Thesis, University of Bonn, Bonn, Germany, 2017.

9. Nero, B.F. Urban green space dynamics and socio- environmental inequity: Multi-resolution and spatiotemporal data analysis of Kumasi, Ghana. *Int. J. Remote Sens.* **2017**, *38*, 6993–7020. [CrossRef]

10. Cumming, G.S.; Buerkert, A.; Hoffmann, E.M.; Schlecht, E.; Von Cramon-Taubadel, S.; Tscharntke, T. Implications of agricultural transitions and urbanization for ecosystem services. *Nature* **2014**, *515*, 50–57. [CrossRef] [PubMed]

11. Codjoe, S.N.A.; Okutu, D.; Abu, M. Urban Household Characteristics and Dietary Diversity: An Analysis of Food Security in Accra, Ghana. *Food Nutr. Bull.* **2016**, *37*, 202–218. [CrossRef] [PubMed]

12. Malabo Montpellier Panel. Nourished: How Africa Can Build a Future Free from Hunger and Malnutrition. Available online: http://www.ifpri.org/publication/nourished-how-africa-can-build-future-free-hunger-and-malnutrition (accessed on 12 September 2018).

13. Walsh, C.M.; van Rooyen, F.C. Household food security and hunger in rural and urban communities in the Free State Province, South Africa. *Ecol. Food Nutr.* **2015**, *54*, 118–137. [CrossRef] [PubMed]

14. Kengne, A.P.; Bentham, J.; Zhou, B.; Peer, N.; Matsha, T.E.; Bixby, H.; Di Cesare, M.; Hajifathalian, K.; Lu, Y.; Taddai, C.; et al. Trends in obesity and diabetes across Africa from 1980 to 2014: An analysis of pooled population-based studies. *Int. J. Epidemiol.* **2017**, *46*, 1421–1432. [CrossRef]

15. FARA. *Study of Growth, Structural Change and Total Factor Productivity in Eight African Countries*; Pattern Draw Ltd.: Accra, Ghana, 2016.

16. Food and Agriculture Organization (FAO). *Towards food Security and Improved Nutrition: Increasing the Contribution of Forests and Trees*; FAO: Rome, Italy, 2013.

17. Arnold, M.; Powell, B.; Shanley, P.; Sunderland, T.C.H. EDITORIAL: Forests, biodiversity and food security. *Int. For. Rev.* **2011**, *13*, 259–264. [CrossRef]

18. Nganje, M. Linking forest to food security in Africa: Lessons and how to capture forest contributions to semi-urban and and urban food security. *Nat. Fauna* **2014**, *28*, 12–16.

19. Clark, K.H. Urban Food Forestry—Low-Hanging Fruit for Improving Urban food Security? Lund University: Lund, Sweden, 2011; Volume 46.

20. Clark, K.H.; Nicholas, K.A. Introducing urban food forestry: A multifunctional approach to increase food security and provide ecosystem services. *Landsc. Ecol.* **2013**, *28*, 1649–1669. [CrossRef]

21. Larinde, S.; Oladele, A. Edible Fruit Trees Diversity in a Peri-Urban Centre: Implications for Food Security and Urban Greening. *J. Environ. Ecol.* **2014**, *5*, 234–248. [CrossRef]

22. Aworinde, D.O.; Erinoso, S.M.; Ogundairo, B.O.; Olanloye, A.O. Assessment of plants grown and maintained in home gardens in Odeda area Southwestern Nigeria. *J. Hortic. For.* **2013**, *5*, 29–36. [CrossRef]

23. Ickowitz, A.; Powell, B.; Salim, M.A.; Sunderland, T.C.H. Dietary quality and tree cover in Africa. *Glob. Environ. Chang.* **2014**, *24*, 287–294. [CrossRef]

24. United Nations. *World Urbanization Prospects*; United Nations: New York, NY, USA, 2014; ISBN 9789211515176.

25. Ghana Statistical Service. *2010 Population and Housing Census Final Results Ghana Statistical Service*; Sakoa Press Limited: Accra, Ghana, 2012.

26. Smith, L.; Ruel, M.; Ndiaye, A. Why is Child Malnutrition Lower in Urbanthan in Rural Areas? Evidence from 36 Developing Countries. FCND: Wahington, DC, USA, 2004; Volume 3.

27. Galbete, C.; Nicolaou, M.; Meeks, K.A.; de-Graft Aikins, A.; Addo, J.; Amoah, S.K.; Smeeth, L.; Owusu-Dabo, E.; Klipstein-Grobusch, K.; Bahendeka, S.; et al. Food consumption, nutrient intake, and dietary patterns in Ghanaian migrants in Europe and their compatriots in Ghana. *Food Nutr. Res.* **2017**, *61*, 1341809. [CrossRef] [PubMed]

28. Owusu, J.S.; Colecraft, E.K.; Aryeetey, R.N.; Vaccaro, J.A.; Huffman, F.G. Nutrition Intakes and Nutritional Status of School Age Children in Ghana. *J. Food Res.* **2017**, *6*, 11. [CrossRef]

29. Fobil, J.; May, J.; Kraemer, A. Assessing the Relationship between Socioeconomic Conditions and Urban Environmental Quality in Accra, Ghana. *Int. J. Environ. Res. Public Health* **2010**, *7*, 125–145. [CrossRef] [PubMed]

30. Stow, D.A.; Weeks, J.R.; Toure, S.; Coulter, L.L.; Lippitt, C.D.; Ashcroft, E. Urban Vegetation Cover and Vegetation Change in Accra, Ghana: Connection to Housing Quality. *Prof. Geogr.* **2012**, *65*, 451–465. [CrossRef] [PubMed]

31. Ofori-Sarpong, E.; Annor, J. Rainfall over Accra, 1901–1990. *Weather* **2001**, *56*, 55–62. [CrossRef]

32. Simmering, D.; Addai, S.; Geller, G.; Otte, A. A university campus in peri-urban Accra (Ghana) as a haven for dry-forest species. *Flora Vegetatio Sudano-Sambesica* **2013**, *16*, 10–21.

33. Creswell, J.W. *Steps in Conducting a Scholarly Mixed Methods Study*; DBER Speaker Series; Report No.: Paper 48; University of Nebraska: Lincoln, NE, USA, 2013.

34. Hawthorne, W.; Gyakari, N. *Photoguide for the Forest Trees of Ghana; Trees Spotters's Field Guide for Identifying the Largest Trees*; Oxford Forestry Institute: Oxford, UK, 2006.

35. Oteng-Amoako, A.A. *100 Tropical African Timber Trees from Ghana: Tree Description and Wood Identification with Notes on Distribution, Ecology, Silviculture, Ethnobotany, and Wood Uses*; Forest Research Institute of Ghana: Kumasi, Ghana, 2002.

36. Dionisio, K.L.; Arku, R.E.; Hughes, A.F.; Jose Vallarin, O.; Carmichael, H.; Spengler, J.D.; Agyei-Mensah, S.; Ezzati, M. Air Pollution in Accra Neighborhoods: Spatial, Socioeconomic, and Temporal Patterns. *Environ. Sci. Technol.* **2010**, *44*, 2270–2276. [CrossRef] [PubMed]

37. Arku, R.E.; Vallarino, J.; Dionisio, K.L.; Willis, R.; Choi, H.; Wilson, J.G.; Hemphill, C.; Agyei-Mensah, S.; Spengler, J.D.; Ezzati, M. Characterizing air pollution in two low-income neighborhoods in Accra, Ghana. *Sci. Total Environ.* **2008**, *402*, 217–231. [CrossRef] [PubMed]

38. Heip, C.H.R.; Herman, P.M.J.; Soetaert, K. Indices of diversity and evenness. *Oceanis* **1998**, *24*, 61–87.

39. Magurran, A.E. *Measuring Biological Diversity*; Backwell Publishing: Oxford, UK, 2004.

40. Agbelade, A.D.; Onyekwelu, J.C.; Oyun, M.B. Tree Species Richness, Diversity, and Vegetation Index for Federal Capital Territory, Abuja, Nigeria. *Int. J. For. Res.* **2017**, *2017*, 1–12. [CrossRef]

41. Nero, B.F.; Campion, B.B.; Agbo, N.; Callo-Concha, D.; Denich, M. Tree and trait diversity, species coexistence, and diversity- functional relations of green spaces in Kumasi, Ghana. *Procedia Eng.* **2017**, *198*, 99–115. [CrossRef]

42. Raoufou, R.; Kouami, K.; Koffi, A. Woody plant species used in urban forestry in West Africa: Case study in Lomé, capital town of Togo. *J. Hortic. For.* **2011**, *3*, 21–31.

43. Nero, B.F.; Callo-Concha, D.; Denich, M. Structure, composition and diversity of the tree community of Kumasi, Ghana. *Forest* **2017**, *9*, 519. [CrossRef]

44. Kühn, I.; Brandl, R.; Klotz, S. The flora of German cities is naturally species rich. *Evol. Ecol. Res.* **2004**, *6*, 749–764. [CrossRef]

45. Mcdonald, R.I.; Marcotullio, P.J.; Guneralp, B. Urbanization, Biodiversity and Ecosystem Services: Challenges and Opportunities. In *Urbanization, Biodiversity and Ecosystem Services: Challenges and Opportunities. A Global Assessment*; Elmqvist, T., Fragkias, M., Goodness, J., Güneralp, B., Marcotullio, P.J., McDonald, R.I., Parnell, S., Schewenius, M., Sendstad, M., Seto, K.C., et al., Eds.; Springer: Berlin, Germany, 2013; pp. 609–628, ISBN 978-94-007-7087-4.

46. Fuwape, J.A.; Onyekwelu, J.C. Urban Forest Development in West Africa: Benefits and Challenges. *J. Biodivers. Ecol. Sci.* **2011**, *1*, 77–94.

47. Eichemberg, M.T.; Christina, M.; Amorozo, D.M.; Moura, C. De Species composition and plant use in old urban homegardens in Rio Claro, Southeast of Brazil. *Acta Bot. Bras.* **2009**, *23*, 1057–1075. [CrossRef]

48. Agbogidi, O.M.; Adolor, E.B. Home gardens in the maintenance of biological diversity. *Appl. Sci.* **2013**, *1*, 19–25.

49. Cilliers, S.; Cilliers, J.; Lubbe, R.; Siebert, S. Ecosystem services of urban green spaces in African countries—Perspectives and challenges. *Urban Ecosyst.* **2013**, *16*, 681–702. [CrossRef]
50. Hope, D.; Gries, C.; Zhu, W.; Fagan, W.F.; Redman, C.L.; Grimm, N.B.; Nelson, A.L.; Martin, C.; Kinzig, A.; Nelsonll, A.L. Socioeconomics drive plant diversity. *Proc. Natl. Acad. Sci. USA* **2003**, *100*, 8788–8792. [CrossRef] [PubMed]
51. Kinzig, A.P.; Warren, P.S.; Martin, C.; Hope, D.; Katti, M. The effects of human socioeconomic status and cultural characteristics on urban patterns of biodivertsity. *Ecol. Soc.* **2005**, *10*. [CrossRef]
52. Food and Agriculture Organization (FAO). *Improving Nutrition through Home Gardening*; FAO: Rome, Italy, 1996.
53. Danquah, A.O.; Amoah, A.N.; Steiner-Asiedu, M.; Opare-Obisaw, C. Nutritional Status of Participating and Non-participating Pupils in the Ghana School Feeding Programme. *J. Food Res.* **2012**, *1*, 263. [CrossRef]
54. Global Panel on Agriculture and Food Systems for Nutrition. *Food Systems and Diets: Food Systems and Diets*; Global Panel on Agriculture and Food Systems for Nutrition: Accra, Ghana, 2016.
55. Sy, M.; Baguian, H.; Gahi, N. Multiple use of green spaces in Bobo-Dioulasso, Burkina Faso. *Urban Agric. Mag.* **2014**, *27*, 33.
56. Elmqvist, T.; Setala, H.; Handel, S.N.; Van Der Ploeg, S.; Aronson, J.; Blignaut, J.N.; Gomez-Baggethun, E.; Nowak, D.J.; Kronenberg, J.; Groot, R. Benefits of restoring ecosystem services in urban areas. *Curr. Opin. Environ. Sustain.* **2015**, *14*, 101–108. [CrossRef]
57. Bolund, P.; Hunhammer, S. Ecosystem services in urban areas. *Ecol. Econ.* **1999**, *29*, 293–301. [CrossRef]
58. Tzoulas, K.; Korpela, K.; Venn, S.; Yli-Pelkonen, V.; Kaźmierczak, A.; Niemela, J.; James, P. Promoting ecosystem and human health in urban areas using Green Infrastructure: A literature review. *Landsc. Urban Plan.* **2007**, *81*, 167–178. [CrossRef]
59. Hosek, L. Urban Forestry in Africa—Insights from a Literature Review on the Benefits and Services of Urban Trees. In Proceedings of the Trees, People and Built Environment II Conference, Edgbaston, UK, 2–3 April 2014; pp. 43–53.
60. Popoola, L.; Ajewole, O. Public perceptions of urban forests in ibadan, nigeria: Implications for environmental conservation. *Arboric. J.* **2001**, *25*, 1–22. [CrossRef]
61. Users, F.; Schroeder, H.W. Perceptions and preferences of urban forest users. *J. Arboric.* **1990**, *16*, 58–61.
62. Yeboah, A.K. *The Interrelationship between Urban Forest and Residential Area Classes—Case Study of the Ayawaso Submetro Districts in the Accra Metropolis, Ghana*; University of Eastern Finland: Joensuu, Finland, 2016.
63. Mills, J.R.; Cunningham, P.; Donovan, G.H. Urban forests and social inequality in the Pacific Northwest. *Urban For. Urban Green.* **2016**, *16*, 188–196. [CrossRef]
64. Frimpong, J. Planning Regimes in Accra, Ghana. Master's Thesis, University of Waterloo, Waterloo, ON, Canada, 2017.
65. Johnston, M.; Shimada, L.D. Urban forestry in a multicultural society. *J. Arboric.* **2004**, *30*, 185–192.
66. Etshekape, P.G.; Atangana, A.R.; Khasa, D.P. Tree planting in urban and peri-urban of Kinshasa: Survey of factors facilitating agroforestry adoption. *Urban For. Urban Green.* **2018**, *30*, 12–23. [CrossRef]
67. Faleyimu, O. Public perceptions of Urban forests in Okitipupa Nigeria: Implications for Environmental Conservation. *J. Appl. Sci. Environ. Manag.* **2014**, *18*, 469–478.
68. Faleyimu, O.; Akinyemi, M. Socio Economic Assessment of Urban Forestry Respondents' income in Okiti Pupa, Ondo State, Nigeria. *J. Appl. Sci. Environ. Manag.* **2014**, *18*, 603–607. [CrossRef]
69. Donovan, G.H.; Mills, J. Environmental justice and factors that influence participation in tree planting programs in Portland, Oregon, U.S. *Arboric. Urban For.* **2014**, *40*, 70–77.

 © 2018 by the authors. Licensee MDPI, Basel, Switzerland. This article is an open access article distributed under the terms and conditions of the Creative Commons Attribution (CC BY) license (http://creativecommons.org/licenses/by/4.0/).

Article

Transdisciplinary Research: Collaborative Leadership and Empowerment Towards Sustainability of Push–Pull Technology

Isaac Mbeche Nyang'au [1,2,*], Girma Kelboro [1], Anna-Katharina Hornidge [3,4], Charles A. O. Midega [2] and Christian Borgemeister [1]

[1] Center for Development Research (ZEF), University of Bonn, Genscherallee 3, D-53113 Bonn, Germany; gmensuro@uni-bonn.de (G.K.); cb@uni-bonn.de (C.B.)
[2] International Centre of Insect Physiology and Ecology (ICIPE), P.O. Box 30772-0100 Nairobi, Kenya; cmidega@icipe.org
[3] Institute of Sociology, University of Bremen, 28334 Bremen, Germany; anna-katharina.hornidge@leibniz-zmt.de
[4] Leibniz Center for Tropical Marine Research (ZMT), Fahrenheitstr 6, 28359 Bremen, Germany
* Correspondence: imbeche@icipe.org; Tel.: +254-799615194

Received: 4 April 2018; Accepted: 4 July 2018; Published: 9 July 2018

Abstract: A transdisciplinary research approach requires that different scientists with their discipline-specific theories, concepts and methods find ways to work together with other societal players to address a real-life problem. In this study, the push–pull technology (PPT) was used as a boundary object to enable interactions among stakeholders across science-practice boundaries engaged in the control of stemborer pest in maize crops in Bako Tibe, Jimma Arjo and Yayu *Woredas* in Ethiopia between August 2014 and April 2015. The PPT is a biological mechanism developed by researchers for the control of stemborer pests and Striga weed in maize crop. It involves inter-cropping maize with a stemborer moth-repellent silverleaf or greenleaf Desmodium (push), and planting an attractive trap crop, Napier or Brachiaria grass (pull), around it. The on-farm implementation of PPT was used to provide an opportunity for collaboration, interaction and learning among stakeholders including researchers from the Ethiopian Institute of Agricultural Research and practitioners from the Ministry of Agriculture and smallholder farmers/traders. The research was implemented following the transdisciplinary action research and the data collected using mixed methods approach, including key informant interviews, focus group discussions, workshops, on-farm practical demonstrations and participant observations. The findings show that collaborative leadership provides a chance for the stakeholders to participate in the technology learning and decision making, by enabling them to jointly contribute skills towards development, refinement and adaptation of the PPT. In situations where there are conflicts, they are embraced and converted to opportunities for in-depth learning, finding solutions and adaptation of the innovation processes rather than being sources of contradictions or misunderstandings. The leadership roles of the farmers enabled them to reflect on their own practices and draw on scientific explanations from researchers. It also enabled them to take the lead in new technology implementation and information sharing with fellow farmers and other stakeholders in a free and easy manner. Although the perennial nature of the companion crops in the PPT provides opportunities for continuous stakeholder interaction and learning, it requires a personally committed leadership and formal institutional engagements for the sustainability of the activities, which span several cropping seasons. Market forces and the involvement of the private sector also play a role as shown from the involvement of individual farmers and traders in Desmodium and Brachiaria seed production, collection and distribution during the PPT implementation.

Keywords: collaboration; leadership; push–pull technology; sustainability; transdisciplinary research; Ethiopia

---

1. Introduction

The uptake of academic research into policy and practice requires the input and active participation of other societal stakeholders. In transdisciplinary research, participation of non-academic or lay persons bring a large range of knowledge and expertise which is vital to the research process where it is framed and conducted with flexibility [1]. It is almost impossible to ignore transdisciplinarity when engaging in sustainability research and vice versa. These approaches are aimed at generating knowledge to solve societal challenges such as poverty, climate change leading to drought or less regular rainfall, increased incidences of pests, diseases and weeds, rising food prices, conflicts, etc. [2]. These challenges cannot be resolved by individuals, one community or a country alone. Addressing them requires joint efforts from multiple disciplines, groups, and organizations. That means the challenges require new ways of knowledge generation and decision making through the involvement of diverse stakeholders including those from outside academia [ibid.]. The process is promising when these approaches have clearly set out goals and competent management to facilitate creativity, innovation and management of conflicts which may exist or arise out of the stakeholder engagement [3]. Participation of non-academics contributes to democratizing the research process, thereby enabling production of better and socially more robust research outputs [4–7]. The knowledge production processes become democratized to the extent that the dominant and non-dominant actors have—at least temporarily—equal access and ability to contribute towards solving societal challenges [8].

Transdisciplinary research aims at establishing a form of collaboration which empowers stakeholders to influence the research, to question and to modify the dominant structures which guide epistemic processes [9]. In other words, the process enhances mutual learning for knowledge production between researchers and other stakeholders, and to bring about the empowerment of the participants either through education on areas of interest or to have their agenda and perspectives integrated in the knowledge production process [10]. Such an authentic and inclusive participation empowers the stakeholders to become better agents of change in their communities from problem identification to finding solutions [11,12]. This model of research in sustainability studies embraces a sequence of stakeholder engagement and involvement from informing to consulting, to collaborating and to empowerment where practitioners are able to implement the research findings [13,14]. In this case, empowerment is seen as a process of providing stakeholders with the space to make decisions right from research planning to the use of the findings. Such an empowerment is bound to take place when it is linked with leadership and innovation [15].

Transdisciplinary leadership comes in handy when stakeholder engagement and interaction is mediated towards fruitful collaborations at the level of boundary-crossing [16,17]. Although the diverse sets of stakeholders are brought together by a common societal problem, in practice they are still ascribed to different disciplinary, organizational and institutional silos [18]. Leadership with collaborative approach plays an important role in trying to overcome some of these boundaries across different agencies or agency units [19–23]. It is recommended that, to make progress in tackling the common problems, all these stakeholders take on the leadership challenge of building shared-power arrangements [13,16]. Such empowerment and leadership should not be motivated by personal power needs, but the potential through which stakeholder collaboration can be used to address common societal problems for mutual benefits [24]. A collaborative leader, therefore, has a responsibility to provide guidance, coordinate the transdisciplinary research process and ensure that the other stakeholder groups participate in making decisions and taking actions in a democratic atmosphere to address the problem at hand. According to Gray, such a leader is equated with a transformative leader who can combine cognitive, structural, and processual tasks:

The task of effective cognitive leadership is to provide a vision and commitment that link and motivate the scientific researchers to step beyond their disciplinary silos, relax old assumptions, and search for creative frame-breaking solution. Effective structural leadership adds value by breaking the gaps existing between science and practice through building bridges. Effective processual leadership task encourages trust and turns potentially destructive conflicts into constructive and participatory interactions. ([18], p. 9)

Studies on transdisciplinary collaboration show that there is progress in connecting and integrating the knowledge with action to support the research agenda, decision making and governance and sustainable use of natural resources and climate change adaptation [14,25–29]. Evaluation of these studies shows that contextual factors either at individual or collective levels greatly contribute to influencing the communicative and collaborative readiness of the teams and chances for success of the transdisciplinary work [23,28]. The presence or absence of institutional support for cross-disciplinary collaboration, the disciplinary specialization and mandates of the research centers, prior cooperation experiences of the research team members in working with other stakeholders, availability of logistical support in cases where the research sites and teams are in distant locations, etc. are some of these factors. The more aligned they are at the beginning of the transdisciplinary research, the greater is the prospect for success in terms of effective and sustainable transdisciplinary collaboration [23]. It is acknowledged that the individuals within the research teams with collaborative management skills appear to make a difference between success and failure in transdisciplinary efforts from their charismatic personality, leadership, knowledge base, broad network and engagement [30]. Such individuals have the ability to anticipate challenges during the transdisciplinary process, learn from failures in case they happen, and be flexible in the process and prepared to adapt to new conditions within the overarching objective of addressing the problem at hand [3]. Collaboration among the stakeholders is instrumental to ensure that their engagement and interactions worked in terms of learning and effective implementation of agricultural innovations. Such innovative processes on joint innovation implementation require leadership with strong advocacy at scientific research and practice levels and, at the same time, willingness to play a central role in the promotion of a free and divergent thinking, risk taking, and readiness to challenge the established methods, theories and practices [30]. That means the transdisciplinary stakeholder collaboration contributes to the empowerment and leadership skill enhancement of especially smallholder farmers so that they can effectively conduct and implement research findings and train others do it [14] To illustrate and show this in practice, this paper is using the experiences on the introduction and implementation of a new technology dubbed as push–pull technology (PPT) in the Oromia region of Ethiopia.

In Ethiopia, extension staffs tend to serve as agents of centralized political power in their respective *Woredas* or *Kebeles*. In the course of their work, they are mostly engaged in delivering political messages and undertaking tasks on behalf of the ruling party, such as collecting taxes under the cover of extension work [31]. However, few initiatives have been used to facilitate collaboration among the researchers and other stakeholders in the strengthening of agricultural innovation and networks in the country. For example, it has been shown that a strong partnership between national research and extension systems with regional administrations, cooperative unions, the private sector and farmers can be effectively put to work [32]. This was done during the promotion of high-value and market-oriented grain legumes (i.e., common beans, chickpeas and lentils). The so-called Agriculture Development-Led Industrialization (ADLI) economic growth strategy is spearheaded by the Government of Ethiopia to enhance crop and livestock productivity, increase the commercialization of smallholder surpluses, enhance accessibility of credit to smallholders and facilitate grassroot participation in development projects for poverty reduction [33]. The participatory extension system within the ADLI framework is aimed at building the capacity of the smallholder farmers to solve their production challenges, raise their income and develop management skills to improve their lives by using their own local resources. It has been reported that, through this extension approach, farmers have been trained on standard land preparation and management practices, the benefits of applying fertilizer, usage of improved seeds, cultural practices in control of pests, etc. The approach has seen about 1% expansion of cultivated area, and, with favorable weather conditions, agricultural production rose by 5.4% and grain yield increased from <1 to 2 t/ha in 2013/2014 [34]. However, this is merely rhetoric. To sustain such production levels and achieve food and nutrition security in the country, the Ethiopian agriculture sector require continuous investment in research and extension for the generation, promotion and adoption of new and production enhancing technologies [35].

The transdisciplinary research approach is seen in principle as an opportunity for empowerment by bestowing confidence in the farmers and making their views, knowledge and skills shared with other farmers, but also researchers and extension professionals on an equal basis. The new skills learned and knowledge generated from such interactions are meant to inform further research and practice. The paper reports the findings of our study which addressed the questions of how collaborative leadership and empowerment work in transdisciplinary research in practice in the context of smallholder farmers and what to consider to enhance the processes.

1.1. The Operationalization of Key Terms Used in This Study

Agricultural Innovation refers to the development or adoption of new agricultural technologies, concepts or ideas, and/or successful exploitation of new ideas. It also embraces the totality and interaction of actors involved in the development of the ideas [36] (p. 7).

Boundary objects refers to those objects which are both plastic enough to adapt to local needs and constraints of the several parties employing them, yet robust enough to maintain a common identity across sites. This is an analytical concept of those scientific objects which both inhabit several intersecting social worlds and satisfy the informational requirements of each of them [37] (p. 393).

Transdisciplinary Research (TDR) refers to the researchers and other societal players working together by integrating their knowledge and perspectives into a shared problem definition to solve a real-life problem [23,38,39]. In this paper, TDR is operationalized to refer to joint efforts of research scientists consisting of entomologists, chemical ecologists, weed scientists, social scientists, and other stakeholders consisting of farmers, traders, extension service providers (public or private sectors), input suppliers, Non-Governmental Organizations (NGOs), community-based organizations and farmer groups/associations working together in the PPT implementation.

Push–Pull Technology (PPT) refers to a cropping strategy for controlling stemborer pests and the parasitic *Striga* weed in maize where it is intercropped with a stemborer moth repellent fodder legume, Desmodium (the push) and with an attractant trap plant, such as Napier grass (the pull) planted around the intercropped cereal [40]. In this paper, the PPT is operationalized as a boundary object.

Stakeholders refers to a person, group or organization who have interest or concern and are affected by the innovation or their activities strongly affect the performance of the innovation [39]. In this paper, stakeholders include researchers, farmers, traders, extension service providers (public or private sectors), input suppliers, NGOs, community-based organizations and farmer groups or associations.

1.2. The push–pull Technology

In Ethiopia, based on the Central Statistical Agency [41], maize accounts for 28% of the total cereal production compared to sorghum (22%), and teff (20%) [41]. Smallholder farmers in Ethiopia account for >90% of the total agricultural production and 95% of total maize area with an average yield of about 2.5 t/ha [32,42] which is well below the yield potential of over 5 t/ha [43]. Higher productivity in the maize sector could propel Ethiopia's food production and would enable the country to reduce its national food deficit and keep pace with a growing population. However, production is severely constrained by stemborer pests and parasitic weeds, particularly Striga, and low soil fertility [44,45]. Stemborers alone can result in significant yield losses ranging 10–80% of the total maize yield, depending on pest population density and phenological stage of the crop at infestation [46]. Stemborer control is, however, difficult, largely as a result of the cryptic and nocturnal habits of moths, and protection provided by host stem for immature pest stages [47]. The control measures mostly advised by research and extension services to smallholder farmers are based on the use of synthetic insecticides, which is not only harmful to the environment but often too expensive for smallholder farmers [48]. The push–pull technology offers an ecological approach for effective pest and weed management based on a combined use of inter- and trap cropping systems where stemborers are driven away from maize crops by push plants (push) and attracted by and trapped on trap plants (pull), and where the parasitic Striga weeds are effectively controlled through root exudates from the Desmodium plants [49].

In the PPT strategy, a cereal crop such as maize is intercropped with a stemborer moth repellent fodder legume, Desmodium (the push), together with an attractant trap plant, Napier/Brachiaria grass (the pull), planted around the maize–legume intercrop (Figure 1). The green leaf volatiles produced by Desmodium repel stemborer moths while those produced by the trap grasses attract them [40]. In addition, Napier or Brachiaria grasses greatly reduce the development of the stemborer larvae and hence the majority of them die before reaching adulthood [50]. Desmodium root exudates also suppress and eliminate Striga, leading to significantly enhanced maize/cereal yields [51]. Studies have identified the mechanisms by which Desmodium suppresses Striga, with an allelopathic effect being the most important one [52]. Desmodium roots produce a blend of chemical compounds, some of which stimulate Striga seeds to germinate while others inhibit lateral growth of Striga roots, thereby hindering their attachment to maize roots, i.e., suicidal germination [52–54]. Striga emergence is thus suppressed, with an in situ reduction of the Striga soil seed bank]. In addition, the PPT is beneficial in crop–livestock integration through fodder provision (by the grass and legume companion plants), soil improvement, agro-ecosystems resilience through crop intensification, income generation and meeting food security demands of smallholder farmers [55]. Currently about 200,000 farmers are practicing the PPT in East Africa [56]. Although the PPT concurrently addresses both the *Striga* weed and stemborer pests, Striga was not considered as a serious threat by the farmers in the study areas. The PPT implementation (Table 1) was undertaken on these areas where stemborer pests was identified as the most challenging in maize crop production constraint.

Table 1. Study area and location of the push-pull technology (PPT) demonstration plots.

Study Area and Location of the PPT Demonstration Plots							
Zone	*Woreda*	*Kebele*	**AEZ**	**Location of the Plots**	**Altitude**	**Longitude (N)**	**Latitude (E)**
West Shawa	Bako Tibe	Dembi Gobu	Mid-land	Gibe river	1660	09°09.173′	037°02.131′
		Seden Kite	Mid-land	Leku river	1648	09°05.331′	037°09.847′
East Wollega	Jima Arjo	Wayu Kumba	High land	Nageso river	1990	08°46.442′	036°31.616′
		Wayu Kumba	High land	Nageso river	1986	08°46.582′	036°32.554′
Ilu-Ababora	Yayu	Jame Shono	High land	Sky-sky	1904	08°21.353′	035°36.584′
		Jame Shono	High land	Jame-Bone	1870	08°21.351′	035°57.736′

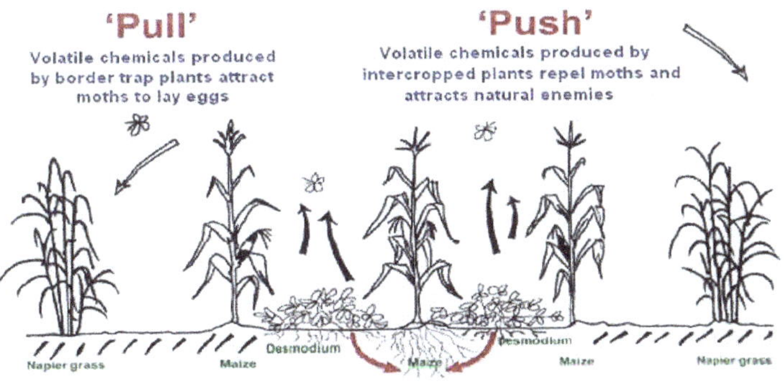

Figure 1. How push–pull technology works [56].

2. Methodology: Transdisciplinary Action Research

The implementation of the study was based on transdisciplinary action research (TDR) [23,25] and drawing on the experiences of the Follow the Innovation (FTI) approach which was developed by the Center for Development Research (ZEF) in the context of Uzbekistan's agriculture [57,58]. The action research design is an interactive, collaborative and rigorous iterative process which engages practitioners in managing systematic enquiries to improve their practices and better realize their desired outcomes and contribute effectively towards the longer-term goals [59,60]. It is a systematic (Figure 2) and participatory paradigm that supports transdisciplinary reflective learning, communication and co-operation between the different stakeholders in the learning and knowledge production for addressing the real-life problems, developing innovative solutions and incorporating them in new action strategies [23,25,61].

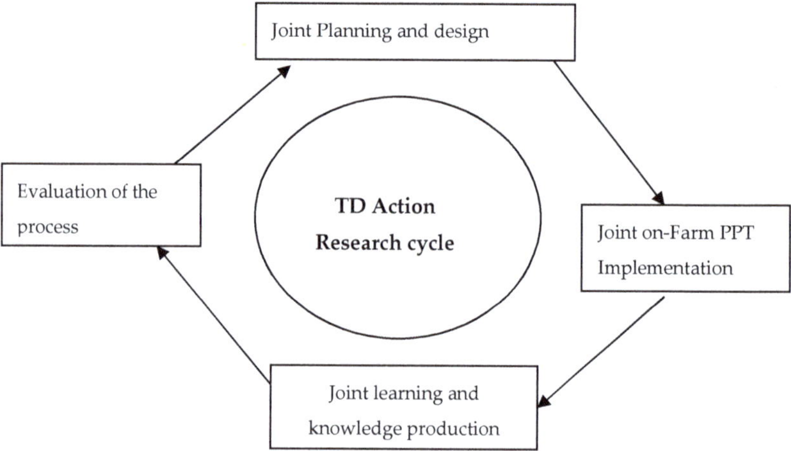

Figure 2. Transdisciplinary (TD) action research cycle.

As a joint process, the transdisciplinary action research was used as a strategy to enable learning among the farmers, researchers and other stakeholders, to share their knowledge and experiences towards the production of locally adaptable knowledge [56,62] and, in this particular case, on the control of stemborer pests and parasitic weeds in maize. The team of researchers consisted of an entomologist, chemical ecologist, weed scientist, and social scientist, whereas other stakeholders included farmers, agro-dealers, public extension service providers and smallholder farmers. The TDR is premised on good communication between researchers and other stakeholders to ensure that knowledge sharing, and learning takes place with eventual empowerment and development of new knowledge or practices. The PPT implementation followed the stepwise practical on-farm activities of maize crop agronomic and management practices; land preparation, planting, weeding, harvesting of either as green maize cobs for roasting or dry cobs for grains, harvesting Desmodium and Brachiaria plants as fodder for livestock and seeds. After the harvesting of the maize, the process involved off-cropping season activities mainly on the process evaluation of the learning outcomes and any impacts achieved.

2.1. Research Design and Data Collection Methods

2.1.1. Research Design

The research process started with stakeholder identification and organizing planning workshops at the regional, *Woreda* and farmer levels. The identified stakeholders were selected based on their professional and organizational mandates, research and disciplinary specializations, or because they were in positions of influencing decision making or had interest in finding a lasting solution to the stemborer problems in cereal crops such as maize. Some of the stakeholders were researchers/agronomists from the University of Jimma, Ethiopian Institute of Agricultural Research (EIAR) and the Jimma Plant Health Clinic. Other stakeholders were crop protection experts from the Federal Ministry of Agriculture (MoA), agronomists, extensionists and crop protection experts from the Oromia Regional Bureau of Agriculture and zonal agricultural offices, journalists (Fana Radio and Oromia Television) and the maize growing smallholder farmers.

The first workshop was conducted at the regional level, whereas the follow-up workshops were conducted at the *Woreda* and farmer levels. During these workshop, the discussions centered on how the use of the PPT as an innovation can be used to address the problem of stemborer pests and also as an opportunity for collaboration and joint learning by exploiting the technical expertise, knowledge and the experiences from the stakeholders (research, farmers and extension service providers). Although the PPT concurrently address both Striga weed and stemborer pests, Striga was not considered as serious threat by the farmers in the study areas. The practical PPT implementation, field observations, joint learning and evaluation took place at farm level and involved participatory collaboration and engagement of the transdisciplinary research team in undertaking the PPT activities. Facilitation of the process by the researchers played a key role of ensuring that there was a balanced participation during discussions and in managing potential biases or conflicts. This was aimed at ensuring that knowledge and concerns of all the stakeholders were consolidated during the implementation of the research. The facilitation considered the consent to participate and confidentiality of the information shared by the stakeholders.

2.1.2. Data Collection

In the data collection process, different qualitative methods were used. The process engaged 37 Key Informant (KI) interviews, 16 Focus Group Discussions (FGD), 2 stakeholder workshops, on-farm practical demonstration and Participant Observations (PO) to assure triangulation. The study lasted for eight months from August 2014 to April 2015. The KI interviewees were mainly researchers, extension and administration officers from the ministries and farmer leaders. The FGD participants were the PPT farmers and their neighbors. The PO were based on the researchers' observation during on-farm practical demonstrations and workshop discussions to document observations on the attitudes, behaviors and actions of the participants.

The interviews were guided and focused on: knowledge and the background information on the stemborer pests' problem in maize crop and effect on yield; type of pest control measures commonly applied and their effectiveness; how the PPT management was similar or different from local farming practices; and new opportunities, practices and knowledge gained and challenges faced in the course of the PPT implementation. The implementation of PPT in terms of its sustainability was discussed. The data were transcribed into themes which were analyzed and interpreted in relation with the research objective. The analysis was guided by an innovation systems approach [36,63–66] by exploring how different actors collaborate and work together during technology implementation; and manage boundaries by integrating their knowledge bases through joint learning and communication. The approach helped to draw the attention of different stakeholders who contributed to the PPT implementation, by focusing on their roles and responsibilities, actions and interactions.

2.2. The Study Area

The study was implemented in three *Woredas* in the Oromia Region of western Ethiopia, namely Bako Tibe, Jimma Arjo and Yayu (Figure 3). These sites were identified in Ethiopia using a GIS based "hot-spot" approach by African and German partners of the BiomassWeb project (https://biomassweb.org). The *Woredas* are located in the western parts of Oromia and were selected as the potential "maize growing basket" of Ethiopia [67,68] in which the "livelihood of local communities (…) traditionally stands on household-based subsistence agriculture, extensive use of forests and cultivation on considerably small plots of agricultural land with an average of one hectare of cropland mainly for the cultivation of staple cereals such as maize, *teff* and wheat" [69] (p. 10). The soils in the areas are severely degraded due to nutrient depletion and/or poor in organic matter from continuous mono-cropping and unsustainable farming practices. Insect pests, principally stemborers, are major cereal production constraints [70–72]. According to Hurni [73], the area is characterized predominantly by bi-modal rainfall, with short rains in March and April, and long rains from June to October, with a distinct dry season extending usually from November to February. The dominant soil types are Nitosols with fertile alluvial soils in valley bottoms and depressions. Major crops, in order of importance, are maize, *teff*, pepper, sorghum, millet and pulses. In Yayu, coffee is an important cash crop. The farming system is mixed crop–livestock based.

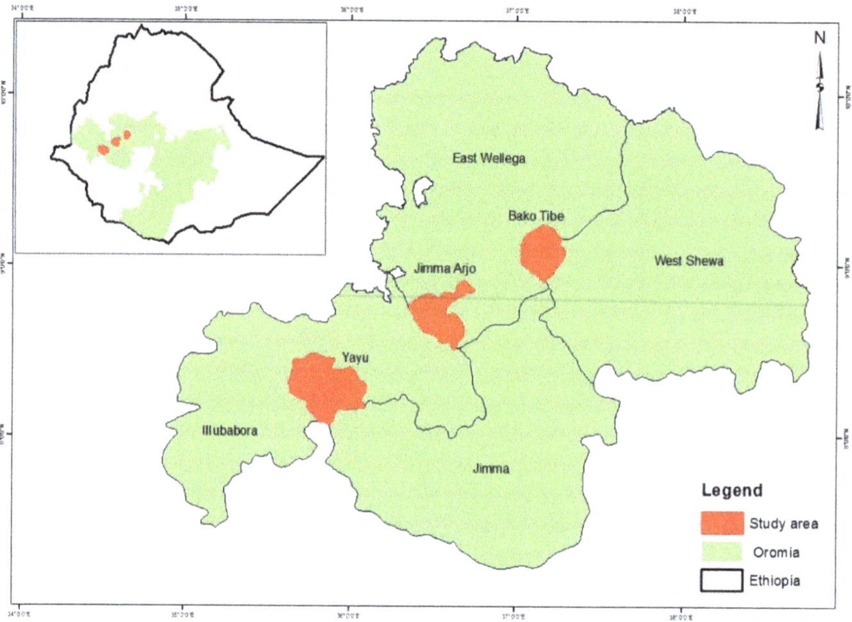

Figure 3. Map of the study area in western Oromia, Ethiopia.

3. Findings

3.1. The Benefits

3.1.1. Transdisciplinary Collaboration as an Opportunity for Farmers Empowerment to Take Up Leadership Roles

The PPT planning and implementation process was undertaken based on a joint and collaborative environment during workshops and on-farm discussions and activities, where the researchers and other stakeholders interacted, contributed and applied their knowledge and practices on how the PPT activities could address the problem of stemborer pests and other challenges such as availability of fodder for livestock, soil and water conservation. During the processes, the researchers' contributions were driven by their scientific knowledge mainly from laboratory and on station research experiments. Farmers and extension staff contributions were based on their experiential and day-to-day encounters with the stemborer pests and other farming challenges at farm level. In the conventional research–extension–farmers linkages, the farmers are regarded the recipients of the already tested and ready-to-use knowledge or practices. However, joint stakeholder implementation of the PPT was a different approach involving the collaboration between researchers and practitioners. In this approach, the farmer had an active a role to play in setting the agenda during planning and implementation activities.

Several efforts have been made in Ethiopia to bridge the research–practice gap through the introduction of new technologies [74]. However, in a situation where there are different levels of participation and the farmers are at the receiving end, whereas the researchers and extension professional are controlling resources for technology implementation, it is not easy to eliminate such a gap [74,75]. With such type of relationship, the farmers believe that, whenever these technologies are introduced, their participation is limited to following and adhering to instructions provided [75]. Under such circumstances, farmers' trust is low whenever new initiatives are introduced by the people whom they believe to have a hidden agenda or acting on behalf "state authorities or other agencies". However, in a situation where there are efforts to build trust in a democratic atmosphere, this is empowerment; the farmers are able to take new technologies with the seriousness they deserve. For example, the sustainability of the PPT is not only dependent on its effective performance in the control of stemborer pests, but also on market forces from the demand and supply of Desmodium and Brachiaria seeds and other inputs. In addition, it is dependent on leadership initiatives from the micro level, i.e., at the household level where the head ensures that all members of the household participate and their capacities are built with the PPT-based knowledge. The farmers can organize and allocate their farm families in terms of roles and responsibilities on carrying out the PPT activities. The PPT was implemented as an enterprise where the different family members were empowered to exploit the various benefits and opportunities which come with it, e.g., from Desmodium and Brachiaria seed production, process and marketing as a source of income generation of livestock fodder for milk production. This was also in anticipation that one of the family members is incapacitated or moves on to other activities, and the rest of the family members can continue with the technology implementation. This sort of leadership at family level was built from the confidence derived by the farmers, whereby their role is recognized by both research and extension professionals. The farmers' leadership was demonstrated by their ability to initiate, conduct experiments and take decisions with confidence regarding the PPT implementation activities. For example, researchers recommend the establishment of the PPT using line planting where the seeds are sown in holes or drills using a straight line. However, the farmers preferred to use ox-drawn ploughs to drill the lines (Figure 4). These lines were not straight as required or recommended by the research or extension, but the farmers, thriving in a participatory and democratic learning process, could make decisions based on valid reasons on what they knew works best for them. This was different from previous relationships where they were dictated by the Development Agents (DAs) or local administration and researchers on what to do during on-farm trials or implementation of new technologies. This time, they could use their own measurements and tools:

> The joint implementation of the PPT enhances participatory leadership in its approach whereby it enables everybody from farmers to researchers to participate in the technology. Thus, enabling all the participants (i.e., stakeholders) to develop and contribute skills to manage its implementation and also push the research agenda . . . Through this approach new players are attracted into the research process and contribute to good results, e.g., the private sector, youths and women . . . . all the players involved are able to learn new skills on PPT openly from each other and by drawing on own experiences. (Researcher, April 2015, EIAR, Bako center)

Figure 4. Pictures of farmers, extension officers and researchers' joint sessions of establishing the PPT.

The effort to change research and extension services provision is in line to what was recommended by Deneke and Gulti [74], i.e., linear relation should be replaced by systems which are iterative, interlinked and overlapping where by the role of the farmers is not that of recipients of technologies developed elsewhere but to empower them to have confidence and trust to contribute to decision making together with other stakeholders. The stakeholder collaboration process was organized in such a way that the on-farm implementation process was undertaken in a participatory manner by the team consisting of the farmers, the Development Agents (DAs), *Woreda*/District level extension professionals and researchers. The on-farm PPT activities created a platform for these stakeholder groups to interact and learn various topics related to stemborer pest control, soil and water conservation and fodder production. Such a joint working relationship was an opportunity for each stakeholder to introduce and contribute skills on the management and implementation of the PPT and other related emergent enterprises or benefits:

> The DAs, researchers, visitors from overseas, Germany and Kenya, have been with us from the start to date . . . Team work always has positive outcomes . . . . The team has been around from sensitization, training, planning for all the implementation stages to now . . . This has formed a continuous interaction process for learning and appreciating the efforts of each other. (DA, Bako Tibe, March 2014)

The farmers were confident with the promoters of the technology. This was based on the reputation of the PPT which they heard; it has been researched vastly and there were many experiences to draw from other countries in the eastern African region such as Kenya, Uganda and Tanzania. They share similar circumstances with farmers from these countries who are also faced by the same stemborer challenges. They were flexible and at ease to make decisions about the PPT from the beginning. In their own circumstances as Ethiopian farmers, 2–3 cropping seasons with the PPT was sufficient to enable them to have their own concrete experiences. At the time, they were optimistic:

> icipe Director General is Ethiopian and she is supporting this technology . . . . She is fully aware of its benefits, so we shall fully embrace it and make it a model. What she thought for her country and population to come out of poverty and challenge of being faced with pests, erosion, and fodder. We are receiving this information through the extension and researchers at the local level. (The PPT farmer, Bako Tibe, April 2014)

In addition to cereal crops, farmers plant horticultural crops for household consumption and for the local market as a source of cash income. These crops are affected by similar pest problems such as nematodes, armyworms, aphids. etc. Out of curiosity to find a solution to the horticultural pest problems as well, a farmer raised a question whether Desmodium could help in addressing the problems he has with tomatoes and cabbages on his farm. This was an interesting question for the researchers and extension alike. The farmer took the initiative to learn about it. This was based on what he observed for the first time, i.e., the maize crops had less infestation with stemborer pests when compared to the previous maize cropping seasons. Contrary to the conventional wisdom of waiting for an answer from the specialists, he offered to run the trial experimentation by growing tomatoes on 25 m by 25 m plot intercropped with Desmodium. Out of this plot, he planned to keep records on the observations made and he was willing to share with researchers and extension staff his own findings. Such a bold gesture from the farmer was attributed to confidence and freeness gained from the joint nature of interaction with researchers and the extension staff. This also was a boost to the other farmers to believe that their ideas and simple experiments can contribute to generating new knowledge in a research engagement process. Such a research initiative is an example of how through the empowerment and supporting farmers' ideas can contribute to generation of new knowledge in pest control. The farmers experimentation, combined with the laboratory-based methods, can potentially reduce the time usually needed to generate results to address the immediate need or real-life problems.

Weeds compete with the maize crops for nutrients and moisture. It is assumed that the farmers should weed their maize even when the crops have matured. However, the farmers had a different perspective. At the later stages of maize growth when they start flowering or forming grains, the farmers stopped weeding. From their point of view, the weeds together with maize stalk leftovers was fodder for their livestock after harvesting the maize cobs. At the same time, weeds retain soil moisture due to sufficient soil cover. With the introduction of the PPT, some farmers in the area saw similarity with weed retention on-farm. They were able to connect with what they already knew and what the PPT had in store for them. Instead of previously leaving weeds to grow with their crops, introducing Desmodium as a companion crop seemed a novel idea. It addresses the stemborer problem and provides additional fodder for livestock. This helped to dispel the rumors which were initially spreading in the study areas that Desmodium was a weed being introduced by researchers and extension staff:

> The overgrown weeds are fodder for livestock. Immediately after the harvest of maize, animals are left to freely roam and feed on "weeds" plus the maize stovers. . . . Now with PPT, Desmodium does the same. Besides, it is a perennial crop which can be cut back and fed to livestock. It is a permanent source of fodder . . . this is good. (the PPT farmer, during FGD interviews, March 2014)

3.1.2. Farmers Leadership in Push–Pull Technology and Related Activities

A few farmers managed to overcome the negatively rumored messages. These are a group of farmers who went ahead to implement the PPT despite these rumors. They eventually became the lead farmers and as leaders to their community through sharing and extending information to fellow farmers during events such as field days and face to face encounters with fellow farmers leaders [76,77] while at the same time playing a key role as key stakeholders in the research process. They embraced the leadership role for fellow farmers to emulate [78,79]. They were able to make decisions regarding the PPT based on their own intuition and experiences on what they think is of best interest to their community. This statement was echoed by the Ministry of Agriculture extension expert:

> The PPT has become like a "school feed", produces a lot of knowledge, food to feed us humans and livestock ... most farmers resist new technologies, but the model farmers accepted to lead the implementation of the PPT for the first time . . . . Other farmers from other Woredas will learn from their example by attending such field days' events which is an important platform for learning and experience sharing. (Extension expert, Bako Tibet, during the PPT farmers' field day Sedan Kite Kebele, March 2014)

The PPT learning plots were established and placed under the management of the lead farmers. This gave them an opportunity to experience the first-hand gains of the PPT on their farms and knowledge about the technology as compared to fellow farmers who were visiting to learn. The day-to-day interaction with research and extension experts, new farmers and inquiries, elevated their quest to search and learn more information about the technology. This increased their level of expertise, not only on PPT, but other management skills such as decision-making and communication skills from their greater contacts beyond their immediate surroundings [78]. Their contact with other stakeholders such as the researchers and extension personnel during the planning and implementation, monitoring and evaluation of the PPT activities, visitors from other countries was a learning journey for them. The information and new knowledge gained by these farmers was to be freely shared with their fellow farmers through social events such as community meetings, market places and even churches. Their social standing in the community through undertaking such new initiatives was elevated. This is an indication that all the farmers are potential innovators from the beginning if they get the chance to take an active role to participate in learning and implementing new initiatives. To be able to illustrate this aspect in detail, here is an example of one of the PPT host farmer.

The PPT host farmer *Kebele* A is an illustration contrary to what is commonly thought, that lead farmers should be well established with good social standing and well to do in the society in terms of assets and landed property or formally educated. However, this particular case is a farmer aged about 38 years without formal education, no family and land of his own. He lives with his elder brother's family and among other extended family members. He takes care of his extended family's large herd of livestock of over 30 animals consisting of dairy cattle, oxen, sheep, goats and donkeys. He rented the land from his brother to plant maize and accepted to participate in the implementation of the PPT. He is the most active member of the family even though he has no family of his own. From his active farming and other social activities, he is respected by the community members. He was selected for participation based on his interests as the farmer to take up new initiatives and the respect he had earned himself from fellow farmers in the village.

> Innovative/model farmer is the one who is ready to accept, learn lessons, pass lessons to other farmers, conduct own research and share with fellow farmers and researchers how technology works on reducing the losses due to stemborers pests. (Small scale irrigation and drainage expert, Bako Tibe, April 2015)

During further discussion with the irrigation expert, he had the same farmer in mind, from *Kebele* A; he commented that the respective farmer was always willing to offer his rented farm for the trial of new technologies. He is a lead farmer and risk taker and previously had participated in the establishment of trial plots of other new technologies such as tree nursery for agroforestry and Napier grass nursey for fodder production in his community. PPT is among these initiatives he is keen to try and bring to fruition by working closely with the researchers and extension staff. "Apart from the immediate benefits such fodder, reduced pest infestations and soil conservation properties, PPT is a promising technology" the farmer commented. "He is fully in it and looking forward into the future on what the technology has to offer to him", commented a fellow farmer in the community.

> The technology is interesting for farmers, researchers and student learning. It addresses the complex challenge of stemborers simply (. . .) and has attracted many schoolchildren from around the village who are interested to learn how the PPT mechanism works in controlling pests in such a simple way. This is science-simplified and in the hands of farmers. This is because I am able to talk about science-based practice and practically implementing it in my own farm. (Interview with PPT farmer, Bako Tibe, March 2015)

This particular farmer was able to talk easily and with confidence about the new PPT because he had understood how it works and it was practically existing and implemented on his farm. He took it upon himself to become a leader in information sharing with fellow farmers and school going students who had interest in learning the technology. The "teaching" aspect is what motivated and made him pay attention during interaction with researchers and extension staff to learn in great detail and later to respond with accurate information to questions raised by those seeking to know more about the PPT. Combining his own intuition with local knowledge and understanding of crop–pests interactions and practical observations, he was able to talk about the PPT freely and easily. This, according to him, was indeed science demystified [59] and gave him an opportunity to become a leader. Apart from stemborer challenges, he was motivated to participate in the PPT to address other problems of soil erosion and mono-cropping. These practices contribute to soil fertility losses in the area and the PPT was seen as a novel idea which was fitting and has the potential to address these challenges. The majority of fellow farmers have no experience or knowledge on cereal crops intercropping practices and are dependent on cereal–livestock production systems. They are constrained by perennial fodder shortages. To these farmers, the PPT was an opportunity to learn and share wide ranging knowledge and practices.

> Apart from pest control, food and fodder production to environmental conservation, the PPT is easy to understand and use despite the underlying scientific principles at play. The resource-constrained farmers are not able to afford expensive farming inputs aimed at increasing productivity such as fertilizers and pesticides. (Interview the PPT farmer, Bako Tibe, March 2015)

Another example is a farmer, from the neighboring village B who had actively participated as lead farmer in other technologies previously introduced in the village, thus his social standing was already elevated. Whenever new opportunities arose, he still stands out to be selected by fellow farmers and the local administration as their farmer leader.

> I have participated in other extension packages such as tomato, onion, and maize production. I am ready to make the PPT a model and aesthetic farm in my village. I previously learned from Sasakwa Global2000 farm extension model plot, so am determined. I have become a lead farmer through training and actively trying extension packages. Previously I was selected by the Kebele administrator as a model farmer in the village. I am happy with being model farmer, it is rewarding from the certificates of participation and sense of satisfaction either socially or economically. . . . . I have managed to invest in a new house, bought a mule and cart from increased farming income from being a model farmer. (Interview with the PPT farmer, Bako Tibe, March 2015)

The selection of farmer leaders was undertaken by the other farmers who had trust, confidence and were comfortable to have one of their own to take lead and whom they could learn from. Their elevated social status after participation in new technologies was a motivating factor for other farmers with similar backgrounds to emulate. Although it may seem to be pointing to replicating the "farmer opinion leadership" where an individual farmer is believed to influence other farmers' attitude and opinion [78], farmer participation in the PPT was different. It involved participation of other stakeholders (academic and non-academic researchers) in sharing of their diverse forms of practices, expertise and knowledge on stemborer pest control and their input and contribution to the research process was a result of joint efforts.

3.2. The Challenges

3.2.1. The Spread of Negative Rumors Regarding Push–Pull Technology

The concept of intercropping Desmodium with maize as an effective cover crop was not as well received as one would have expected. Some of the farmers were reluctant to embrace the PPT either due to cultural orientation or personal attributes such as fears, not trusting new information or technologies or being swayed by negatively rumored information. For example, in Yayu *Woreda*, some farmers were thinking that Desmodium was weed and wild plants being introduced into their farms to become an unnecessary competition with food crops or to be just a nuisance. According to some of these farmers, Desmodium was annoying. It gets stuck on clothes when walking through or weeding their maize fields. Some of them knew Desmodium as a wild plant growing on the farm hedges or in the forest and could not comprehend the fact that it can be intercropped with maize. At the same time, freely roaming grazing livestock after harvesting of maize is a common practice in the study area. Therefore, to some of the farmers, researchers and extension professionals, the introduction of PPT was not likely to work. In particular, they had doubts on how the PPT could be managed during the off-season when there is no growing of maize. That means, the perennial PPT companion crops will be destroyed and would need to replanted during the subsequent seasons. However, such doubts were expected because the PPT was being introduced into the area for the first time. To the promoters of the technology, the emergence of negative rumors and misinformation are some of the key areas of concern which had to be understood and addressed. These issues turned out to be the key learning points or resources for detailed learning about the technology [79].

> I had many questions in my mind on the future and sustainability of this technology. Now slowly the questions are being answered (. . .) such as roaming of livestock immediately after harvesting maize, preparing land for planting maize in the next season in the presence of Desmodium and Brachiaria grass (. . .). Because of the interactions we have had together with other stakeholders over time, I have cleared some of my doubts. (PPT Follower farmer, FGD in Bako area, March 2014)

3.2.2. Emergence of New Conflicts as a Result of the Push-pull Technology Implementation

The introduction and implementation of the PPT in the study area had mixed blessings. It not only brought new knowledge on managing stemborers and other benefits, but also new dimensions of conflicts which were not expected nor existed before.

> Initially we had an informal agreement on water use and allocation mainly for irrigation purposes. Now with the introduction of the PPT, some farmers who didn't understand the technology started questioning why water was diverted to these plots . . . This caused us to have irrigation water use conflicts, forcing us to irrigate PPT plots at night with the aid of torch. (Interview with the PPT Farmer, sedan kite Kebele, Bako Tibe, March 2015)

The implementation of the PPT in the area had contributed to the emergence of water use conflicts which had not been experienced before. The conflict was as a result of using irrigation water in the PPT plots as a supplement to low rainfalls in the area during the period of the study. The result may be seen as a negative outcome, but it brought the problem to the attention of the Ministry of Water and Irrigation for an intervention. Previously, water use agreements in the area were informally set up among the farmers who were practicing horticultural farming and other users along the Gibbe River. However, with the introduction of the PPT, it generated mixed reactions on water use allocation which was in high demand from many users along the river. According to the horticultural farmers and other users, there was no water allocation for the PPT. That means the allocation was causing unnecessary demand on the already "scarce resource". Despite the allegations, the PPT farmers attributed these conflicts to jealousy from their fellow farmers. This was because the maize crops in the PPT farms were outperforming the horticultural crops in terms of productivity and market value prospects. Previously, without the PPT, these farmers were planting maize using the same water and such conflicts were not witnessed. To prevent the resultant conflict from escalating or recurring in the future, the officials from the Ministry of Water and Irrigation who were part of the research team, together with other stakeholders, convened a joint meeting and agreed to have a formalized water use arrangement. This was a bottom up approach of formalization through stakeholder mobilization and engagement. The agreements were put on paper with a clear outline on who, when, where and what amount of water is allowed to be used for irrigation along the river considering other users downstream on the Gibbe River, which cuts across many *Woredas* in the region. This is in concurrence with Gray [18], whereby the absence of conflict resolution is a detriment to any form of transdisciplinary collaboration which requires integrating of diverse aims and interests among the stakeholders. In such circumstances, leadership with process skills to manage and facilitate conflict resolution plays an important role towards the success or failure of transdisciplinary efforts [18,30].

"My wife was skeptical and lost interest in the technology. She felt that I have been misled and sold out the land without her knowledge," the lead farmer commented. To this particular farmer's wife, this appeared as a sort of land grabbing where she could lose access to her land on which she previously could grow crops—and choose them freely on an annual basis—for her family's livelihood. Land is an important resource in terms of investment and food production opportunities for the socio-economic and environmental wellbeing of the smallholder farmers. Thus, when the same land is targeted to be acquired by outsiders who are either individuals or companies to serve their interests at the expense of its farmers, it can be seen as a form of land grabbing. The introduction of the PPT can be easily be equated with "land grabbing" which take place through introduction of commodity crops such as sugar cane, tea, coffee, Jatropha, etc. by multinational companies. This is because the crops are perennial, i.e., they are planted once and remain on the farm over the years. In most instances, the owners of these lands are poorly informed of the consequences of the investments and growing certain crops on their farms. There are several advantages highlighted by the proponents of such land transactions, e.g., food production, technology transfer, job creation and infrastructure development in those areas. They argue that such investments on land will enhance yield and close productivity gaps through the supply of high capital agricultural inputs such as irrigation and other technologies [80]. With awareness creation and information availability either through media or civil society groups, the farmers are seemingly alarmed with new technologies which are tailored or based on land investments. The PPT was linked with such examples. It was introduced as a new technology with perennial companion crops planted under smallholder farming systems. At the same time, it was being promoted by researchers from other countries.

3.2.3. Inconsistence of the Push-pull Technology with Locally Established and Traditional Farming Practices

The joint implementation of the PPT by researchers and other stakeholders was linked with introduction of new information which was inconsistent with their common practices, thus creating an uncomfortable psychological state due to uncertainty from what the technology had in store for them and famers were not sure about that. It was seen as source of conflicts by challenging the locally entrenched practices by the farmers and to even the extension and research officials. For example, maize is grown as a mono crop. Intercropping maize with an additional crop was not a common practice among the farmers. The agricultural extension staff had no experience on how to go about changing this practice. That means, introducing the PPT was directly conflicting with this already entrenched and established culture of mono-cropping. At the same time the farmers and extension officials were accustomed to the culture of conventional and linear technology transfer as the main source of extension messages. The participatory processes of stakeholder planning and implementing activities on a joint basis was seen by these officials as a source of conflict to their "usual business environment and ways of doing things". The farmers are usually relegated to the receiving end and they are only supposed to implement what have been instructed by the extension or by researchers. The off-season Brachiaria and Desmodium crops management introduced by the PPT required that the farmers had to fence off their farms or guard their farms from freely roaming livestock, which was conflicting with the age-old traditional practice of allowing livestock to freely graze.

However, this can be overcome by planning and working together by combining different ways of knowing, doing and learning among different social actors [10]. This stems from the fact that such differences as technical competencies, cultural beliefs, social status in the community and language barriers may contribute to a mistaken understanding and meaning, thereby leading to distorted or even unheeded messages [81]. Therefore, effective communication and coordination between such individuals can be used to overcome cognitive dissonance [82]. It is also specified that the PPT activities should last at least 2–3 maize cropping seasons for effective performance and learning to take place [51]. This gives sufficient time and allows stakeholders involved to become familiar with each other and adjust their local practices and beliefs to new technology and even the ways of working. This means that the organization and leadership of the TD research process provides the direction towards successful interaction, fruitful collaboration and learning. With the notable example of Ethiopia, the organization and implementation of the PPT requires institutional changes such as embracing participation by both non-traditional players such as private sector actors and youthful farmers who are showing interest in commercial or enterprise-based farming activities. This will broaden their collective learning beyond traditional communities and approaches of academic researchers and state-run extension agencies. This view was echoed by the key informant:

> What we have learned so far never existed before on intercropping cereals with Desmodium ... The PPT implementation process brings on board other new players such as the farm input sellers, youths and women to contribute to agriculture ... During dry season, farmers plant tomatoes, onions and cabbage as horticultural crops using furrow irrigation. This is a new custom of planting maize with Desmodium and Brachiaria grass during dry time. (Ministry of Agriculture extension expert, Bako Tibe, April 2015)

The emergence of conflicts as a result of new ideas from the PPT, became embraced and used as an opportunity for in-depth learning about the technology and also to find solutions and their possible adaptation. According to Suchman [83], these kind of situations require boundary crossing and indeed effective leadership to manage the transitions. During discussions on the sort of leadership required among the stakeholder teams that will enable the successful learning of PPT despite situations where it is seen as a potential source of conflicts with local practices, it was suggested that:

> There is need for flexibility among the collaborative stakeholders. The actors have to understand the needs of each other, their potential strengths and capacities. Although understanding the needs of the people and working on their minds is challenging .... but, the leaders should be experts who are committed with a guided vision long into the future. (Agriculture extension expert, Bako Tibe Woreda, April, 2015)

3.2.4. Lack of Formal Stakeholders' Memorandum of Understanding

There was no formal Memorandum of Understanding (MOU) among the stakeholders with details on collaborative working arrangements between the researchers and extension staffs from the different ministries and departments. The working arrangements were based on individual commitments by the staffs from the different departments and ministries. However, between partner organizations, it was not always smooth running. For example, some of the research and extension staff were asking for payments to participate. They were seeing the PPT implementation not as part of their mandate or their acknowledgement was unspecified. Some of the extension personnel and researchers felt that participating in the PPT implementation was conflicting with their day-to-day activities unless it was formally arranged. The MOU concerns were necessitated by the fact that previous initiatives on joint stakeholder engagements had not yielded tangible results but were limited to boardroom agreements. For example, the Agricultural Development Partners' Linkages Advisory Council (ADPLAC) was meant to strengthen linkages among researchers, extension staff and farmers. However, this had not been generally operationalized due to challenges from lack of actor commitment, financial limitations, staff time to frequent transfers turn overs. Therefore, it was hoped that collaboration cemented with a MOU in the implementation PPT activities with specific stakeholders could be a better way to learn how to deal with this kind of challenges.

4. Discussion

The implementation of PPT is associated with stemborer pests control, Striga seed bank depletion in the soil and/or the provision of fodder for livestock. This is based on the immediate results or the products farmers experience or benefit from practicing the PPT. However, beyond these benefits, there is a challenge on its sustainability. Apart from the durability and effective performance of the companion cropping, the sustainability of the PPT as an innovation is dependent on the strength of relationships and social learning processes which occur among the collaborating stakeholders, type and nature of new enterprises and activities which emerge or are created during its implementation. Equally, the sustainability depends on whether these activities will continue being implemented or practiced by the farmers over the successive cropping seasons.

4.1. Opportunities for Interactions

The Transdisciplinary action research approach and the practical on-farm nature of the PPT implementation provided frequent interactions and continuous learning based on its workings and even challenges for both researchers and farmers stakeholders. This was evidenced by their comments:

> The PPT is a unique technology. I think it is sustainable based on its integrated nature, and continuous learning on step by step basis (…) There are many questions which can be raised and answered with the introduction of the PPT on the farm. The major one being, it is addressing the serious challenge of stemborer pests in maize production (…), we can have the best breeder seed, best fertilizer, enough rainfall and well-prepared field, but without a sustainable, affordable pest control, still maize productivity will be affected or reduced. You may have all the capital, but if no effective pest control, you still have low crop yields. In the future, this technology has lots of potentials. (Interview with the Ethiopian Institute of Agricultural Researcher (EIAR) April 2015)

> The PPT is a useful technology for the new generation of farmers. This is because it lasts longer in the field and provides opportunities for continuous learning and interaction […] and the farmers have started to understand how environmental factors contribute to pest problems and how the same can be used on their management. (Interview with the PPT farmer, Bako Tibe, March 2015)

"I had many questions in my mind on the sustainability of this technology. Now, slowly, the questions are being answered" a follower farmer commented during FGD. Farmers like to participate in new technologies which can give them flexibility to improve in their farming enterprises or maximum returns on their investments. For example, the minimum average plot size recommended for PPT is about 600 m^2. In the study area, due to population increase, the land is scarce, and most farmers own less than 2 ha of land. The PPT plots require management practices which farmers can acquire with time, and when convinced with the performance results, they can expand to the entire farm or if not convinced, they can drop the technology. This makes the practice of PPT a learning journey. The benefits, such as fodder and stemborer control, are witnessed from two or more maize cropping seasons [51]. This requires patience to get full benefits and even to make the adoption and adaptation decisions. During the interaction process among the stakeholders, the PPT provided opportunities for learning about other farming methods and related enterprises, e.g., farmers were able to learn how to plant maize under irrigation. In the past, farmers in the study area have been growing horticultural crops using irrigation. With irrigation the PPT farmers were able to produce maize for food and additional fodder for their livestock. During the dry season, most farmers do not participate in cereal crop farming and let their livestock roam freely scavenging for scant fodder in the harvested fields. This not only affects the production levels of the livestock, but also contributes to environmental degradation through overgrazing and destruction of vegetational cover. The PPT had shown the potential that these negative environmental conditions can be addressed and even reversed:

> Land for grazing is reducing and slowly in-door feeding is gaining currency in the district. Ethiopia has highest number of cattle in the region, but the quality and productivity is poor due to low quality feeding. Therefore, the PPT is part of the solution in terms of complementary fodder production for livestock in the area. (Interview, Ministry of Agriculture, livestock production expert, Bako Tibe, April 2015)

The application of PPT as a method for stemborer pest control had not been used previously by the MOA officials in the study area. Despite this, as decision makers, their commitment and policy support were needed from grassroot to national levels of administration. The commitment shown from farmers on learning and their willingness to use the PPT was an encouragement for the officials to provide the necessary support. This is what culminated in the inclusion of the PPT as part of MOA's extension and soil and water conservation programs. The PPT was included as part of the Ministry's Integrated Pest Management (IPM) program and also in the soil and water conservation measures. This was because it was complementing the already existing efforts. For example, the MOA officials encouraged farmers to plant Brachiaria grass along the farm borders and slopes and intercropping maize with Desmodium to reduce surface water run-off and prevent soil erosion. In the past, the same program promoted the use of vetiver grass as a cover crop; however, due to the turf nature of its leaves, it is not preferred for harvesting as fodder for livestock. Brachiaria grass was preferred to vetiver grass, which has same growth and conservation properties, but the soft leaves make it easy for feeding livestock.

The soil and water conservation such as terracing, planting grasses on the hillsides or planting trees are conducted as mass campaigns mainly with long term objectives. In contrast, the PPT was done based on personalized contact with individual farmers and the conservation benefits have both immediate and long-term returns, e.g., Brachiaria grass providing fodder and stabilizing the soil on the hillsides. Even though the technology was proven as a potential platform upon which both immediate and long-term farming enterprises could be built, the question on its sustainability was still lingering in the minds of the stakeholders' team. To address the sustainability concerns, the team raised and discussed, among other issues, the potential linkages with new market opportunities as well as the PPT as content for training agriculture students in the technical and vocational centers.

4.2. Promoting Push-pull Technology Based New Market Opportunities

The implementation of the PPT came with new market opportunities for the farmers and other stakeholders to participate, i.e., in commercially-oriented farming. Some of these farming enterprises include dairy, beef (fattening), commercial seed production and agro-dealership for the PPT inputs (such as Desmodium and Brachiaria seed material). The coming together of several actors creates a market place, a network of the PPT practitioners with a broad-based social capital to draw from as well as for learning and dissemination of the PPT. As a learning platform, the PPT can be used to promote other technologies which have a linkage to it. For example, during the off season, the PPT plots had sufficient soil and residual moisture which farmers used to grow chickpeas. The high demand and good market prices for chickpeas in Addis Ababa and other big cities was a strong incentive for other farmers to participate in the PPT. This was because, during the dry season, only the PPT farms in the area had some residual moisture due to soil cover provided by Desmodium. This was an incentive:

> For the sustainability of this technology, we need to supply seeds for planting and target young farmers and develop local structures on seed supply … … The PPT is setting up the pace for accepting other new technologies. We have to use the existing platforms for us to get maximum benefits and, for its sustainability, we have to ensure seed source from own production. (Interview, Women and youth affair affairs expert, Bako Tibe, April 2014)

Sufficient production and distribution of *Desmodium* and *Brachiaria* seeds forms the basic component for sustainable expansion of the technology. The involvement of farmers in the PPT seed collection and production is a very important practice towards achieving this goal. This can only be achieved if the farmers and traders will have the interest to invest in seed production and distribution.

> Bako research center has no mandate to produce and sell the Desmodium seeds; we are only engaged in conducting research and producing breeder seeds. However, we can provide leadership on the distribution of the produced seed, i.e., support seed production and

coordinate the distribution, but the government authorities should provide guidelines on production and distribution. (Interview, EIAR researcher Bako center, November 2014)

Farmers in the area know Desmodium as new crop which came from the research centers, it is not an indigenous crop. It was introduced as a fodder and cover crop for livestock and in coffee farms. Farmers at the same time had no knowledge on seed production, processing and selling as a source of income. This is an opportunity for learning and to meet the demands of the new and emerging Desmodium seed market. (Interview, EIAR researcher Matuu Center, March 2014)

There is significant production potential for silver leaf and a modest quantity for green leaf Desmodium from the Yayu and Jimma areas adjoining forests and along the farm hedges. This is an opportunity for the youths and women farmer groups to participate in the PPT through Desmodium seed collection as a response to the demand and market for supplying the seeds for the new PPT farmers in the country. For instance, 3 tones of Desmodium seeds were bought by *icipe* in March/April 2015 from the farmers, Metu and Jimma Research Centers, and agro-dealers in Jimma. This was a good incentive for the producers to collect more seeds and an opportunity to generate cash income for the farmers. The market linkage was associated as a measure for sustainability of PPT. The private agro-dealers have an existing network of seed collection and distribution and Desmodium is one of the seeds collected and traded. The successful promotion and expansion of PPT will increase their market penetration with Desmodium seeds and this has significant influence on the sustainability of the PPT.

4.3. Applying the Push-Pull Technology Knowledge as Training Content

The PPT content is suitable as a curriculum material not only for training farmers, but also for training agricultural college students who are training to become DAs. The DAs training takes two years and are posted in the village levels as agricultural extension officers. Their participation with tutors during farmers' field days was an opportunity for them to link theory with practical realities on the farmers' fields. This stimulated their interest to learn more with the intention to apply the knowledge when they graduate. They learned that the PPT touches on key sectors of their training: crops, livestock and natural resources management. This was noted by one of the students who participated during the farmers' field days:

The PPT is simple to implement, it is cheap, uses less expensive inputs, it is very important for the livelihood of the farmers. Bako College is a center of excellence for training DAs in plant science and animal health. Including the PPT in our curriculum will be an opportunity for information transfer to the students who will finally become Das . . . .and will transfer the same knowledge to our farmers. (Interview, student Bako College, April 2015)

Bako technical and vocational training college is mandated to train frontline extension staffs, who are otherwise called DAs. They are employed immediately after graduating to work as village extension staff. During an interview with one of the tutors who attended farmers' field days, he said:

In our college, we provide DA's training in two types. That is occupation-based and project-based. The former one is based on Ethiopian occupational standard prepared by Ministry of Education. For the second one, the projects are prepared based on the competencies or topics selected from list of occupations like crop production, animal production and natural resource conservation. The project-based training contains competence, entrepreneurship and technology. So, having our curriculum content with technologies like the PPT is important for our students who will eventually teach farmers. (Interview, Bako college tutor, April 2015)

This statement was backed up by another tutor from the college:

The PPT field day provides an opportunity for the teachers to learn and have experience with the new technology. Now with practical knowledge, it is easy to read and understand and transfer to learners who will become Das . . . . This is a perfect opportunity for introducing the technology as a tool for training and for learning by various disciplines . . . . and seems to touch all the departments . . . from technology, trade, crops to livestock/health etc. (Comment by Bako college tutor during farmers' field day, April 2015)

4.4. Mutual Trust in Transdisciplinary Stakeholder Interaction

The critical basis for stakeholder collaboration and cooperation is trust which is built and maintained through regular meetings, openness, transparency and offering something of immediate importance to the stakeholders [58]. The PPT implementation process follows the maize crop phenology and covering several cropping seasons. The stakeholder interaction during these seasons has the potential to generate mutual relationship and agreements among the stakeholders involved. Sometimes, it can also lead to disagreements. In normal situations, for such interactions to come to a point of developing trust requires some time and sustained interactions [84]. During the PPT implementation over successive seasons, the stakeholders have an opportunity to frequently interact and learn over common and emerging issues related not only to stemborer pest control and other farming enterprises, but also to form strong relationships among the stakeholders.

Individual stakeholders must know each other well enough to be able to interact productively [83]. They must trust each other and also have no doubt on their willingness to contribute to the common enterprises of their community. This makes them feel comfortable to openly discuss and speak truthfully to each other in an attempt to address their common challenges. In Ethiopia, for example, as an outsider, the first author observed that there is a culture of benevolent trust among the people, especially in the rural areas. This means that nobody is willing to share information which can lead to some harm on a neighbor. They may prefer to keep useful information rather than share what is injurious to their neighbors. This culture makes them share information on new technologies or opportunities which are meant to directly address their real-life challenges now and in the future. Achieving the expected increase in maize production by 4–5 t/ha and provision of high quality livestock fodder which are associated with introduction of the PPT, is an incentive of immediate benefit for the farmers and other stakeholders to develop interest and share information regarding the technology. However, at the same time, if these farmers do not trust the source of information, they could easily give up on its benefits. The trust is even higher if the information comes from a competent source, i.e., the farmers know that the person is knowledgeable about a given subject. Such a culture of appreciating new information from competent sources and sharing it freely is an entry process to the learning journey. "Farmers easily trust new technologies if it is from competent contacts or sources in the community such as lead farmers, extension staff and researchers whom it is assumed have already tested or put into use before introducing or recommending it to us" commented a PPT follower farmer during an FGD. There may be histories of mistrust, personal differences or communication barriers between farmers, researcher and extension services providers in the past [31]. However, the TDR approach and the need to address common challenges of stemborer pests contributed to some extent to improved ties between the different stakeholders to learn and implement PPT together [85]

Therefore, the researchers, extension staff and smallholder farmers started to work together from the point of trusting each other's contributions. Although building such trust requires time which is longer than the time taken in this study, some "sort of ties" in an evolving learning journey of the PPT implementation was initiated:

I learnt that the Ethiopian farmers are interested to take up and adopt the PPT more than Kenyan and Ugandan farmers. It will take a short time to achieve the high adoption numbers due to the fact that we trust new information from competent sources so long as it addresses the problem at hand. (Comment by follower farmer who participated in exchange visit to Kenya and Uganda, FGD, May 2015)

This observation was based on the fact that, the PPT technology has been promoted in Kenya and Uganda for longer time than in Ethiopia. The reasons cited for low adoption rate was related to the commitments by the research and extension systems of these countries which had not taken up the technology at policy levels and promotion being mainly research driven. The Ethiopian MOA officials requested for the introduction of the PPT into the country from the *icipe* researchers with the support of donors in the year 2012. This signifies the high level of trust the MOA officials have on research and also their commitment for new information to address the problem of stemborers. However, this dependence on research and extension officers as the main source of information and knowledge on new technologies is not sustainable in the long-run. For example, every year, the government sets aside 30 days dedicated for soil and water conservation efforts. The farmers are compelled to participate in digging terraces, and planting soil cover crops such as vetiver grasses on the slope areas. In most instances, this has degenerated to mistrusts, whereby the farmers do not trust the intention of the government officials, whether from research centers or extension services. This is also linked to some of the cases where the DAs are used to collect "taxes" on behalf of the government. The farmers are forced to cooperate with the political power needs. This is completely conflicting with agricultural extension activities where farmers are supposed to be treated with freedom and trusting by the extension agents.

Farmers often have some apprehension vis-à-vis new technologies whenever introduced in their localities. Connecting with the previous experiences during the SG2000 technology promotion, the farmers were not sure what the PPT had in store for them. This was attributed partly to the lack of enough information at the initial stages of implementation. However, through continuous interactions with researchers and extension personnel, they were slowly appreciating the technology and learning in detail how it functions. For example, the farmers had noticed that with their current maize crop, the perennial pest problem of stemborer and birds had decreased:

> I haven't seen stemborer attack this time round in my maize field . . . . This is contrary to during the normal cropping season, the stemborer always infest stems and birds always attack the maize tassel . . . . I think birds are searching for stemborer eggs on the maize tassel, hence breaking it . . . I hope this Desmodium protects maize from stemborer and from birds . . . . I will continue to observe this trend in the coming seasons. In future I will make some conclusion. (The PPT Farmer interview, April 2015, Dembi Gobu, Bako Tibe)

However, this observation did not last long. Towards the end of season, the birds started attacking the maize crop, but the effect was less, because the cobs had already matured and it was difficult for the birds to feed on mature maize cobs. This was an advantage to the farmer. This keenness of the farmers to make observation and try to make sense of it either by conducting own research or engaging researchers for further discussions was a significant step towards appreciating the technology. This was not only in addressing the pest problem, but also an opportunity for learning by seeking answers to "new observations" made. For example, why did the birds not recognize the maize earlier? Was maize made invisible as a result of intercropping with Desmodium? Some of these questions were raised by the farmers based on the observations made for the first time and during the first cropping season. However, for certainty, they decided to continue to follow on this observation for some time and during subsequent seasons. They hoped that after 2–3 years of continuous cropping, they would have formed an opinion based on the long-term observations on the performance of the technology and even make recommendations for changes or adaptations. In the meantime, it came out from the discussions that there is no scientific method so far for scaring or controlling birds. Locally, the farmers use a cassette tape ribbon which they wind along the borders of the maize field. The hissing sound and reflection produced by the ribbon due to the winds blowing is used to scare off the birds. However, this is only effective for small plots of land of <0.5 ha. On the larger fields, they rely on manually scaring the birds away mainly by throwing stones into the maize plantation. This lasts for about two weeks, after which the maize cobs mature and the birds cannot manage to feed on them. However, during off-season crop, the population of birds is too high and feeding on the grain can be intense within a very short period, causing huge losses to the farmers.

5. Conclusions

The article shows that effective collaborative leadership coupled with social empowerment provides a chance, especially for the farmer stakeholders, to participate in the learning and decision-making processes by enabling them to contribute skills towards the development, refinement and adaptation of Push–Pull Technology. The personal interest among the researchers and other stakeholders is linked with the newness and first-time introduction of the technology to the study area to partake in its emergent opportunities and relationships. Although the perennial nature of cropping in the PPT provides opportunities for empowerment through continuous interaction and learning, it requires committed leadership and institutional engagements starting from the initial stages of planning and implementation, involving the researchers and other stakeholders. In the transdisciplinary research process, leadership roles taken by farmers is empowering in terms of their ability to reflect on their own practices and drawing on scientific explanations from researchers. It also enables them to take lead in new technology implementation and information sharing in a free and easy manner with fellow farmers and other stakeholders. Even though the implementation process was undertaken jointly; the different stakeholders represented their respective interest groups. That means that collaboration with "personally" committed leadership is most likely to contribute to the sustainability of the PPT activities when the personal interests are met first or understood well. The immediate PPT benefits and assurances will sustain the stakeholder interests and gain their trust for continued interaction and learning or follow-up on the PPT. Market forces and the involvement of the private sector players also have a role to play in this, as shown by the interests of individual farmers and traders in Desmodium and Brachiaria seed production, collection and distribution.

The findings show that, in a situation where there are conflicts, either of professional or disciplinary nature or practice-based, they should be embraced and become opportunities for in-depth learning, finding solutions and adaptation, rather than being sources of contradictions or misunderstandings about the new technology. That means new challenges which come with such a new technology should be addressed and developed in situ and used as the essential points of learning for further research, thus refining and improving its practical use.

The sustainability of the PPT activities and the transdisciplinary processes require longer term evaluation studies. Consequently, a follow-up to this study should be considered to document the changes which occurred. This is based on the fact that the introduction of the PPT, although seems positive and progressive, was not immediately accepted by all the stakeholder groups.

Author Contributions: G.K. and A.-K.H. supported in designing the study, and actively participated in its implementation and provided guidance in data analysis. C.A.O.M. and C.B. contributed to the provision of materials and logistics for field research and the data collection. I.M.N. conceived the study, collected and analyzed the data, and wrote—supported by the co-authors—the article.

Funding: The authors highly appreciate the scholarships and financial support received from The German Ministry for Economic Cooperation and Development (BMZ) through the BiomassWeb project (https://biomassweb.org) component of the International Center for Insect Physiology and Ecology (*icipe*) for the entire study.

Acknowledgments: Special thanks go to the farmers, Ministry of Agriculture officials and researchers from the Center for Development research (ZEF), University of Bonn and the Ethiopian Institute of Agricultural Research (EIAR) who provided guidance and technical support for the study.

Conflicts of Interest: No conflict of interest was reported by the authors.

References

1. Bracken, L.J.; Bulkeley, H.A.; Whitman, G. Transdisciplinary research: Understanding the stakeholder perspective. *J. Environ. Plan. Manag.* **2015**, *58*, 1291–1308. [CrossRef]
2. Lang, D.J.; Wiek, A.; Bergmann, M.; Stauffacher, M.; Martens, P.; Moll, P.; Swilling, M.; Thomas, C.J. Transdisciplinary research in sustainability science: Practice, principles, and challenges. *Sustain. Sci.* **2012**, *7*, 25–43. [CrossRef]

3. Wright Morton, L.; Eigenbrode, S.D.; Martin, T.A. Architectures of adaptive integration in large collaborative projects. *Ecol. Soc.* **2015**, *20*. [CrossRef]

4. Hirsch Hadorn, G.; Hoffmann-Riem, H.; Biber-Klemm, S.; Grossenbacher-Mansuy, W.; Joye, D.; Pohl, C.; Wiesmann, U.; Zemp, E. (Eds.) *Handbook of Transdisciplinary Research, Springer: Dordrecht, Netherlands*; Springer: Berlin, Germany, 2008; ISBN 978-1-4020-6700-6. Available online: https://www.springer.com/gp/book/9781402066986 (accessed on 4 April 2018).

5. Hoppe, R. Rethinking the Science-Policy Nexus: From Knowledge Utilization and Science Technology Studies to Types of Boundary Arrangements. *Poiesis Prax.* **2005**, *3*, 199–215. [CrossRef]

6. Nowotny, H.; Scott, P.; Gibbons, M. Re-thinking science. In *Re-Thinking Science: Knowledge and the Public in an Age of Uncertainty*; Polity Press: Cambridge, UK, 2001.

7. Stauffacher, M.; Flüeler, T.; Krütli, P.; Scholz, R. Analytic and Dynamic Approach to Collaboration: A Transdisciplinary Case Study on Sustainable Landscape Development in a Swiss Prealpine Region. *Syst. Pract. Action Res.* **2008**, *21*, 409–422. [CrossRef]

8. Bunders-Aelen, J.G.F.; Broerse, J.E.W.; Keil, F.; Pohl, C.; Scholz, R.W.; Zweekhorst, M.B.M. How can transdisciplinary research contribute to knowledge democracy. In *Knowledge Democracy—Consequences for Science, Politics and Media*; In't Veld, R., Ed.; Springer: Heidelberg, Germany, 2010; pp. 125–152.

9. Popa, F.; Guillermin, M.; Dedeurwaerdere, T. A pragmatist approach to transdisciplinarity in sustainability research: From complex systems theory to reflexive science. *Futures* **2015**, *65*, 45–56. [CrossRef]

10. Siebenhüner, B. Social learning and sustainability science: Which role can stakeholder participation play? *Int. J. Sustain. Dev.* **2004**, *7*, 146–163. [CrossRef]

11. Mobjörk, M. Centrum för Urbana och Regionala Studier. In *Crossing Boundaries: The Framing of Transdisciplinarity*; Örebro Universitet: Örebro, Sweden, 2009.

12. Mobjörk, M. Consulting versus participatory transdisciplinarity: A refined classification of transdisciplinary research. *Futures* **2010**, *42*, 866–873. [CrossRef]

13. Harris, F.; Lyon, F. Transdisciplinary Environmental Research: A Review of Approaches to Knowledge Co-production. Available online: http://www.thenexusnetwork.org/wp-content/uploads/2014/08/Harris-and-Lyon_pg.pdf (accessed on 8 July 2018).

14. Reid, R.S.; Nkedianye, D.; Said, M.Y.; Kaelo, D.; Neselle, M.; Makui, O.; Onetu, L.; Kiruswa, S.; Kamuaro, N.O.; Kristjanson, P.; et al. Evolution of models to support community and policy action with science: Balancing pastoral livelihoods and wildlife conservation in savannas of East Africa. *Proc. Natl. Acad. Sci. USA* **2016**, *113*, 4579–4584. [CrossRef] [PubMed]

15. Schäpke, N.; Omann, I.; Wittmayer, J.M.; van Steenbergen, F.; Mock, M. Linking transitions to sustainability: A study of the societal effects of transition management. *Sustainability* **2017**, *9*, 737. [CrossRef]

16. Crosby, B.C.; Bryson, J.M. *Leadership for the Common Good: Tackling Public Problems in a Shared-Power World*, 2nd ed.; Jossey-Bass: San Francisco, CA, USA, 2005.

17. Bushe, G.R. *Clear Leadership: Sustaining Real Collaboration and Partnership at Work*; Davies-Black Pub: Mountain View, CA, USA, 2009.

18. Gray, B. Enhancing Transdisciplinary Research through Collaborative Leadership. *Am. J. Prev. Med.* **2008**, *35*, 124–132. [CrossRef] [PubMed]

19. Archer, D.; Cameron, A. *Collaborative Leadership: How to Succeed in an Interconnected World*; Elsevier/Butterworth-Heinemann: Amsterdam, The Netherland, 2009.

20. Chrislip, D.D. *The Collaborative Leadership Fieldbook: A Guide for Citizens and Civic Leaders*; Jossey-Bass: San Francisco, CA, USA, 2002.

21. Frydman, B.; Wilson, I.; Wyer, J. *The Power of Collaborative Leadership: Lessons for the Learning Organization*; Butterworth-Heinemann: Boston, MA, USA, 2000.

22. Linden, R.M. *Working across Boundaries: Making Collaboration Work in Government and Nonprofit Organizations*, 1st ed.; Jossey-Bass: San Francisco, CA, USA, 2002.

23. Stokols, D. Toward a Science of Transdisciplinary Action Research. *Am. J. Commun. Psychol.* **2006**, *38*, 79–93. [CrossRef] [PubMed]

24. Pedrosa, A. Motivating stakeholders for co-created innovation. *Open Sour. Bus. Resour.* **2009**. Available online: http://timreview.ca/article/311 (accessed on 4 April 2018).

25. Siarta, S.; Blochb, R.; Knierima, A.; Bachingerb, J. Development of Agricultural Innovations in Organic Agriculture to Adapt to Climate Change–Results from a Transdisciplinary R&D Project in North-eastern Germany. In Proceedings of the 10th European IFSA Symposium, International Farming Systems Association, Aarhus, Denmark, 1–4 July 2012.

26. Stokols, D.; Fuqua, J.; Gress, J.; Harvey, R.; Phillips, K.; Baezconde-Garbanati, L.; Unger, J.; Palmer, P.; Clark, M.A.; Colby, S.M.; et al. Evaluating transdisciplinary science. *Nicot. Tob. Res.* **2003**, *5*, S21–S39. [CrossRef] [PubMed]

27. Pohl, C. What is progress in transdisciplinary research? *Futures* **2011**, *43*, 618–626. [CrossRef]

28. Siebenhüner, B.; Romina, R.; Franz, E. Social Learning Research in Ecological Economics: A Survey. *Environ. Sci. Policy* **2016**, *55*, 116–126. [CrossRef]

29. Godemann, J. Knowledge Integration: A Key Challenge for Transdisciplinary Cooperation. *Environ. Educ. Res.* **2008**, *6*, 625–641. [CrossRef]

30. EU SCAR. *Agricultural Knowledge and Innovation Systems in Transition—A Reflection Paper*; Standing Committee on Agricultural Research (SCAR); Collaborative Working Group on Agricultural Knowledge and Innovation Systems (AKIS); European Commission & Directorate-General for Research and Innovation: Brussels, Belgium, 2012.

31. Berhanu, K.; Poulton, C. The political economy of agricultural extension policy in Ethiopia: Economic growth and political control. *Dev. Policy Rev.* **2014**, *32*, 197–213. [CrossRef]

32. Abate, T.; Shiferaw, B.; Gebeyehu, S.; Amsalu, B.; Negash, K.; Assefa, K.; Million, E.; Aliye, S.; Shiferaw, B.; Abate, T.; et al. A systems and partnership approach to agricultural research for development: Lessons from Ethiopia. *Outlook Agric. J.* **2011**, *40*, 213–220. [CrossRef]

33. Ministry of Finance and Economic Development (MoFED). *Federal Democratic Republic of Ethiopia: Sustainable Development and Poverty Reduction Program*; MoFED: Addis Ababa, Ethiopia, 2002.

34. United Nations Development Programme (UNDP). *African Economic Outlook 2015: Regional Development and Spatial Inclusion*; OECD Publishing: Paris, France, 2015.

35. Elias, A.; Nohmi, M.; Yasunobu, K.; Ishida, A. Effect of Agricultural Extension Program on Smallholders' Farm Productivity: Evidence from Three Peasant Associations in the Highlands of Ethiopia. *J. Agric. Sci.* **2013**, *5*, 163. [CrossRef]

36. World Bank. *Enhancing Agricultural Innovation: How to Go Beyond the Strengthening of Research Systems*; The International Bank for Reconstruction and Development: Washington, DC, USA, 2006.

37. Star, S.L.; Griesemer, J.R. Institutional Ecology, Translations and Boundary Objects: Amateurs and Professional in Berkeley's Museum of Vertebrate Zoology, 1907-39. *Soc. Stud. Sci.* **1989**, *19*, 387–420. [CrossRef]

38. Rosenfield, P. The potential of transdisiplinary research for sustaining and extending linkages between the health and social sciences. *Soc. Sci. Med.* **1992**, *35*, 1343–1357. [CrossRef]

39. Ul-Hassan, M.; Hornidge, A.-K.; van Veldhuizen, L.; Akramkhanov, A.; Rudenko, I.; Djanibekov, N. *Participatory Testing and Adaptation of Agricultural Innovations in Uzbekistan-Guidelines for Researchers and Practitioners*; Center for Development Research (ZEF), University of Bonn: Bonn, Germany, 2011.

40. Chamberlain, K.; Khan, Z.R.; Pickett, J.A.; Toshova, T.; Wadhams, L.J. Diel periodicity in the production of green leaf volatiles by wild and cultivated host plants of stemborer moths, *Chilo partellus* and *Busseola fusca*. *J. Chem. Ecol.* **2006**, *32*, 565–577. [CrossRef] [PubMed]

41. Demeke, M. *Analysis of Incentives and Disincentives for Maize in Ethiopia*; Technical Notes Series; MAFAP, FAO: Rome, Italy, 2012.

42. Statistical Central Agency (CSA). *Agricultural Sample Survey 2010/2011 (September–December 2010)*; Report on Area and Production of Major Crops; CSA: Addis Ababa, Ethiopia, 2011.

43. Smale, M.; Byerlee, D.; Jayne, T. Maize revolutions in sub-Saharan Africa. In *An African Green Revolution*; Springer: Berlin, Germany, 2013; pp. 165–195. Available online: http://link.springer.com/chapter/10.1007/978-94-007-5760-8_8 (accessed on 4 April 2018).

44. Mantel, S.; Van Engelen, V.W.P. Assessment of the impact of water erosion on productivity of maize in Kenya: An integrated modelling approach. *Land Degrad. Dev.* **1999**, *10*, 577–592. [CrossRef]

45. Oswald, A. Striga control Technologies and their dissemination. *Crop Prot.* **2005**, *24*, 333–342. [CrossRef]

46. Kfir, R.; Overholt, W.A.; Khan, Z.R.; Polaszek, A. Biology and management of economically important lepidopteran cereal stemborers in Africa. *Annu. Rev. Entomol.* **2002**, *47*, 701–731. [CrossRef] [PubMed]

47. Midega, C.A.O.; Bruce, T.J.A.; Pickett, J.A.; Pittchar, J.O.; Murage, A.; Khan, Z.R. Climate-adapted companion cropping increases agricultural productivity in East Africa. *Field Crops Res.* **2015**, *180*, 118–125. [CrossRef]
48. Midega, C.A.O.; Pittchar, J.; Salifu, D.; Pickett, J.A.; Khan, Z.R. Effects of mulching, N-fertilization and intercropping with *Desmodium uncinatum* on *Striga hermonthica* infestation in maize. *Crop Prot.* **2013**, *44*, 44–49. [CrossRef]
49. Midega, C.A.; Wasonga, C.J.; Hooper, A.M.; Pickett, J.A.; Khan, Z.R. Drought-Tolerant Desmodium Species Effectively Suppress Parasitic Striga Weed and Improve Cereal Grain Yields in Western Kenya. *Crop Prot.* **2017**, *98*, 94–101. [CrossRef] [PubMed]
50. Khan, Z.R.; Midega, C.A.O.; Wadhams, L.J.; Pickett, J.A.; Mumuni, A. Evaluation of Napier grass (*Pennisetum purpureum*) varieties for use as trap plants for the management of African stemborer (*Busseola fusca*) in a push-pull strategy. *Entomol. Exp. Appl.* **2007**, *124*, 201–211. [CrossRef]
51. Khan, Z.R.; Pickett, J.A.; Wadhams, L.J.; Hassanali, A.; Midega, C.A.O. Combined control of *Striga hermonthica* and stem borers by maize-*Desmodium* spp. intercrops. *Crop Prot.* **2006**, *25*, 989–995. [CrossRef]
52. Tsanuo, M.K.; Hassanali, A.; Hooper, A.M.; Khan, Z.R.; Kaberia, F.; Pickett, J.A.; Wadhams, L.J. Isoflavanones from the allelopathic aqueous root exudates of *Desmodium uncinatum*. *Phytochemistry* **2003**, *64*, 265–273. [CrossRef]
53. Khan, Z.R.; Pickett, J.A.; Van den Berg, J.; Wadhams, L.J.; Woodcock, C.M. Exploiting chemical ecology and species diversity: stem borer and striga control in maize in Kenya. *Pest Manag. Sci.* **2000**, *56*, 957–962. [CrossRef]
54. Khan, Z.R.; Hassanali, A.; Overholt, W.; Khamis, T.M.; Hooper, A.M.; Pickett, A.J.; Wadhams, L.J.; Woodcock, C.M. Control of witchweed *Striga hermonthica* by intercropping with *Desmodium* spp., and the mechanism defined as allelopathic. *J. Chem. Ecol.* **2002**, *28*, 1871–1885. [CrossRef] [PubMed]
55. Khan, Z.R.; Midega, C.A.O.; Amudavi, D.M.; Njuguna, E.M.; Wanyama, J.W.; Pickett, J.A. Economic performance of the "Push-pull" technology for stemborer and Striga control in smallholder farming systems in western Kenya. *Crop Prot.* **2008**, *27*, 1084–1097. [CrossRef]
56. Habermann, B.; Misganaw, B.; Peloschek, F.; Dessalegn, Y.; Yihenew, G. *Inter-and Transdisciplinary Research Methods in Rural Transformation Case Studies in Northern Ethiopia*; BOKU: Vienna, Austria, 2013. [CrossRef]
57. Hornidge, A.-K.; Ul-Hassan, M.; Mollinga, P.P. 'Follow the Innovation'—A joint experimentation and learning approach to transdisciplinary innovation research. *ZEF Work. Paper Ser.* **2009**, *39*. [CrossRef]
58. Hornidge, A.-K.; Ul-Hassan, M.; Mollinga, P.P. Transdisciplinary innovation research in Uzbekistan—One year of 'Follow-the-Innovation'. *Dev. Pract.* **2011**, *21*, 834–847. [CrossRef]
59. Allen, W.; Ogilvie, S.; Blackie, H.; Smith, D.; Sam, S.; Doherty, J.; McKenzie, D.; Ataria, J.; Shapiro, L.; MacKay, J.; et al. Bridging Disciplines, Knowledge Systems and Cultures in Pest Management. *Environ. Manag.* **2014**, *53*, 429–440. [CrossRef] [PubMed]
60. Kohler, T.; Fueller, J.; Matzler, K.; Stieger, D. Co-creation in virtual worlds: The design of the user experience. *MIS Q.* **2011**, *35*, 773–788. [CrossRef]
61. Bortz, J.; Döring, N. *Forschungsmethoden und Evaluation für Human-und Sozialwis-Senschaftler*; Springer: Berlin/Heidelberg, Germany, 2003.
62. Miller, T.R.; Baird, T.D.; Littlefield, C.M.; Kofinas, G.; Chapin, F.; Redman, C.L. Epistemological pluralism: Reorganizing interdisciplinary research. *Ecol. Soc.* **2008**, *13*, 46. [CrossRef]
63. Freeman, R.E. *Strategic Management: A Stakeholder Approach*; Pitman: Boston, MA, USA, 1984.
64. Lundvall, B.-Å. (Ed.) *National Systems of Innovation: Towards a Theory of Innovation and Interactive Learning*; Frances Pinter: London, UK, 1992.
65. Nelson, R.R. (Ed.) *National Innovation Systems: A Comparative Analysis*; Oxford University Press: Oxford, UK, 1993.
66. Edquist, C. *Systems of Innovation: Technologies, Institutions and Organizations*; Pinter Publishers/Cassell Academic: London, UK, 1997.
67. Abate, T.; Shiferaw, B.; Menkir, A.; Wegary, D.; Kebede, Y.; Tesfaye, K.; Kassie, M.; Bogale, G.; Tadesse, B.; Keno, T. Factors that transformed maize productivity in Ethiopia. *Food Secur.* **2015**, *7*, 965–981. [CrossRef]
68. Abdissa, G.; Ethiopian Agricultural Research Organization; International Maize and Wheat Improvement Center (Eds.) *Farmers' Maize Seed Systems in Western Oromia, Ethiopia*; International Maize and Wheat Improvement Center: Addis Ababa, Ethiopia, 2001.

69. Stellmacher, T.; Grote, U. *Forest Coffee Certification in Ethiopia: Economic Boon or Ecological Bane?* ZEF Working Paper Series No. 76; University of Bonn, Center for Development Research (ZEF): Bonn, Germany, 2011.

70. Assefa, G.A. Development of maize stalk borer, *Busseola fusca* (Fuller), in wild host plants in Ethiopia. *J. Appl. Entomol.* **1998**, *106*, 390–395.

71. Belay, D.; Foster, J.E. Efficacies of habitat management techniques in managing maize stemborers in Ethiopia. *Crop Prot.* **2010**, *29*, 422–428. [CrossRef]

72. Getu, E.; Overholt, W.A.; Kairu, E.; Omwega, C.O. Status of stemborers and their management in Ethiopia. In Proceedings of the Integrated Pest Management, Kampala, Uganda, 8–12 September 2002.

73. Hurni, H. Agroecological Belts of Ethiopia. Explanatory Notes on Three Maps at a Scale of 1, 1. Soil Conservation Research Programme, Research Report. *J. Technol. Glob.* **1998**, *2*, 279–287.

74. Deneke, T.T.; Gulti, D. Agricultural Research and Extension Linkages in the Amhara Region, Ethiopia. In *Technological and Institutional Innovations for Marginalized Smallholders in Agricultural Development*; Gatzweiler, F.W., von Braun, J., Eds.; Springer: Cham, Switzerland, 2016; pp. 113–124.

75. Wenger, E.; McDermott, R.; Snyder, W. *Cultivating Communities of Practice: A Guide to Managing Knowledge*; Harvard Business School Press: Boston, MA, USA, 2002.

76. Amudavi, D.M.; Khan, Z.R.; Wanyama, J.M.; Midega, C.A.O.; Pittchar, J.; Hassanali, A.; Pickett, J.A. Evaluation of farmers' field days as a dissemination tool for push-pull technology in Western Kenya. *Crop Prot.* **2009**, *28*, 225–235. [CrossRef]

77. Amudavi, D.M.; Khan, Z.R.; Wanyama, J.M.; Midega, C.A.O.; Pittchar, J.; Nyangau, I.M.; Hassanali, A.; Pickett, J.A. Assessment of technical efficiency of farmer teachers in the uptake and dissemination of push–pull technology in Western Kenya. *Crop Prot.* **2009**, *28*, 987–996. [CrossRef]

78. Rogers, E.M. *Diffusion of Innovations*, 5th ed.; Free Press: New York, NY, USA, 2003.

79. Akkerman, S.F.; Bakker, A. Boundary Crossing and Boundary Objects. *Rev. Educ. Res.* **2011**, *81*, 132–169. [CrossRef]

80. Anseeuw, W.; Boche, M.; Breu, T.; Giger, M.; Lay, J.; Messerli, P. *Transnational Land Deals for Agriculture in the Global South: Analytical Report Based on the Land Matrix Database*; Geographica Bernensia; CDE/CIRAD/GIGA: Bern/Montpellier/Hamburg, Germany, 2012.

81. Medlin, B.D. The Factors that May Influence a Faculty Member's Decision to Adopt Electronic Technologies in Instruction. Ph.D. Thesis, Virginia Polytechnic Institute and State University, Blacksburg, VA, USA, 2001.

82. Cross, R.L.; Parker, A.; Sasson, L. (Eds.) *Networks in the Knowledge Economy*; Oxford University Press: New York, NY, USA, 2003.

83. Suchman, L. Working relations of technology production and use. *Comput. Supported Cooperative Work* **1994**, *2*, 21–39. [CrossRef]

84. Cundill, G.; Roux, D.J.; Parker, J.N. Nurturing communities of practice for transdisciplinary research. *Ecol. Soc.* **2015**, *20*, 22. [CrossRef]

85. Berger-González, M.; Stauffacher, M.; Zinsstag, J.; Edwards, P.; Krütli, P. Transdisciplinary Research on Cancer-Healing Systems between Biomedicine and the Maya of Guatemala: A Tool for Reciprocal Reflexivity in a Multi-Epistemological Setting. *Qual. Health Res.* **2016**, *1*, 77–91. [CrossRef] [PubMed]

© 2018 by the authors. Licensee MDPI, Basel, Switzerland. This article is an open access article distributed under the terms and conditions of the Creative Commons Attribution (CC BY) license (http://creativecommons.org/licenses/by/4.0/).

 *sustainability*

Article

The Potential of Plantain Residues for the Ghanaian Bioeconomy—Assessing the Current Fiber Value Web

Tim K. Loos [1,2,*], **Marlene Hoppe** [3], **Beloved M. Dzomeku** [4] and **Lilli Scheiterle** [2]

1   Food Security Center, University of Hohenheim, Wollgrasweg 43, 70599 Stuttgart, Germany
2   Institute of Agricultural Sciences in the Tropics (Hans-Ruthenberg-Institute), University of Hohenheim, Wollgrasweg 43, 70599 Stuttgart, Germany; lilli.scheiterle@uni-hohenheim.de
3   C.S.P. Consulting und Service für Pflanzliche Rohstoffe GmbH, Schnorrstraße 70, 01069 Dresden, Germany; marlenehoppe@web.de
4   CSIR-Crops Research Institute, P.O. Box 3785 Fumesua-Kumasi, Ghana; bmdzomeku@gmail.com
*   Correspondence: timloos@uni-hohenheim.de; Tel.: +49-711-459-22548

Received: 27 September 2018; Accepted: 13 December 2018; Published: 18 December 2018

Abstract: An essential part in the concept of any emerging bioeconomy includes the sustainable use of biomass as a resource for industrial raw materials. Focusing on the increasing demand for natural fibers, it will be necessary to identify alternative sources without compromising food security. Here, untapped potential lies in the use of plantain residues. Yet, it is unclear how or whether this can be activated. This article investigates the current situation in Ghana as a major plantain producer in Africa. Based on data collected with participatory tools, expert interviews, and group discussions, we (i) assess predominant plantain production structures, (ii) derive a stakeholder network map identifying institutional challenges, and (iii) discuss the potential starting points for linking the supply side with the national or international fiber market. Results indicate that there is substantial interest of private enterprises for high quality fibers. Despite traditional knowledge, after fruit harvest the fiber rich plantain pseudostems usually remain in the field. From an institutional point of view, key stakeholders and structures exist that could boost the establishment of a sustainable plantain based fiber value web. Key to such an endeavor, however, would be to pilot activities, including technology transfer of suitable innovations from other countries.

Keywords: plantain residues; fiber; value web; bioeconomy; Ghana

---

1. Introduction

Plantain is an essential crop for many rural households in the southern regions of Ghana. With more than 90% of the cultivated area belonging to smallholder farmers [1], it not only provides staple food to the population, but also serves as a key source of rural income which is reflected in contributing about 13% to the agricultural gross domestic product [2,3]. It is increasingly becoming an integral part of the agro-food industry [1]. Next to Cameroon and Uganda, Ghana is the main producer of plantain fruit in Africa. In West Africa, it is the largest producer with annual production of 3.95 million tons harvested from almost 363,000 ha in 2105/2016 [3,4]. The yield potential of plantain under rain-fed condition is estimated at 38 mt/ha with a current yield of about 11 mt/ha [3,5].

While marketing structures of fresh plantain fruit are very well established, the utilization of other parts of the plant are practically limited to the use of fresh leaves as animal feed, of green leaves for wrapping food, like *Kaakle*, or of dried leaves as wrappings for traditional dishes like *Kenkey*. There are also niches where traditional non-food uses are sometimes practiced. For example, the sap from the pseudostem is used as a dye for the traditional textile industry [6], or plant extracts being used for medicinal purposes, in particular for its blood coagulant properties. Despite all the traditional knowledge, other uses widely disappeared in most communities. Especially the use of the fiber-rich parts, like the stalk and pseudostem, lost their importance, which resulted in large quantities of plant residues being left on the fields after harvest.

Although there is high potential of plantain and other fiber-rich plants, the Ghanaian textile industry only processes cotton. Over the past couple of years, the cotton sector seems to fade away due to unauthorized and cheap imported textiles from e.g., China, India, and some African countries, and due to imported second hand clothing from Europe [7,8]. Relying on cotton as the sole raw material for manufacturing textiles and apparel is risky. Cotton grows in the northern part of the country, making transportation difficult, transaction costs high, and available produce often limited [9].

Against the background of such bottlenecks, the concept of a bio-economy with cascading uses, efficient utilization of plant by-products, and interlinkages between different value chains resulting in a value web, appears rather attractive for Ghana, not only for the textile sector, but generally for the whole economy. Focusing again on the increasing demand for natural fiber for textiles, it will be necessary to identify alternative plant-based sources that do not compromise food security. A major untapped potential in Ghana lies in the use of plantain residues as a source for natural fiber.

In contrast to other countries, like Uganda, India, or the Philippines, where banana fiber is already being used for fabrics, paper, etc. [10,11], the abundance of plantains in Ghana remains overseen when it comes to its potential contribution to the development of a sustainable bioeconomy. However, it is unclear which institutional, logistic, or socio-economic structures hinder the use of plantain residues in this context.

Therefore, this study (i) assessed the predominant production structures of plantain, (ii) derived a fiber (textile) focused value web for plantain and cotton, (iii) identified bottlenecks regarding institutional settings, and (iv) discussed a possible pathway to initiate the use of plantain fiber as part of the bioeconomy.

After a short reflection on plantain as a potential source of fiber and the current uses of plantain in Ghana, we describe the underlying conceptual framework that guided this assessment. Subsequently, a reflection on the applied methods and tools are given, and the empirical results are presented. The following discussion of the findings led to the conclusions and some implications for potential starting points to kick-start the development of plantain fiber-based value generation in Ghana. Thus, the results may contribute to an action-oriented follow-up project linking the identified stakeholders and initiating collaborations or a pilot project between local and possibly international stakeholders.

2. Background and Underlying Concepts

2.1. Reflecting on the Bioeconomy

Historically, biomass has been an intrinsic part of every economy. With the intensive focus on fossil fuels and the importance of the petro-chemical industry, the use of biomass as raw material inputs has lost its importance. Due to the expectation of a "peak-oil" scenario in the mid-21st century, the concept of a bioeconomy, i.e., an economy that is based on renewable biomass, was developed so as to offer a solution. Despite novel oil extraction techniques that allow for new resource exploitation, thus leading to substantial price drops, the strategy remains high on the political agenda, because of its strong contribution to achieving the set of environmental and climate change targets [12].

The concept of the bioeconomy has gained international importance and an increasing number of countries developed bioeconomy strategies. Each national strategy provides a country-specific, systematic framework with the aim to guide the design and development of a sustainable economy that is grounded on sound knowledge and is based on the use of biological and renewable resources. There is no single definition for the term, but at the Global Bioeconomy Summit in 2018 it was defined from a global perspective in rather general terms as: "the knowledge-based production and utilization of biological resources, innovative biological processes and principles to sustainably provide goods and services across all economic sectors" [12] (p. 4). Further, the "sustainable bioeconomy can help respond to societal aspirations and needs through its links to several relevant SDGs including poverty reduction, food security, access to water, energy and education, and sustainable innovation, production and consumption" [13] (p. 8).

In the past decade, the use of biomass for non-food proposes strongly affected prices and political debates (e.g., food vs. fuel), and global trade. Its repercussions were observed to be rather dramatic during the economic crisis in 2008/09. Therefore, it is of global interest to use scarce natural resources more efficiently and sustainably. This is particularly relevant when looking at the development of new materials and processes. This implies that an increased use of biomass as an industrial primary material should not happen at the expense of food and nutrition security, the environment, or social dimension [14,15].

The traditional sectors of the bioeconomy are not only linked to agriculture and forestry, but also they include the processing and service industry for both food and non-food materials (e.g., paper, textiles, construction building blocks, chemicals, etc.). In addition, it is required to develop new or optimize existing technologies with bio-based processes, and to more efficiently use available products and industrial applications [12].

Currently, 50 countries have developed national bioeconomy strategies. The targets and pathways to achieve a sustainable economic system differ among the countries, especially because of their different factor conditions. Technologically advanced and land-scarce nations focus more on bio-, nano-, and information technology, whereas countries with high net-biomass-productivity have a comparative advantage in providing primary products that are essential for further processing [16,17]. To transform the economic system towards a sustainable bioeconomy, efficiency gains are inevitable. This implies that processing needs to be based on fewer inputs, with less negative externalities (e.g., environmental pollution and greenhouse gas emissions). Furthermore, the holistic use of plants and the cascading use and/or re-use of products and commodities during their lifespan are relevant [15]. On the production side, indigenous and locally adapted crops may play an important role, as they often achieve good yields without the need of high external inputs. Examples in recent literature address the case of sugarcane in Brazil [18] and the role of cassava in Ghana [19].

Regarding the Ghanaian bioeconomy, plantain represents an optimal crop with untapped potential for various reasons. Firstly, plantain is a locally grown crop, which is well established and adapted to the local pedo-climatic conditions. Secondly, it is in the first place a food crop of which less than half of the plant's total biomass is harvested. The remaining crop parts (leaves, pseudostem, roots) are very rarely used leaving an unexploited potential. Thirdly, it is competitive in its resource use as compared to, for example, cotton. Furthermore, due to its extensive biomass production and its large number of traditionally known uses, it has a strong potential to be easily accepted by the local population as a sustainable biomass source for the bioeconomy. Picking fiber as the focus of this study, a closer look reveals that plantain may prove superior to other fiber crops, like cotton, which is grown under sub-optimal environmental conditions demanding a lot of water and external inputs. It may be expected that fiber sourced from plantain residues comply with the overall aspirations of the bioeconomy by fulfilling the aims of a sustainably grown, processed, and manufactured end product.

Natural fibers find application in various sectors in the industry. For example, in Brazil, they are used to reinforce recycled polypropylene to strengthen the material and are expected to replace glass fibers and mineral fillers in the plastic industry [20]. Another example is the textile and clothing sector with its growing demand over the past decades. Regarding the different physical properties of different natural fibers, it may be required to refine the raw material mechanically or with certain chemicals or enzymes. Next to clothing, other promising uses of fiber include non-woven technical textiles, composite material in the automotive sector, components in building materials, like light-weight construction materials, composite in polymers, or molded pulp applications.

2.2. Theoretical Basis

To better understand how plantain residues, especially the underutilized pseudostems, can serve as an industrial raw material in the emerging bio-economy of Ghana, we combine three methodological approaches in the conceptual framework. First, the biomass-based value web extends the traditional idea of a more linear value chain through a systems approach. When looking at product flow, information flow, or the value generation in a multi-dimensional way, linkages, synergies, and leakages between different value chains may be identified, addressed, and consequently may lead to productivity increases of the whole system or network [15,21]. A crucial part of the web approach comprises (i) the idea of a full use of biomass, e.g., crops, including so far disposed of plant parts and (ii) the aim of cascading use of products, which may be recycling or upcycling, too. Hence, tapping into plantain residues to open new value generation strategies provides a good example of a value web view.

As for the complexity of the value web with interwoven value chains, a functioning system will require different factor conditions and structures to be in place. As a second method, the Porter's diamond model [22] is adapted so as to identify determinants and preconditions that are needed for a sustainable and productive bio-economy. In detail, six elements are included: (i) factor conditions referring to the available resources, (ii) demand conditions reflecting marketing opportunities, (iii) related and supporting industries providing physical, technical, or intellectual inputs, (iv) firm strategy, structure, and rivalry, i.e., the way national companies are set up and operated, (v) government determining the institutional and political setting, and (vi) chances or shocks that may influence the development. All elements are interlinked, influence each other, and may also be practically linked in one or more ways to the biomass-based value web idea.

The third concept applied is the National Innovation System (NIS). It allows for assessing which key elements (~stakeholders) and which relationships (~linkages) exist and they are important for generating and diffusing new knowledge and innovations that are useful for the economy [23]. Established and functioning interactions within and between science, technology, economy, policy, and culture combined with complementarities between regions and the whole country are needed for national development [24]. Combining and linking these three theoretical models allows us to identify key stakeholders and to interpret their role in fostering the inclusion of plantain fiber into the value web and bioeconomy.

3. Data and Methods

This article originates from a demand-driven approach to explore the potential of national or international demand for example biomass products for industrial purposes that can be produced under comparative advantages in Africa. Plantain fiber was identified as a high potential commodity for a sustainable textile industry. Against this background, the supply side structures in Ghana, as well as the gaps, challenges, and opportunities in the value web are assessed in this study.

3.1. Research Design

In order to address the first objective of describing predominant plantain production structures, a thorough literature review was conducted. During on-site fieldwork, the preliminary findings where confirmed by experts and validated in several field visits. The review also allowed for identifying research institutes and experts in the Ghanaian plantain context, thereby providing the starting point for addressing the second objective, i.e., determining the plantain-cotton-value web. In a first step, key informants from the Council for Scientific and Industrial Research (CSIR), particularly the Food Research Institute (CSIR-FRI), and the Crops Research Institute (CSIR-CRI), where interviewed and asked to name relevant stakeholders in the plantain value chain. Similar to a snowball-sampling approach, the identified stakeholders, e.g., GRATIS foundation, ministries, farmer groups, etc. were visited, interviewed, and asked to suggest additional respondents. This design enabled a constantly growing picture of the plantain-cotton network.

Fieldwork and data collection took place during April and May, 2016. Next to the interviews, farmers' fields, weaving workshops, and textile factories were visited. In the case of plantain, the sites visited are located in Brong-Ahafo, Ashanti, and Eastern Regions. These three regions are located in an agro-ecological zone favorable for plantain, and account for 79% of total plantain production in Ghana [25]. For logistic reasons, the field visits and group discussions were organized along major roads connecting agricultural markets in Techiman, Sunyami, Kumasi, Nyambekyere, Ofinso, and Akosombo. In the case of cotton, the Volta region hosts the few processing centers in the country and it was therefore selected. In addition, it was possible to visit two banana plantations and get an idea of large-scale, export-oriented producers.

Complementing the interviews, focus group discussions and literature review with drawing the plantain value-web enabled the assessment of the current stage and potential future development, like the cascading utilizations of the plantain biomass for food and non-food purposes.

3.2. Research Tools

In order to visualize the actors and linkages in the plantain and cotton value web, a participatory tool that is known as Net-Map was applied. Developed by Schiffer [26], this instrument starts with informing the respondent or responding focus group about the purpose of the research and explaining how the Net-Map exercise works, in order to describe and understand the linkages between actors in the crop sectors. Using a blank Din A2 sheet of paper, the main questions guiding through the exercise for the plantain-cotton sector were:

1. 'Who is involved in the plantain/cotton sector?' Key persons and/or respondent groups were asked to identify stakeholders in the production, marketing, processing, and consumption of plantain (use of cotton), as well as other actors or institutions that may play a role in supporting or otherwise influencing the sector. The actors were categorized by type (e.g., producer, processor, private or public institution, etc.) and noted down on the blank sheet.
2. 'How are the actors linked?/Which linkages exist between the actors?' The interviewees were asked to identify and describe existing (and non-existing) linkages between the different actors. The type of linkage was color-coded and differentiated into (i) knowledge and information flows and (ii) product flows reflecting actual business connections. When drawing the linkages on the paper sheet, the direction of interaction was indicated by one- and double-sided arrows.
3. 'How influential are the actors with respect to initiating and/or developing the use of plantain for fiber?' The interviewees were asked to assess, from their perspective, which role different actors play when it comes to establishing a plantain based fiber value web. Using flat washers, the level of importance or influence could be indicated by allocating different numbers to different actors.

To gather additional information on bottlenecks, challenges, opportunities, etc., the individual network maps were discussed with the respective respondent or group. This allowed for gaining an in-depth understanding about the interviewees' reasoning of why some players are more important than others, how different linkages work (or not work), if there is a chance for plantain fiber, and how they perceive the sector.

With the intention of getting a full picture, the separate plantain Net-Maps were combined into a final map. Acknowledging the biomass-based value web idea and considering that plantain fiber for the textile industry is considered, the map was extended to include national actors from the cotton fiber processing sector. This amendment was possible by including key person interviews of and factory/workshop visits to textile companies and local, traditional weavers.

Furthermore, the mapping exercise was supplemented by common qualitative research tools, including in-depth personal interviews with key persons and experts, and focus group discussions with different stakeholders. Overall, in-depth interviews were carried out with 24 persons (nine key experts), six focus group discussions were organized, and six detailed Net-Maps were drawn. Stakeholders interviewed included plantain farmers, market actors (transporters, traders), scientists at research institutes, and other institutions from the private and public sector. Details are listed in Table A1 in the Appendix A.

Data quality was ensured through the combination of the different methods applied, which guaranteed the triangulation of the results. In-depth interviews follow a pre-defined structure and purpose, and thus differ from a conventional exchange of perceptions, thereby allowing the collection of rich data [26].

4. Results

With plantain at the focus of this study, this section starts with a brief overview of the main plantain production systems that are currently found in Ghana. It then maps out and discusses the links between different institutional structures that are relevant for integrating plantain into the bioeconomy. Connecting the plantain knowledge and information flows, plantain product flows, and cotton product flows allows for the illustration of a potential plantain-cotton value web for Ghana. The developed figures were generated based on information collected with participatory tools and are supplemented by the findings of in-depth interviews.

4.1. Plantain Production Structures and Value Web

About 90% of plantains are grown by smallholder farmers. Of the different farming systems that are described in the literature for West Africa [27], three intercropping strategies are predominantly found in Ghana. Most important is the cocoa-plantain combination. Here, a farmer starts with cultivating plantain as a nurse crop (locally referred to as pioneer crop) providing shade for young cocoa plants. During the initial first three years when the cocoa matures, plantain provides food for home consumption or sale. Once the cocoa is established, plantain is gradually phased out or moved to the borders of the fields [25,28]. In fact, this strategy is heavily supported by a project of the COCOBOD (Ghana Cocoa Board) which aims at providing ten million cocoa seedlings to farmers [29]. The support from the COCOBOD not only includes information on good agronomic practices that are related to cocoa, but also the provision of seedlings to the participating farmers. In general, the project is seen as very beneficial by farmers. They reported to be particularly interested in getting involved in cash crops, like cocoa, even more so when seedlings and equipment are being provided. One drawback appears to be the focus areas. Farmers living too far from the cocoa nurseries were unhappy about an unfair support. Another critical point may be agronomic practices. Several field visits to cocoa-plantations revealed that in a lot of cases, cocoa pods are not harvested properly, left rotting on the stem, or not used for processing (drying, etc.). Farmers mentioned that this lack of interest and commitment is due to the limited access to markets with good cocoa-prices.

The other two production systems are more oriented towards producing (home-)consumption crops. One cropping system is intercropping plantain with cocoyam (taro). According to farmers' statements, this mix provides them with an additional crop that can either be used by themselves in the household, or be sold on the market if there is surplus available or cash income required. From a food security perspective, diversifying the diet by adding roots and tubers to daily meals should have a particularly positive effect on nutritional aspects. To prove this empirically, further studies will be helpful. Cash-crop farmers also stated that the crop mix is often chosen as an alternative to a single nurse crop for cocoa. The advantage here is that, during the early years, two types of food crops can be harvested. Relying on a more diverse crop portfolio is seen as a means of reducing risk and, at the same time, increasing land productivity.

The third cropping system that is regularly encountered is intercropping plantain with cassava. This mix also draws its importance from very popular food preferences, i.e., *Fufu. Fufu* is a national dish made by mashing plantain with cassava. For many rural farmers this is a main staple food consumed on daily basis. However, also in more urban areas, from small food shacks to high-level restaurants, this traditional dish is part of the menu. Not surprisingly, several research activities, small and medium enterprises, and large players in the food processing industry invest in developing and setting up processing lines to produce plantain and cassava flour, so as to increase shelf-life of the perishable fruits and roots, respectively. Packaged flour ready for use is especially relevant for urban areas, like Accra, where supermarkets are becoming more and more important as a source for groceries.

The map of the plantain biomass web illustrated in Figure 1 identifies the cascading uses of plantain biomass in Ghana and potential other products that are relevant for the bioeconomy.

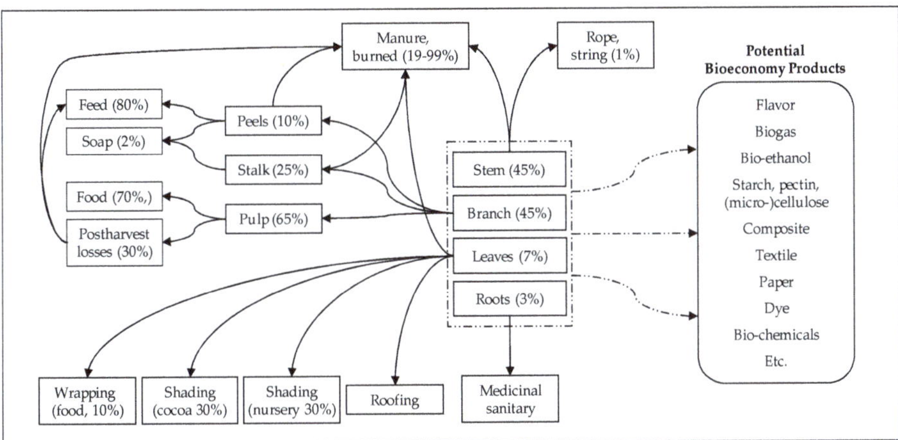

Figure 1. Biomass flows in the plantain value web; Source: own data, additions [30,31]; percentages are estimates based on expert interviews and literature [32].

4.2. Mapping the Plantain-Cotton Biomass Web

Drawing on the theoretical considerations of a biomass-based value web as a cornerstone of a sustainable bioeconomy, the plantain and cotton fiber related institutional structures and networks that are assessed in this paper include three components. Firstly, knowledge and information flows relevant for plantain (and cotton) production, processing, and marketing provide the foundation for an up-to-date technological and innovative development processes. Secondly, the plantain product flow identifies currently produced and traded products, as well as the respective stakeholders involved. Thirdly, when assessing the potential of using plantain pseudostem residues for fiber production, it seems essential to also understand the cotton product flows, as the main natural fiber currently used.

4.2.1. Plantain Knowledge Flows

Starting with the national innovation system perspective, Figure 2 shows stakeholders, linkages, and flows of knowledge and information about plantain production, use, and processing. Relevant in Ghana are actors from research, politics, extension services, and non-governmental organizations (NGOs). In general, two types of information are generated and/or provided. Smallholder producers are supported with suggestions on how to improve production practices targeting an increased fruit yield. Involved are the government extension service through the Ministry of Food and Agriculture (MoFA), the Council for Scientific and Industrial Research–Crops Research Institute (CSIR-CRI), NGOs, and to some extent other research institutions, including universities. Offered are e.g., group trainings or field visits with individual advice on topics like plantain propagation, greenhouse multiplication, and general agronomic techniques. When considering knowledge generation, the most active are universities and the CSIR-CRI, which may also collaborate, and in some cases, exchange information. Regarding food processing, the CSIR–Food Research Institute (CSIR-FRI) seems to be the driving force in developing and introducing processed plantain products. By providing technical advice to private, mainly large enterprises, the mandate is to bring *Fufu* powder, plantain chips, and other products with long shelf life into supermarkets so as to improve supply and access to key food commodities to the urban population.

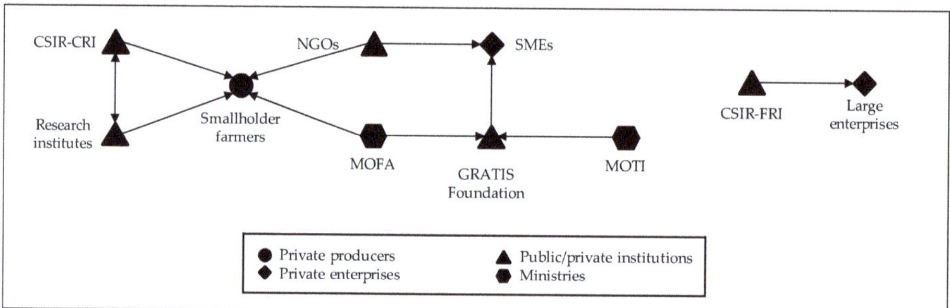

Figure 2. Visualization of plantain related information and knowledge flows; Source: own data.

Special attention should be given to the GRATIS Foundation of Ghana. Initiated as a 'spinoff' of a project carried out by the Ministry for Trade and Industry (MOTI), the goal of the foundation is to provide a platform for the structured vocational training of young adults to be (agricultural) mechanics that are capable of opening their own workshop, and to produce and market technical equipment, like cassava grinders, etc. The network of GRATIS-workshops is supported by two ministries and it frequently collaborates with small and medium enterprises in need of machinery equipment. What appears to be missing is a direct link with the research side (CSIR institutes, universities), thus making it likely that innovations and new technologies face a time lag regarding dissemination.

4.2.2. Plantain Product Flows

At the focal point of a plantain fiber value web is, of course, plantain production and product flow. As indicated in the short literature reflection and observed during fieldwork data collection, there is currently hardly any use of plantain fiber in Ghana. Although traditional uses, like ropes, sanitary pads, or medicinal purposes, are reported, it seems that its actual use widely disappeared among farmers, traders, and other people. In contrast to several other countries, Ghana appears to be unaware of the potential of plantain pseudostems as a natural fiber source for national and international processing including e.g., the textile industry. Thus, Figure 3 rather visualizes the current flows of fresh fruits and leaves. Yet, the stakeholders could easily pick up steps, like bulking, transportation, or marketing of plantain fiber or fiber bundles, and therefore may serve as a blueprint when making use of the whole plantain biomass.

Two producer structures were identified during fieldwork: Small-scale plantain production by smallholder farmers in the Southern regions of Ghana; and large export-oriented plantations of banana in the South-Eastern regions. Although this paper focuses on plantain, banana or generally all *Musa* species may have comparable potentials as fiber sources. Especially, large-scale plantation enterprises can be assumed to have logistics, management, bulk production capacities, etc. that may be expanded to also consider fiber production.

Despite some bottlenecks in marketing [33], the marketing channels for plantain fruit are very well established. Fruit bunches find their way to consumers via traders at the village level, market women, aggregators and wholesalers, and retailers. Particularly important for markets in Ghana are female trader groups that are headed by 'market queens' who manage the marketing. While older research claims them to control prices and market access [34,35], more recent findings suggest that they are essential for the value chains as types of collective action and credit groups [36,37].

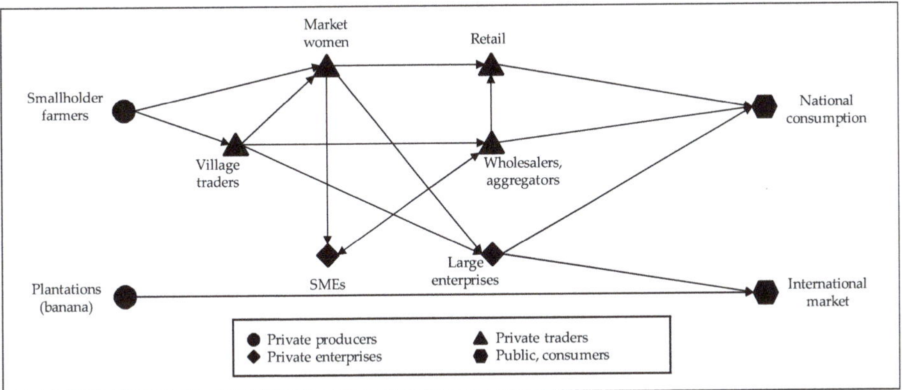

Figure 3. Visualization of plantain product flow; Source: own data.

Small, medium, and large enterprises processing plantain fruits or using leaves as wrappings when preparing food dishes either get their raw material from local village traders, or from market women who usually can provide larger quantities. Regarding plantain, constraints mentioned frequently are the perishable nature of the fruit as well as its bulkiness. It seems difficult to optimize transportation logistics and match demand with available produce when many traders, aggregators, and wholesalers compete. Especially the seasonality of plantain availability leads to price fluctuation.

Plantain fiber would be a new commodity added to the plantain product flows. From a generic value chain perspective, fiber bundles are extracted, dried, and baled close to its cultivation area. This implies that either producers, producer groups, or private start-ups are needed to enter the network between the initial production and local village traders. As farmer groups already exist and Ghana's business environment appears very vibrant, this bottleneck may be addressed if appropriate support measures are put into place. The further downstream transportation and marketing could easily be taken over by the established traders who in many cases deal with several different commodities. In order to use the fibers, for example, in the textile industry, further processing steps, like spinning, weaving, or apparel manufacturing may be carried out by other companies that are yet to be established. Especially, if the finally targeted product is supposed to show properties like cotton, more complex processing like chemical or enzymatic treatment will be needed to reduce the high stiffness of banana fiber. A recent study discusses the mechanical properties of banana and other fibers, and shows that it is possible to produce a textile grade of banana fibers [38]. The study also discusses options of mixing banana with other natural fibers, like wool, flax, hemp, or cotton. Currently, banana based rayon, often referred to as banana-silk, is already widely available on the market as yarn, fabric, clothing, or other goods, and is offered by e.g., Frabjous Fibers (USA), ZS Fabrics (USA, India), or Green Banana Paper (Micronesia). Considering merely the textile sector is falling short of the many other potential uses of plantain residues or by-products, some requiring sophisticated technologies, others low-tech solutions.

4.2.3. Cotton Product Flows

Emphasizing and concentrating on the textile and clothing sector as a continuously growing market worldwide, the national production and processing of cotton might appear as a key sector in the Ghanaian bioeconomy. Figure 4 presents the cotton stakeholder map for Ghana. National cotton production takes place in the Guinea Savanna zone in the northern part of the country. The produce may pass through village traders and wholesalers to the few national large-scale processing factories that are located in the Volta region. Ginning, spinning, weaving, and finishing fabrics all takes place in the factories. The final steps in manufacturing textile apparel generally takes place in small and medium enterprises before entering the retail market. The in-depth interviews with textile companies revealed that the more important source of cotton is the import from China. The input is said to be cheaper, more accessible, and reliable due to the closer location to the harbor, more abundant, and of similar quality to the national or regional supply. The real challenge, however, was mentioned to be the cheap imported clothing that gradually displaces nationally produced apparel. There seems to be double the burden of depending on imports of cotton bales (mainly from China) for further processing, and having to compete with the imported end products. Therefore, the interviewed managers of the textile factories were keen on learning about alternative natural fibers that may prove competitive in the near future.

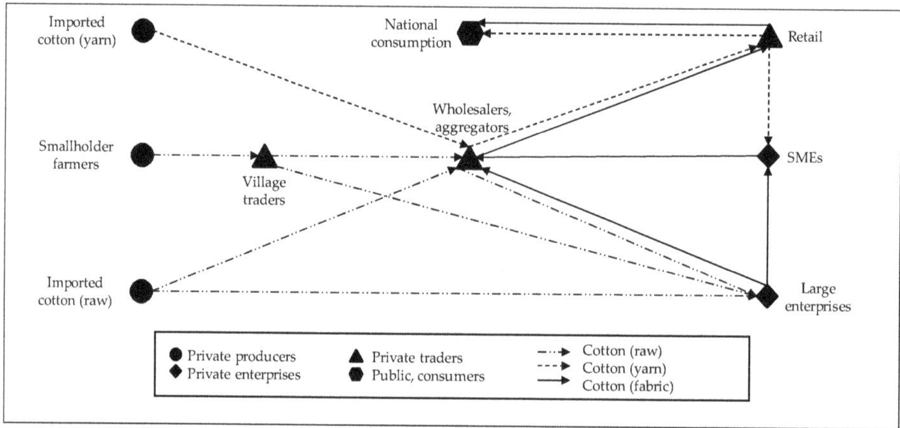

Figure 4. Visualization of cotton product flow (simplified); Source: own data.

An exception in the Ghanaian textile sector can be found in *Kente* cloth and fashion. *Kente* are traditional, colorful clothing carrying different meanings and are worn at special occasions. Using hand operated looms, small-scale weavers produce the required number of cloth-pieces (each ca. 10 × 10 cm to 20 × 20 cm), which then are sewn together to get the whole expressive garment. The employment arrangements among the interviewed weavers varies: self-employment, working in loose groups, working as part of an association, or being employed. However, all of these use similar production techniques and rely on the same input namely imported cotton yarn and/or thread purchased on the local market. In the end, they claimed not to really care about the original source of fiber used for the yarn, if the required colors are available and weaving properties are guaranteed.

4.2.4. Plantain-Cotton Value Web

In order to fully understand a natural fiber value-web considering both plantain and cotton in Ghana, the information and knowledge network and the two product flow networks are combined in a single Net-Map, as shown in Figure 5. Most relevant from the value-web perspective are those stakeholders being active in different networks, hence forming the linkages between the systems.

Starting with crop production, smallholder farmers are the stakeholder in common. As cotton and plantain grow in different agro-ecological zones, i.e., cotton in the northern part, plantain in the central and southern parts of Ghana, there will not be much overlap, competition, or synergies to be expected. However, with the public extension service as the main source of agronomic advice, there are likely to be similar bottlenecks, like coverage and frequency of support given. During focus group discussions with plantain farmers, farmer groups, and extension officers the lack of enough trained staff, and targeted (re-)training were highlighted several times. In addition, information about new agricultural technologies was reported to not reach extension officers. Innovations only find their way quickly to farmers if they collaborate with research projects.

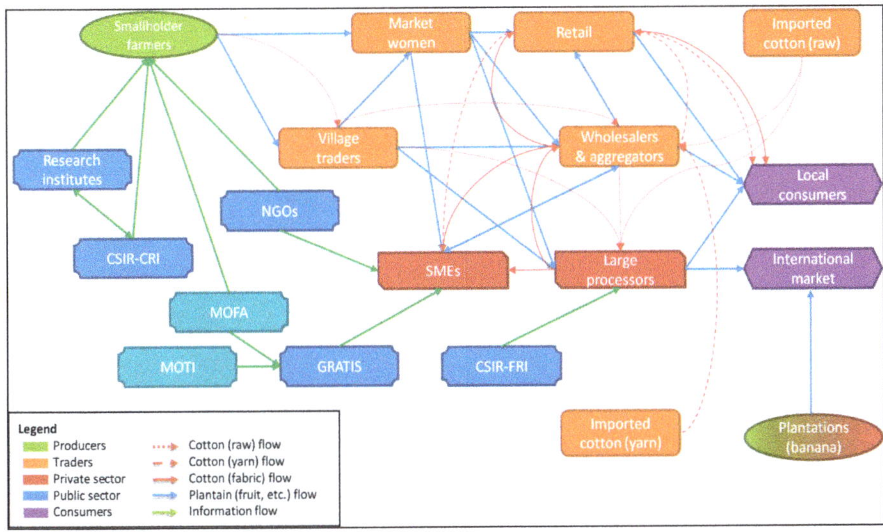

Figure 5. Plantain-cotton value web in Ghana; Note: Currently, only plantain fruit and leaves are used commercially; Source: own data.

Regarding processing, the key to a sustainable downstream value generation based on plantain fiber involves small and medium enterprises (SMEs) or large processors. Currently, both types of private sector actors handle plantain fruit. SMEs are backed by NGOs, and, more importantly, the GRATIS foundation. Each of these knowledge support agents may also expand their field of activity to provide information and technical know-how necessary to establish plantain fiber extraction. The GRATIS workshops may even be involved in designing or adapting suitable equipment for fiber extraction. If financing issues can be overcome, new SMEs may develop taking over processing steps, like ginning, spinning, weaving, and further processing, thereby creating new jobs and opportunities. The large-scale textile companies currently do not operate at full capacity. Thus, with some amendments to processing lines, they could evolve into the most important domestic demand side players utilizing both cotton raw material, plantain fiber raw material, and in the long run possibly other natural fiber sources, like e.g., pineapple, bamboo, rattan, etc.

5. Discussion

This section mainly discusses the empirical findings concerning the institutional structure of plantain with respect to the elements of the Porter model and the National Innovation System. The discourse allows for drawing implications regarding the potential of plantain fiber in Ghana and to develop a possible pathway for initiating fiber use of plantain residues.

5.1. The Plantain Biomass Web

Plantain is one of many species of the *Musa* genus grown worldwide. Whether as sweet fruits or as a staple food with high starch content, the fruit is generally the main part of economic interest. In Ghana, the number of smallholder farming households cultivating and relying on plantain for home consumption and income generation is estimated to be more than 400,000 families [25]. Assuming that the biomass of an average plantain plant is distributed to the bunch branch (45% of the biomass), stem (45%), leaves (7%), and roots (3%), and only the stem and leaves to be regularly used, there is a huge gap in sustainably using the full plant potential in the sense of a cascading and whole plant use envisaged in a functioning bioeconomy.

While the use of by-products is widely applied and established in many other parts of the world, the systematic use and value generation through plantain residue processing is very rare in Africa in general. An exception that was identified in this research is an initiative to start utilizing plantain in Uganda through an innovation platform [11], but the technology and knowledge transfer seems to be stuck in the country. At least there is no evidence that any collaboration with West-African partners were attempted or conducted. Potential uses of banana by-products reported in the literature are manifold (see Section 4.1, Figure 1, [30,31]) and could easily be adapted to any *Musa* species. Depending on different processing techniques, end products may serve as food or as non-food commodities. Looking at the pseudostem, different extraction methods exist to yield fiber and pulp to be used in textiles, paper, or as a composite. Known and established extraction methods include mechanical, chemical, and enzymatic treatment. Though still at a research and development stage, steam explosion may prove as an interesting and effective alternative in the future. In the Philippines, Indonesia, India, and other South-East Asian countries, the fiber of different *Musa* species, e.g., *Musa textilis*, is already being used for e.g., teabags, ropes, sacks, fabrics, and in some cases also textiles for clothing. In many of these countries, the production and demand structures are well established. Both knowledge and technical equipment, like specific machinery required to extract and process raw fiber from banana pseudostems are available.

5.2. Factor Conditions

From a production perspective, two possible strategies stick out. First, the abundant plantain and its current neglect of residue use theoretically may serve as a sufficiently large source of available resources. The crucial challenge is overcoming the small-scale production. With thousands of smallholder farmers operating on little land area, a bulking system would need to be put into place so as to be able to gather adequate quantities of raw material. This could be achieved through forming cooperatives, relying on traders and middlemen, or through implementing outgrower schemes with e.g., contract farming. Here, lessons for an optimal contract design in the context of Ghana's bioeconomy may be drawn from recent findings in the cassava sector [19,39].

When considering that the plantain pseudostems contain large shares of water, an efficient and sustainable use implies the extraction of fiber on, or close to, the production site. With the first processing step near to the production sites, the sap and scraps from extraction may be returned as (fluid) fertilizer to the farms.

Regarding transportation and further processing, the well-established and functioning plantain fruit marketing system may be expanded to include the fiber trade as a regular activity. Already now, traders and transporters fill their trucks with a mix of different marketed food and non-food commodities in order to operate at full loading capacity.

As an alternative to relying on smallholder production, a second approach could be the large-scale, export oriented banana plantations. If they become aware of a potential economic benefit, they may add fiber extraction to their production portfolio. The current practice of utilizing all banana residues as green manure so as to maintain nutrients, is not necessarily compromised if only fiber is extracted. The sap and scraps from extraction may yet be returned on the plantations.

5.3. Demand Conditions

On a large scale, the international demand for fiber is expected to continue to increase over the next decade, especially in the textile, paper, and pulp/composite industry. On the one hand, this is due to higher consumer preferences for natural resource based products. On the other hand, the availability as well as the market price for alternative natural fibers like cotton plays a role. Focusing on Ghana, cheap cotton imports heavily affected national production in a way that almost all raw material is now imported. This is especially important for the few large-scale textile companies that reported to be willing to become independent of imported cotton. As national and regional raw cotton production cannot meet the demand, they would gladly handle alternatives. The prerequisite, of course, would be the physical (e.g., spinning, weaving) and chemical (e.g., dyeing, coloring) properties of such an alternative natural fiber, as well as a competitive price.

Another positive aspect for fostering plantain based fiber production is that these large-scale companies only operate at a fraction of their installed processing capacity. For example, in the interviewed factory about two-thirds of the equipment is not used at the moment. Despite this, if cotton remains economically superior to plantain fiber, there would still be the option of adapting part of the currently vacant equipment to plantain fiber, thereby providing a basis for generating additional marginal return and making the company more profitable overall.

Apart from the large-scale producers, local demand also includes small-scale weavers who currently fully depend on imported yarn. Similar to the factories, they are ready to switch to yarn originating from fiber other than cotton, as long as the handling is not influenced substantially and the price for yarn is not increased.

5.4. Related and Supporting Industries

In terms of supportive structures, especially the knowledge and innovation aspect is rather active in Ghana. Several universities, research institutes, and other initiatives particularly consider plantain. Their focus, however, is mostly on propagation, productivity, and fruit processing. Yet, the institutional structures are in place and they could easily broaden the view to include the use of plantain by-products for e.g., extracting natural fiber from plantain pseudostems. While technical knowledge or equipment is currently not available, a technology transfer is possible from e.g., India, or from the African continent (Uganda). Adaptation, lab testing, or field trials so as to develop suitable equipment could be conducted by universities with a technical focus, research institutes, private consulting companies, or the GRATIS Foundation of Ghana. The latter, and possibly other stakeholders, also have the potential for the upscaling and dissemination of know-how through their vocational training programs. Optimizing, producing, and marketing of machinery could also be organized through their workshop network and marketing system.

5.5. Firm Strategy, Structure, and Rivalry

From a national perspective, the large-scale textile companies that are located in the Volta region are likely to serve as the cornerstone of processing plantain fiber. They all have productive capacity, but they face different challenges. The currently active enterprise operates at about one-third of the installed production capacity, it is equipped with rather old spinning machinery and weaving looms, and it struggles with efficiency issues and competitiveness with cheap imports from e.g., China. When targeting the European Union (EU) as a potential international trading partner, previous research by the authors showed that companies pay specific attention to aspects like working conditions or labor safety, which may need to be improved to some extent before receiving certain certifications requested for by the customers. To meet such international requirements, it may be necessary to invest in technical, managerial, and other improvements for which investments are likely to be needed.

At the time of fieldwork data collection, the employees of a second textile company were on strike, because the management did not pay them for three to six months. In informal discussions with some workers, they claimed that the whole board and the executive management were not interested or were not able to properly run the factory. Unfortunately, it was not possible to meet with the management staff due to the strikers barricading the entrance to the compound. Again, this hints at the need for managerial improvements for the large-scale processors.

Turning to small-scale weavers, the plantain fiber raw material is not of interest. Especially, the traditional *Kente*-weavers use hand-operated looms and they only work with yarn that was bought on the market. Although the interviewed weavers reported to be interested in trying other than imported Chinese cotton yarn, they most likely would still choose the cheaper option for manufacturing. This implies that in order to sustainably include plantain fiber in the Ghanaian (fiber) value web, it is most likely inevitable to also initiate the establishment of spinning facilities in the form of e.g., separate enterprises or to couple the refining process with the first processing step of extracting raw fibers on or close to the primary production sites. This approach may induce competition, thereby fostering efficiency and (economic) sustainability.

5.6. Government and Chance

Considering the high importance of plantain for the Ghanaian farming population as a staple crop to ensure food security and generate income, public interest, activities, and ministerial support mainly focusses around fruit production and processing. Hence, it is not surprising that farmers are supported mainly through the national agricultural extension system. Fostering research and innovations that aim to extend shelf life or to process fruit to e.g., *Fufu* powder is of relevance. Other key supporting industries that were identified in the network analysis also receive attention. Most prominent is the support of GRATIS as an approach to improve the vocational training of rural and/or peri-urban youth, to strengthen the dissemination of agricultural innovations, like cassava peelers, and to set up a network of workshops across the country. Regarding the use of plantain residues, including the pseudostem, no specific policy could be identified during the fieldwork interviews. It seems that this neglect originates from the lacking awareness of the economic potential of plantain fiber for the Ghanaian bioeconomy, starting with the rural farmers, traders, processors, and the retail market.

5.7. Potential Pathway for Utilising Plantain Fiber

Over the past decades it seems that the traditional knowledge and uses of plantain residues is being forgotten. May it be medical uses of roots or using the pseudostem fibers for rope, string, or sanitary pads, as reported by older interviewees. In order to reignite a whole plant biomass use in an efficient and sustainable manner, the first steps could include fiber extraction. Figure 6 presents how the current fruit-oriented value chain could be expanded to include the use of fiber as a step to a biomass web. Required is the collection of residues from the fields. I.e. the pseudostems need to be collected and processed close to the fields to avoid the transportation of unnecessary weight, i.e., nutrient rich water that can be returned to the field as manure. The raw fiber material can be bundled, undergo mechanical, chemical, or enzymatic treatment so as to reach textile component production (yarn, thread, or fabric), before reaching end products (apparel, home furnishings, or components to other end products), and marketing.

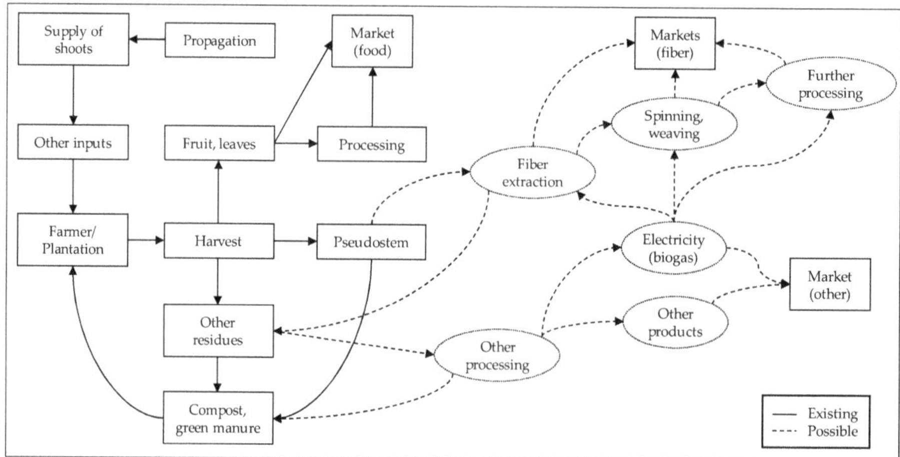

Figure 6. Potential expansion of the plantain value chain to include fiber utilization in a value web; Source: own illustration.

Residues from the fiber extraction process that are not going directly back to the field may be used in other processing lines. For example, the waste could be used as feedstock for (small-scale) biogas production, which in turn could be used for electricity generation, hence linked back to extraction. Depending on other organic waste that could be added as feedstock and the pursued biogas strategy, other uses are easily possible. Low-cost small-scale biogas reactors may serve as a source of energy at the household level; larger installation may be more market-focused.

In any case, the most critical step is likely to be the bulking of sufficient raw fiber bundles, so as to make it economically viable to establish further processing lines. If it is possible to organize the small-scale plantain producers and to overcome this bottleneck, utilizing plantain based fiber may be a promising addition to the Ghanaian bioeconomy. Drawing from the experience with other crops and/or products, suitable approaches may include the formation of cooperatives who also take charge of extracting fibers near to the production sites, outgrower-schemes with contracted plantain farmers, coupling plantain fiber initiatives with the cocoa-promotion-program, or large-scale banana plantations expanding their portfolio by including fiber production. In this context, looking into the experience made in other countries, identifying existing machinery and equipment, piloting fiber extraction in Ghana, and assessing technology transfer appear as useful, action-oriented follow-up research directions.

6. Conclusions and Implications

In this article, we investigate the institutional environment of the Ghanaian plantain sector and assess the potential to tap the abundant, but so far underutilized, plantain residues as a source for natural fibers. The objectives of the research were to (i) describe predominant production structures, (ii) derive a stakeholder network map, respectively, the value web, so as to identify challenges and opportunities regarding institutional settings, and (iii) discussing how this supply side may be linked to the national market or the international demand side.

Required information on the plantain production structure and the connected institutional network system were mainly gathered through qualitative research techniques, including key person interviews, focus-group discussions, and the participatory Net-Map tool. When considering the theoretical concept of a biomass-based value web, the cotton value chain was incorporated in the value web. Currently, cotton is the main source for natural fiber processing actors in the textile sector, who were included as national users of natural fiber. Drawing on the National Innovation System and Porter's diamond model, the potential of initiating and developing a plantain fiber based value generation was discussed.

Starting with the description of plantain production structures, we found three predominant intercropping systems: plantain with cassava as a mixed food production, and both plantain or plantain with cocoyam (taro) as pioneer crops for cocoa in early years. The latter is heavily supported by public initiatives. The majority (90%) of plantain is grown by smallholder farmers. In the research area, i.e., the main plantain producing regions, there are farmer groups or associations that may play a role in bulking produce. Regarding the production, marketing, and processing of fresh plantain, the value chain is very well established. A notable utilization of plantain for fiber or other purposes is not present in Ghana.

Assessing the institutional map and value web revealed that, from a National Innovation System perspective, the required institutions are all available and operating in the country. In detail, this means that specific plantain focused research, extension, and support actors are already intensively in contact and collaborating with smallholder farmers. In addition, general technical knowledge is available and the adaptation of foreign fiber extraction machinery could be carried out easily by e.g., the GRATIS Foundation of Ghana (a 'spinoff' of a project of the Ministry of Trade and Industry). On the processing side, i.e., spinning raw fiber to yarn, the private sector only has experience with cotton, but it shows willingness to experiment with other natural fibers to become more independent of imported raw material and to optimize capacity utilization.

In a broader sense, the findings of this research indicate that Ghana has got the infrastructure and institutional structures that would allow for developing a sustainable commercial use of plantain residues for fiber. What is needed, however, is (re-)discovering and acknowledging the use of plantain residues as a valuable resource, especially by public support institutions, like the agricultural ministry and its extension system. Combining technology transfer and the adaptation of equipment from e.g., India or Uganda with fostering knowledge about plantain's fiber potential appears to be a promising first step to kick-start a new branch in the Ghanaian value web and bioeconomy. Particularly useful would be further action oriented research, like e.g., developing appropriate equipment and a pilot project involving all stakeholders from (plant) fiber to fabric.

Author Contributions: Conceptualization, T.K.L., M.H.; Methodology, T.K.L.; Investigation, all authors; Writing—original draft preparation, T.K.L.; writing—review and editing, all authors.

Funding: This research was funded by the German Federal Ministry of Education and Research (BMBF) within the framework of the GlobE–Research for the Global Food Supply program. It is part of the collaborative project "Improving food security in Africa through increased system productivity of biomass-based value webs." (BiomassWeb); Grant No. 031A258C.

Acknowledgments: The authors are thankful for the assistance of Solomon Darkey during fieldwork activities and to Michael Kwaku (INBAR, Ghana) for the logistical support.

Conflicts of Interest: The authors declare no conflict of interest.

Appendix A

Table A1. Overview of expert and focus group interviews.

Stakeholder Category	Field Visits	Net-Maps	Interviews	Focus Group Discussions
Farmers	3	3	3	3
Plantation	2		2	
Market actors	2		6	
GRATIS foundation	1	1	1	
Cocobod	1		1	
MoFA/government		1		2
Textile SMEs	2		7	
Textile factory	2		2	
Research institutes		1	2	1

Note: Field visits include visits to agricultural fields, markets, workshops, and factories.

References

1. Dzomeku, B.M.; Dankyi, A.A.; Darkey, S.K. Socioeconomic Importance of Plantain Cultivation in Ghana. *J. Anim. Plant Sci.* **2011**, *21*, 269–273.

2. Dankyi, A.A.; Dzomeku, B.M.; Anno-Nyako, F.O.; Adu-Appiah, A.; Antwi, G. Plantain Production Practices in the Ashanti, Brong-Ahafo and Eastern Regions of Ghana. *Asian J. Agric. Res.* **2007**, *1*, 1–9. [CrossRef]

3. Ministry of Food and Agriculture. *Agriculture in Ghana. Facts and Figures, 2015*; Ministry of Food and Agriculture Statistics, Research and Information Directorate (SRID): Accra, Ghana, 2016.

4. FAOSTAT. Available online: http://www.fao.org/faostat/en/#home (accessed on 25 September 2018).

5. Ministry of Food and Agriculture. *Agricultural Sector Progress, Report 2016*; Ministry of Food and Agriculture: Accra, Ghana, 2017.

6. Dzomeku, B.M.; Boateng, O.K. Exploring the potential of banana sap as dye for the Adinkra industry in Ghana. *J. Ghana Sci. Assoc.* **2012**, *14*, 20–24.

7. Asare, I.T. Critical success factors for the revival of the textile sector in Ghana. *Int. J. Bus. Soc. Sci.* **2012**, *3*, 307–310.

8. Adikorley, R. The Textile Industry in Ghana: A Look into Tertiary Textile Education and Its Relevance to the Industry. Master's Thesis, The Patton College of Education of Ohio University, Athens, OH, USA, August 2013.

9. Naab, F.X. An Assessment of the Contribution of Cotton Production to Local Economic Development in Sissala East and West Districts. Master's Thesis, Kwame Nkrumah University of Science and Technology, Kumasi, Ghana, August 2015.
10. NAIP (National Agricultural Innovation Project). *Development of Value Added Products from Banana Pseudostem*; Navsari Progress Report; Navsari Agricultural University: Navsari, India, 2011.
11. Afribanana Products Limited. Available online: http://afribanana.co.ug/index.php (accessed on 25 September 2018).
12. German Bioeconomy Council. *Further Development of the "National Research Strategy BioEconomy 2030"*; Office of the Bioeconomy Council: Berlin, Germany, 2016.
13. German Bioeconomy Council. Global Bioeconomy Summit, Communiqué—Innovation in the Global Bioeconomy for Sustainable and Inclusive Transformation and Wellbeing. In Proceedings of the Global Bioeconomy Summit, Berlin, Germany, 19–20 April 2018.
14. Schneider, R. Tailoring the bioeconomy to food security. *Rural 21* **2014**, *3*, 19–21.
15. Virchow, D.; Beuchelt, T.D.; Kuhn, A.; Denich, M. Biomass-Based Value Webs: A Novel Perspective for Emerging Bioeconomies in Sub-Saharan Africa. In *Technological and Institutional Innovation for Marginalized Smallholders in Agricultural Development*; Gatzweiler, F.W., von Braun, J., Eds.; Springer International Publishing: Basel, Switzerland, 2016; pp. 225–238.
16. German Bioeconomy Council. *Bioeconomy Policy (Part II)—Synopsis of National Strategies around the World*; Office of the Bioeconomy Council: Berlin, Germany, 2015.
17. German Bioeconomy Council. *Bioeconomy Policy (Part III)—Update Report of National Strategies around the World*; Office of the Bioeconomy Council: Berlin, Germany, 2018.
18. Scheiterle, L.; Ulmer, A.; Birner, R.; Pyka, A. From commodity-based value chains to biomass-based value webs: The case of sugarcane in Brazil's bioeconomy. *J. Clean. Prod.* **2018**, *172*, 3851–3863. [CrossRef]
19. Poku, A.-G.; Birner, R.; Gupta, S. Making contract farming arrangements work in Africa's bioeconomy: Evidence from cassava outgrower schemes in Ghana. *Sustainability* **2018**, *10*, 1604. [CrossRef]
20. Leao, A.L.; Casarino, I.; Narine, S.; Sain, M. Innovation under the bioeconomy context. In *Knowledge-Driven Developments in the Bioeconomy: Technological and Economic Perspectives*; Dabbert, S., Lewandowski, I., Weiss, J., Pyka, A., Eds.; Springer: Cham, Switzerland, 2017; pp. 117–141, ISBN 978-3-319-58373-0.
21. Virchow, D.; Beuchelt, T.D.; Denich, M.; Loos, T.K.; Hoppe, M.; Kuhn, A. The value web approach—So that the South can also benefit from the bioeconomy. *Rural 21* **2014**, *48*, 16–18.
22. Porter, M.E. *The Comparative Advantage of Nations*; Free Press: New York, NY, USA, 1990.
23. Lundvall, B.-Å. *National Innovation Systems: Toward a Theory of Innovation and Interactive Learning*; Pinter: London, UK, 1992.
24. Freeman, C. Continental, national and sub-national innovation systems—Complementarity and economic growth. *Res. Policy* **2002**, *31*, 191–211. [CrossRef]
25. Context Network. Multi Crop Value Chain Phase II: Ghana-Plantain. Commissioned Report. 2014. Available online: https://www.agriknowledge.org/downloads/dr26xx473 (accessed on 24 September 2018).
26. Schiffer, E. *Net-Map Toolbox: Influence Mapping of Social Networks*; International Food Policy Research Institute: Washington, DC, USA, 2007.
27. Gilbert, N. *Researching Social Life*, 3rd ed.; Sage: London, UK, 2008.
28. Cauthen, J.; Jones, D.; Gugerty, M.K.; Anderson, L. *Banana and Plantain Value Chain: West Africa*; Evans School Policy Analysis and Research (EPAR) Brief; Evans School Policy Analysis and Research: Seattle, WA, USA, 2013; Volume 239.
29. Anonymous; (COCOBOD, Sunyani, Brong-Ahafo, Ghana). Personal communication, 2016.
30. Padam, B.S.; Tin, H.S.; Chye, F.Y.; Sbdullah, M.I. Banana by-products: An under-utilized renewable food biomass with great potential. *J. Food Sci. Technol.* **2012**, *51*, 3527–3545. [CrossRef] [PubMed]
31. Uma, S.; Kalpana, S.; Sathiamoorthy, S.; Kumar, V. Evaluation of commercial cultivars of banana (*Musa* spp.) for their suitability for the fibre industry. *Plant Genet. Resour. Newsl.* **2005**, *142*, 29–35.
32. Adeniji, T.A.; Tenkouano, A.; Ezurike, J.N.; Ariyo, C.O.; Vroh-Bi, I. Value-adding post-harvest processing of cooking bananas (*Musa* spp. AAB and ABB genome groups). *Afr. J. Biotechnol.* **2010**, *9*, 9135–9141.
33. Mensah-Bonsu, A.; Agyeiwaa-Afrane, A.; Kuwornu, J.K.M. Efficiency of the plantain marketing system in Ghana: A co-integration analysis. *J. Dev. Agric. Econ.* **2011**, *3*, 593–601.

34. Banful, A.B. Production of plantain, an economic prospect for food security in Ghana. In *Bananas and Food Security, Proceedings of the INIBAP International Symposium, Douala, Cameroon, 10–14 November 1998*; Picq, C., Fouré, E., Frison, E., Eds.; INIBAP: Montpellier, France, 1998; pp. 151–160.
35. Katila, S. Between Democracy and Dictatorship: The Market Queen Institution of Ghanian Daily Markets. In *State, Market and Organizational Form*; Bugra, A., Üsdiken, B., Eds.; Gruyter Studies in Organization: Berlin, Germany, 1997; Volume 80, pp. 269–288, ISBN 978-3-11-080073-9.
36. Scheiterle, L. The Role of Institutions and Networks in Developing the Bioeconomy: Case Studies from Ghana and Brazil. Ph.D. Thesis, University of Hohenheim, Stuttgart, Germany, November 2017.
37. Scheiterle, L.; Birner, R. Gender, knowledge and power: A case study of Market Queens in Ghana. In Proceedings of the 2018 Annual Meeting, Washington, DC, USA, 5–7 August 2018; Agricultural and Applied Economics Association: Milwaukee, WI, USA, 2018.
38. Ortega, Z.; Morón, M.; Monzón, M.D.; Badalló, P.; Paz, R. Production of banana fiber yarns for technical textile reinforced composites. *Materials* **2016**, *9*, 370. [CrossRef] [PubMed]
39. Poku, A.-G. *The New Wave of Contract Farming in Ghana: The Role of Contract Design in Facilitating Sustainable Outgrower Schemes*; BiomassWeb Policy Brief; Center for Development Research: Bonn, Germany, 2018; Volume 2.

© 2018 by the authors. Licensee MDPI, Basel, Switzerland. This article is an open access article distributed under the terms and conditions of the Creative Commons Attribution (CC BY) license (http://creativecommons.org/licenses/by/4.0/).

 *sustainability*

Article

Biogas Potential of Coffee Processing Waste in Ethiopia

Bilhate Chala [1,*], Hans Oechsner [2], Sajid Latif [1] and Joachim Müller [1]

[1] Institute of Agricultural Engineering, Tropics and Subtropics Group(440e), University of Hohenheim, 70599 Stuttgart, Germany; s.latif@uni-hohenheim.de (S.L.); joachim.mueller@uni-hohenheim.de (J.M.)

[2] State Institute of Agricultural Engineering and Bioenergy, University of Hohenheim, 70599 Stuttgart, Germany; hans.oechsner@uni-hohenheim.de

* Correspondence: info440e@uni-hohenheim.de or bilhat@gmail.com; Tel.: +49-711-459-24759

Received: 27 June 2018; Accepted: 28 July 2018; Published: 31 July 2018

Abstract: Primary coffee processing is performed following the dry method or wet method. The dry method generates husk as a by-product, while the wet method generates pulp, parchment, mucilage, and waste water. In this study, characterization, as well as the potential of husk, pulp, parchment, and mucilage for methane production were examined in biochemical methane potential assays performed at 37 °C. Pulp, husk, and mucilage had similar cellulose contents (32%). The lignin contents in pulp and husk were 15.5% and 17.5%, respectively. Mucilage had the lowest hemicellulose (0.8%) and lignin (5%) contents. The parchment showed substantially higher lignin (32%) and neutral detergent fiber (96%) contents. The mean specific methane yields from husk, pulp, parchment, and mucilage were 159.4 ± 1.8, 244.7 ± 6.4, 31.1 ± 2.0, and 294.5 ± 9.6 L kg^{-1} VS, respectively. The anaerobic performance of parchment was very low, and therefore was found not to be suitable for anaerobic fermentation. It was estimated that, in Ethiopia, anaerobic digestion of husk, pulp, and mucilage could generate as much as 68×10^6 m^3 methane per year, which could be converted to 238,000 MWh of electricity and 273,000 MWh of thermal energy in combined heat and power units. Coffee processing facilities can utilize both electricity and thermal energy for their own productive purposes.

Keywords: husk; pulp; parchment; mucilage; methane; renewable energy

1. Introduction

Coffee production is a livelihood for about 125 million people worldwide, particularly from developing countries [1]. Ethiopia is known to be the origin of and gene pool for coffee Arabica [2]. In the last decade, Ethiopia has been the largest coffee producer in Africa, and it remains fifth in the world, contributing a share of about 4.5% to the world production. Annual coffee production increased from 273,400 Mg in 2007 to 469,091 Mg in 2016, while the cultivation area increased from 407,147 ha to 700,475 ha (Table 1). The annual green bean production has increased in the last 4 consecutive years, but productivity (yield per harvest area) has declined in the same period. The amount of coffee by-products is directly related to coffee production. Coffee is Ethiopia's leading export commodity; and the livelihood of more than a million households depends on coffee production [3]. Coffee made up about 24% of the country's total export earnings for the fiscal period 2012/13 [4]. Ethiopian coffee is produced under forest, semi-forest, garden, and plantation production systems contributing 10, 35, 50, and 5% of the country's coffee production, respectively. Thus, about 95% of Ethiopia's coffee is produced by small holder farmers [2,4].

Table 1. Area, production, export, and yield of coffee in Ethiopia.

Year	Area (ha)	Yield (kg/ha)	Green Beans Production (Mg)	Green Beans Export (Mg)	Export Value ($\times 1000$ USD)
2007	407,147	671.5	273,400	158,467	417,323
2008	391,296	665.1	260,239	179,283	561,511
2009	395,003	672.1	265,469	129,833	365,689
2010	498,618	743.2	370,569	211,840	676,517
2011	515,882	730.4	376,823	159,135	844,555
2012	528,571	521.3	275,530	203,652	887,549
2013	538,466	728.0	392,006	218,937	803,965
2014	561,762	747.6	419,980	238,631	1,023,691
2015	653,910	698.9	457,014	234,218	1,018,149
2016	700,475	669.7	469,091	159,712	714,885

Adapted from www.faostat.org database [5].

In Ethiopia, the main regional states involved in coffee production are Oromia, Southern Nations Nationalities & People (SNNP), and Gambella. As of 2013/14, there were 1026 wet and 696 dry milling stations in these regions. About 60% of the wet milling and 79% of dry milling stations were located in SNNP and Oromia regions, respectively. Coffee milling stations are owned by private companies, cooperatives, or states. About 75% of the wet and 96% of the dry mills are owned by private firms. Cooperatives own 23% of wet and 3% of dry stations (Table 2) [5]. Unlike the dry milling stations, the majority of the wet mills are not connected to the electric grid; they are run on diesel engines for pulping and further processing steps. Replacing fossil fuels by bioenergy could be an option with ecologic and economic advantages.

Table 2. Wet and dry milling stations in Ethiopia (Ethiopian Ministry of Agriculture, coffee, tea, and spices directorate, 2015).

Regional State	Wet Milling Stations				Dry Milling Stations				Grand Total
	Private	Cooperative	State	Total	Private	Cooperative	State	Total	
Oromia	297	95	15	407	524	20	6	550	957
SNNP	470	146		616	136	4	–	140	756
Gambella	3	–	–	3	6	–	–	6	9
Total	770	241	15	1026	666	24	6	696	1722

Coffee cherries are collected from the coffee trees by selective harvesting (picking only the ripe fruits by hand) or strip harvesting (fruits striped at once with different maturity levels). The cherries are then processed to green beans following either the dry method or the wet method (Figure 1). In the wet method, the cherries undergo pulping, fermentation/washing, and peeling/polishing one after the other before producing the green beans in 7–12 days. The respective by-products are pulp (43% *w*/*w*), mucilage (12% *w*/*w*), and parchment (6.1% *w*/*w*) on the intrinsic fresh weight basis of coffee cherries [6]. On the other hand, the dry method uses hulling after drying the cherries to produce green beans. Husk is the major single by-product of the dry method. It represents about 50% *w*/*w* of the dry cherry. The dry method often requires up to 4 weeks from harvest to green beans [7]. Despite the higher water requirement (about 80 L kg^{-1} green beans) and disposal problems, green coffee beans from wet processing have a superior aroma and are sold with higher premiums [8]. Coffee by-products are dumped within the hosting community with virtually no economic benefits, thus causing severe environmental problems. The by-products are known to be rich in organic pollutants (e.g., proteins, sugars, and pectin), tannins, and phenolic compounds harmful to plants, humans, and aquatic biota [9,10]. Globally, coffee processing generates about 15×10^6 Mg (dry weight basis) of coffee residues, of which 9.4×10^6 Mg is pulp [11].

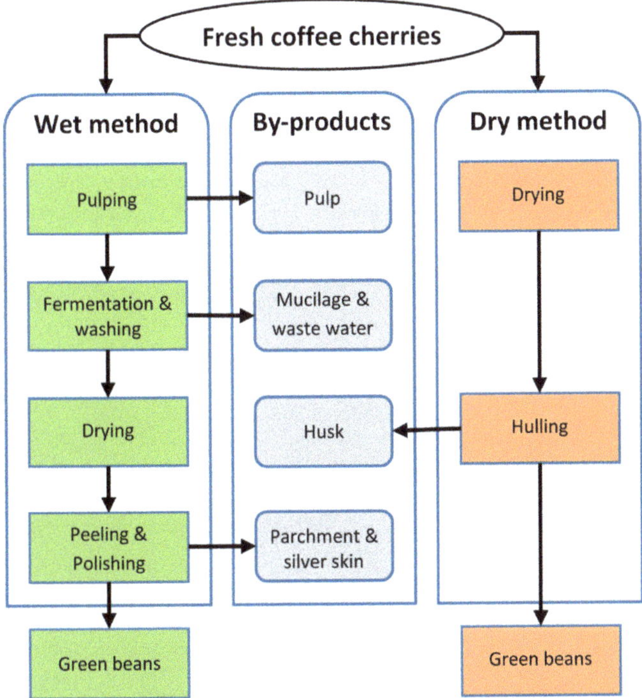

Figure 1. Primary coffee processing pathways to produce green beans following the dry and wet method.

Biogas technology was introduced to Ethiopia five decades ago. The technology is envisaged as a de-centralized energy source for improved livelihood and a source of organic fertilizer. The Ethiopian national biogas program installed about 8,000 household biogas units (4 to 10 m^3) in 2009–2013 [12,13]. Most of the biogas digesters are fed with cow dung or latrine waste. By-products of coffee processing, which are currently discarded as waste, could be an alternative feedstock.

The rates and extent of bio-degradation are crucial in anaerobic fermentation of agricultural residues, which in turn depend on lignocellulose contents and properties [14]. Higher cell contents (protein, carbohydrates, and fats) tend to ferment easily and result in higher methane production [15]. The complex interaction of hemicellulose and lignin often leads to a reduced cellulose hydrolysis. However, in a first approximation, lignin content in substrates has been proven to be a good predictor of methane potential from agro-industrial wastes [15,16].

The biogas yield from coffee processing by-products has been investigated by several researchers, but the bio-methane yield potentials reported vary substantially [17]. Ulsido, et al. [18] found a methane production potential of 132 L kg^{-1} VS from husk, while Kivaisi et al. [19] reported 650 L kg^{-1} VS of *Robusta* and 730 L kg^{-1} VS of *Arabica* mixed solid wastes (husk and pulp). Similarly, Baier and Schleiss [20] determined a biogas potential of 380 L kg^{-1} VS (57–66% methane) and 900 L kg^{-1} VS of pulp using batch assay and semi-continuous digesters, respectively. Adams and Dougan [21] reported the suitability of anaerobic conversion of coffee pulp and the waste water for waste treatment, and the generation of useful fuel, as high as 66 m^3 per ton of pulp digested. Variation in the bio-methane potentials of substrates could be attributed to the variety of coffee waste investigated and mode of fermentation. Chanakya and De Alwis [1] summarized several research and field trials on coffee processing wastes. Different types of reactors (batch, CSTR, UASB, and BIBR) were tested at varying scales of operation.

On the other hand, Franca and Oliveira [7] pointed out that, despite the bio-methane potential, coffee by-products are comparable to common agricultural residues, and that the lack of scientific studies hinders the wider utilization of the by-products for technical or economic reasons. Jayachandra, et al. [22] reported that the acidic pH and polyphenols in the coffee husk makes it resistant to anaerobic conversion. However, the treatment of husk with thermophilic fungus (*Mycotypha*) resulted in suitable pH levels for anaerobic conversion and thus increased the bio-methane yield considerably. Ensiled coffee pulp/husk presents an ideal method to preserve the substrate for longer periods of use in anaerobic digesters and reduces caffeine content (13–63%), total polyphenols (28–70%), and condensed polyphenols (51–81%) [23]. Seasonal availability of coffee by-products could be a limitation for continuous anaerobic fermentation.

Summary of bio-methane yields from typical anaerobic substrates is listed in Table 3. Municipal wastes tend to produce more methane than other substrate types. Despite frequent use of animal manure as biogas substrate, the methane production rate was relatively low.

Table 3. Methane yield potential of different substrates common to anaerobic digesters.

Substrates	Methane Yield (L kg^{-1} VS)	Reference
Basic substrates: farm manures		
Cattle	130–300	[24]
Pig	210–320	[24]
Poultry	250–400	[24]
Agricultural products		
Straw	71–240	[24]
Maize silage	320–400	[24]
Grass	286–324	[25]
Sunflower	235–347	[25]
Agro industrial wastes		
Potato pulp	250–400	[24]
Vegetable waste	400	[24]
Brewer grains	370–390	[24]
Municipal wastes		
Bio-wastes	200–600	[24]
Rumen content (slaughterhouse waste)	160–400	[24]
Kitchen waste	350–600	[24]
Sewage sludge	250–350	[26]

In 2015, Ethiopia had an annual electricity production of 10.08 TWh. The overall installed capacity was 2.7 GW, in which hydro, fossil fuel, and other renewables contributed 79.5%, 7.5%, and 13% of the installed capacity, respectively [27]. As of 2016, 85.4% of the urban and 26.5% of the rural population of Ethiopia had access to electricity [28].

To the authors' knowledge, no comprehensive study on the characterization and bio-methane potential of coffee husk, pulp, parchment, and mucilage has been published in the scientific literature. Electrical and thermal potentials (in Ethiopian context) of the bio-methane from husk, pulp, and mucilage were not reported either.

The objective of this study was to examine the physico-chemical characteristics and determine the anaerobic bio-methane potential of coffee husk, pulp, parchment, and mucilage generated from primary coffee processing in Ethiopia. Furthermore, the potential of the bio-methane from husk, pulp, and mucilage to generate electrical and thermal energy was estimated.

2. Materials and Methods

2.1. Raw Materials

Coffee husk, pulp, parchment, and mucilage were collected from Gomma-2 estate coffee farm (7°55′16.32″ N, 36°37′06.62″ E) about 400 km South-West of Addis Ababa, Ethiopia. The pulp was collected right after pulping of the cherries, from consecutive processing days of a month. The pulp then was spread on a plastic sheet for 3–4 days of sun drying. The pulp was shuffled in regular intervals, in order to dry evenly. The mucilage was fetched from a second fermentation pit, poured on a plastic sheet, and left to dry under the sun for 5–6 days. The parchment and husk were collected from the same farm, without the need for further drying. All samples underwent a size reduction to pass 1 mm pores, packed in polyethylene bags, and shipped to the University of Hohenheim, Germany. The samples were stored in a cool and dry place until further use.

2.2. Inoculum

The inoculum was obtained from the State Institute of Agricultural Engineering and Bioenergy, University of Hohenheim. It consists of dairy manure and energy crops pre-digested in laboratory reactors, cultivated under the institute's standard procedure for bio-methane potential (BMP) assays [29]. Prior to the BMP assay, the inoculum was digested at 37 °C to degas and allow temperature adaptation of the anaerobic bacteria for further experimental assays. The inoculum had a total solids and volatile solids content of 5.0% and 61.6% (of TS), respectively.

2.3. Chemical Analysis

Moisture, volatile matter, and ash contents of samples were determined according to DIN EN 14774-3:2010-02, DIN EN 15148:2010-03 and DIN EN 14775:2010-04, respectively. Neutral detergent fiber (NDF), acid detergent fiber (ADF), acid detergent lignin (ADL), and crude fiber contents were determined by AOAC Official Method 973.18 (FibreBag Analysis System, Gerhardt, Königswinter, Germany). Sugar (sucrose, glucose, and fructose) and acid (lactic acid, formic acid, acetic acid, and propionic acid) contents of the samples were determined with HPLC (Bischoff, Leonberg, Germany) equipped with RI detector. Samples were separated on a Bio-Rad Aminex HPLC organic acid column (HPX—87H 300 × 7.8 mm^2) at 35 °C with a mobile phase of 0.02 N H_2SO_4, at a rate of 0.6 mL min^{-1} and pressure 6.0 MPa. Sugar and acid contents were analyzed at the laboratory of the State Institute of Agricultural Engineering and Bioenergy, University of Hohenheim. Elemental composition of the samples was analyzed by the State Institute of Agricultural Chemistry, University of Hohenheim, and by the laboratory of Schaumann Bioenergy GmbH.

2.4. Anaerobic Batch Digestion Tests

The Hohenheim Biogas Yield Test (HBT) is a laboratory batch method used to determine the methane and biogas yield potential of substrates [30], which is mentioned in the German Engineers Association (VDI) Guideline 4630 [31], as one of six recommended methods for BMP assays. The trial was carried out using 100 mL calibrated glass syringes as fermenters (Figure 2). The syringes were fitted to a motorized rotor to enable continuous mixing of the samples. The fermentation was operated for duration of 35 days at a temperature of 37 °C. At the beginning of the process, 30 g of inoculum was placed inside a glass syringe with 0.4 g of ground and dried samples [29,31]. Reference samples (hay and concentrate) of known methane yield were used in order to verify the quality of the inoculum and the reliability of the digestion process [29]. The blank (zero variant) fermenter was filled with about 50 g of inoculum alone (Table 4). Biogas and methane yields were obtained by deducting the proportional biogas and methane yields from the zero variant. The biogas volume was measured according to the volume difference before and after emptying the gas inside the syringe. While emptying the syringes, biogas was injected into the methane sensor (Advanced Gasmitter, Pronova, Berlin, Germany), which was calibrated with ambient air and gas (~60% CH_4). The fermenter temperature (mostly 2 °C less than the digestion temperature) and ambient air pressure were measured during the gas measurement and used to normalize the gas yield at STP (101.325 kPa, 273 K). The specific methane yield (L CH_4 kg^{-1} VS added) is the net normalized methane yield divided by the amount of organic dry matter of substrate added.

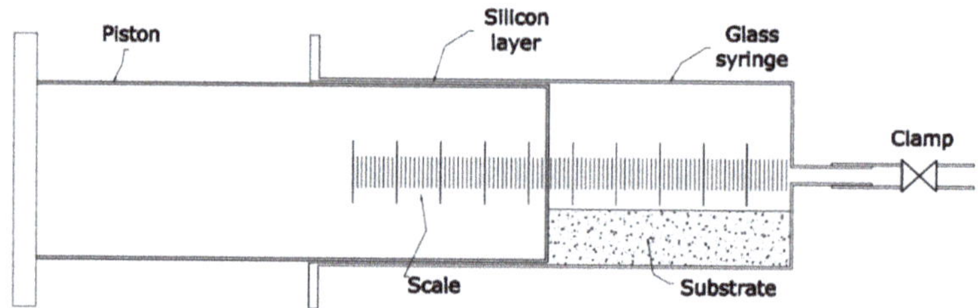

Figure 2. Glass syringe to determine the bio-methane potential in HBT anaerobic batch assay (adapted from Mittweg, et al. [29]).

Table 4. Inoculum and substrate inputs in HBT batch assay.

Variant	Inoculum (mg)	Substrate (mg)	TS (% FM)	VS (% TS)
Blank	50,000	—	4.99	61.55
Hay standard	30,000	400	90.56	89.99
Pulp	30,000	400	91.62	88.26
Husk	30,000	400	92.65	93.09
Parchment	30,000	400	95.44	99.59
Mucilage	30,000	400	94.38	85.99

The degree of degradation (η_{oDM}) was evaluated according to the mass ratio of gas produced m_{biogas} to volatile solids loaded (Equation (1)).

$$\eta_{oDM} = \frac{m_{biogas}}{m \cdot c_{VS}} \tag{1}$$

in which m is the dry mass of substrate added and c_{VS} is the concentration of volatile solids in the substrate.

The molar masses of methane (M_{CH_4}) and carbon dioxide (M_{CO_2}) were used to determine the mass of the biogas (Equation (2)) based on their respective concentrations in the biogas, assuming the biogas was composed of methane and carbon dioxide only [31,32].

$$m_{biogas} = V_{biogas} \cdot \left(\frac{M_{CO_2} \cdot c_{CO_2}}{22.4 \cdot 100} + \frac{M_{CH_4} \cdot c_{CH_4}}{22.4 \cdot 100} \right) \tag{2}$$

in which V_{biogas} is the volume of biogas measured, c_{CO_2} is the concentration of carbon dioxide in the biogas, and c_{CH_4} is the concentration of methane in the biogas.

The energy recovery (ER) rate was determined by the ratio of calorific values of the methane generated to the calorific value of input substrates in the batch assay. The gross energy (GE) of substrates can be estimated using results from Weende feed analysis and applying the Equation (3) [33].

$$GE = \left(0.0239 \text{ MJ g}^{-1} \cdot XP + 0.0398 \text{ MJ g}^{-1} \cdot XL + 0.0201 \text{ MJ g}^{-1} \cdot XF + 0.0175 \text{ MJ g}^{-1} \cdot NfE \right) \tag{3}$$

in which XP is crude protein, XL is crude fat, XF is crude fiber, NfE is nitrogen free extract (NfE) (g kg^{-1}, dry mass basis), and GE is gross energy (MJ kg^{-1}). The specific energy content of component parameters has certain differences, as shown in the fixed coefficients of the equation.

The theoretical energy potential of by-products from coffee processing was calculated under the following scenario: (1) that all the husk, mucilage, and pulp generated in a harvest season is recovered and utilized in biogas digesters (the parchment was excluded from calculation), (2) the electric conversion efficiency in a CHP is 35% electricity and 40% thermal energy that could be used for coffee drying, and (3) the lower methane heating value of 9.968 kWh m^{-3} is used to calculate the primary energy yield.

Statistical analysis was made using Microsoft Excel (2013) to determine the mean and standard deviation of SMY of substrates and other parameters related to physico-chemical characterization.

3. Results and Discussion

3.1. Chemical Composition

There were certain differences in the chemical composition between coffee husk, pulp, parchment, and mucilage (Table 5). The parchment showed the highest proportion (76.9%) of crude fiber contents, followed by husk (39.9%), pulp (24.8%), and mucilage (19.4%). NDF, ADF, and ADL contents in parchment were also considerably higher than in husk, pulp, and mucilage. Non-fiber carbohydrates (sugar, starch and pectin) in mucilage were about 28.8%, while husk and pulp showed comparable NFC of 16.2% and 17.6%, respectively. Hemicellulose contents in parchment, husk, pulp, and mucilage were 20%, 15%, 8%, and 1%, respectively. However, cellulose contents in pulp, husk, and mucilage were roughly the same (32%), while the parchment showed higher values (45%). The lignin in the parchment was about two times higher than for husk and pulp, each. Parchment exhibited the highest lignin content (32%) among the by-products. Samples were also examined for their sugar (sucrose, glucose, and fructose) and organic acid (lactic, formic, acetic, and propionic) contents (Table 5). The highest sugar content was obtained from husk, followed by pulp and mucilage. Fructose was the main component in the sugar analysis. Mucilage had the highest content in organic acids, while husk and pulp contained little amounts of the acid fractions. Lactic acid was the main component in the acid analysis. There was neither detectable sugar nor organic acids in the parchment. Carbon to nitrogen ratio (C:N) in husks showed an ideal value for anaerobic digestion of 25. Mucilage, however, showed lower than the recommended optimum value of 20–30, which might need co-digestion with other substrates with a fairly higher C:N ratio. The parchment, on the other hand, showed the highest C:N ratio (190) [34].

Table 5. Chemical composition of husk, pulp, parchment, and mucilage; values on % DM basis, unless stated. n = 3, mean ± std.

Substrate	DM (% FM)	VS	XA	XP	XL	XF	NfE	NDF	ADF	ADL	NFC	GE (MJ kg^{-1})	C:NRatio	Sugars *	Organic Acids **
Husk	93.5 ± 0.0	92.9 ± 0.1	7.2 ± 0.05	11.1 ± 0.0	1.5 ± 0.0	39.9 ± 0.1	40.4 ± 0.0	64.0 ± 1.1	49.5 ± 0.0	17.5 ± 1.6	16.2 ± 1.1	18.8 ± 0.0	24.9	4.45	1.22
Pulp	92.3 ± 0.1	88.3 ± 0.2	11.7 ± 0.2	14.1 ± 0.0	1.0 ± 0.1	24.8 ± 1.2	48.5 ± 0.0	55.6 ± 1.4	47.1 ± 0.1	15.5 ± 1.6	17.6 ± 1.4	17.4 ± 0.0	18.4	1.73	0.13
Parchment	96.0 ± 0.1	99.6 ± 0.1	0.45 ± 0.0	1.6 ± 0.0	0.9 ± 0.1	76.9 ± 0.3	20.1 ± 0.0	96.8 ± 0.3	76.9 ± 0.2	32.2 ± 0.0	0.3 ± 0.3	19.7 ± 0.1	190	0	0
Mucilage	94.8 ± 0.0	85.1 ± 0.1	14.9 ± 0.1	18.5 ± 0.0	0.1 ± 0.1	19.4 ± 0.4	47.1 ± 0.0	37.7 ± 1.2	36.9 ± 0.6	5.0 ± 0.3	28.8 ± 1.2	17.7 ± 0.1	14	0.46	5.19

FM = fresh matter; DM = dry matter; VS = volatile solid; XA = crude ash; XP = crude protein; XL = crude fat; XF = crude fibre; NfE = nitrogen free extract; NDF = neutral detergent fibre; ADF = acid detergent fibre; ADL = acid detergent lignin; NFC = non-fiber carbohydrate; GE = gross energy * sucrose, glucose, and fructose; ** lactic, formic, acetic, and propionic acid; NFC = 100-NDF-XA-XL-XP.

3.2. Elemental Analysis

The elemental analysis of husk, pulp, parchment, and mucilage showed that the substrates might exhibit a deficiency in some important trace elements required for optimal and stable biogas production, as illustrated in Table 6, compared to the range of values suggested by Oechsner, et al. [35]. Mucilage was deficient in Mo, Se, and W, pulp in Mn and Co., and husk in all analyzed trace elements besides Co. The deficiency of Mo in mucilage, pulp, and husk was as high as 70%, 80%, and 90%, respectively, compared to minimum trace element requirements. Ni and Se were also deficient with 60% and 55%, respectively, both in husk and pulp. Fe, which is required in a higher amount than other trace elements, was available 29% in husk and 65% in pulp compared to the minimum requirements. Shortage of trace elements often causes a decline in biogas production due to the loss of a stable digestion process and eventually leads to a lower substrate feeding rate [36–39]. Full scale anaerobic digestion of coffee by-products thus requires supplementation of trace elements through co-digestion with animal manure or commercial formulas.

Table 6. Elemental composition of the husk, pulp, parchment, and mucilage in terms of manganese (Mn), zinc (Zn), cobalt (Co.), molybdenum (Mo), iron (Fe), nickel (Ni), selenium (Se), and tungsten (W) [mg kg^{-1} DM].

		Mn	Zn	Co.	Mo	Fe	Ni	Se	W
Husk		83	4.4	0.5	0.1	440	1.2	0.09	<0.05
Pulp		159	21.7	0.5	0.2	969	1.2	0.09	<0.05
Mucilage		149	125.5	1.0	0.3	1719	3.7	0.12	<0.05
Parchment		12	6.8	0.1	<0.05	47	0.1	<0.05	<0.05
Optimum	Min.	100	30	0.4	1.0	1500	3.0	0.20	0.1
values	Max.	1500	300	5.0	6.0	3000	16.0	2.00	30.0

3.3. HBT Analysis

The mean SMY from husk, pulp, parchment, and mucilage was examined at 37 °C for 35 days following the HBT batch assay protocol (Table 7). The SMY from husk was 159 L kg^{-1} VS, and the average methane content of biogas was 56.8%. The energy recovery rate of husk was 34%, which indicates a fairly low anaerobic performance. Mönch-Tegeder, et al. [33] demonstrated similar results from batch fermentation of horse dung. The pulp showed SMY of 245 L kg^{-1} VS with a methane quality of 51.5%. Parchment exhibited the lowest SMY of 31 L kg^{-1} VS. The result suggested that coffee parchment is not suitable for anaerobic conversion. The highest recalcitrant content in the parchment could be attributed to the lowest SMY. Previous research demonstrated that a higher lignin content was one of the main reasons for lower bio-methane conversion [26,40–42]. Mucilage yielded SMY of 294 L kg^{-1} VS with about 55.4% of methane from the biogas. Thus, the mucilage showed the highest SMY methane yield among the coffee by-products, followed by pulp and husk. The degradability of pulp and mucilage was 63% and 68%, respectively. Higher SMY recovered from mucilage among all examined substrates can be attributed to higher soluble cell-contents (62.3%), as well as lower lignin (5%) and hemicellulose (0.8%) contents. Moreover, the organic acid content of mucilage was much higher than that of the other substrates [43]. Furthermore, the SMY and quality of the biogas from husk, pulp, and mucilage was comparable to common agro-industrial wastes and some energy crops [25] as depicted in Table 3. Khan, et al. [44] determined a SMY of 256 L kg^{-1} VS–349 L kg^{-1} VS from different fractions of banana waste, and Haag, et al. [45] demonstrated SMY 228 L kg^{-1} VS–261 L kg^{-1} VS from different varieties of cup plants (*Silphium perfoliatum*) under batch assay.

Table 7. Specific methane yield (SMY, mean ± STD, n = 6), degradability, and energy recovery rate of husk, pulp, and mucilage from HBT batch assay.

Substrate	SMY (L kg^{-1} VS)	Methane Content (%)	Degradability (%)	Methane Energy (MJ kg^{-1} VS)	Energy Recovery
Husk	159.4 ± 1.8	51.5	35.3	6.33	33.7%
Pulp	244.7 ± 6.4	56.8	63.0	9.75	56.1%
Parchment	31.1 ± 2.0	84.2	3.4	1.12	5.7%
Mucilage	294.5 ± 9.6	55.5	68.0	11.7	66.1%

The energy recovery rate ranges from 4.6% (parchment) to 66% (mucilage), while pulp and husk showed 56% and 33%, respectively. The lowest recovery rate was expected from parchment due to the very high recalcitrant contents in the biomass. Since the ash content of the parchment is very low, it might be suitable for other types of energy conversion technologies, like pelleting and briquetting. The husk is basically a combination of pulp, mucilage, and parchment; hence, it reflects intermediate values of the components.

3.4. Energy Potential of Coffee by-Products in Ethiopia

About 70% of Ethiopian coffee is produced following the dry method and 30% by the wet method. The average green coffee bean production for the past 3 harvest years (2014–2016) was 448,695 Mg year^{-1} (Table 1). For each unit weight of green coffee beans produced, 0.6 kg of pulp and 0.103 kg of mucilage (DM basis) are generated from the wet method, and 0.933 kg of husk from the dry method. This translates to an average generation of 312,060 Mg DM and 289,748 Mg VS of husk, 85,976 Mg DM and 75,925 Mg VS of pulp, and 14,764 Mg DM and 12,564 Mg VS of mucilage a year. Considering the SMY obtained at the end of the digestion period (35 days) in the BMP assay, the energy potential per year from husk, pulp, and mucilage was estimated to be 160,729 MWh, 64,898 MWh and 12,887 MWh, respectively (Table 8). Since the pulp and mucilage are both available in a single wet processing station, their respective energy potential could be combined. The aggregate methane estimate can produce 238,000 MWh of electricity and 272,586 MWh of thermal energy, provided a CHP unit (total conversion efficiency 75%) is applied. The diesel equivalent of the total bio-methane estimated from the husk, pulp, and mucilage was 68,365 m^3, which costs about 40.3 million USD, based on the current fuel retail price in Ethiopia. The bio-methane could displace fossil fuel, which is often used by coffee processing facilities. The diesel is mainly used to run pulping machines and pump water for coffee processing. In Ethiopia, the average diesel consumption by farmers' cooperatives to process 1 Mg of fresh coffee cherry into parchment coffee was 2.8 L. However, in large estate coffee farms, the consumption was very low (0.9 L Mg^{-1} fresh cherry). Furthermore, the thermal energy from the CHP units could be utilized in drying parchment coffee or fresh cherries. Therefore, it would be a supplement for open-air drying and reduce labor costs and weather risks faced by traditional coffee drying. Fischer, et al. [17] estimated a potential of 18MW$_{el}$ from anaerobic fermentation for the Kenyan coffee sector, with an installed capacity of 50 kW$_{el}$ for coffee cooperatives and 250 kW$_{el}$ for large estate farms.

Table 8. Energy potential of coffee husk, pulp, and mucilage from one-year harvest.

Substrate	SMY ($m^3\ Mg^{-1}\ VS$)	Residues Production Ratio (RPR) $kg\ VS\ kg^{-1}\ GB$ [a]	Biomass Yield ($Mg\ VS\ year^{-1}$)	Methane Yield ($m^3\ year^{-1}$)	CHP Production (MWh/year)		Diesel [b] Equivalent (m^3)	Saved Fuel Cost [c] (USD)
					Electricity ($MWh_{el}\ year^{-1}$)	Heat ($MWh_{Th}\ year^{-1}$)		
Husk	159	0.923	289,748	46,069,906	160,729	183,690	46,070	27,181,244
Pulp	245	0.564	75,925	18,601,671	64,898	74,169	18,602	10,974,986
Mucilage	294	0.093	12,564	3,693,742	12,887	14,728	3,694	2,179,308

[a] Green coffee beans and [b] diesel calorific value = 10.25 kWh/L. [c] The diesel cost was estimated as 0.59 \$/L based on the current retail price in Ethiopia.

4. Conclusions

Ethiopia's coffee processing sector generates a huge amount of by-products, both in liquid (mucilage) and solid (pulp, parchment, and husk) forms. The current electricity shortage coupled with the environmental issues associated with coffee waste disposal make anaerobic conversion technology a benign intervention. Coffee by-products are rich in lignocellulosic contents with different proportions of cell contents and cell wall (cellulose, hemicellulose, and lignin) contents. The by-products exhibited promising bio-methane potential (BMP) comparable to common agro-industrial residues and some energy crops. The specific methane yield was highest in mucilage followed by pulp and husk. The BMP from parchment was very low, thus implying that it is not suitable for anaerobic conversion. It was estimated that anaerobic fermentation of coffee processing by-products generated by the Ethiopian coffee sector has significant potential to generate methane as high as 68×10^6 m^3 per year, which can produce 238,000 MWh$_{el}$ of electricity and 272,586 MWh$_{th}$ thermal energy. Most coffee processing facilities could recover bio-methane from their own by-products, which would be sufficient to run the processing activities. Seasonal availability of coffee by-products, particularly from the wet method, could limit a year-round utilization. Ensiling, however, enables longer storage periods. Policy instruments like the feed-in tariff (FIT) are suggested to encourage private investors to produce electricity/thermal energy from coffee by-products and also promote renewable energy production. The FIT offers long-term (15–25 years) guaranteed purchase contract and cost-based compensation to renewable energy producers. Further research on the performance of coffee by-products in combination with other biomass sources should be investigated.

Nomenclatures

Symbol	Definition
η_{VS}	Degree of volatile solids degradation
c_{VS}	Concentration of volatile solids in dry substrate
m_{biogas}	Mass of biogas
M_{CH_4}	Molar mass of methane
M_{CO_2}	Molar mass of carbon dioxide
c_{CH_4}	Concentration of methane in the biogas
c_{CO_2}	Concentration of carbon dioxide in the biogas
GE	Gross energy of substrate
SMY	Specific methane yield

Author Contributions: Conceptualization, B.C. and J.M.; Methodology, B.C., H.O., and J.M.; Investigation, B.C. and H.O.; Writing—Original Draft Preparation, B.C.; Writing—Review & Editing, B.C. and J.M.; Project Administration, S.L.

Funding: This research was funded by the German Federal Ministry of Education and Research (BMBF) within the framework of the GlobE program in conjunction with the BiomassWeb project, grant number 031A258F.

Acknowledgments: The authors are grateful to the State Institute of Agricultural Engineering and Bioenergy and the laboratory staff members (University of Hohenheim) for the batch assay trials and physico-chemical analysis. In addition, we thank Schaumann Bioenergy, GmbH (Germany) for conducting elemental sample analyses. Authors are indebted to Stiftung Fiat Panis, Germany for financial support. Authors are thankful to Sabine Nugent for language editing.

Conflicts of Interest: The authors declare no conflict of interest.

References

1. Chanakya, H.N.; De Alwis, A.A.P. Environmental issues and management in primary coffee processing. *Process Saf. Environ.* **2004**, *82*, 291–300. [CrossRef]
2. Labouisse, J.-P.; Bellachew, B.; Kotecha, S.; Bertrand, B. Current status of coffee (coffea arabica l.) genetic resources in ethiopia: Implications for conservation. *Genet. Resour. Crop Evol.* **2008**, *55*, 1079–1093. [CrossRef]
3. Petit, N. Ethiopia's coffee sector: A bitter or better future? *J. Agrar. Chang.* **2007**, *7*, 225–263. [CrossRef]

4.	Minten, B.; Tamru, S.; Kuma, T.; Nyarko, Y. *Structure and Performance of Ethiopia's Coffee Export Sector*; International Food Policy Research Institute: Washington, DC, USA, 2014; Volume 66.

5.	Food Agriculture Organization of the United Nation. *Faostat Statistics Database*; Food Agriculture Organization of the United Nation: Rome, Italy, 2018.

6.	Gadhamshetty, V.; Arudchelvam, Y.; Nirmalakhandan, N.; Johnson, D.C. Modeling dark fermentation for biohydrogen production: Adm1-based model vs. Gompertz model. *Int. J. Hydrog. Energy* **2010**, *35*, 479–490. [CrossRef]

7.	Franca, A.S.; Oliveira, L.S. Coffee processing solid wastes: Current uses and future perspectives. In *Agricultural Wastes*; Ashworth, G.S., Azevedo, P., Eds.; Nova Science Publishers, Inc.: Hauppauge, NY, USA, 2009; pp. 155–190.

8.	Esquivel, P.; Jiménez, V.M. Functional properties of coffee and coffee by-products. *Food Res. Int.* **2012**, *46*, 488–495. [CrossRef]

9.	Beyene, A.; Kassahun, Y.; Addis, T.; Assefa, F.; Amsalu, A.; Legesse, W.; Kloos, H.; Triest, L. The impact of traditional coffee processing on river water quality in ethiopia and the urgency of adopting sound environmental practices. *Environ. Monit. Assess.* **2012**, *184*, 7053–7063. [CrossRef] [PubMed]

10.	Shemekite, F.; Gómez-Brandón, M.; Franke-Whittle, I.H.; Praehauser, B.; Insam, H.; Assefa, F. Coffee husk composting: An investigation of the process using molecular and non-molecular tools. *Waste Manag.* **2014**, *34*, 642–652. [CrossRef] [PubMed]

11.	Bakker, R. *Availability of Lignocellulosic Feedstocks for Lactic Acid Production-Feedstock Availability, Lactic Acid Production Potential and Selection criteria*; Wageningen UR-Food & Biobased Research: Wageningen, The Netherlands, 2013.

12.	Mengistu, M.G.; Simane, B.; Eshete, G.; Workneh, T.S. A review on biogas technology and its contributions to sustainable rural livelihood in ethiopia. *Renew. Sustain. Energy Rev.* **2015**, *48*, 306–316. [CrossRef]

13.	Gwavuya, S.G.; Abele, S.; Barfuss, I.; Zeller, M.; Müller, J. Household energy economics in rural ethiopia: A cost-benefit analysis of biogas energy. *Renew. Energy* **2012**, *48*, 202–209. [CrossRef]

14.	Tong, X.; Smith, L.H.; McCarty, P.L. Methane fermentation of selected lignocellulosic materials. *Biomass* **1990**, *21*, 239–255. [CrossRef]

15.	Triolo, J.M.; Sommer, S.G.; Møller, H.B.; Weisbjerg, M.R.; Jiang, X.Y. A new algorithm to characterize biodegradability of biomass during anaerobic digestion: Influence of lignin concentration on methane production potential. *Bioresour. Technol.* **2011**, *102*, 9395–9402. [CrossRef] [PubMed]

16.	Mussatto, S.I.; Fernandes, M.; Milagres, A.M.F.; Roberto, I.C. Effect of hemicellulose and lignin on enzymatic hydrolysis of cellulose from brewer's spent grain. *Enzyme Microb. Technol.* **2008**, *43*, 124–129. [CrossRef]

17.	Fischer, E.; Schmidt, T.; Hora, S.; Geirsdorf, J.; Stinner, W.; Scholwin, F. *Agro-Industrial Biogas in Kenya: Potentials, Estimates for Tariffs, Policy and Business Recommendations*; German International Cooperation (GIZ): Berlin, Germany, 2010.

18.	Ulsido, M.D.; Zeleke, G.; Li, M. Biogas potential assessment from a coffee husk: An option for solid waste management in gidabo watershed of ethiopia. *Eng. Rural Dev.* **2016**, 1348–1354.

19.	Kivaisi, A.K.; Rubindamayugi, M.S.T. The potential of agro-industrial residues for production of biogas and electricity in tanzania. *Renew. Energy* **1996**, *9*, 917–921. [CrossRef]

20.	Baier, U.; Schleiss, K. Greenhouse gas emission reduction through anaerobic digestion of coffee pulp. In Proceedings of the 4th International Symposium Anaerobic Digestion of Solid Waste, Copenhagen, Denmark, 31 August–2 September 2005.

21.	Adams, M.R.; Dougan, J. Waste products. In *Coffee: Volume 2: Technology*; Clarke, R.J., Macrae, R., Eds.; Springer Netherlands: Dordrecht, The Netherlands, 1987; pp. 257–291.

22.	Jayachandra, T.; Venugopal, C.; Anu Appaiah, K.A. Utilization of phytotoxic agro waste—Coffee cherry husk through pretreatment by the ascomycetes fungi mycotypha for biomethanation. *Energy Sustain. Dev.* **2011**, *15*, 104–108. [CrossRef]

23.	Porres, C.; Alvarez, D.; Calzada, J. Caffeine reduction in coffee pulp through silage. *Biotechnol. Adv.* **1993**, *11*, 519–523. [CrossRef]

24.	Jungbluth, T.; Büscher, W.; Krause, M. *Technik Tierhaltung*; Eugen Ulmer: Stuttgart, Germany, 2017; Volume 2641.

25.	Weiland, P. Biogas production: Current state and perspectives. *Appl. Microbiol. Biotechnol.* **2010**, *85*, 849–860. [CrossRef] [PubMed]

26. Roati, C.; Fiore, S.; Ruffino, B.; Marchese, F.; Novarino, D.; Zanetti, M. Preliminary evaluation of the potential biogas production of food-processing industrial wastes. *Am. J. Environ. Sci.* **2012**, *8*, 291.

27. *The World Factbook: 2018*; Central Intelligence Agency: Washington, DC, USA, 2018.

28. WorldBank. *Worlddatabank:Ethiopia*; WorldBank: Washington, DC, USA, 2018.

29. Mittweg, G.; Oechsner, H.; Hahn, V.; Lemmer, A.; Reinhardt-Hanisch, A. Repeatability of a laboratory batch method to determine the specific biogas and methane yields. *Eng. Life Sci.* **2012**, *12*, 270–278. [CrossRef]

30. Helffrich, D.; Oechsner, H. The hohenheim biogas yield test: Comparison of different laboratory techniques for the digestion of biomass. *Landtechnik* **2003**, *58*, 148–149.

31. VDI. Vdi 4630—Fermentation of organic materials, characterisation of substrate, sampling, collection of material data, fermentation tests [1872]vdi gesellschaft energietechnik. In *VDI Handbuch Energietechnik*; Beuth Verlag GmbH: Berlin, Germany, 2006; pp. 44–59.

32. Lindner, J.; Zielonka, S.; Oechsner, H.; Lemmer, A. Effect of different ph-values on process parameters in two-phase anaerobic digestion of high-solid substrates. *Environ. Technol.* **2015**, *36*, 198–207. [CrossRef] [PubMed]

33. Mönch-Tegeder, M.; Lemmer, A.; Oechsner, H.; Jungbluth, T. Investigation of the methane potential of horse manure. *Agric. Eng. Int. CIGR J.* **2013**, *15*, 161–172.

34. Mao, C.; Feng, Y.; Wang, X.; Ren, G. Review on research achievements of biogas from anaerobic digestion. *Renew. Sustain. Energy Rev.* **2015**, *45*, 540–555. [CrossRef]

35. Oechsner, H.; Lemmer, A.; Ramhold, D.; Mathies, E.; Mayrhuber, E.; Preissler, D. Method for Producing Biogas in Controlled Concentrations of Trace Elements. U.S. Patent Application DE200850002359, 29 May 2008.

36. Schattauer, A.; Abdoun, E.; Weiland, P.; Plöchl, M.; Heiermann, M. Abundance of trace elements in demonstration biogas plants. *Biosyst. Eng.* **2011**, *108*, 57–65. [CrossRef]

37. Facchin, V.; Cavinato, C.; Fatone, F.; Pavan, P.; Cecchi, F.; Bolzonella, D. Effect of trace element supplementation on the mesophilic anaerobic digestion of foodwaste in batch trials: The influence of inoculum origin. *Biochem. Eng. J.* **2013**, *70*, 71–77. [CrossRef]

38. Zhang, L.; Lee, Y.W.; Jahng, D. Anaerobic co-digestion of food waste and piggery wastewater: Focusing on the role of trace elements. *Bioresour. Technol.* **2011**, *102*, 5048–5059. [CrossRef] [PubMed]

39. Chala, B.; Oechsner, H.; Fritz, T.; Latif, S.; Müller, J. Increasing the loading rate of continuous stirred tank reactor for coffee husk and pulp: Effect of trace elements supplement. *Eng. Life Sci.* **2017**. [CrossRef]

40. Chen, Y.; Cheng, J.J.; Creamer, K.S. Inhibition of anaerobic digestion process: A review. *Bioresour. Technol.* **2008**, *99*, 4044–4064. [CrossRef] [PubMed]

41. Triolo, J.M.; Pedersen, L.; Qu, H.; Sommer, S.G. Biochemical methane potential and anaerobic biodegradability of non-herbaceous and herbaceous phytomass in biogas production. *Bioresour. Technol.* **2012**, *125*, 226–232. [CrossRef] [PubMed]

42. Rojas-Sossa, J.P.; Murillo-Roos, M.; Uribe, L.; Uribe-Lorio, L.; Marsh, T.; Larsen, N.; Chen, R.; Miranda, A.; Solís, K.; Rodriguez, W.; et al. Effects of coffee processing residues on anaerobic microorganisms and corresponding digestion performance. *Bioresour. Technol.* **2017**, *245*, 714–723. [CrossRef] [PubMed]

43. Thomsen, S.T.; Spliid, H.; Østergård, H. Statistical prediction of biomethane potentials based on the composition of lignocellulosic biomass. *Bioresour. Technol.* **2014**, *154*, 80–86. [CrossRef] [PubMed]

44. Khan, M.T.; Brulé, M.; Maurer, C.; Argyropoulos, D.; Müller, J.; Oechsner, H. Batch anaerobic digestion of banana waste-energy potential and modelling of methane production kinetics. *Agric. Eng. Int. CIGR J.* **2016**, *18*, 110–128.

45. Haag, N.L.; Nägele, H.-J.; Reiss, K.; Biertümpfel, A.; Oechsner, H. Methane formation potential of cup plant (silphium perfoliatum). *Biomass Bioenergy* **2015**, *75*, 126–133. [CrossRef]

© 2018 by the authors. Licensee MDPI, Basel, Switzerland. This article is an open access article distributed under the terms and conditions of the Creative Commons Attribution (CC BY) license (http://creativecommons.org/licenses/by/4.0/).

Article

Phytotoxicity of Corncob Biochar before and after Heat Treatment and Washing

Kiatkamjon Intani *, Sajid Latif, Md. Shafiqul Islam and Joachim Müller

University of Hohenheim, Institute of Agricultural Engineering, Tropics and Subtropics Group (440e), Stuttgart 70599, Germany; s.latif@uni-hohenheim.de (S.L.); msislamdebd@yahoo.com (M.S.I.); joachim.mueller@uni-hohenheim.de (J.M.)

* Correspondence: k_intani@uni-hohenheim.de or info440e@uni-hohenheim.de; Tel.: +49-711-459-23114

Received: 27 September 2018; Accepted: 18 December 2018; Published: 21 December 2018

Abstract: Biochar from crop residues such as corncobs can be used for soil amendment, but its negative effects have also been reported. This study aims to evaluate the phytotoxic effects of different biochar treatments and application rates on cress (*Lepidium sativum*). Corncob biochar was produced via slow pyrolysis without using purging gas. Biochar treatments included fresh biochar (FB), dried biochar (DB), washed biochar (WB), and biochar water extract (WE). Biochar application rates of 10, 20, and 30 t/ha were investigated. Significant phytotoxic effects of biochar were observed on germination rates, shoot length, fresh weight, and dry matter content, while severe toxic effects were identified in FB and WE treatments. Germination rate after 48 h (GR_{48}) decreased with the increase of biochar application rates in all treatments. The observed order of performance of the biochar treatments for germination, shoot length, and shoot fresh weight for every biochar application rate was WB>DB>WE>FB, while it was the reverse order for the shoot dry matter content. WB treatment showed the best performance in reducing the phytotoxicity of biochar. The mitigation of the phytotoxicity in fresh corncob biochar by washing and heat treatment was found to be a simple and effective method.

Keywords: biochar; crop residue; corncob; germination; phytotoxicity; self-purging pyrolysis; soil amendment

---

1. Introduction

Corncob is biomass residue, which is available and underused in many countries including Ethiopia [1,2]. It was reported in a previous study that in the corn belt areas of Ethiopia the average annual corn stover (i.e., cobs, husks, leaves and stalks) production per household was 2.09 tons [3]. In developing countries, corn is still harvested manually. The cobs with kernels are picked, dried, and stored. After shelling the kernels, the corncobs are available for other uses without extra effort for the collection and transportation. Therefore, corncobs have a potential to be processed into biochar for both energy and soil improvement via a slow pyrolysis process. This can improve the resource use efficiency along the biomass-based value webs. Corncob biochar could have an innovative use as propagation substrate in agricultural production systems [4,5].

In general, slow pyrolysis process requires a purging gas (e.g., nitrogen) to obtain oxygen-limited condition in the system [6]. The unaffordability and poor accessibility of pure nitrogen could hinder the establishment of biochar systems for smallholders in developing countries [7]. There are limited numbers of studies on pyrolysis without using nitrogen as purging gas (self-purging pyrolysis) [8,9]. Therefore, the need exists for studies on the possibility of producing biochar via slow pyrolysis without using purging gas. To address the existing knowledge gap, the production and characterization of corncob biochar from a self-purging pyrolysis reactor was carried out in previous studies of the authors [10,11]. It was proven that a self-purging pyrolysis reactor effectively produced corncob biochar with acceptable quality. However, in order to avoid negative effects of corncob biochar on plants and soils, phytotoxicity tests should be carried out before the biochar will be applied. Cress seeds are widely used for germination and toxicity tests due to their specific characteristics such as rapid growth and response, affordability, and hydroponic characteristics [12–15]. Cress seeds were also selected by Rombolà et al. [16] for a study to identify the phytotoxicity of biochars derived from poultry litter and corn stalk. To compare the results with previous studies, cress seeds were used in the present study for the germination test.

In recent years, biochar has gained substantial attention for its potential use as carbon sequestration, soil improver and biofuel. The concept of using biochar as a soil improver was established based on the research on Amazonia Dark Earths (Terra Preta) [17]. It was proved by several studies that agriculture, environment, and energy sectors can benefit from various applications of biochar [18,19]. However, increasing numbers of studies reported negative effects and dubious benefits of biochar for agricultural applications and carbon sequestration under field conditions [20,21]. More research on biochar derived from different biomass feedstocks is required before applying biochar to different types of soils under various climate conditions [22]. Moreover, the existence of potential contaminants in biochar was also reported in previous studies. Polycyclic aromatic hydrocarbons (PAHs), volatile organic compounds (VOCs), heavy metals (Cd, Cr, Cu, Ni, Pb, and Zn), tars, furans, and dioxins were the conventional pollutants that could be present in biochar [23,24]. These pollutants were commonly formed during pyrolysis and might exist at a low concentration with less bioavailability since they could be absorbed by the biochar [23–26]. A few studies on the composition and impact of residual tars and VOCs were carried out [27–29]. It was determined that biochar containing leachable organic compounds showed positive effects on corn productivity [30], while negative effects on aquatic microorganisms were observed [27]. The contamination of biochar during pyrolysis was caused by the re-condensation of the VOCs, which condensed into liquid products. These liquids were the degradation products of lignin, cellulose, and hemicellulose during the pyrolysis process [31], which contained acetone, acetic acid, methanol, and furfural. The other substances such as alcohols, aldehydes, furans, ketones, organic acids, and phenols were among the main compounds found in the liquids. Furthermore, PAHs could also be found in the liquid products from the pyrolysis process [31,32]. In the case of biochar toxicity, many studies focused on PAHs and heavy metals [33]. VOCs such as acetone and benzene in biochar were found to have severe phytotoxic effects [29,33]. In the presence of transition metals such as Cr, Cu, Fe, Pb, and Zn in the biomass feedstock, free radicals were generated during pyrolysis [25]. These free radicals reacted with water and formed reactive oxygen species, which could have negative effects on seed germination, root-shoot elongation and plant growth [25]. Liao et al. [25] also reported that free radicals in biochar had more detrimental effects on seed germination than the PAHs and heavy metals. Salinity in biochar was also found to have a negative effect on seed germination and seedling growth due to osmotic stress [34,35]. The biochar salinity was usually estimated by electrical conductivity (EC) [36]. It was evident that salt stress induced by high salinity in biochar significantly inhibited seed germination [16]. Moreover, the alkaline nature of biochar indicated by a high pH value might lead to an insolubilization of plant nutrients [34,37].

Previous studies were carried out to identify methods to reduce biochar phytotoxicity [37,38]. It was reported that washing biochar with water or an organic solvent significantly reduced its toxicity. In this washing process, an aqueous extract was separated from the solid biochar [37,39]. It should be noted that VOCs in biochars derived from different feedstocks at high temperatures (800–860 °C) did not inhibit germination of barley [37]. It was also evident that biochar produced at a higher temperature had less phytotoxic effects than biochar produced at a lower temperature [6]. VOCs in biochar could be reduced by handling, processing, and long-term storage [38,40]. A previous study by Kołtowski and Oleszczuk [41] showed that biochar toxicity was also reduced by drying the biochar at 100 to 300 °C for 24 h. However, biochar toxicity and phytotoxicity mitigation methods have not yet been widely studied on biochar produced in a self-purging pyrolysis. Therefore, the need exists to assess the toxicity of the fresh corncob biochar before any agricultural applications. Furthermore, it is also necessary to identify whether simple washing and heat treatments can adequately reduce the phytotoxicity of corncob biochar produced in a self-purging pyrolysis reactor. After washing biochar, the biochar water extract might have high environmental risks. Thus, the phytotoxicity of the biochar water extract should be also assessed, which will provide useful information for a proper management of the water extract. This can avoid the contamination to soils, ground water, and the ecosystems.

It should be noted that while biochar might offer potential benefits, there are also risks arising from its applications. Therefore, biochar properties must be carefully evaluated before field applications. Different treatments such as heat treatment and washing could be options for the reduction of the phytotoxicity in biochar. In this study, the physiochemical properties of the corncob biochar were determined. The objective of this study was to investigate the phytotoxic effects of different biochar treatments on seed germination of cress (*Lepidium sativum*). Fresh biochar (FB), dried biochar (DB), washed biochar (WB), and biochar water extract (WE) with the application rates of 10, 20, and 30 t/ha were investigated. The phytotoxic effects on germination rates, shoot length, fresh weight, and dry matter content of cress seedlings were identified.

2. Materials and Methods

2.1. Biomass and Biochar Preparation

Corncobs were provided by the experimental station of the University of Hohenheim (Heidfeldhof, 48°42′56″ N 9°11′29″ E, 396 m a.s.l.). The corncobs were collected from cultivars being used for animal feed. The cobs without kernels were collected manually after being discarded from a combine harvester. The biomass was dried at the experimental station and stored in a wooden box with a perforated floor.

A self-purging pyrolysis reactor was used for the pyrolysis experiments as presented in previous studies [10,11]. The reactor was constructed of stainless steel with inner dimensions of 22.6, 15.6, and 9.8 cm in length, width, and height, respectively. Various sizes of holes on the lid of the reactor were used as pyrolysis gas outlets. In the present study, a hole with a diameter of 6 mm was selected. The pyrolysis reactor was placed in a pyrolysis chamber during the experiments. A gas outlet pipe was attached to the pyrolysis chamber with inner dimensions of 25, 18, and 10 cm (length, width, and height). The pyrolysis unit was heated in an ashing furnace (LVT 15/11/P330, Nabertherm GmbH, Lilienthal, Germany).

In total 3.7 kg of whole corncobs without size reduction were pyrolyzed in 8 batches. The corncob biomass was heated from 100 to 450 °C at a heating rate of 10 °C/min. The pyrolysis temperature was kept at 450 °C for 60 min without using a purging gas. After removing the pyrolysis reactor from the furnace, it was covered with aluminum foil to avoid further oxidation and cooled down for 30 min at room temperature. Those operating conditions have been chosen to obtain corncob biochar with a low volatile matter content, high fixed carbon content, high heating value, high pH, high EC, and high biochar yield [11].

2.2. Characterization of Corncob Biochar

The biochar yield was calculated on air-dried basis (ad), as follows:

$$Y_{biochar,ad} = 100 \times \frac{M_{biochar}}{M_{biomass}} \tag{1}$$

where $Y_{biochar,ad}$ is the air-dried basis yield of biochar (%), $M_{biochar}$ represents the mass of produced biochar (g), and $M_{biomass}$ represents the total mass of biomass (g).

Biochar was ground using a coffee mill (CM3260, GRUNDIG Intermedia GmbH, Neu-Isenburg, Germany) and sieved to obtain a particle size of <2 mm. The biochar samples were thoroughly mixed in a plastic container before random samples were taken for the analyses of physicochemical properties and germination test. The analyses of the physicochemical properties of the biochar were done in triplicate. The oven drying method based on DIN 51718 [42] was used to determine the moisture content (MC) of the biochar. The analysis of volatile matter content (VM) was carried out according to the standard methods DIN 51720 [43], while the ash content (AC) of the biochar was determined based on DIN 51719 [44]. The calculation of the fixed carbon content (FC) was done using the following equation:

$$FC = 100 - VM - AC \tag{2}$$

The pH values were measured in a biochar solution. Approximately 0.5 g of biochar and 5 mL of 0.01 mol/L $CaCl_2$ solution were mixed. After shaking the biochar solution for an hour, the pH values were determined using a pH meter (HQ40D, Hach Company, Colorado, USA). The EC was determined in the biochar water extract (1:10 w/v) using an electrical conductivity meter (Cond 315i/SET, Xylem Analytics Germany Sales GmbH & Co. KG, Weilheim, Germany).

The elemental analyzer (vario Max CNS, Elementar Analysensysteme GmbH, Langenselbold, Germany) was used to measure the C, N, and S content. Inductively coupled plasma optical emission spectrometry (ICP-OES) (Vista-PRO, Varian Inc., California, USA) was employed to estimate the major extractable mineral cations such as Al, Ca, Fe, Mg, Mn, Na, Zn, and P. Whereas, an inductively coupled plasma mass spectrometry (ICP-MS) (NexION 300X, PerkinElmer Inc., Massachusetts, USA) was used to determine the water extractable trace elements such as Cd, Co, Cr, Cu, Ni, Pb, Rb, and Sr. The effective cation exchange capacity (ECEC) was determined by simultaneous ICP-OES applying 1 M ammonium chloride extraction based on the handbook forest analytics (HFA) method A3.2.1.8 [45]. The ECEC was calculated by summing the exchangeable cations of Al, Ca, Fe, K, Mg, Mn, and Na.

2.3. Treatments of Corncob Biochar

Phytotoxic potential was determined before and after heat treatment and washing. Untreated biochar further on is referred to as FB. Heat treatment was applied to reduce VOCs and PAHs [41], and to change physicochemical properties such as hydrophobicity and hydrophilicity [38]. For this purpose biomass was oven dried according to DIN 51,718 [42] at 105 ($\pm$2)°C for 24 h and then placed in a desiccator for 30 min. The samples were stored in airtight containers until use. The heat-treated biochar is further on referred to as DB. To reduce water-soluble toxic compounds in the solid phase [37,39], biochar was also washed. The fresh biochar was mixed with 24 mL of de-ionized water and manually shaken for 1 min. The biochar suspension was shaken in the lab shaker (Type 1083, Gesellschaft für Labortechnik mbH, Burgwedel, Germany) for 24 h at room temperature. Subsequently, the solid and liquid fractions in the biochar suspension were separated using the filter paper (Filter discs, Grade 388, Pore size 10–15 μm, Sartorius Stedim Biotech GmbH, Göttingen, Germany) and vacuum pump (N 035.3 AN. 18, KNF Neuberger GmbH, Freiburg, Germany). As a result, WB and biochar WE were obtained and used for the germination test. The WE treatment was used to identify the potential environmental risks and contamination, which could be caused by the leaching effect of fresh biochar. For the control (CON), sand without the addition of biochar was taken.

2.4. Germination Test

Cress seeds were acquired from Bruno Nebelung GmbH, Everswinkel, Germany. The company reported a germination rate of 99% for the seeds. A pre-test revealed a germination rate of 97%.

Plastic mini greenhouses each holding one planting tray of 57 × 38 × 18 cm, (Maximus Complete, Romberg GmbH & Co. KG, Ellerau, Germany) were set for the germination tests. The mini greenhouse was equipped with two illumination units (24 watt, 6400 Kelvin). A heat mat (50 × 30 cm, 30 watt) was used to control the temperature in the mini greenhouse. A control unit was employed to regulate the temperature and lighting (HortiSwitch Thermostat, Romberg GmbH & Co. KG, Ellerau, Germany).

Sand with a particle size of 0.6–1.4 mm in diameter was used. The sand was disinfected in the pyrolysis reactor at 600 °C for 3 h.

The germination test was carried out using a block design. The experiment was performed with three and four replications for treatments and control (CON), respectively. Two mini greenhouses were assigned as block 1 and block 2. A seedling tray with 20 germination pots was placed in each of the mini greenhouse. The size of each germination pot was 5 cm × 4 cm. A total of 16 germination pots were used for the replications in this experiment as described in Table 1. Three application rates for biochar were selected, namely 2, 4, and 6 g of biochar per pot representing 10, 20, and 30 t/ha, respectively. The application rates were determined according to the assumption that the biochar was incorporated into the soil over a depth of 50 mm [46]. Each germination pot of the mini greenhouse was filled with a mixture of 100 ($\pm$ 0.05) g disinfected sand, 24 mL de-ionized water and 2, 4, and 6 g of biochars FB, DB, and WB as shown in Table 1. Also, the biochar WE was tested by mixing 100 ($\pm$ 0.05) g disinfected sand with 24 mL WE, where 2, 4, and 6 g of biochar were washed with de-ionized water. This corresponds to 83, 167 and 250 g biochar washed in 1 L water.

Table 1. Experimental set-up.

Treatments		Block 1 and 2		Biochar	Extract
		Replications	Seeds per replication	(g)	(g/L)
Control	(CON)	4	5	-	-
Fresh biochar	(FB)	3	5	2, 4, 6	-
Dried biochar	(DB)	3	5	2, 4, 6	-
Washed biochar	(WB)	3	5	2, 4, 6	-
Biochar water extract	(WE)	3	5	-	83, 167, 250

The sand, biochar, and de-ionized water or WE were filled into a plastic bottle and manually shaken for 45 seconds before filling the germination pot with the mixture. The moisture content in the substrate was maintained at 85% of the maximum water holding capacity (WHC) of the sand [47].

There were three runs for the whole experiment, in which the three application rates were performed separately. The germination period lasted 7 days under 24 h continuous lighting, 5500–7500 lux [48] and controlled temperatures of 23 ($\pm$ 2)°C. The distance between the lamp (T5HO 24 watt, PAR white 2400K, Romberg GmbH & Co. KG, Ellerau, Germany) and the top of the germination pot was 25 cm. Total numbers of germinated seeds were counted after 24 h and 48 h. The normal and abnormal seedling emergence and above-ground biomass (fresh and dry shoot weight, shoot length) were evaluated [49].

Germination rate (GR) was defined as a percentage of the germinated seeds to the total seeds within 24 h (GR_{24}) and 48 h (GR_{48}). A seed was considered as germinated when there was visible root development [35]. The above-ground biomass of cress was collected. The shoot length (mm) was measured using a digital slide caliper (M 823–160, Brüder Mannesmann Werkzeuge GmbH, Remscheid, Germany). The fresh weight of the shoot biomass was measured in grams using a digital balance (ED224S, Sartorius AG, Göttingen, Germany). Subsequently, the biomass was dried in an oven at 105 ($\pm$ 2)°C for 24 h and the dry weight was recorded. Finally, the MC and dry matter content (DM) of the shoot biomass were calculated.

2.5. Statistical Analysis

Generalized linear mixed model (GLMM) using the GLIMMIX procedure in statistical analysis system (SAS) (Ver. 9.3, SAS Inst., Cary, NC, USA) was employed for the statistical analysis. The analysis of variance (ANOVA) and the comparison of means using least significant difference (LSD) were evaluated at 95% confidence level ($p < 0.05$).

3. Results and Discussion

3.1. Characteristics of Corncob Biochar

The biochar yield was 26.42 ± 0.91%. The results of the proximate analysis, pH, and EC of the biochar are presented in Table 2. The biochar particle distribution is also shown in Table 2. The biochar yield in our study (26.42%) was similar to the results in a previous study, which reported biochar yields of 25–27% [6]. This confirmed that the pyrolysis of corncob without using purging gas produced similar amounts of solid products. Özçimen and Ersoy-Meriçboyu [50] reported that the biochar yield decreased linearly with an increase in the purging gas flow rate. This might be explained by the removal of the volatile matters from the feedstock, which reduced the secondary reaction such as re-polymerization and re-condensation. However, the impact of the purging gas on the biochar yield was not highly significant. On the other hand, there might be a risk of high VM in biochar in the absence of purging gas during pyrolysis, which might increase the phytotoxic substances [10,11].

Table 2. Proximate analysis, pH, EC, and particle size distribution of the corncob biochar.

Parameters	Unit	Value
Moisture content (MC)	wt.% db [a]	1.18 ± 0.10
Volatile matter content (VM)	wt.% db	14.48 ± 0.21
Ash content (AC)	wt.% db	6.52 ± 0.06
Fixed carbon content (FC)	wt.% daf [b]	78.98 ± 0.29
pH	-	10.34 ± 0.04
Electrical conductivity (EC)	mS/cm	7.21 ± 0.05
Particle size		
CBP [c] > 2 mm	%	0.39
2 mm ≥ CBP > 1 mm	%	2.30
1 mm ≥ CBP > 850 μm	%	2.12
850 μm ≥ CBP > 600 μm	%	7.96
600 μm ≥ CBP > 500 μm	%	5.74
500 μm ≥ CBP	%	80.15
Processing loss	%	1.35

Value presented as mean ± standard deviation; [a] Weight percentage on dry basis; [b] Weight percentage on dry ash free basis; [c] CBP stands for corncob biochar particle.

There were various sizes of biochar particles in the corncob biochar, and more than 80% of the particles were smaller than 500 μm (Table 2). Although biochar particle size distribution did not show any effect on the toxic effect of the biochar [51], a smaller particle size could increase the surface area and water holding capacity. The proximate analysis of biochar indicated that there was 14.48% and 78.98% of VM and FC, respectively (Table 2). These results were in agreement with the results in the previous study [6]. It was reported that the AC, VM, and FC were 1.2–71.8%, 14.9–34.9%, and 36.4–85.0%, respectively, in various biochars produced via slow pyrolysis process at 450 °C using different feedstocks [6,52]. The VM of the biochar in this study was 14.48%, which was below the threshold of VM of biochar for soil application (20%) [53]. This implied that a self-purging pyrolysis reactor was able to produce biochar with an acceptable amount of volatile matter. The AC and pH value of the biochar in the present study were also in agreement with the results from previous studies [6,33,52]. The biochar had a mean pH value of 10.34, which indicated a high alkalinity. The alkalinity generally results from the salts present in the ash. The high pH value of biochar might be a result of the high K content (25,602 mg/kg) in the biochar [54], which was the highest elemental content in the corncob biochar. Salt stress on plants might be exposed by the high AC of the biochar [38]. Biochar with a high pH could be applied to soils that need a liming effect [55]. However, the negative impacts of salt stress on seed germination were reported in previous studies [38,56]. Hoekstra et al. [35] reported that salinity might negatively affect germination and plant growth, especially at the early stage of the germination period. Generally, the effects of the salt stress were irrelevant in biochar water extract with an EC value of ≤ 2.0 mS/cm [35]. The EC value of the biochar was 7.21 mS/cm as shown in Table 2, thus the negative effects of the salt stress on germination and seedling growth could not be disregarded in this study.

The results of the mineral content, trace element content, ultimate analysis, and cation exchange capacity of biochar are described in Tables 3 and 4. The major mineral contents in the biochar were K, P, Mg, Ca, Zn, Na, Mn, Fe, and Al, respectively. Consequently, the analysis of the effective cation exchange capacity showed the highest exchangeable K^+ of 62.7 $cmol_c/kg$. The ECEC was 71.1 $cmol_c/kg$. The addition of biochar to soils has a potential to enhance the soil cation exchange capacity, as reported in a previous study [57]. The major heavy metals in the biochar were Zn, Mn, Fe, Cu, Cr, Ni, Pb, As, Cd, and Co, respectively. The content of the heavy metals was less than 52.9 mg/kg. In this study, the heavy metal content (Zn, Fe, Cu, Cr, Pb, and Cd) in biochar was relatively low compared to the results (up to 214.05 mg/kg) from rice straw, wheat straw, and corn stalk biochar in previous studies [25,58]. The ultimate analysis indicated that the biochar had a high C (77.6%), but low N content (65.21%). This resulted in a high C/N ratio (65.21), which could lead to a short-term N immobilization after incorporating biochar into soils [59]. Based on the elemental properties of the corncob biochar, it was observed that the biochar had the potential to provide additional macronutrients such as K, P, C, N, Mg, and Ca to plants. In addition, micronutrients such as Zn, Na, Fe, Mn, and Cu could have positive effects on plant growth, especially when biochar was applied in small quantities [25].

Table 3. Major mineral content, trace element content and ultimate analysis of corncob biomass and biochar.

Element	Biomass	Biochar
Mineral content (mg/kg) [a]		
Al	97.5	6.00
Ca	285	787
Fe	61.9	20.5
K	6.38	25,602
Mg	373	1268
Mn	8.00	22.1
Na	18.9	33.7
P	430	2930
Zn	33.1	52.9
Trace element (mg/kg) [a]		
As	<0.100	<0.0500
Cd	<0.150	0.0300
Co	<0.150	0.0100
Cr	8.42	1.77
Cu	5.81	9.32
Ni	2.61	<0.150
Pb	0.500	0.0600
Rb	0.970	21.4
Sr	1.06	2.06
Ultimate analysis (wt. %) [b]		
C	45.7	77.6
N	0.611	1.19
S	0.0580	<0.100
C/N	74.8	65.2

[a] Values on dry basis; [b] Weight percentage on dry basis.

Table 4. Exchangeable cations and effective cation exchange capacity (ECEC) of corncob biochar.

Exchangeable Cations	Concentrations ($cmol_c$/kg)
Al^{3+}	0
Ca^{2+}	1.80
Fe^{3+}	0
K^+	62.7
Mg^{2+}	6.60
Mn^{2+}	0
Na^+	0
ECEC	71.1

3.2. Effects of Corncob Biochar on Cress Seed Germination

The germination rates (GR_{24} and GR_{48}) are shown in Figure 1. At a biochar application rate of 10 t/ha, biochar treatments did not have a significant effect ($p = 0.4750$) on the GR_{48}. All of the biochar treatments (FB, DB, WB, and WE) resulted in similar germination rates (GR_{48}) as in the CON. In case of GR_{24}, it was observed that the biochar treatments also did not significantly delay ($p = 0.9191$) the germination of cress seeds. Therefore, FB, DB, WB, WE, and CON showed similar values of GR_{24}. This result implied that a biochar application rate of 10 t/ha or lower did not significantly reduce the germination rate of cress. At an application rate of 20 t/ha, biochar treatments had a significant effect ($p = 0.0335$) on the GR_{48}. The CON had the highest GR_{48} of 95% followed by WB (93%). It was observed from GR_{24} that the biochar treatments had a significant impact ($p = 0.0216$) on cress germination. Compared to the CON, FB, and WE reduced GR_{24}, while DB and WB increased GR_{24}. It was evident that WB performed as good as CON on GR_{48} and even better than CON on GR_{24}, which indicated that washing biochar successfully reduced its toxicity at an application rate of 20 t/ha. At an application rate of 30 t/ha, the effect of biochar treatments on GR_{48} was significant with p-values of <0.0001 but was not significant for GR_{24} ($p = 0.1105$). FB and WE completely inhibited the germination of cress after 24 h ($GR_{24} = 0\%$). The GR_{48} in WB treatments was not significantly different compared to the value in the CON. CON and DB treatments were significantly different ($p < 0.05$) with a GR_{48} of 95 and 73%.

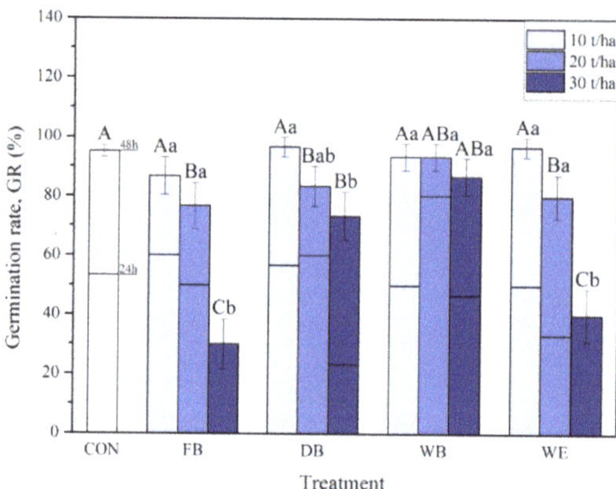

Figure 1. Germination rates (GR) under different biochar application rates (10, 20, and 30 t/ha) and treatments (Control: CON, Fresh biochar: FB, Dried biochar: DB, Washed biochar: WB and Biochar water extract: WE). The first stacked column indicates the germination rate after 24 h (GR_{24}), while the second stacked column shows the germination rate between 24–48 h. The sum of stacked columns represents the germination rate after 48 h (GR_{48}) Different capital letters show significant differences in GR_{48} between biochar treatments at same application rate ($p < 0.05$), and different small letters show significant differences in GR_{48} between biochar application rates of same biochar treatment ($p < 0.05$). Standard errors of GR_{48} are represented by the bars (n = 6 for the treatments, and n = 24 for the CON).

The application rate of biochar had a significant effect on GR_{48} in FB and WE treatments with p-values of <0.0018, and 0.0033. This result was in agreement with the findings from previous research describing a significant influence of the biochar application rate on seed germination [37,38,60]. On the contrary, the application rate did not significantly impact the GR_{48} in DB ($p = 0.1267$) and WB ($p = 0.5986$) treatments. This indicated that washing biochar did sufficiently reduce the toxicity even at an application rate of 30 t/ha, when the germination rate was considered as an indicator. As shown in Figure 1, the germination rate (GR_{48}) was significantly different in FB and WE treatments with increasing application rates from 20 to 30 t/ha, indicating a high toxicity in the fresh biochar and biochar water extract. While the application rate of biochar increased from 10 to 30 t/ha, the GR_{48} decreased from 93 to 87, 97 to 73, 97 to 40, and 87 to 30% in WB, DB, WE, and FB treatments, respectively. WB treatments showed the best performance in reducing biochar toxicity, when germination was used as an indicator. Apart from WB treatments, DB also showed a good potential to reduce the germination inhibition induced by biochar toxicity.

The germination rate can be considered as the most sensitive indicator for the toxicity in different biochar treatments, which was confirmed by a previous study [33]. Based on the results of this study, corncob biochar could be used at a low application rate (10 t/ha) without a significant phytotoxic effect on seed germination. In a previous study, it was reported that biochar from date palm leaflets did not result in toxic effect on lettuce germination [61]. It was reported that the addition of biochar at low application rates (10–50 t/ha) increased germination of wheat [60]. At an application rate of more than 20 t/ha, biochar treatments have to be applied to reduce the phytotoxicity in biochar and negative effects on germination. Washing and heat treatment of biochar showed promising results in reducing negative effects of biochar on seed germination. FB and WE showed acute and severe toxic effects on germination. A strong negative effect of WE on germination was also reported by Buss and Mašek [33]. Therefore, after washing corncob biochar, the biochar water extract must be managed properly to avoid contamination to the environment. Water-soluble toxic substances such as PAHs, phenols, organic acids, ketones, and alcohols in WE can result in high phytotoxicity on seed germination [62]. In the FB treatments, cress seeds were in direct contact with biochar and were also exposed to water-soluble compounds and volatiles. Therefore, FB resulted in high levels of toxicity on seed germination. In this study, the germination test was carried out in closed mini greenhouses, therefore the seeds were fully exposed to volatiles. These volatiles accumulate in the air, which could inhibit the germination without direct contact to cress seeds. It was reported that the germination of spring barley decreased due to the volatiles released from hydrochar [37]. Volatile toxic substances in biochar also demonstrated negative effects on germination, which resulted in a delayed germination in cress [33]. A pH value of 5.5–7.5 was found to be preferable for seed germination [63]. The high pH value (10.34) of the biochar might be responsible for germination inhibition [54]. In addition, the high AC in biochar could also induce salt stress, which might also have reduced the germination rate. It was suggested that water-soluble compounds in biochar could have played a predominant role in biochar phytotoxicity [27]. In the FB, DB, and WB treatments, biochar was mixed with water, in which toxic substances could have been released. Therefore, there was a high risk of germination inhibition by the water-soluble compounds.

3.3. Effects of Corncob Biochar on Shoot Length

Figure 2 shows the effect of biochar treatments and application rates on the shoot length. At a biochar application rate of 10 t/ha, biochar treatments had a highly significant effect ($p < 0.0001$) on the shoot length. The CON had the highest shoot length of 33.01 mm. Among the treatments, the shoot length was highest in WB (23.28 mm), but it was not significantly higher than the shoot length in DB. On the other hand, FB and WE significantly reduced the shoot length compared to DB and WB. At a biochar application rate of 20 t/ha, the effect of biochar treatments on the shoot length was also highly significant ($p < 0.0001$). In WB, the shoot length (21.90 mm) increased significantly compared to FB, DB, and WE. As expected, the lowest shoot length was found in FB (7.93 mm). However, the shoot lengths among FB, DB, and WE treatments were not significantly different. This implies that the phytotoxic effect on the shoot length was significantly reduced by washing the biochar. In case of an application rate of 30 t/ha, biochar treatments also had a significant impact ($p < 0.0001$) on the shoot length. WB treatment showed the highest potential in reducing the phytotoxicity, which resulted in significantly higher shoot lengths (11.57 mm) compared to the other biochar treatments (FB, DB, and WE). This indicated that at a high application rate washing biochar with de-ionized water could still reduce the phytotoxicity to some extent.

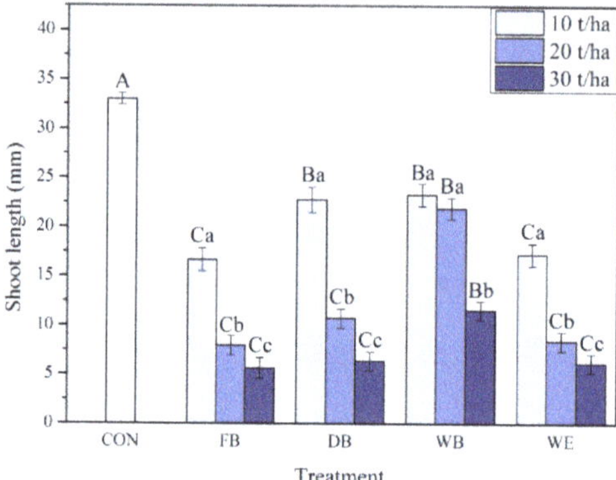

Figure 2. Shoot length of cress under different biochar application rates (10, 20, and 30 t/ha) and treatments (Control: CON, Fresh biochar: FB, Dried biochar: DB, Washed biochar: WB and Biochar water extract: WE). Different capital letters show significant differences between biochar treatments at same application rate (p <0.05), and different small letters show significant differences between biochar application rates of same biochar treatment (p <0.05). Standard errors are represented by the bars (n = 6 for the treatments, and n = 24 for the CON).

The effect of biochar application rates on the shoot length was highly significant (p < 0.0001) in all of the biochar treatments (FB, DB, WB, and WE), which was in agreement with the results from a previous study [25]. The shoot length decreased significantly with increasing application rates from 10–20 and 20–30 t/ha in FB, DB, and WE. It should be noted, that increasing application rates from 10–20 t/ha did not significantly reduce the shoot length in WB (only 6% reduction). However, increasing the application rate from 20–30 t/ha resulted in a significant reduction (51%) of the shoot length in WB treatments.

The shoot length was found to be mainly affected by the delayed germination [33]. The delayed germination reduced the time for plant growth, which might result in reduced shoot lengths. The shoot length proved to be a good indicator to evaluate the effect of biochar treatments and application rates. This could be supported by the fact that all of the treatments and application rates showed significant differences in the shoot lengths compared to the CON. The shoot length was significantly reduced by the addition of biochar produced from peanut hull at 500 °C [38]. Contrarily, it was reported that biochar derived from date palm leaflets promoted the shoot length of lettuce [61]. However, the germination test in the previous study was carried out on filter papers, which could lead to the different outcomes compared to the germination test on biochar-sand mixtures in our study. The strong negative effect of WE treatments on the shoot length was supported by similar findings from a previous research, where rice straw biochar significantly reduced the shoot length of corn, rice, and wheat [25].

3.4. Effects of Corncob Biochar on Shoot Fresh Weight

The effect of biochar treatments on shoot fresh weight is illustrated in Figure 3. The biochar treatments had a significant impact (p < 0.0001) on shoot fresh weight in all biochar application rates (10, 20, and 30 t/ha). At a biochar application rate of 10 t/ha, the DB and WB treatments resulted in similar shoot fresh weight compared to the CON. The FB had the most negative effect on the shoot fresh weight among the treatments. The WB treatment still showed the highest shoot fresh weight among the biochar treatments at an application rate of 20 t/ha, but was significantly reduced compared to the CON. The FB, DB, and WE had significantly lower shoot fresh weights than that of WB. In CON, WB, DB, WE, and FB treatments, the shoot fresh weight was 0.054, 0.050, 0.032, 0.027, and 0.027 g, respectively. At a 30 t/ha application rate, there was a significant reduction of the shoot fresh weight in the WB compared to the CON (0.054 g). However, the shoot fresh weight in the WB (0.032 g) was significantly higher than in the FB, DB, and WE treatments. The FB, DB, and WE treatments resulted in a similar shoot fresh weight of 0.011–0.014 g.

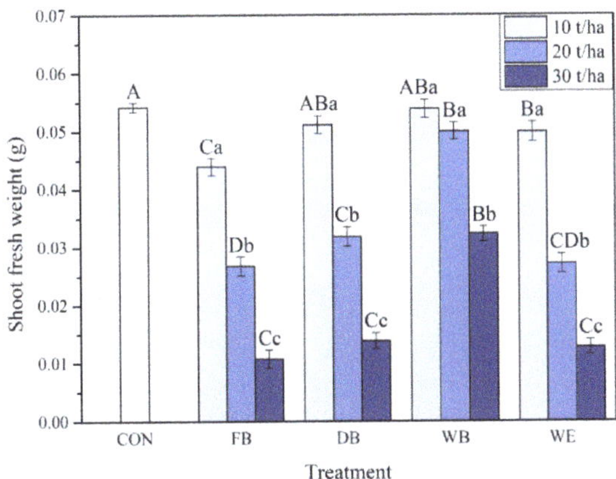

Figure 3. Shoot fresh weight of cress under different biochar application rates (10, 20, and 30 t/ha) and treatments (Control: CON, Fresh biochar: FB, Dried biochar: DB, Washed biochar: WB and Biochar water extract: WE). Different capital letters show significant differences between biochar treatments at same application rate (p <0.05), and different small letters show significant differences between biochar application rates of same biochar treatment (p <0.05). Standard errors are represented by the bars (n = 6 for the treatments, and n = 24 for the CON).

The biochar application rate significantly influenced (p < 0.0001) the shoot fresh weight in all biochar treatments, including FB, DB, WB, and WE. The reduction of the shoot fresh weight was highly significant (about 50% reduction) with increasing application rates from 10 to 20 and 20 to 30 t/ha in FB, DB, and WE treatments. The shoot fresh weight was not significantly reduced (only 7% reduction), when the application rate increased from 10 to 20 t/ha in WB treatments. This result indicated that washing biochar could adequately reduce the phytotoxic effect on shoot fresh weight up to an application rate of 20 t/ha. However, at an application rate of 30 t/ha the shoot fresh weight was significantly reduced by 36%. In general, the WB treatments showed the highest shoot fresh weight among the biochar treatments for every application rate investigated in this study.

Contrary to the findings in our study, it was determined that adding corncob biochar significantly increased the fresh weight of wheat in sandy soil [64]. A previous study also showed that biochar derived from argan shells increased fresh biomass weight of salad and barley [65]. In addition, it was reported that peanut hull biochar did not have a significant influence on the fresh weight of barley after germination [38]. In agreement with the results in our study, peanut hull biochar significantly reduced the fresh weight of cress seedlings, when the seeds were germinated without extra irrigation [38]. This implied that biochar from different feedstocks and pyrolysis conditions could have various effects on seed germination and seedling growth [54]. Therefore, potential toxic effects of each biochar need to be determined before its application in agriculture.

3.5. Effects of Corncob Biochar on Dry Matter Content of the Shoot Biomass

Figure 4 describes the treatment and application rate effect of biochar on the dry matter content of the shoot biomass. The treatment effect was highly significant (p < 0.0001) for all biochar application rates. At a biochar application rate of 10 t/ha, the FB treatment had a significantly higher DM than other treatments and the CON, while the CON, DB, WB, and WE had comparable DM contents. At a 20 t/ha application rate, FB and WE performed significantly better than other treatments and the CON, which resulted in DM contents of 35.34 and 34.46%, respectively. At a 30 t/ha application rate, the FB, DB, and WE had comparable DM contents, which were significantly higher than the DM content in the WB and CON. The DM content in FB, WE, DB, WB, and CON was 37.37, 35.82, 35.35, 29.48, and 21.10%, respectively. It should be noted that for all application rates the CON had the lowest DM content compared to the biochar treatments.

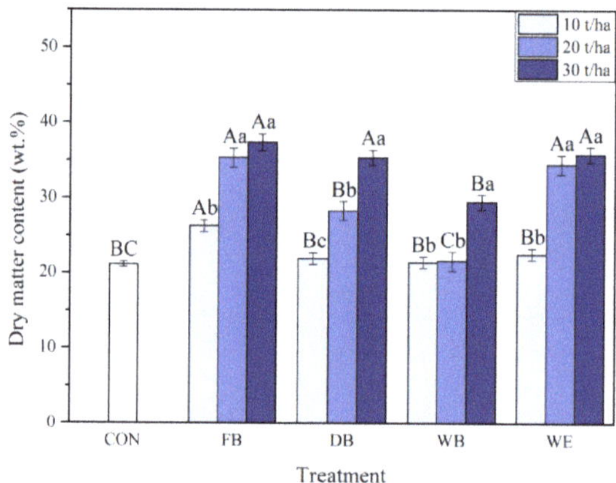

Figure 4. Dry matter content of the shoot biomass under different biochar application rates (10, 20, and 30 t/ha) and treatments (Control: CON, Fresh biochar: FB, Dried biochar: DB, Washed biochar: WB and Biochar water extract: WE). Different capital letters show significant differences between biochar treatments at same application rate (p <0.05), and different small letters show significant differences between biochar application rates of same biochar treatment (p <0.05). Standard errors are represented by the bars (n = 6 for the treatments, and n = 24 for the CON).

It was determined that the biochar application rate had a significant impact on the DM content in all biochar treatments. The p-values were $\leq$0.0004. In the FB treatment, the DM content did not increase significantly, when the application rate increased from 20 to 30 t/ha. This might be due to the phytotoxic effect at high application rates in FB treatments. In DB treatments, the DM content significantly increased with increasing biochar application rates. Only the DM content at 30 t/ha was significantly higher than the other application rates in WB treatments. The DM content in WB treatments increased by 0.17 and 7.89% with increasing application rates from 10 to 20 and 20 to 30 t/ha, respectively. In case of WE treatments, the DM content did not significantly increase with increasing application rates of 20 to 30 t/ha, which could be explained by a high phytotoxic effect at high application rates. Among the biochar treatments, WB had the lowest DM content. This implies that the DM content could be promoted by the mineral nutrients. Since the nutrients leached away in the WB treatment, the DM content of the shoot biomass was negatively affected by the nutrient deficiency.

In general, the DM content of shoot biomass increased in biochar treatments compared to the CON. A positive effect of biochar on shoot dry weight was also reported in a previous study [66]. This can be explained by the addition of mineral nutrients from biochar ash [67]. Water extract from corn stover biochar was also found to increase the shoot dry weight of tomato seedlings [68].

3.6. Phytotoxic Compounds in Biochar

There was only a limited amount of research on the phytotoxic effects of biochar. It was reported that various phytotoxic compounds were responsible for the negative effects on germination and seedling growth. Conventional toxic compounds in biochar include toxic metals, tars, PAH, furans, and dioxins [23,24]. Leachable organic compounds in biochar were found to have negative effects on soil, plants, and the environment [27]. In addition, other potential toxic compounds in biochar were free radicals, which were generated during a pyrolysis process. Negative effects on seed germination, such as germination inhibition and delayed shoot growth were observed in biochar with ample free radicals [25]. A previous study carried out by Liao et al. [25] showed that heavy metals and PAHs had less negative impacts on seed germination compared to the free radicals. It was found that mobile organic compounds in biochar resulted in high phytotoxic effects on cress seed germination, and shoot and root length, while PAHs in biochar might not be responsible for the toxicity on seed germination, since the mobility of the PAHs was relatively low [33]. Furthermore, salt stress and water-soluble toxic organic compounds could lead to high toxic effects on seed germination [56]. Therefore, one or a few of the above-mentioned elements could have a negative effect on germination and seedling growth in cress. Further research is needed to identify the phytotoxic substances in biochar from a self-purging pyrolysis reactor in order to ensure a safe and sustainable application in agriculture.

3.7. Effects of Biochar Treatments and Application Rates

A significant negative effect of FB treatments on germination was identified in cress. The negative impact on the germination rate drastically increased with increasing application rates. Therefore, the raw corncob biochar should be tested for its phytotoxicity and treated to reduce the toxicity. As shown in Figure 4, the shoot biomass in FB treatments had the highest DM content (i.e., the lowest MC). This might be explained by the positive effect of nutrients in FB on the biomass production, which led to a high DM content. The high level of water-soluble toxic compounds in FB might also damage the plasma membrane of cress cells, which reduced the water uptake ability of cress seedlings resulting in a low MC in the shoot biomass. It was reported that a direct contact of FB with seeds intensified the phytotoxic effect of biochar [33]. Nevertheless, it was observed that FB showed some positive effects on the dry matter content of the shoot biomass, which might be associated with the additional P and K from biochar. In the present study, high amounts of K and P were identified in the corncob biochar. Moreover, the effective cation exchange capacity of K^+ was also considerably high, which could lead to a better retention capacity of the cation against leaching. FB was hydrophobic and had a lower capacity to absorb water. The hydrophobicity of FB could be reduced by an aging and weathering process [69,70]. A short-term application in this study showed, that FB had a negative effect on germination rate, shoot length, and shoot fresh weight, but the results might be different in a long-term application. In general, the most significant phytotoxic effect on the germination and seedling growth of cress was observed in FB treatments. This could be associated with the low water holding capacity and the high phytotoxicity in FB treatments. The results of FB treatments in this study were in agreement with the findings from a previous study, where the effect of FB suspension was studied [16]. Therefore, it is necessary to perform biochar aging and weathering treatments such as washing and heat treatment to reduce the phytotoxicity.

The WE treatments showed a less negative impact on the germination rate and seedling growth of cress compared to the FB treatments. The germination rate, shoot length, and shoot fresh weight decreased sharply with increasing biochar application rates, while the dry matter content increased. In a previous study, the negative effect of the water extract from macadamia nut shell biochar produced at 430 °C on the germination of radish and corn seeds was identified [59]. The water extract of biochars from varying feedstocks was found to reduce seedling growth in corn [39]. A negative effect of water extract from pine biochar produced at 450 °C was also observed on the growth of aquatic photosynthetic microorganisms [27]. Water-soluble toxic compounds in biochar were found to reduce the germination rate in cress [33]. Rombolà et al. [16] reported that the germination rate of cress in biochar water extract was similar to that of fresh biochar suspensions, which was in agreement with the findings in our study. It was evident that the water-soluble phytotoxic compounds decreased the germination rate and shoot growth in cress. Therefore, after washing biochar the water extract generated from this process must be handled with care and disposed of properly without contaminating the environment. Future research could be carried out to identify the methods to eliminate the toxicity in the water extract and its potential uses.

It was reported that the phytotoxicity in biochar could be reduced by handling, processing, and storage [33]. The initial property of biochar could be changed from hydrophobic to hydrophilic by heat treatment. The reduction of VOCs in biochar can also be achieved by heat treatment. Nevertheless, the drying process might not be able to eliminate all of the water-soluble toxic substances in biochar. Therefore, there might be less phytotoxicity in DB treatments, but some toxic compounds still remain. In comparison, DB treatments should have higher water holding capacity and lower phytotoxicity than FB and WE treatments. The findings of this study indicated that the phytotoxicity in biochar was significantly reduced in the DB treatments at an application rate of 10 t/ha. However, the DB treatments had a slightly higher shoot length and shoot fresh weight than FB and WE treatments. It was reported in a recent study that drying of biochar for 24 h at different temperatures ranging from 100 to 300 °C significantly reduced the phytotoxicity [41]. In our study, a drying temperature of 105 ($\pm$2) °C might not be sufficient to remove all of the volatile matters in the biochar. It was reported that drying biochar at 100 °C for 24 h reduced PAHs by 33.8–88.1% [41].

The WB treatment showed the best result for the germination rate, shoot length, and shoot fresh weight of cress. This implied that washing biochar had the best performance in reducing biochar phytotoxicity. Water-soluble toxic compounds were leached away after washing the biochar with de-ionized water. This result was in agreement with the findings from previous studies stating that the phytotoxicity was significantly reduced by washing biochar with water or an organic solvent [37,39]. The amount of VOCs in biochar was also reduced by repeated leaching [33]. It was evident that washing biochar sufficiently mitigated negative effects of biochar phytotoxicity on plant growth [71]. Rombolà et al. [16] also reported that biochar washed with ethanol showed a better performance than FB and WE on the seed germination of cress. It was found that the volatile matters in biochar did not have a significant toxic effect in a gas exchangeable environment, but they could lead to high toxicity in a closed environment [33,37] like the condition of the mini greenhouses in our study. Therefore, it can be expected that the WB treatments could have a better performance in unsealed environment.

According to the results in this study, the germination rate, shoot length, shoot fresh weight, and dry matter content of the shoot biomass proved to be sensitive indicators of the cress response to phytotoxic substances in biochar. These indicators were significantly influenced by the biochar application rate, especially at the application rate of 30 t/ha or higher. It was also reported in previous studies that the biochar application rate had a significant impact on germination and seedling growth [25,68]. Therefore, it is very important to identify suitable application rates of biochar for various agricultural applications.

In this study, low-cost equipment for germination tests was used to identify the phytotoxic effects of biochar on seed germination. The results indicated that the equipment and settings were suitable to effectively identify the phytotoxic effects of different biochar treatments and application rates. Future research on the phytotoxicity of biochar should focus on the effect of biochar on the root biomass, which will increase the knowledge of biochar-root interactions.

4. Conclusions

Significant phytotoxic effects of biochar on the germination rate, shoot length, shoot fresh weight, and shoot dry matter content of cress were identified. Untreated corncob biochar (FB) and the WE showed severe toxic effects, revealing high soil and environmental risks, while the DB and WB showed less toxicity. The observed order of performance of the biochar treatments for germination, shoot length, and shoot fresh weight at every biochar application rate investigated in this study was WB>DB>WE>FB. When the biochar application rate increased from 10 to 30 t/ha, in the WB, DB, WE, and FB treatments, the germination rate after 48 h (GR_{48}) decreased by 7, 23, 57, and 57%, respectively. With increasing application rates from 10 to 30 t/ha, the shoot fresh weight also decreased by 41, 72, 74, and 75%, in the WB, DB, WE, and FB treatments, respectively. It was concluded that washing (WB) was adequate to significantly reduce the phytotoxicity in biochar. However, the effective mitigation of the phytotoxicity by WB treatments was only confirmed up to a biochar application rate of 20 t/ha.

Simple heat treatment (DB) also significantly reduced the phytotoxicity, but only up to an application rate of 10 t/ha. Therefore, the selection of an economical biochar treatment should depend on biochar properties, application rates, and specific requirements of the application. Based on the findings of this study, it can be concluded that washing and heat treatment are effective in reducing the phytotoxicity of the corncob biochar. Combining washing and heat treatments could further reduce the biochar toxicity.

Author Contributions: Conceptualization, K.I. and S.L.; Methodology, K.I., S.L. and M.S.I.; Software, K.I. and M.S.I.; Validation, K.I., S.L. and M.S.I.; Formal Analysis, K.I. and M.S.I.; Investigation, K.I. and M.S.I.; Resources, K.I., S.L. and J.M.; Data Curation, K.I. and M.S.I.; Writing-Original Draft Preparation, K.I.; Writing-Review and Editing, K.I., S.L. and J.M.; Visualization, K.I. and M.S.I.; Supervision, K.I., S.L. and J.M.; Project Administration, K.I. and S.L.; Funding Acquisition, K.I., S.L. and J.M.

Funding: This research was funded by the Food Security Center of the Universität Hohenheim, the foundation fiat panis, the German Academic Exchange Service (DAAD), the Federal Ministry of Economic Cooperation and Development (BMZ) and the German Federal Ministry of Education and Research (BMBF). This research is the result of the project BiomassWeb WP 5.1 (Project No. 031A258F). The APC was funded by the project BiomassWeb.

Acknowledgments: The authors gratefully thank Jens Hartung, University of Hohenheim, Institute of Crop Science, Biostatistics Group (340c) for his advice on the statistical analysis. The authors would like to thank Sabine Nugent for proofreading and language editing of this manuscript.

Conflicts of Interest: The authors declare no conflict of interest. The funders had no role in the design of the study; in the collection, analyses, or interpretation of data; in the writing of the manuscript, and in the decision to publish the results.

References

1. Liu, X.; Zhang, Y.; Li, Z.; Feng, R.; Zhang, Y. Characterization of corncob-derived biochar and pyrolysis kinetics in comparison with corn stalk and sawdust. *Bioresour. Technol.* **2014**, *170*, 76–82. [CrossRef] [PubMed]
2. Budai, A.; Wang, L.; Gronli, M.; Strand, L.T.; Antal, M.J.; Abiven, S.; Dieguez-Alonso, A.; Anca-Couce, A.; Rasse, D.P. Surface properties and chemical composition of corncob and miscanthus biochars: Effects of production temperature and method. *J. Agric. Food Chem.* **2014**, *62*, 3791–3799. [CrossRef] [PubMed]
3. Gebremedhin, B.; Fernandez-Rivera, S.; Hassena, M.; Mwangi, W.; Ahmed, S. *Maize and Livestock: Their Inter-Linked Roles in Meeting Human Needs in Ethiopia*; ILRI Publications: Nairobi, Kenya, 2007.
4. Hoover, B.K. Herbaceous perennial seed germination and seedling growth in biochar-amended propagation substrates. *HortScience* **2018**, *53*, 236–241. [CrossRef]
5. Dumroese, R.K.; Heiskanen, J.; Englund, K.; Tervahauta, A. Pelleted biochar: Chemical and physical properties show potential use as a substrate in container nurseries. *Biomass Bioenergy* **2011**, *35*, 2018–2027. [CrossRef]
6. Ronsse, F.; van Hecke, S.; Dickinson, D.; Prins, W. Production and characterization of slow pyrolysis biochar: Influence of feedstock type and pyrolysis conditions. *GCB Bioenergy* **2013**, *5*, 104–115. [CrossRef]
7. Scholz, S.M.; Sembres, T.; Roberts, K.; Whitman, T.; Wilson, K.; Lehmann, J. *Biochar Systems for Smallholders in Developing Countries: Leveraging Current Knowledge and Exploring Future Potential for Climate-Smart Agriculture*; The World Bank: Washington, DC, USA, 2014.
8. Chen, T.; Liu, R.; Scott, N.R. Characterization of energy carriers obtained from the pyrolysis of white ash, switchgrass and corn stover - Biochar, syngas and bio-oil. *Fuel Process. Technol.* **2016**, *142*, 124–134. [CrossRef]
9. Meng, J.; Wang, L.; Liu, X.; Wu, J.; Brookes, P.C.; Xu, J. Physicochemical properties of biochar produced from aerobically composted swine manure and its potential use as an environmental amendment. *Bioresour. Technol.* **2013**, *142*, 641–646. [CrossRef] [PubMed]
10. Intani, K.; Latif, S.; Kabir, A.K.M.R.; Müller, J. Effect of self-purging pyrolysis on yield of biochar from maize cobs, husks and leaves. *Bioresour. Technol.* **2016**, *218*, 541–551. [CrossRef]
11. Intani, K.; Latif, S.; Cao, Z.; Müller, J. Characterisation of biochar from maize residues produced in a self-purging pyrolysis reactor. *Bioresour. Technol.* **2018**, *265*, 224–235. [CrossRef]
12. Pare, T.; Gregorich, E.G.; Dinel, H. Effects of stockpiled and composted manures on germination and initial growth of cress (Lepidium sativum). *Biol. Agric. Hortic.* **1997**, *14*, 1–11. [CrossRef]
13. Gehringer, M.M.; Kewada, V.; Coates, N.; Downing, T.G. The use of Lepidium sativum in a plant bioassay system for the detection of microcystin-LR. *Toxicon* **2003**, *41*, 871–876. [CrossRef]

14. Moser, B.R.; Shah, S.N.; Winkler-moser, J.K.; Vaughn, S.F.; Evangelista, R.L. Composition and physical properties of cress (Lepidium sativum L.) and field pennycress (Thlaspi arvense L.) oils. *Ind. Crops Prod.* **2009**, *30*, 199–205. [CrossRef]

15. Diwakar, B.T.; Dutta, P.K.; Lokesh, B.R.; Naidu, K.A. Physicochemical properties of garden cress (lepidium sativum l.) seed oil. *JAOCS, J. Am. Oil Chem. Soc.* **2010**, *87*, 539–548. [CrossRef]

16. Rombolà, A.G.; Marisi, G.; Torri, C.; Fabbri, D.; Buscaroli, A.; Ghidotti, M.; Hornung, A. Relationships between Chemical Characteristics and Phytotoxicity of Biochar from Poultry Litter Pyrolysis. *J. Agric. Food Chem.* **2015**, *63*, 6660–6667. [CrossRef] [PubMed]

17. Bezerra, J.; Turnhout, E.; Vasquez, I.M.; Rittl, T.F.; Arts, B.; Kuyper, T.W. The promises of the Amazonian soil: Shifts in discourses of Terra Preta and biochar. *J. Environ. Policy Plan.* **2016**, *0*, 1–13. [CrossRef]

18. Lehmann, J.; Rillig, M.C.; Thies, J.; Masiello, C.A.; Hockaday, W.C.; Crowley, D. Biochar effects on soil biota—A review. *Soil Biol. Biochem.* **2011**, *43*, 1812–1836. [CrossRef]

19. Lehmann, J.; Joseph, S. *Biochar for Environmental Management: Science, Technology and Implementation*, 2nd ed.; Routledge: Abingdon-on-Thames, UK, 2015; ISBN 9780415704151.

20. Wardle, D.; Nilsson, M.-C.; Zackrisson, O.; Lehmann, J.; Sohi, S. Comment on "Fire-Derived Charcoal Causes Loss of Forest Humus". *Science* **2008**, *321*, 1295. [CrossRef]

21. Sagrilo, E.; Rittl, T.F.; Hoffland, E.; Alves, B.J.R.; Mehl, H.U.; Kuyper, T.W. Rapid decomposition of traditionally produced biochar in an Oxisol under savannah in Northeastern Brazil. *Geoderma Reg.* **2015**, *6*, 1–6. [CrossRef]

22. Aller, M.F. Biochar properties: Transport, fate, and impact. *Crit. Rev. Environ. Sci. Technol.* **2016**, *46*, 1183–1296. [CrossRef]

23. Hale, S.E.; Lehmann, J.; Rutherford, D.; Zimmerman, A.R.; Bachmann, R.T.; Shitumbanuma, V.; O'Toole, A.; Sundqvist, K.L.; Arp, H.P.H.; Cornelissen, G. Quantifying the total and bioavailable polycyclic aromatic hydrocarbons and dioxins in biochars. *Environ. Sci. Technol.* **2012**, *46*, 2830–2838. [CrossRef]

24. Freddo, A.; Cai, C.; Reid, B.J. Environmental contextualisation of potential toxic elements and polycyclic aromatic hydrocarbons in biochar. *Environ. Pollut.* **2012**, *171*, 18–24. [CrossRef] [PubMed]

25. Liao, S.; Pan, B.; Li, H.; Zhang, D.; Xing, B. Detecting free radicals in biochars and determining their ability to inhibit the germination and growth of corn, wheat and rice seedlings. *Environ. Sci. Technol.* **2014**, *48*, 8581–8587. [CrossRef] [PubMed]

26. Oh, T.K.; Shinogi, Y.; Chikushi, J.; Lee, Y.H.; Choi, B. Effect of Aqueous Extract of Biochar on Germination and Seedling Growth of Lettuce (Lactuca sativa L.). *J. Fac. Agric. Kyushu Univ.* **2012**, *57*, 55–60.

27. Smith, C.R.; Buzan, E.M.; Lee, J.W. Potential impact of biochar water-extractable substances on environmental sustainability. *ACS Sustain. Chem. Eng.* **2013**, *1*, 118–126. [CrossRef]

28. Yang, H.; Kudo, S.; Hazeyama, S.; Norinaga, K.; Mašek, O.; Hayashi, J.I. Detailed analysis of residual volatiles in chars from the pyrolysis of biomass and lignite. *Energy Fuels* **2013**, *27*, 3209–3223. [CrossRef]

29. Spokas, K.A.; Novak, J.M.; Stewart, C.E.; Cantrell, K.B.; Uchimiya, M.; DuSaire, M.G.; Ro, K.S. Qualitative analysis of volatile organic compounds on biochar. *Chemosphere* **2011**, *85*, 869–882. [CrossRef] [PubMed]

30. Elad, Y.; Cytryn, E.; Meller Harel, Y.; Lew, B.; Graber, E.R. The biochar effect: Plant resistance to biotic stress. *Phytopathol. Mediterr.* **2011**, *50*, 335–349. [CrossRef]

31. Cordella, M.; Torri, C.; Adamiano, A.; Fabbri, D.; Barontini, F.; Cozzani, V. Bio-oils from biomass slow pyrolysis: A chemical and toxicological screening. *J. Hazard. Mater.* **2012**, *231–232*, 26–35. [CrossRef]

32. Sfetsas, T.; Michailof, C.; Lappas, A.; Li, Q.; Kneale, B. Qualitative and quantitative analysis of pyrolysis oil by gas chromatography with flame ionization detection and comprehensive two-dimensional gas chromatography with time-of-flight mass spectrometry. *J. Chromatogr. A* **2011**, *1218*, 3317–3325. [CrossRef]

33. Buss, W.; Mašek, O. Mobile organic compounds in biochar - A potential source of contamination - Phytotoxic effects on cress seed (Lepidiumsativum) germination. *J. Environ. Manag.* **2014**, *137*, 111–119. [CrossRef]

34. Fornes, F.; Belda, R.M. Acidification with nitric acid improves chemical characteristics and reduces phytotoxicity of alkaline chars. *J. Environ. Manag.* **2017**, *191*, 237–243. [CrossRef] [PubMed]

35. Hoekstra, N.J.; Bosker, T.; Lantinga, E.A. Effects of cattle dung from farms with different feeding strategies on germination and initial root growth of cress (Lepidium sativum L.). *Agric. Ecosyst. Environ.* **2002**, *93*, 189–196. [CrossRef]

36. Buss, W.; Graham, M.C.; Shepherd, J.G.; Mašek, O. Suitability of marginal biomass-derived biochars for soil amendment. *Sci. Total Environ.* **2016**, *547*, 314–322. [CrossRef] [PubMed]

37. Bargmann, I.; Rillig, M.C.C.; Buss, W.; Kruse, A.; Kuecke, M. Hydrochar and biochar effects on germination of spring barley. *J. Agron. Crop Sci.* **2013**, *199*, 360–373. [CrossRef]

38. Busch, D.; Kammann, C.; Grünhage, L.; Müller, C. Simple Biotoxicity Tests for Evaluation of Carbonaceous Soil Additives: Establishment and Reproducibility of Four Test Procedures. *J. Environ. Qual.* **2012**, *41*, 1023. [CrossRef] [PubMed]

39. Rogovska, N.; Laird, D.; Cruse, R.M.; Trabue, S.; Heaton, E. Germination Tests for Assessing Biochar Quality. *J. Environ. Qual.* **2012**, *41*, 1014. [CrossRef]

40. Spokas, K.A.; Cantrell, K.B.; Novak, J.M.; Archer, D.W.; Ippolito, J.A.; Collins, H.P.; Boateng, A.A.; Lima, I.M.; Lamb, M.C.; McAloon, A.J.; et al. Biochar: A Synthesis of Its Agronomic Impact beyond Carbon Sequestration. *J. Environ. Qual.* **2012**, *41*, 973. [CrossRef]

41. Kołtowski, M.; Oleszczuk, P. Toxicity of biochars after polycyclic aromatic hydrocarbons removal by thermal treatment. *Ecol. Eng.* **2015**, *75*, 79–85. [CrossRef]

42. DIN 51718. *Testing of Solid Fuels—Determination of the Water Content and the Moisture of Analysis Sample*; Deutsches Institut für Normung e.V., Ed.; Beuth Verlag GmbH: Berlin, Germany, 2002.

43. DIN 51720. *Testing of Solid Fuels—Determination of Volatile Matter Content*; Deutsches Institut für Normung e.V., Ed.; Beuth Verlag GmbH: Berlin, Germany, 2001.

44. DIN 51719. *Testing of Solid Fuels—Solid Mineral Fuels-Determination of Ash Content*; Deutsches Institut für Normung e.V., Ed.; Beuth Verlag GmbH: Berlin, Germany, 1997.

45. König, N.; Blum, U.; Symossek, F.; Bussian, B.; Ellinghaus, R.; Furtmann, K.; Gärtner, A.; Gutwasser, F.; Hauenstein, M.; Kiesling, G. *Handbuch forstliche Analytik*; Bundesministerium für Ernährung und Landwirtschaft: Bonn, Germany, 2005.

46. van Zwieten, L.; Kimber, S.; Morris, S.; Chan, K.Y.; Downie, A.; Rust, J.; Joseph, S.; Cowie, A. Effects of biochar from slow pyrolysis of papermill waste on agronomic performance and soil fertility. *Plant Soil* **2010**, *327*, 235–246. [CrossRef]

47. Buss, W.; Kammann, C.; Koyro, H.-W. Biochar Reduces Copper Toxicity in Willd. in a Sandy Soil. *J. Environ. Qual.* **2012**, *41*, 1157. [CrossRef]

48. Müller, K.; Tintelnot, S.; Leubner-Metzger, G. Endosperm-limited Brassicaceae Seed Germination: Abscisic Acid Inhibits Embryo-induced Endosperm Weakening of Lepidium sativum (cress) and Endosperm Rupture of Cress and Arabidopsis thaliana. *Plant Cell Physiol.* **2006**, *47*, 864–877. [CrossRef] [PubMed]

49. OECD. *Test No. 208: Terrestrial Plant Test: Seedling Emergence and Seedling Growth Test*; OECD Publishing: Paris, France, 2006.

50. Özçimen, D.; Ersoy-Meriçboyu, A. A study on the carbonization of grapeseed and chestnut shell. *Fuel Process. Technol.* **2008**, *89*, 1041–1046. [CrossRef]

51. Oleszczuk, P.; Rycaj, M.; Lehmann, J.; Cornelissen, G. Influence of activated carbon and biochar on phytotoxicity of air-dried sewage sludges to Lepidium sativum. *Ecotoxicol. Environ. Saf.* **2012**, *80*, 321–326. [CrossRef] [PubMed]

52. Crombie, K.; Mašek, O.; Sohi, S.P.; Brownsort, P.; Cross, A. The effect of pyrolysis conditions on biochar stability as determined by three methods. *GCB Bioenergy* **2013**, *5*, 122–131. [CrossRef]

53. Brewer, C.E.; Unger, R.; Schmidt-Rohr, K.; Brown, R.C. Criteria to Select Biochars for Field Studies based on Biochar Chemical Properties. *Bioenergy Res.* **2011**, *4*, 312–323. [CrossRef]

54. Buss, W.; Graham, M.C.; Shepherd, J.G.; Mašek, O. Risks and benefits of marginal biomass-derived biochars for plant growth. *Sci. Total Environ.* **2016**, *569–570*, 496–506. [CrossRef] [PubMed]

55. Mukome, F.N.D.; Zhang, X.; Silva, L.C.R.; Six, J.; Parikh, S.J. Use of chemical and physical characteristics to investigate trends in biochar feedstocks. *J. Agric. Food Chem.* **2013**, *61*, 2196–2204. [CrossRef]

56. Gell, K.; van Groenigen, J.W.; Cayuela, M.L. Residues of bioenergy production chains as soil amendments: Immediate and temporal phytotoxicity. *J. Hazard. Mater.* **2011**, *186*, 2017–2025. [CrossRef] [PubMed]

57. Zhao, B.; Xu, R.; Ma, F.; Li, Y.; Wang, L. Effects of biochars derived from chicken manure and rape straw on speciation and phytoavailability of Cd to maize in artificially contaminated loess soil. *J. Environ. Manag.* **2016**, *184*, 569–574. [CrossRef]

58. Shen, X.; Huang, D.-Y.; Ren, X.-F.; Zhu, H.-H.; Wang, S.; Xu, C.; He, Y.-B.; Luo, Z.-C.; Zhu, Q.-H. Phytoavailability of Cd and Pb in crop straw biochar-amended soil is related to the heavy metal content of both biochar and soil. *J. Environ. Manag.* **2016**, *168*, 245–251. [CrossRef]

59. Deenik, J.L.; McClellan, T.; Uehara, G.; Antal, M.J.; Campbell, S.; Antal, M.J.; Sonia, C. Charcoal volatile matter content influences plant growth and soil nitrogen transformations. *Soil Sci. Soc. Am. J.* **2010**, *74*, 1259–1270. [CrossRef]

60. Solaiman, Z.M.; Murphy, D.V.; Abbott, L.K. Biochars influence seed germination and early growth of seedlings. *Plant Soil* **2012**, *353*, 273–287. [CrossRef]

61. Al-Wabel, M.I.; Rafique, M.I.; Ahmad, M.; Ahmad, M.; Hussain, A.; Usman, A.R.A. Pyrolytic and hydrothermal carbonization of date palm leaflets: Characteristics and ecotoxicological effects on seed germination of lettuce. *Saudi J. Biol. Sci.* **2018**. [CrossRef]

62. Buss, W.; Mašek, O.; Graham, M.; Wüst, D. Inherent organic compounds in biochar-Their content, composition and potential toxic effects. *J. Environ. Manag.* **2015**, *156*, 150–157. [CrossRef]

63. Shoemaker, C.A.; Carlson, W.H. pH Affects Seed Germination of Eight Bedding Plant Species. *HORTSCIENCE* **1990**, *25*, 762–764.

64. Amin, A.E.-E.A.Z. Impact of Corn Cob Biochar on Potassium Status and Wheat Growth in a Calcareous Sandy Soil. *Commun. Soil Sci. Plant Anal.* **2016**, *47*, 2026–2033. [CrossRef]

65. Bouqbis, L.; Daoud, S.; Koyro, H.W.; Kammann, C.I.; Ainlhout, F.Z.; Harrouni, M.C. Phytotoxic effects of argan shell biochar on salad and barley germination. *Agric. Nat. Resour.* **2017**, *51*, 247–252. [CrossRef]

66. Olszyk, D.M.; Shiroyama, T.; Novak, J.M.; Johnson, M.G. A rapid-test for screening biochar effects on seed germination. *Commun. Soil Sci. Plant Anal.* **2018**, *49*, 2025–2041. [CrossRef]

67. Deenik, J.L.; Cooney, M.J. The potential benefits and limitations of corn cob and sewage sludge biochars in an infertile Oxisol. *Sustainability* **2016**, *8*, 131. [CrossRef]

68. Li, Y.; Shen, F.; Guo, H.; Wang, Z.; Yang, G.; Wang, L.; Zhang, Y.; Zeng, Y.; Deng, S. Phytotoxicity assessment on corn stover biochar, derived from fast pyrolysis, based on seed germination, early growth, and potential plant cell damage. *Environ. Sci. Pollut. Res.* **2015**, *22*, 9534–9543. [CrossRef]

69. Cheng, C.-H.H.; Lehmann, J.; Thies, J.E.; Burton, S.D.; Engelhard, M.H. Oxidation of black carbon by biotic and abiotic processes. *Org. Geochem.* **2006**, *37*, 1477–1488. [CrossRef]

70. Lehmann, J.; Joseph, S.; Thies, J.E.; Rillig, M.C.; Lehmann, J.; Joseph, S. *Biochar for Environmental Management*; Earthscan: London, UK, 2009; Volume 1, ISBN 9781844076581.

71. Di Lonardo, S.; Vaccari, F.P.; Baronti, S.; Capuana, M.; Bacci, L.; Sabatini, F.; Lambardi, M.; Miglietta, F. Biochar successfully replaces activated charcoal for in vitro culture of two white poplar clones reducing ethylene concentration. *Plant Growth Regul.* **2013**, *69*, 43–50. [CrossRef]

 © 2018 by the authors. Licensee MDPI, Basel, Switzerland. This article is an open access article distributed under the terms and conditions of the Creative Commons Attribution (CC BY) license (http://creativecommons.org/licenses/by/4.0/).

 *sustainability* 

Review

Comparing Characteristics of Root, Flour and Starch of Biofortified Yellow-Flesh and White-Flesh Cassava Variants, and Sustainability Considerations: A Review

Oluwatoyin Ayetigbo [1,*], Sajid Latif [1], Adebayo Abass [2] and Joachim Müller [1]

[1] Institute of Agricultural Engineering, Tropics and Subtropics Group, Faculty of Agricultural Science, University of Hohenheim, Garbenstraße 9, 70599 Stuttgart, Germany; s.latif@uni-hohenheim.de (S.L.); joachim.mueller@uni-hohenheim.de (J.M.)

[2] International Institute of Tropical Agriculture (IITA), 25, Light Industrial Area, Mikocheni B, P.O. Box 34441, Dar es Salaam, Tanzania; a.abass@cgiar.org

* Correspondence: ayetigbo_oluwatoyin@uni-hohenheim.de; Tel.: +49-15214626004

Received: 26 June 2018; Accepted: 22 August 2018; Published: 30 August 2018

Abstract: Cassava is a significant food security and industrial crop, contributing as food, feed and industrial biomass in Africa, Asia and South America. Breeding efforts have led to the development of cassava variants having desirable traits such as increased root, flour, and starch yield, reduced toxicity, reduced pest/disease susceptibility and improved nutrient contents. Prominent among those breeding efforts is the development of colored-flesh cassava variants, especially biofortified yellow-fleshed ones, with increased pro-vitamin A carotenoids, compared to the white-flesh variants. The concept of sustainability in adoption of biofortified yellow-flesh cassava and its products cannot be fully grasped without some detailed information on its properties and how these variants compare to those of the white-flesh cassava. Flour and starch are highly profitable food products derived from cassava. Cassava roots can be visually distinguished based on flesh color and other physical properties, just as their flours and starches can be differentiated by their macro- and micro-properties. The few subtle differences that exist between cassava variants are identified and exploited by consumers and industry. Although white-flesh variants are still widely cultivated, value addition offered by biofortified yellow-flesh variants may strengthen acceptance and widespread cultivation among farmers, and, possibly, cultivation of biofortified yellow-flesh variants may outpace that of white-flesh variants in the future. This review compares properties of cassava root, flour, and starch from white-flesh and biofortified yellow-flesh variants. It also states the factors affecting the chemical, functional, and physicochemical properties; relationships between the physicochemical and functional properties; effects of processing on the nutritional properties; and practical considerations for sustaining adoption of the biofortified yellow-flesh cassava.

Keywords: yellow cassava; sustainability; cassava variants; cassava processing; carotenoids retention; amylose

---

1. Introduction

Cassava is a staple food crop for more than half a billion people in the tropical and subtropical regions of the world and mainly used as food, feed and industrial raw material [1–4]. It is the sixth most important commercially cultivated food crop after wheat, maize, potato, rice and barley [2]. Cassava is a highly resilient crop that can withstand stress from drought and dry poor soils while still giving acceptable yields [2]. In addition, cassava is easy to cultivate and its roots can stay reserved in the soil for several months [2] when the farmer's storage space is limited, thereby creating opportunity for extended harvest and sustained availability. It therefore fits description as a *survival* crop that can potentially secure food supply [4] and sustain livelihoods of large populations in difficult times.

In the regions where cassava is cultivated, the flesh color is traditionally white [1]. This is conventionally important to produce "white" starch or high quality flour. However, other colored-flesh cassava variants (yellow, orange, cream, and red) have emerged to challenge this supposition. In this article, the term "variant" generally refers to phenotypes, genotypes, progenies, accessions, varieties, clones, or cultivars—as the case may be in the referred literature—while yellow-flesh cassava and biofortified yellow-flesh cassava all refer to same thing. Flesh color and culinary quality have therefore become vital in the selection of cassava for food [1]. White-flesh (and other dull colored) variants have negligible carotenoid content compared to yellow-flesh variants [2–5]. Variants with deeper color intensity have a higher carotene content [6]. Visual differences also exist between root and flour of white-flesh and yellow-flesh variants, but not for the starch.

Sub-Saharan Africa, Asia, and South America are the largest cassava producers in the world. Yellow-flesh cassava, and other biofortified cassava have considerable potential in alleviating food insecurity in developing countries [4]. Currently, issues remain about widespread acceptance, commercial cultivation, and consumption of the yellow-flesh cassava variants despite their immediate nutritional advantage over white-flesh cassava variants. Some of the challenges have arisen from poor understanding of nutritional benefits of the colored-flesh cassava, misinformation about the nature of development of the variants as genetically modified crops, unwillingness of farmers to change cultivation pattern, and weak governmental commitment to propagation and dissemination for public awareness. These challenges could be stumbling blocks in the sustenance of cultivating yellow-flesh and other colored-flesh cassava. This is particularly unexpected since crops such as maize, potato and cocoyam have recorded relatively huge successes in this regard in the past.

Cassava generally has location-specific properties which have been discussed in the literature. At locations in Southern China, variants that are now commonly cultivated for starch include five white-flesh variants and two yellow-flesh variants (Table 1) [7]. Indonesia also has different variants cultivated for food [8]. Among those are yellow-flesh (Mentega) and white-flesh (Adira 2 and Darul Hidayah) variants (Table 1). A notable difference between these Indonesian variants is that all yellow-flesh variants are sweet, while white-flesh ones are either bitter or sweet. The bitter taste may not only be due to the high content of toxic cyanogenic glucosides (HCN > 100 mg/kg) in the variants, but also due to high levels of antioxidants. Hence, the yellow-flesh variant is preferred for local food preparation in Indonesia [8]. In a bid to sustain production of improved cassava variants in Nigeria, six yellow-flesh variants were released under the IITA-HarvestPlus Project between 2011 and 2014 [9,10]. The first set of released variants had a β-carotene content of 6–8 µg/g on fresh weight basis, while the second set of variants introduced had an average β-carotene content of about 10 µg/g on fresh weight basis. Several clones of yellow-flesh variants have been under investigation to select the most suitable traits for release. A total carotenoid content of almost 25 µg/g has been attained in South American variants [11,12]. Sustained efforts are ongoing to develop variants with up to 15 µg/g of β-carotene [13]. The acceptability of yellow-flesh variants is promising among farmers and consumers, as they are more stable after harvest than their white-flesh counterparts [14]. The increasing marketability of yellow-fleshed varieties, e.g. Narayanakappa in India [1,15] and NR 07/0326 in Nigeria [16], is documented due to the culinary value and the quality of flour and baked goods that they can yield. From a nutritional point of view, yellow-flesh variants have been used to partially meet the recommended vitamin A requirements of cassava-consuming populations. For example, in Kenya [17], where every year 23,500 children die due to a deficiency of micronutrients, and where many school-aged children suffer from sub-clinical vitamin A deficiency [18], yellow-flesh cassava was used to combat these deficiencies. Adding biofortified yellow-flesh cassava to the lunch of the schoolchildren successfully improved nutrient adequacy, but additional nutritional supplement was recommended. In addition, majority of the respondents in the organoleptics aspect of the study preferred the taste, color, and texture of yellow-flesh cassava to the white-flesh one [17].

Table 1. Studies conducted on various white-flesh, and yellow-flesh cassava variants in different locations.

Location	Institution	Flesh Color	Variant(s)	Use	Properties of Interest	Reference
Asia						
India	CTCRI	White	T.P white, T.P brown, I-4, Raman kappa	Food	Good cooking quality	[19]
		Yellow	Zoen kunnu, Chirakkarode, Narayana kappa; I-2, -3, -5, -6; Sree Visakham, NTA, Ambakkadan, Kaliyan, Kandhari padarppan; Narukku I,II,III; Kalikalan		Good cooking quality, moderate or high carotenoids content	
Indonesia	LIPI	White	Adira 2 and 4, Darul Hidayah, Gatot Kaca, Menti, Perelek, Aceh Utara, Manihot, Ketan, Maneot, Gebang	Food, feed, starch, Flour, pellets	Sweet, bitter	[8]
		Yellow	Adira 1, Tim-tim 29, Mangkler, Tim-tim 40, Mentega, Mentega Aceh Besar	food, starch	Sweet	
South China	CATAS	White	SC 5, SC 6, SC 7, SC 8, and SC 205	Starch, biofuel	low/high MC *,high SC *, low/high HCN	[7]
		Yellow	SC 9, SC 10		high DM *, low/high SC *, low/high HCN	
America						
USA	USDA-ARS-PWA	White	Unspecified cultivar (from Las Montanas supermarket)	Food	negligible carotenoids content	[20]
		Yellow	GM905-69 (from CIAT)		High carotenoids content retainable in blood plasma	
Africa						
Nigeria	IITA	White	TME1, 30572, 91/02324	Flour	White clones	[21]
		Yellow	01/1115, 01/1224, 01/1235, 01/1273, 01/1277, 01/1331, 01/1335, 01/1368, 01/1371, 01/1412, 01/1413, 01/1442, 01/1610, 01/1646, 01/1649, 01/1662, 01/1663, 90/01554, 94/0006, 94/0330, 95/0379, 98/2132, 00/0028, 00/0093, 01/1172, 01/1181, 01/1206, 01/1231, 01/1296, 01/1380, 01/1404, 01/1417, 01/1423, 01/1551, 01/1560, 01/1635, 01/1659, 99/2987, 99/7578, Z97/0474		Clones of varying β-carotene content	
Nigeria	NRCRI	White	AR-37-108, TMS 30572, CR-36-5, 02/0007, CR-12-45, NR03/0155, NR02/0018, AR 1-82, 03/0174	Gari, flour	High starch, dry matter	[22]
		Yellow	UMUCASS 36, 37, 38, and 98/2132		High moisture, amylose carotenoids,	
Nigeria	NRCRI	White, Yellow	UMUCASS 36, 37, 38, and TMS 30572	Fufu, flour, mash	Fermented mash and dough formation and acceptability	[23]
Nigeria	NRCRI	White	TMS 30572	Flour, starch, bread, cake, chinchin, strips, salad cream	High starch content	[16]
		Yellow	NR 07/0427, 07/0432, 07/0326, 07/0506, 07/0497, 07/0499		High starch and carotenoids	

Note: * Location and/or genotype dependent; CATAS, Chinese Academy of Tropical Agricultural Sciences; LIPI, Indonesian Institute of Sciences; IITA, International Institute of Tropical Agriculture; CIAT, International Center for Tropical Agriculture; CTCRI, Central Tuber Crops Research Institute; USDA-ARS-PWA, Western Human Nutrition Research Center; NRCRI, National Root Crops Research Institute; DM, dry matter; MC, water content; SC, starch content, dry basis; HCN, hydrocyanide conten.

Cassava flour is a valuable product obtained from cassava roots after processing. Generally, to produce the flour, cassava root is peeled, washed, chipped, milled, pressed to expel most of the toxic liquor, dried, fine-milled, and sieved. It is relatively cheap to produce traditionally. Industry-grade high quality cassava flour, however, requires improved processing inputs, which may add to the costs. Due to its special properties such as its clear appearance, low off-flavor tendency and ideal viscosity, it is regarded as a vital ingredient in the food industry [21].

Cassava starch is a highly suitable material for food and industrial use. It is edible, non-toxic, and functionally important in the food and non-food sectors of industry. Briefly, cassava starch is produced in sequence by peeling, chipping, wet milling, sieving, sedimenting, decanting, drying and pulverizing. Cassava has a high proportion (65–80%) of starch [24,25], which is low in contaminants compared to other botanical starches [26,27]. Cassava starch embodies positive characteristics such as high paste clarity, relatively good stability to retrogradation, low protein clog/complex, swelling capacity, and good texture [27], which makes it suitable for use in many foods. For instance, better quality bread has been made when cassava starch is included in baking [28]. Therefore, it is highly desirable to select certain variants of cassava as industrial starch source, depending on their inherent characteristics [7].

It is worth mentioning that, while cassava flour consists mostly of starch, the presence of relatively higher fiber, protein, minerals and vitamins contents in the flour compared to the starch could confer certain differences in their properties.

Several review articles have been written on the properties of cassava root starches (in native and modified forms) especially from white-flesh variants [26,29,30]. In comparison, there has been little focus on the properties of root, flour, and starch of yellow-flesh cassava variants. It is therefore necessary to compile more information on yellow-flesh (and other colored-flesh) variants, since there is an increasing interest in the value-added, biofortified cassava variants and products derived from them. Due to issues with relatively restricted acceptance, cultivation and consumption of yellow-flesh cassava, it is pertinent to study if they are comparable to, or offer any nutritional, functional, or physicochemical advantages over the white-flesh cassava variants in a bid to argue for continued sustenance of efforts to improve its level of acceptance and adoption by the agro-allied industry and the public at large. A positive outcome could encourage public-private investments in breeding and adoption programs in Sub-Saharan Africa. Therefore, the comparison of properties of cassava root, flour, and starch is discussed from a number of literature sources, based on differing flesh colors. For instance, nutritional and chemical properties such as protein, carotenoids, minerals, starch and amylose contents are discussed. Physicochemical and functional properties such as water and oil absorption capacity, swelling power, solubility, and pasting characteristics are also discussed. Factors that influence the differences in properties and effects of processing on the variants are considered. In addition, some peculiar challenges and possible solutions in adoption of the biofortified yellow cassava in Africa are discussed with regard to issues of sustainability. Recommendations based on the review findings are made.

2. Methodology

This work uses an extensive review of several scientific, technical and economic literature on analysis of variants of white-flesh and biofortified yellow-flesh cassava focused on the properties of the cassava variants and how they compare to one another. Most of the studies considered were very recent (2000–2018), with very few older studies published decades ago (1959–1999). Authors sourced information from original research articles, review articles, proceedings of conferences, books, working papers and own research. No personal communication sources were consulted, and very few online sites were consulted. Data reported were based on direct values published in the literature and deductions from illustrations. While literature is fairly available on sustainability and the adoption of many improved white-flesh cassava variants with regards to disease resistance, yield, and other agronomic attributes, very few studies have reported the subject of sustainability and adoption of biofortified yellow-flesh cassava with focus on post-harvest quality. Even fewer have focused on detailed comparison of the several technical properties of white-flesh and yellow-flesh cassava variants in locations around the world with a view to arguing for its sustained cultivation, use and adoption in Sub-Saharan Africa.

3. Chemical Properties

The literature classifies cassava root as a high calorie food [4,31] with a high percentage of carbohydrates (80–90% dry basis) consisting almost entirely of starch [32]. Nonetheless, cassava root is relatively poor in other nutrients such as proteins, lipids, and vitamins [17]. Cassava roots generally have a high moisture content, which can differ with variants. A reliable comparison of the moisture contents between variants is possible with dry basis measurements. Most literature reports higher values of moisture contents in yellow-flesh variants. Moisture content of the respective root, flour and starch of yellow-flesh cassava variants is reported to be higher [16], marginally higher [21] or significantly higher [33] than for those of white-fleshed variants (Table 2). Likewise, Ukenye et al. [22] reported a higher dry matter content for white-flesh variants compared to yellow-flesh variants, and dry matter of starch extracted from white-flesh variants (89.04–96.41%) was significantly higher than that from yellow-flesh variants (88.47%) [33]. An association between the concentration of carotenoids and the moisture content of cassava and its products may be possible considering that, in numerous publications, the average moisture content of cassava roots, flour and starch usually ranks in the color sequence: yellow = orange > cream > white [16,19–22,33]. Despite the relatively higher amount of moisture in yellow-flesh cassava roots, they tend to store better after harvest than their white-flesh counterparts, perhaps due to the additional anti-oxidative effect of carotenoids present [6]. This beneficial property could be appreciated by farmers, thus encouraging sustainable post-harvest practices and processing. A safe moisture limit for starch storage on the international market is 13% [34–36].

The poor nutrients density of cassava [37–39] is exemplified in its low protein content. Compared to other roots and tubers, cassava roots have a low protein content of about 1–3% on dry basis [37], and are particularly poor in sulfur-rich amino acids. Acidic and basic amino acids such as glutamic, aspartic and arginine are, however, relatively plentiful in cassava roots [32]. Because of this, cassava diet from the roots has to be supplemented by other protein sources. Nonetheless, Montagnac et al. [4] reported a number of attempts to improve protein in cassava by biofortification and post-harvest processing, with some recorded successes. Protein content (Table 2) was significantly higher in the flour of white-flesh cassava variants than in the flour of yellow-flesh variants [21]. In cassava roots reported by Ukenye et al. [22], the protein content was not significantly different between variants. Protein may be lost during processing of cassava root into flour and starch. Hence, protein content in cassava could be variant-dependent.

The fiber content of cassava roots depends on the variant and age at harvest [4]. A relatively higher fiber content of 0.62–4.92% has been found in roots of white-flesh cassava variants than in yellow-flesh and cream-flesh variants [22]. Likewise, crude fiber in cassava flour is generally higher in white-flesh variants than yellow-flesh ones [16]. The difference in fiber content contributes to the higher dry matter in white-flesh variants than yellow-flesh variants. The isolation procedure undergone during starch extrication rids cassava starch of most fiber. Residual fiber influences texture and in vitro digestibility of cassava starch and flour. Cassava starch is characteristically low in fiber (0.10–0.15%) and lipids (0.11–0.22%) [38,39]. About half the lipids in cassava roots are non-polar, or in glycolipid forms (especially galactose diglycerides), but fatty acids such as oleates and palmitates are more commonly found [32,40]. Lipids may be involved in the retention of the lipid-soluble carotenoids in yellow-flesh cassava.

Table 2. Proximate composition of cassava root, flour and starch from white-flesh and yellow-flesh variants.

Location	Flesh Color	Variant(s)	MC (%)	Protein (%)	Lipids (%)	Fiber	Ash (%)	NFE	Reference
Root									
South China	White	5 [a]	57–72	-	-	-	-	-	[7]
	Yellow	2 [a]	56.4–68.5	-	-	-	-	-	
USA	White	2 [b]	-	-	-	-	-	-	[20]
	Yellow/cream-orange	2 [b]	59.1 [α]	-	-	-	-	-	
Nigeria	White	9 [c]	57.46–75.91	2.07–7.92	0.02–3.66	0.62–4.92	0.33–1.1	89.09–95.69	[22]
	Yellow	4 [c]	66.19–73.49	2.75–8.10	0.29–3.2	1.07–2.14	0.55–1.04	89.50–91.52	
Nigeria	White	1 [d]	71.27	-	-	-	-	-	[16]
	Yellow	6 [d]	72.05–75.26	-	-	-	-	-	
India	White	4 [e]	-	2.0–2.9	-	-	-	-	[19]
	Yellow, orange	43 [e]	58.5–81.7	1.5–3.1	-	-	-	-	
Flour									
Nigeria	White	3 [f]	10.90–11.85	0.81–1.26	-	-	0.74–1.43	-	[21]
	Yellow	40 [f]	10.78–12.72	0.56–0.96	-	-	0.77–1.43	-	
Nigeria	White	1 [d]	-	-	-	1.63	1.15	-	[16]
	Yellow	6 [d]	-	-	-	0.62–1.74	0.93–1.85	-	
Starch									
South China	White	5 [a]	-	-	-	-	-	-	[7]
	Yellow	2 [a]	-	-	-	-	-	-	
Indonesia	White	15 [g]	-	-	-	-	-	-	[8]
	Yellow	2 [g]							
Nigeria	White	39 [h]	3.59–10.96	0.23–0.7	-	-	0.03–0.49	-	[33]
	Yellow	1 [h]	11.53	0.23	-	-	0.16	-	

Note: MC, moisture content (wet basis); NFE, Nitrogen-free extract or Total carbohydrate; [α] from separate lots after transport, storage, freezing, and thawing; [a], eight locations, three replicates, completely randomized; [b], two genotypes, two lots, one processing method, randomized tests; [c], 16 cultivars, proximate analyses on dry matter basis; [d], seven varieties, seven processed products; [e], 43 yellow-orange flesh local clones and accessions, 8 cream-flesh clones, 4 white-flesh local clones, 10th month harvest; [f], two sets, four replicates, completely randomized, three drying methods; [g], 11 months at harvest, selected from 71 local and improved genotypes; [h], 36 CMD-resistant and 4 non CMD-resistant varieties.

Cassava is not particularly rich in all mineral nutrients; hence, diets based on cassava alone may not fulfill adequate mineral nutritional requirement in humans. Chavez et al. [6] analyzed 20 variants of cassava collected from several core clones and quantified the major minerals present. Average content of zinc, iron, calcium, magnesium, sodium, potassium and sulfur (dry basis) were 6.4 mg/kg, 9.6 mg/kg, 590 mg/kg, 1153 mg/kg, 66.4 mg/kg, 8903 mg/kg, and 273 mg/kg, respectively. Other minerals were only found in negligible quantities. Total phosphorus content in cassava is as low as 70–120 mg/kg of root [41]. An average phosphorus content of 1284 mg/kg has been reported for cassava roots [6], and the phosphate content did not vary by flesh color. Phosphate in starches of seven yellow-flesh and white-flesh Indonesian variants were similar, and had negligible amounts (23.5–25.3 nmol/mg) of phosphorus [8], attached mostly at C-3 and C-6 positions on anhydro-glucose units. In other tuber crops, especially potato, the phosphate content is relatively high, and influences a number of physicochemical and functional properties [26]. However, the use of cassava in industrial food and non-food use is almost unrivalled. Due to its comparably superior solubility and paste clarity, low gelatinization temperature and low retrogradation tendency [26], the industrial use of cassava in food applications, textile processing and in confectioneries endears it to the industry. In addition, cassava flour and starch are more popular, cheaper, and more available in commercial quantities than those of potato across regions of Sub-Saharan Africa.

Other micronutrients of more recent significance and interest in cassava are pro-vitamin A carotenoids and vitamin C. Vitamin C, which is important for mineral absorption in the gut, is found in relatively higher amounts than carotenoids in fresh cassava. For instance, among more than 500 cassava root lines evaluated by Chavez et al. [6], vitamin C averaged 9.5 mg/100 g in fresh cassava although it is much more susceptible to losses during processing than carotenes. Minimal processing of cassava by methods susceptible to oxidation is recommended, if some vitamin C is to be retained. However, much of the processing techniques used for converting cassava to edible, safe food cannot guarantee its retention. Carotenoids are important for healthy body metabolism and disease prevention. Of the carotenoids found in yellow-flesh cassava roots, β-carotenes are present in higher concentrations than other carotenoids involved in the biosynthesis of vitamin A [42]. They are important micronutrients in global health issues, particularly in developing countries [43–45]. Many studies have measured carotenoids in cassava (Table 3) with significant differences existing between the variants. The Indian yellow-flesh variant Narayanakappa had lower total carotenoids and β-carotene (3.1 µg/g, and 2.3 µg/g, respectively) contents than three other orange-flesh variants [1]. The higher color intensity of orange-fleshed variants could indicate higher concentrations in carotenoids than the yellow-flesh cassava. The concentrations differed by variant when some processing techniques were employed [15]. Total carotenoids of roots of three yellow-flesh cassava variants from Nigeria ranged between 2.6–7.3 µg/g (average 4.9 µg/g), and varied significantly depending on the variant [46]. These values are close to those reported (6.26–7.76 µg/g) for other variants of yellow-flesh cassava by Omodamiro et al. [23], but considerably higher than that of white-flesh variants (0.35 µg/g). Again, root and flour from six yellow-flesh cassava variants were about three to six folds richer in carotenoids than from white-flesh variants [16]. Thus, the value addition offered by yellow-flesh cassava over white-flesh cassava as a vector for biofortification cannot be overemphasized. Ceballos et al. [47] determined total carotenoid content of six variants of yellow-flesh cassava from Colombia as 8.32–16.40 µg/g (wb), which consists of 70% all-trans β-carotene and 5% each of the isomers 9-cis and 13-cis β-carotene. The proximal parts of the roots had a higher concentration of total carotenoids, β-carotenes, and higher dry matter than central and distal parts, signifying inhomogeneous distribution of carotenoids in cassava, and could be informative in preferred choice of parts for consumption. Visual inspection of cassava root color intensity could be used as a casual, non-empirical indicator for carotenoid concentration as conducted by Ukenye et al. [22], where color intensity and carotenoid content ranked: yellow-flesh > cream-flesh > white-flesh.

Table 3. Chemical composition of cassava root, flour and starch from white-flesh and yellow-flesh variants.

Location	Flesh Color	Variant(s)	Starch Content (dw) (%)	Sugars (%)	Amylose (%)	Carotenoids µg/g (fw)	β-Carotene µg/g (fw)	HCN (mg/kg)	Reference
Root									
South China	White	5 [a]	63–83.2		-			47.4–304.7	[7]
	Yellow	2 [a]	55.1–76.5		-	-	-	68–265	
USA	White	2 [d]	-	-	-	-	0.99	5.27	[20]
	Yellow/Cream-orange	2 [d]	-	-	-	-	8, 21.1 [α]	280, 5.51 [α]	
Nigeria	White	9 [e]	13.47–30.97 *	-	9.79–22.88 [µ]	0.51–2.29	-	-	[22]
	Yellow	4 [e]	14.29–20.00 *	-	13.31–24.48 [µ]	3.31–4.79	-	-	
Nigeria	White	1 [g]	29.13 *	-	-	~0.53	-	-	[16]
	Yellow	6 [g]	11.69–29.16 *	-	-	~1.2–3.88	-	-	
India	White	4 [h]	-	-	-	-	-	-	[18]
	Yellow, orange	43 [h]	-	-	-	2.0–13.6	0.7–11.1	-	
Flour									
Nigeria	White	3 [c]	70.48–82.42	2.25–3.79	18.18–20.29 [µ]	-	-	-	[19]
	Yellow	40 [c]	67.08–81.18	1.71–5.66	15.71–22.25 [µ]	-	-	-	
Nigeria	White	1 [g]	61.05 *	-	-	~0.15	-	-	[16]
	Yellow	6 [g]	23.18–53.56 *	-	-	~0.6–0.88	-	-	
Starch									
South China	White	5 [a]	-	-	13.5–22.5	-	-	-	[7]
	Yellow	2 [a]	-	-	15–25	-	-	-	
Indonesia	White	15 [b]			16.8–21.3 [Ω]	-	-	-	[8]
	Yellow	2 [b]			17.4–20.2 [Ω]	-	-	-	
Nigeria	White	39 [f]	60.34–86.79	0.51–3.56	15.24–30.20 [µ]	-	-	-	[33]
	Yellow	1 [f]	69.90	0.97	28.61 [µ]	-	-	-	

Note: Fw, fresh weight; dw, dry weight; *, starch by wet sedimentation; [α], from separate lots after transport, storage, freezing, and thawing; [Ω], amylose by enzymatic method; [µ], amylose by iodine-binding spectrophotometry; [a], eight locations, three replicates, completely randomized; [b], 11 months at harvest, selected from 71 local and improved genotypes; [c], two sets, four replicates, completely randomized, three drying methods; [d], two genotypes, two lots, one processing method, randomized tests; [e], 16 cultivars, proximate analyses on dry matter basis; [f], 36 CMD-resistant and 4 non CMD-resistant varieties; [g], seven varieties, seven processed products; [h], 43 yellow-orange flesh local clones and accessions, 8 cream-flesh clones, 4 white-flesh local clones, 10th month harvest.

On a dry basis, total carotenoids and β carotene contents quantified for three cassava roots [48] were 9.06–21.95 μg/g and 7.16–13.50 μg/g, respectively, with HPLC chromatograms revealing the presence of isomeric forms of carotene: 13-cis-β- carotene and 9-cis-β-carotene in relatively lower quantities. A similar study was conducted by Oliveira et al. [49] for 12 bitter yellow-flesh cassava roots, where total carotenoids and β-carotene (isomeric forms: all-E-β carotene, 13-Z-β carotene and 9-Z-β carotene) ranged between 1.97–16.33 μg/g and 1.37–7.66 μg/g, respectively, while the total carotenoid content for flour of five other bitter yellow-flesh variants was 3.65–18.92 μg/g. Thus, carotenoids in cassava can be unstable during processing, and converted to other more stable forms. Two genes have been thought to be implicated in defining concentration of carotenes found in several cassava variants, one coding for transportation of products of precursors, and the other for accumulation [6,50]. These genes could be manipulated for breeding purposes in developing cassava of increasingly higher carotenoids content. Carotenoid concentration in cassava can be considered a qualitative trait, determined by a few genes, and not readily repressed or promoted due to effects of the environment [51]. The study [51], which describes the individual and interactive effects of genotype (G), year of cultivation (Y), and location (L) on the fresh yield and total carotenoid quality of 24 yellow-flesh and 3 white-flesh variants in Nigeria, showed that G had the strongest impact on total carotenoids, while the location was the decisive factor for high fresh weight yields. Each factor had a significant effect on both qualities, but Y and interaction effect G-L-Y were only partially significant. Interaction effects Y-L and G-L were strongly significant, but G-Y was not significant. Interacting principal component analysis (IPCA) was helpful in selecting the genotypes and locations with best strengths and minimal compromise for the targeted qualities.

Amylose and amylopectin play crucial compositional and functional roles in cassava starch, influencing properties such as crystallinity, gelatinization, retrogradation, gelling, and pasting. Cassava starch with low amylose has higher crystallinity corresponding to a reduced amorphous band [52], and high-amylose starch retrogrades relatively easily [53]. Amylopectin molecular structure, as determined by degree of branching, molecular weight, and chain length can be influenced by the activity of starch branching enzymes of different polymorphic forms [54]. When determined from the roots directly, amylose content of yellow-flesh cultivars was similar to that of white-flesh variants [22]. A contrasting observation was made for other roots/tubers, such as sweet potatoes [55] and yam bean (*Pachyrhizus tuberosus*) tuber, where amylose content in yellow-flesh variants was significantly higher than in white-flesh variants [56]. Significant difference ($p < 0.01$) was found between amylose of cassava flour of 40 yellow-flesh (15.71–22.25%) variants and three white-flesh (18.18–20.29%) variants from Nigeria [21], when studied for effect of drying method and variety. The difference in amylose may have partly contributed to the higher paste peak viscosity of flours from most yellow-flesh variants compared to white-fleshed ones in the study. Amylopectin plays a more significant role, however, in the pasting properties of cassava [57]. Amylose content of cassava starch varies widely between 14–24% [41], but starches of waxy (amylopectin-rich) variants may contain as low as 0–3.4% amylose [58]. Amylose of starches of seven variants from Southern China ranged from 13.5–24.65%, varying significantly with location, environment, and variant [7]. Among these variants, the yellow-flesh ones had similar amylose content to white-flesh ones (Table 3). A similar trend was found in other works [33,59,60], where similar methods were used for amylose determination. Again, amylose in starches of 17 Indonesian variants was similar regardless of flesh color, ranging 17.1–21.3% [8]. No difference was found between the amylose content in starches of yellow-flesh variants (17.4–20.2%) and white-flesh variants (16.8–21.3%). Any variations in values of amylose content reported could be due to different methods of determination employed.

Starch content of cassava can be determined chemically or enzymatically, but starch yield is the amount of starch physically recoverable from cassava root. Total starch content (Table 3) of flour from 40 yellow-flesh cassava variants (67.08–81.18%) was similar to that of flour of three white-flesh variants (70.48–82.42%) from Nigeria, when studied for effect of drying method and variety [21]. Starch yield from four yellow-flesh cassava variants was lower than starch yield from white and cream-flesh variants, with total carbohydrates following the same trend [22]. Again, starch extracted from six roots and flours of recently released yellow-flesh variants was lower on the average in comparison to that from white-flesh variants [16]. These reports indicate that the starch content of white-flesh cassava variants is significantly higher than that of yellow-flesh variants, although genotypic differences [21] and age can also cause differences in starch contents between these variants. This might reveal that there is possibly a reduced activity of granule bound starch synthase (GBBS), soluble starch synthase (SSS) and starch branching enzymes (SBE) for starch synthesis [61] in yellow-flesh variants.

Sugars such as sucrose, glucose, fructose and maltose have been quantified in cassava roots [62]. Overall, about 17% sucrose has been found in sweet variants of cassava roots [31]. Anggraini et al. [8] found all yellow-flesh variants studied were sweet, while white-flesh variants were either bitter or sweet depending on the variant. The amount of compounds responsible for the bitterness, such as tannins and cyanogens, was significantly lower in yellow-flesh roots than in white-flesh roots [7] and may influence taste. In contrast, similar or even higher amounts of sugars have been found in some white-flesh variants compared to yellow-flesh variants [33]. In fact, Oliveira et al. [49] has reported 17 known bitter yellow-flesh cassava variants cultivated in Brazil. Total sugars were found to be significantly higher among flours of yellow-flesh cassava variants (Table 3) than flours of white-flesh cassava variants in the study by Maziya-Dixon et al. [21]. Hence, this might be the plausible reason for mild to sweet taste among most yellow-flesh variants. These findings imply that the differences in variants and cultivation conditions could be strong factors influencing sugar composition in cassava.

The ash content is an important component of cassava, and is an indication of the mineral richness and non-volatiles content of cassava [4]. Total non-incinerable matter (ash) of cassava was reported to be similar between both white-flesh and yellow-flesh variants [22]. Ash content reported for cassava flour and starch made from white cassava were within similar range as for yellow-flesh variants [16,21,33]. Among these cassava materials, starch had the lowest ash content. Cassava leaves, however, have a significantly higher ash content than the roots [4]. Processing of cassava has been reported to significantly reduce ash content of the roots, with a similar trend for minerals [4]. Hence, severe processing techniques such as those involving application of high temperature and chemicals and excessive fermentation, washing and milling treatments could significantly reduce ash in cassava-dominated diets.

4. Physicochemical and Functional Properties

Cassava roots have different physiological and functional parts in various colors (Figure 1). Peel colors of light, and dark brown; cortex colors of pink, yellow, purple, light brown, cream, and white; flesh colors of yellow, white, and red; and root shapes of conical, cylindrical and irregular have been reported [7,8,63,64]. Water content, starch content, dry matter, cyanogen content, and tannins of roots vary with genotype, location and environmental conditions [7], and can influence physicochemical and functional properties of cassava. Due to relatively lower dry matter of cassava variants with higher β-carotene contents such as the yellow-flesh cassava [65], improving the dry matter content in yellow-flesh cassava roots is important to cassava breeders in reducing processing losses of pro-vitamin A [47]. Reducing processing losses could enhance the sustainable adoption of yellow cassava by farmers and food processors. Reducing cassava processing losses primarily depends on root yield. Generally, root yield is considered as a polygenic trait [66,67], resulting from a number of genes expressed based on the environment in which the roots are cultivated.

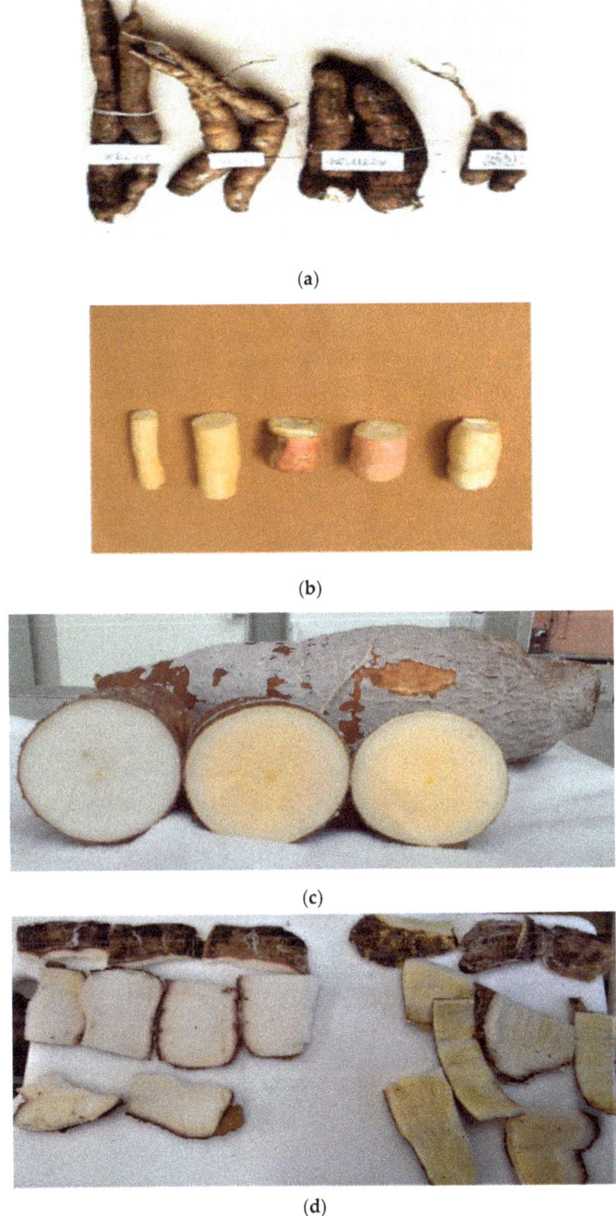

Figure 1. Visual differences in: (**a**) cassava root size and shape; (**b**) cassava cortex color; (**c**) flesh color; and (**d**) peel color. Source: Anggraini et al. [8]. Permissions for (**a**,**b**) granted by Prof (Dr) Richard Visser; (**c**,**d**) from own source.

Physicochemical and functional properties of cassava include morphology of its starch, water and oil absorption, surface color, density, swelling power and solubility, gel freeze–thaw stability, paste clarity, least gelation concentration, and pasting.

4.1. Morphology and Crystallinity of Starch

Morphology of cassava starch plays an important role in its functionality. Morphological characteristics of cassava starch are shown in Table 4. Light and compound microscopy, as well as electron microscopy, has been used to study morphological properties of isolated cassava starch [68]. Authors have also investigated morphology of starch in cassava flour (Figure 2). Starch granules of seven Southern China variants grown in eight different locations had essentially similar shapes and sizes among the white and yellow-flesh variants, such as round, oval, and truncated shapes as well as a wide range of dimension (5–40 µm) [7]. All the granules were of unimodal distribution, with sizes from 10 to 15 µm more frequently occurring than others. Granules with sizes above 30 µm were the fewest. Hence, small-l and medium-sized granules form the bulk class of granule types found in cassava starches. In the work of Anggraini et al. [8], the average starch granule size of a yellow-flesh variant was similar to that of white-flesh variants. In other studies on starches, bimodal distribution was reported [33,69] for 39 white-flesh cassava variants and 40 variants of white and yellow-flesh cassava, consisting entirely of medium sized granules with sizes ranging 12.5–22.50 µm and 12.5–24.17 µm, respectively. Tri-modal distribution has also been reported for nine Ugandan white-flesh cassava starches [69], with medium-sized granules occurring more frequently. Therefore, whenever starch granule size is of paramount interest in industrial applications (such as in paper, textile, and pharmaceutical industries) both types of variants could be used. X-ray diffraction studies of cassava starches revealed A-type crystallinity of about 35% [70,71]. Crystalline pattern of cassava starch has been reviewed by Zhu [30]. Cassava starches with higher crystallinity tend to have a higher peak viscosity, pasting temperature, and gelatinization temperature [71], as they may require more energy to melt starch crystallites of their granules, a factor mainly determined by the nature of amylopectin present [52,53]. Since the literature is sparse on crystallinity of cassava starches from yellow-flesh variants, this is an area requiring more research to compare with starches of white-flesh variants.

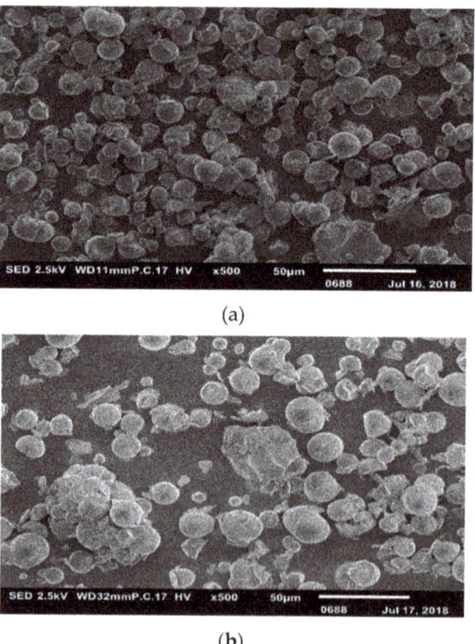

Figure 2. Remission scanning electron micrographs of starch granules showing shape and size of granules, and cell wall fragments in: (**a**) white cassava flour; and (**b**) yellow cassava flour. Source: authors' own source.

Table 4. Morphology and crystallinity of starch of white-flesh and yellow-flesh cassava variants.

Location	Flesh Color	Variant(s)	Shapes	MGS (μm)	GSI (μm)	GMS (μm)	MS (μm)	S, M, L (%)	Reference
South China	White	5 [a]	Round, oval, truncated	-	5–40	10–15	-	33,63,3 *	[7]
	Yellow	2 [a]	Round, oval, truncated	-	5–40	10–15	-	33,63,3 *	
Indonesia	White	6 [b]	-	7.3–9.5	-	ca. 3.2–4.1	7.8–10.8	-	[8]
	Yellow	1 [b]	-	8.0	-	3.5	8.5	-	
Nigeria	White,	40 [c]	Kettledrum, round with indentation on one side	-	12.50–24.17	15.83, 16.67 *	-	0,100,0 *	[33]
	Yellow	1 [c]	Round	20	-	-	-	-	

Note: MGS, mean granule size; GSI, granule size interval; GMS, granule modal size(s); MS, median size; S,M,L (%), percentage of small, medium, and large granules (separated respectively by comma) classified according to Singh, Mc Carthy and Singh [72] and Lindeboom, Chang and Tylera [73]; *, estimated from graph/table; [a], cultivars in eight locations, three replicates, completely randomized; [b], 11 months at harvest, selected from 71 local and improved genotypes; [c], 36 CMD-resistant and 4 non CMD-resistant varieties.

4.2. Surface Color

Color is an important organoleptic quality attribute, which is considered in the acceptance of crop products [74]. Generally, yellow-flesh cassava looks more attractive than white-flesh cassava. Apart from β-carotene, other carotenoids, such as lutein, contribute to the overall color of colored-flesh cassava, as seen in red-flesh Brazilian variants [75]. Cassava flour made from cassava root retains the color of the root flesh. A reason could be that carotenoids are color-active compounds that are lipid-soluble and the color retention in flour could be due to the complex with mucilage and latex [26] as well as starch–lipid, fiber–lipid, and protein–lipid interactions. Hence, cassava flour retains a more intense color, while cassava starch made from cassava root is relatively purer in state, and has less intense color. Few studies report the color of cassava root, flour, or starch and how they relate to other properties. The Commission Internationale de l'Eclairage (CIE) L*, a*, b* color system was employed to measure tri-stimulus surface color of starch from several white-flesh and yellow-flesh variants of cassava. The L*, a* and b* values are measure of extent of lightness, redness-greenness and yellowness-blueness of the surface color of a material [69]. L* values between 83.97 and 93.17 were recorded for several white-flesh and yellow-flesh cassava starches [33]. These values were close to those reported by Eke et al. [69] for similar variants, with average L* values of 87.66–93.73 for two consecutive seasons. Ladeira et al. [76] reported L* values of 72.43–81.19, a* values of −0.9 to −1.63, and b* values of 14.08–16.29 for three cream-fleshed Brazilian cassava roots. Similarly, L* values of 83.65–87.22, a* values of −0.24 to 0.33, and b* values of 3.57–5.02 were reported for starches of these three cassava roots. These values are quite different from L* values of 93.93–97.92, a* values of 0.12–0.23, and b* values of 0.78–1.30 for cassava starch, and L* values of 93.85–95.80, a* values of −0.32 to −0.97, and b* values of 6.07–10.22 for cassava flour of five Ghanaian white-flesh variants [77]. Thus, starches from cassava have higher L* values and lower b* values compared to the corresponding roots and flour, regardless of the flesh color of the variants they have been isolated from. The assumption that starches from colored-flesh cassava variants are probably colored could not be supported because the starches are much lighter and have much lower extent of yellowness than the roots and flours. Moorthy [26] reported the color of isolated cassava starch as white, unlike from some other tuber crops. For instance, yellow color in yellow-flesh yam starch has been reported [78]. A more complete color grading from measurements of other color parameters (a*, b*, H*, ΔC, E and %W) is not reported in most works. The relationship between color and physicochemical properties is not explored in detail. Onitilo et al. [33] found that a strong significant relationship exists between color and swelling power, solubility, and water absorption capacity of cassava starches, but no reason was given for this. An explanation could be some of the subtle physicochemical characteristics that differentiate white-flesh and yellow-flesh variants from one another as well as genetics and cultivation conditions (location, environment, climate/season, weeding, fertilizing, precipitation, etc.).

4.3. Density and Flow Properties

Loose bulk density (LBD) of high-quality cassava flour from six yellow-flesh variants was similar to that from white-fleshed ones, but packed bulk density (PBD) was higher on the average for flour of white-flesh variants (1.42 g/mL) [16]. Packed bulk density (0.695–0.703 g/mL) did not differ significantly between cassava starch extracted from three yellow-flesh variants [60], but was lower than that reported by Agunbiade and Ighodaro [79] for white-flesh variants. A strong positive relationship was found between PBD and peak time for pasting of the starches. It could be reasoned that starches of high PBD have a low void volume, high surface area to volume ratio, and could be arrayed tightly, making disintegration of inter-granular integrity more difficult, thus, requiring a relatively longer time to reach peak viscosity. Longer peak time translates to extra production costs for the food industry, but could also be an advantage functionally, because starches resistant to rapid peak time can be useful in preserving structural integrity of foods. Granular integrity is an important element in the ability of starchy foods to absorb water [41]. Significant negative correlation has been found between PBD and water absorption of yam starches [78].

Starches of white-flesh and yellow-flesh variants of cassava have a similar dispersibility and reconstitution index, and may require only little agitation energy for reconstitution. Addition of surfactants can greatly influence dispersibility. Awoyale et al. [60] reported 80–86% dispersibility for yellow-flesh cassava variants, similar to those of white-flesh variants as reported by Onitilo et al. [33].

4.4. Water and Oil Absorption

Water absorption and oil absorption capacity entail ability of components of cassava to bind water and oil at hydrophilic and hydrophobic sites, respectively. Water and oil absorption are relevant because some major differences between white and yellow cassava variants that cassava breeders find intriguing are the relatively higher dry matter and starch in white-flesh cassava than yellow-flesh cassava [47,65], which influence water and oil absorption. Hence, the ability of the dry matter, starch, and other components to bind with water or oil films is important in studying how these interactions influence properties of cassava flour from a macroscopic point of view. Starch and other components can also form emulsions with oil and water films. In food use, oil absorption capacity of cassava starch and flour contribute to ensuring stable and uniform pastes and emulsions are formed in production of local confections and baked goods, and as anti-sticking material during cooking of pasta or frying of fishes as commonly practices in rural areas in some West African countries. In non-food use, these properties are important in formation of emulsion in paints, in textile sizing and shaping for local fashion or ceremonial wears, in soluble adhesives or gums and in release of active ingredients of drugs in vivo. This contribution to the local industry is important in sustaining livelihoods as it has created jobs for many unskilled and unemployed youths. On average, starches of three yellow-flesh variants of cassava absorbed 79.56% of their weight of water, indicating good water absorption capacity [60]. Significant differences among the variants indicated differences in granular interaction with water molecules. The WAC reported by Ikegwu et al. [59] for starches of 13 white-flesh variants ranged between 59.75–68.02%, but Onitilo et al. [33] reported WAC of 86.83–127.54% for 39 white-flesh cassava variants. This implies that WAC of both variant types are variant-dependent. Cassava starch has a fairly high oil absorption capacity as reported by Ikegwu et al. [59] and Omodamiro et al. [80]. Little, however, is reported on the oil absorption for flour and starch of white-flesh, and yellow-flesh cassava.

4.5. Swelling Power (SP) and Solubility

Starch and flour swell when their molecular components absorb and bind water by hydrogen bonds. The swelling power of cassava starch is functionally beneficial in use as thickener in the food industry, and its solubility, in use as degradable excipients in drug delivery systems. Thickening is required in soups, gravies, baby foods, and breakfast gruels. In a study of Indonesian cassava variants [8], starch suspension of a yellow-flesh variant at 84 °C, swelled above thrice its volume at 64 °C; similar to starch from white-flesh cassava variants, which also swelled about two to three times its initial volume at these temperatures (Table 5). The cassava starches swell increasingly with temperature across all variants, and had higher swelling performance than corn starch (4–18 mL/g), but lower than potato starch (42–168 mL/g). Hence, cassava starch or flour is well suited as food thickener. Starches of higher mean granule size are deemed to have higher SP [81,82], but the contrary was argued for potato and sweet potato starches [83–85]. Three improved yellow-flesh cassava variants from Nigeria had starches whose mean SP (7.46%) and solubility (1.67%) were low at 60 °C [60], signifying low interface between the starch chains in amorphous and crystalline regions.

Table 5. Physicochemical and functional properties of starch from white-flesh and yellow-flesh cassava variants.

Location	Flesh Color	Variant(s)	PC (%)	FTS (%)	SP or SI $^\Omega$	SOL $^\Omega$ (%)	WAC (%)	LGC (%)	Dispersibility (%)	Reference
South China	White	5 [a]	23.2–39.6 *	39.6–96.7	-	-	-	-	-	[7]
	Yellow	2 [a]	23.5–30 *	46–95	-	-	-	-		
Indonesia	White	7 [b]	63.1 $^\alpha$	-	16–32 mL/g, 40–60 mL/g (64 and 84 °C)	-	-	-	-	[8]
	Yellow	1 [b]	67.6 $^\alpha$	-	18 mL/g, 56 mL/g (64 and 84 °C)	-	-	-	-	
Nigeria	White	39 [c]	-	-	9.04–16.9% (80 °C)	1.03–47.07 (80 °C)	86.83–127.54	2–4.67	82–89.5	[33]
	Yellow	1 [c]	-	-	11.53% (80 °C)	13.46 (80 °C)	92.34	2.67	84	

Note: PC, paste clarity; FTS, freeze thaw stability; SP, swelling power; SI, swelling index; SOL, solubility; WAC, water absorption capacity; OAC, oil absorption capacity; LGC, least concentration for gel formation; *, location dependent; $^\alpha$, average estimated from charts; $^\Omega$, swelling power (at respective temperatures in parenthesis separated by comma); [a], eight locations, three replicates, completely randomized; [b], 11 months at harvest, selected from 71 local and improved genotypes; [c], 36 CMD-resistant and 4 non CMD-resistant varieties.

Swelling power is affected by factors such as intrinsic inter-granule binding forces, granule morphology, and amylose–amylopectin ratio [29,86]. Comparison of swelling power between white-flesh and yellow-flesh variants essentially depends on the genetic character of the variant studied. For instance, while white-flesh variants had higher swelling power [33,87] and solubility [33,88] than some yellow-fleshed ones, other starches of white-flesh variants with lower SP have been reported [8]. The amylose of starch introduces a dilution effect and restricts the capability for swelling [89]. Yellow-flesh cassava variants have been found to have similar amylose, but lower dry matter content, as discussed earlier, which could be the reason for their comparably lower swelling power. Modification of starch by chemical and physical means could significantly alter swelling and solubility properties. Little is mentioned in the literature on SP and solubility of cassava flour from white-flesh and yellow-flesh variants, but similar trend is expected as with cassava starch.

4.6. Paste Clarity (PC)

Paste clarity of cassava is important in products requiring translucence or gloss, such as gumdrops, jellies, and paints. Cassava starch generally has a high paste clarity (Table 5). Paste clarity of starches of seven variants from Southern China ranged between 23.2–39.6% [7]. Starch paste clarity of a yellow-flesh variant (SC 10) was not different to paste clarity of starches of white-flesh variants, but one yellow-flesh variant (SC 9) showed comparably lower clarity. For Indonesian variants [8], starch paste clarity of a yellow-flesh variant (Mentega) was considerably higher than that of two white-flesh variants (Gatot Kaca and Ketan). Hence, differences in paste clarity of cassava starch may not be argued from the flesh color perspective. All the cassava starch pastes became twice as turbid after six days of storage. Clarity of tuber and root starches is higher than clarity in cereal starches and the clarity of cassava starch paste ranks next only to potato starch [90]. Because of this, and its inexpensiveness, cassava starch has a long history of industrial use as glazing agent and thickener in the production of textiles and soup sauces respectively. Resultant swelling due to amylopectin leads to a viscous gel of high clarity. Polymer concentration and molecular weight have been correlated with turbidity (or clarity) of the starch [91].

4.7. Least Concentration for Gel Formation (LGC)

Low concentration of starch is usually required for gel formation. Therefore, LGC refers to the least amount of starch required to form a stable gel. This property is relevant to the breakfast meal and confections food industry, and in pharmaceutic excipients. Starch concentration of 2.67% was required to form a stable gel for yellow-flesh variants, and concentrations of 2–4.67% were required for starches of white-flesh variants [33]. Awoyale et al. [60] also revealed that LGC of 4.01–4.06 was required for gel formation in three biofortified yellow-flesh cassava starches. Hence, yellow-flesh and white-flesh variants are quite similar in LGC. Instead, variations in amylose–amylopectin ratio, crystallinity of starch, degree of branching, and molecular weight ratio play larger roles in influencing LGC.

4.8. Freeze–Thaw Gel Stability

When starches gelatinize by heating in water suspension, they lose their ordered crystalline structure and on cooling, reordering occurs, but not to the precise initial crystalline state [29]. Subjecting starch gel to cycles of periodic heating and cooling often results in deterioration of hydrogen bonding with water. There is a loss of hydrogen linkages resulting in readily exuded water molecules. The freeze–thaw stability of starch gels from seven southern Chinese cassava variants during storage was 39.6–96.7% [7]. Freeze–thaw (FT) characteristics of starch of yellow-flesh tubers among these variants varied significantly from those of the white-fleshed ones, a result connected to the starch amylose–amylopectin compositions. Freeze–thaw property was affected by location and environment, in addition to genotypic differences, although, in most locations, the yellow-flesh ones had lower FT values, and thus were less susceptible to effects of freeze–thaw cycles. The cassava starches all had poor freeze–thaw properties. For this reason, the use of cassava starch in foods requiring cold storage is restricted, unless the starch is modified. Little literature is available on FT properties of starch or flour from colored-flesh variants.

4.9. Textural Properties

Textural properties measure strain–stress relationships of a material when a defined magnitude of force is exerted to test material strength. For starch gels, such properties include elasticity, cohesiveness, hardness, and adhesiveness. Elasticity is a measure of extent to which starch gel can be extended before disintegration, and is representative of extent of chemical bond stretching and bending among its constituents. Cohesiveness and adhesiveness are measures of intra-molecular bond strength and resistance to disintegration within the gel. Hardness of starch gel is a measure of amount of compressional force that the homogeneous starch gel can withstand before mechanical disintegration. Starch gel (5.3% *w/v*) from white-flesh and yellow-flesh variants had closely related values of elasticity (0.52–0.54), cohesiveness (0.57–0.81), adhesiveness (1.27–2.27 mJ), and gel hardness (4.7–7.6 N) [8]. The cassava starches formed more adhesive, but softer gel than potato and cereal starches. Adhesive property of cassava starch gel is useful in paper and cardboard manufacturing, while its elastic nature could find relevance in confectioneries, such as soft-centered jellies, gums, and baked goods. Textural properties of starch gels are connected to amylose–amylopectin interactions; intricate networking of both constituent polymers could improve gel strength.

4.10. Pasting Properties

Pasting properties of cassava is important in studying the behavior of cassava starch or flour suspension during regulated heating, holding and cooling temperature regimes. Some pasting properties of cassava can differ significantly among variants, location, and cultivation conditions. Other pasting properties are similar between the cassava variants, and this advantage can be helpful in considering the use of yellow-flesh cassava starch or flour as substitute for white-flesh cassava starch or flour in food, and thereby enhance acceptance by industry in a bid to encourage sustained production by farmers who are less informed of the similarities. Comparing pasting properties of cassava starch or flour in the literature is complicated due to different paste concentration, method of determination, variant analyzed, and temperature regime used. Therefore, varying outcomes of paste viscosity comparisons occur, and are context-specific. Some pasting properties differ significantly between flour or starch from yellow-flesh and white-flesh variants (Table 6). For instance, most starch pasting properties were significantly higher for starches of yellow-flesh cassava [60] than for those of white-flesh cassava at similar or even higher paste concentration of white-flesh cassava starch [59,69] using similar method.

Pasting temperature (PT) of starch or flour is the temperature at which a sudden rise in viscosity is first noted with concurrent swelling. A positive relationship exists between PT, water-binding capacity and gelatinization temperature. High PT relates to a restricted swelling ability [92,93]. Cassava flour from 40 various yellow-flesh variants, however, did not vary significantly in PT to flour from three white-flesh variants [21]. Pasting temperature of starch from seven cassava variants grown in Southern China ranged between 61.7–69.1 °C [7]. The starch pasting temperature of the yellow-flesh ones was similar to that for white-flesh variants. Pasting temperature varied significantly with location and environmental conditions [7]. Pasting temperature of starch of three yellow-flesh variants [60] averaged 76.76 °C, quite higher than the temperature for various white-flesh variants in other works [33,59,68,94]. The differences in amylose content may have influenced this outcome. Amylose acts as a diluent during pasting, and has been cited as being positively related with PT [30]. Presence of non-starch components may insulate, bind, or occlude starch from thermal effects. Low PT of cassava starch has economic advantages in saving energy costs, as less thermal energy will be expended for pasting compared to more temperature-resilient starches from potato [26], rice [95] and yam [96]. Cassava starches readily lose their structural integrity when heated in solution.

Table 6. Pasting properties of flour and starch from white-flesh and yellow-flesh cassava variants.

Location	Flesh Color	Variants	PV (RVU)	TV (RVU)	BV (RVU)	FV (RVU)	SV (RVU)	PT (°C)	Peak time (min.)	Reference
Flour										
Nigeria	White	3 [b]	280.11–352.93	90.40–121.56	159.58–211.43	127.34–173.31	20.60–54.50	64.23–64.73	3.62–3.86	[21]
	Yellow	40 [b]	271.85–471.29	13.24–174.61	177.57–257.71	27.28–240.01	14.04–73.07	64.06–65.21	3.51–3.87	
Starch										
South China	White	5 [a]	187–303 BU	-	-	-	-	61.7–68.5	-	[7]
	Yellow	2 [a]	210–297 BU	-	-	-	-	63–69.1	-	

Note: PV, peak; TV, trough; BV, breakdown; FV, final; SV, setback; PT, pasting temperature; BU, Brabender unit; RVU, Rapid viscosity unit; [a], eight locations, three replicates, completely randomized; [b], two sets, four replicates, completely randomized, three drying methods, 10% *w/v* viscosity paste concentration.

Table 7. Thermal properties of starch from white and colored-flesh cassava variants.

Location	Flesh Color	Variant(s)	Location	T_o (°C)	T_p (°C)	ΔH (J/g)	Reference
Indonesia	White	*Adira 4, Darul Idayah, Gatot Kaca, Roti, Perelek, Gading merah* [a]	Indonesia	63.5—65.6	68.0–69.5	11.1–14.3	[8]
	Yellow	*Mentega* [a]	Indonesia	64.6	69.4	11.5	

Note: [a] 11 months at harvest, selected from 71 local and improved genotypes.

Peak viscosity (PV) is an important rheological property of starchy foods, and reflects the behavior of flour and starch paste under varying shear, temperature and time. Differences in PV of cassava variants have been found in the literature. Of seven cassava variants in Southern China, peak viscosity of the starch of yellow-flesh variants varied widely to that of starches of white-flesh variants [7] depending on location and variant type. For cassava flour of 43 yellow-flesh and white-flesh variants [21], PV was similar between both variant types.

Peak viscosity reported for yellow-flesh cassava starch [60] was higher than for starches of white-flesh cassava studied by Nuwamanya et al. [68], Eke et al. [69] and Ikegwu et al. [59]. Paste concentration and season of harvest may have influenced this outcome, apart from amylose content. An exception is the average PV of starch from three yellow-flesh variants from Nigeria reported as 382.14 RVU (rapid viscosity unit) [60], which was similar to PV of starches of white-flesh variants reported by Onitilo et al. [33] based on similar paste concentration and method of determination.

Other tuber/root crops, for instance, *Pachyhizus tuberosus*, exhibited a similar trend, as there was no significant difference in PV when comparisons were made between yellow-flesh variants and white-flesh ones [56]. The corresponding time required for viscosity to peak (peak time) was not different across variants of different flesh color [21,33]. Mean peak time of starch from three yellow-flesh cassava variants was 4.6 min [60].

Trough viscosity (TV) and breakdown viscosity (BV) are measures of resistance and disintegration of flour or starch, respectively, when exposed to temperature changes during pasting. The TV and BV of flours of white-flesh and yellow-flesh variants were within similar range, as reported by Maziya-Dixon et al. [21]. For starch, a mean BV of 138.69 RVU was recorded for three yellow-flesh variants [60], higher than for all white-flesh variants studied by Ikegwu et al. [59], but lower than for starches of several white-flesh variants studied by Onitilo et al. [33] and Eke et al. [69]. High breakdown viscosity is a disadvantage in many food applications because it results in unevenly distributed viscosity [97].

Setback viscosity (SV) can be considered a measure of tendency for flour or starch to retrograde, due to realignment of amylose molecules. Setback viscosity was within similar range between flours of yellow-flesh and white-flesh variants [21]. The same was not found for starches of several white-flesh variants [33,59,69] which have generally lower SV than starches of yellow-flesh variants [60], with few exceptions.

The FV is characteristic for the final product quality of starch-based foods. FV was within similar range between flours of white-flesh and yellow-flesh variants [21]. Final viscosity of starch of three yellow-flesh variants [60] surpassed those of the white-flesh variants reported so far. Again, variant-specificity influenced the differences in FV for variants studied in both works.

4.11. Gelatinization and Retrogradation

Pure starch crystallites should melt over a narrow range of gelatinization temperatures. Starch of low gelatinization temperature is cost-effective in industrial food production as it limits energy expense. For starch of 17 Indonesian variants analyzed for their gelatinization properties [8], enthalpy (ΔH) of gelatinization of starch from yellow-flesh variants (11.5 J/g) was similar to, or lower than, that of white-flesh variants (Table 7). Gelatinization onset at 63.5–66.1 °C was similar for all variants studied, regardless of flesh color, and was higher than gelatinization onset of wheat (60.2 °C) and potato (61.8 °C) starches in the same study. Gelatinization temperature of cassava starches may, however, be influenced by season of cultivation, location and cultivar [71,81], due to environmental factors and gene-associated factors contributing to the starch structure and composition. Although cassava flour consists mostly of starch, gelatinization temperatures (onset, peak and conclusion) reported for cassava flour was higher than for cassava starch in some studies by Defloor et al. [28,98], due to the presence of non-starch components in cassava flour. Gelatinization characteristics could be used to determine purity or adulteration of starch in the industry. The degree of association between starch crystallites, and crystalline–amorphous interactions may also explain differences in gelatinization temperatures of starch or flour [71].

An index of retrogradation of seven Southern Chinese cassava starches varied between 52.3–100% [7], indicative of high retrogradation tendencies. Retrogradation among starches of both white-flesh and yellow-flesh cassava was not different. Variant differences did not have a significant influence on retrogradation of cassava starches compared to the location and environment of cultivation in the study. Retrogradation is a disadvantage in use of cassava starch since the starch loses its water holding capacity, becomes irreversibly crystalline, unyielding, and undergoes syneresis.

5. Relationships (Correlation or Regression) between Physicochemical and Functional Properties of Cassava

A few studies demonstrate the relationships between physicochemical and functional properties of cassava. Flesh color has been positively correlated with total carotenoids in some works [48,99]. Low dry matter has also been associated with carotenoid contents of cassava variants [65,100]. A common theory that postharvest physiological deterioration (PPD) and shelf storability may be influenced by pro-vitamin A content of yellow cassava roots due to the anti-oxidative nature of the compounds may be true. Postharvest physiological deterioration of cassava includes vascular streaking, tissue softening, rotting, and discoloration resulting from biochemical changes. A report by Chavez et al. [6] correlated PPD of 30 cassava variants with vitamins after six days in storage and revealed there was significant evidence to support this hypothesis, but conclusions could not be drawn since there was a low correlation. It was further established that >0.5 mg/kg carotene ensured that PPD did not result in reduction of carotene nutrients exceeding 30%. Toxicity, as measured by cyanogenic content, varied significantly in different parts of the cassava plant. Concentration of cyanogenic compounds in the root, stem, and leaf of cassava are independent of one another because total cyanogens in those parts are influenced by age at maturity or harvest, precipitation/rainfall, genetics, fertilization, soil type, location, etc. [101–104].

Water absorption capacity of three yellow-flesh variants had a significant negative correlation with breakdown viscosity of the starches. A significant negative linear regression coefficient has been observed between starch granule damage and gelatinization properties of cassava starch, such as enthalpy and onset of gelatinization, especially when grown under less water stress [71].

A study on 79 cassava variants in China revealed no relationship between amylose, peak viscosity and clarity of the starches [105], but a report by Charles et al. [106] found correlations between these properties. The amount of rainfall cassava received during cultivation was negatively correlated to starch yield and dry matter [7], while starch content significantly correlated positively with dry matter (0.54, $p < 0.01$). In the same work, freeze–thaw stability correlated positively to retrogradation (0.355, $p < 0.05$), indicating that cassava starches that are stable to freeze–thaw cycles may not readily retrograde. Starch granule volume, and surface area correlated positively with viscosity (0.350, 0.336, $p < 0.05$).

Pasting properties of cassava as well as other tuber crops have been found to negatively correlate with protein content [107]. Since amylose content is a major composition of cassava starch, there was a positive correlation with pasting temperature and peak time ($r = 0.45$). Higher proportions of amylose in the starch require longer periods to leach out of the granule matrix before complete solubilization and swelling occurs [68,107]. Starch–fiber interaction may have a negative effect on peak viscosity ($r = -0.544$) and swelling power ($r = -0.805$), but a positive correlation with pasting temperature ($r = 0.422$) [68,107]. Swelling of cassava starch is negatively correlated with pasting temperature ($r = -0.629$), but positively correlated with peak viscosity, $r = 0.588$ [68].

6. Effects of Processing on Nutrients in White-Flesh and Biofortified Yellow-Flesh Cassava Roots and Flour

Cassava is safe for consumption only after undergoing appropriate processing, which may include one or a combination of some treatments such as boiling, frying, fermenting, drying, baking, or size-reduction, all of which contribute to reducing cyanogenic glucosides [108]. However, the downside of these processing operations is often a resultant reduction of nutrients, or conversion to other forms than the original nutrients. For instance, production of flour and other foods from cassava often leads to a loss of vital micronutrients, some of which are discussed below.

In a study of 28 clonal variants selected by Chavez et al. [6] from a large genetic pool of cassava, boiling of the roots led to an average reduction of carotene by 34%. In addition, flour produced after oven drying and sun drying lost an average of 44% and 73% carotene, respectively. In India, Narayanakappa, a yellow-fleshed cassava, and three other colored-flesh variants, all of high-carotene content, were processed by boiling, frying, sun drying and oven drying [1,15], and the effect on the concentration of total carotenoids and β-carotene were studied. Destruction of total carotenoids and beta carotene was found to be in the order: sun drying (44–67% and 43–79%) > frying (20–51% and 16–56%) > boiling (16–52% and 19–49%) > oven drying (16–45% and 5–36%) [15]. Boiling intensified the color of the cassava chips, possibly due to starch gelatinization, however, the profound release of carotenoids from cassava matrices could also be a reason [20]. Frying depleted carotenoids, which are known to be fat-soluble. Diminished retention was highest for variant *Acc-3*, which had the highest amounts of carotenoids and β-carotene. Rapid deterioration of carotenoids during exposure to light and heat can be attributed to their sensitivity towards oxidation and isomerization. Similar results have been reported in other studies [6,23,109,110]. Retention of carotenoids after processing is essential in adopting yellow-flesh cassava as a means of combating vitamin A deficiency in affected populations.

Other forms of processing affect retention of carotenoids in cassava. Grated cassava mash, fermented cassava mash, and fermented-cooked cassava dough (fufu) retained 97.68–98.48%, 94.68–96.66%, and 86.42–90.24% of original total carotenoids in yellow-flesh cassava (6.26–7.76 µg/g), respectively [23]. Drying the product from the "odorless" non-fermented method resulted in a significant loss of carotenoids, the severity being higher with sun drying than oven drying. Generally, the acceptability of fufu by sensory panelists, in terms of color, was best with the traditional method [23], because the method enhanced carotenoids retention better than sun drying and oven drying methods.

Evaluation of retained micronutrients, such as zinc, iron and total carotenoids, was made after three yellow-flesh cassava variants TMS 01/1371, 01/1235, and 94/0006 were processed traditionally by four methods: boiling, fermenting (raw fufu), fermenting and cooking (cooked fufu) and fermenting and roasting (gari) [46]. Boiling led to losses of about 4–52% of carotenoids, 3.6–20.6% of iron, and 2.7–21.7% of zinc. Fermentation significantly increased the average carotenoid content of the cassava roots from 4.9 µg/g to 8.64 µg/g according to wet basis measurements. A possible reason could be that, as major compositions of cassava (carbohydrates, moisture, and fiber) reduce by hydrolysis during fermentation, the proportion of other minor compositions such as carotenoids will apparently increase. Dry basis measurements could have given a more accurate trend of what transpired during processing [111]. Fermentation also significantly reduced the average iron and zinc content from 7.47 mg/kg to 7.13 mg/kg, and 8.95 mg/kg to 5.58 mg/kg, respectively. Fermentation leaches minerals due to the acidic nature of fermentate (fufu), and oxidative activities of microbes that use these micronutrients for development and growth. Boiling fermented cassava paste to achieve a doughy consistency further led to a reduction in the average carotenoid (3.64 µg/g) and zinc content (6.23 mg/kg), thus retaining less carotenoids (21.5%) and zinc (34.1%) than the uncooked paste. Only 32.5% iron was retained. The high temperature employed may have resulted in the rapid oxidation, isomerization, and destruction of vitamin A precursors [112–114], and leaching of the minerals. Again, fermenting and subsequent roasting (gari) was reported to increase average carotenoids and iron of three yellow-flesh variants from 4.9 µg/g to 10.6 µg/g, and from 7.5 mg/kg to 8.2 mg/kg. respectively. This is due to the dry nature of gari, and the fresh weight basis by which the determinations were made. Hence, the increase in the carotenoids and iron content was not an increase in actual amounts. There was an average retention of about 45% and 22% of carotenoids and iron. respectively, while 90% of zinc was lost. Heating, increase in surface area, and agitation [115–121] during gari production may have increased carotenoids available for quantification. Chopping (to increase surface area) and brief heating of colored-flesh cassava roots was adjudged to have contributed to the bioavailability of carotenoids and vitamin A by disrupting cell wall and protein–carotenoids complexes [20].

High performance liquid chromatograms [47] revealed that boiling can reduce total carotenoids and all-trans-β-carotenes of yellow-flesh cassava from 32.6 μg/g to 27.2 μg/g and 22.7 μg/g to 15.3 μg/g, respectively, on dry weight basis. The reduction of all-trans β-carotenes has been characterized as isomerization to 13-cis and 15-cis-β-carotene in boiled cassava [47,122]. Dry matter also reduced from 34.4% to 29.7% after boiling for 30 min. However, high retention (86.6%) of total carotenoids was observed in the work [47], and it was postulated that high dry matter in yellow-flesh variants might enhance retention of carotenoids in general.

Individual and interaction effects [48] of variants and processing methods showed significant differences in the retention of β-carotene among yellow-flesh cassava variants. This is an important criterion in the selection of variants to be adopted for vitamin A bio-fortification. A ranking of the processing methods vis-à-vis β-carotene retention was: oven drying (71.9%) > shadow drying (59.2%) > boiling (55.7%) > sun drying (37.9%) > gari production (34.1%). Storage of cassava up to four weeks in any form (flour, chip or root) after sun drying and oven drying resulted in reducing the retained β-carotene by slightly above 50%, depending on the severity of size-reduction (flour > chip > root). Vacuum storage reduced residual β-carotene, possibly due to the permeability of packaging used, and not as a result of the vacuum created. Generally, a reduction of oxygen species in an environment could deter oxidation of carotenoids.

Processing of five bitter yellow-flesh variants of cassava from Brazil [49] to flour reduced total carotenoids by an average of 50%, with further reduction of 33–99% during storage for up to 19 days. By the fourth week in storage, total carotenoids of the variants had completely diminished.

Processing could, in addition, lead to a reduction of cassava toxicity. Breeding to reduce toxicity is an ongoing sustained effort by scientists and breeders in research institutes in Africa and Asia. La Frano et al. [20] determined toxic cyanogenic contents of white-flesh (5.27 ppm) and yellow-flesh (5.51–280 ppm) cassava variants for safety reasons, and found a complete absence after they were processed by boiling into porridge, while 99% and 96% of β-carotene was retained, respectively. Generally, white-flesh cassava variants have a negligible β-carotene content, losing less than 1% during boiling [20], but variants with a high initial carotene content lose it much more readily during processing [6,48]. Furthermore, in the work of Diallo et al. [123], three cassava variants in Senegal were processed into four products—chips, flour, gari, and attieke—and their cyanide concentration (wb) was evaluated for the extent of detoxification. The chips retained 15.1–51.6% of cyanogenic toxicity, but gari, flour and attieke retained as low as 0–1.8%, 0–2.8%, and 1.1–5.4% cyanogens, respectively. These levels were below the allowed toxicity recommendation (<10 ppm) of the Codex Alimentarius Commission [124] for cassava flour, and are thus regarded safe for consumption.

In Nigeria, value-addition has been achieved in using yellow-flesh variants in producing gari instead of the traditional practice ofadding red palm oil to the white cassava granules to improve the vitamin A content of the food. Not only are production costs reduced, but a healthy form of fortification is also acquired by the yellow-flesh variants with yields close to the white-flesh ones. From a culinary and health/nutrition perspective, adding palm oil may not be acceptable in gari due to phase separation in water-based processes as well as for cholesterol-related issues.

Vitamin C retention is very low in processed cassava roots and flour. Only 36.6%, 6.2%, and 0.005% of vitamin C was retained after boiling, oven-drying and sun drying, respectively [6]. Pregnant women, lactating mothers and children [125] need the vital minerals (iron and zinc) and pro-vitamin A to prevent anemia, diarrhea, stunted growth, and defective eyesight. Evidence from trials with American women [20] and Kenyan children [17] has shown that feeding biofortified cassava is more efficient in the bioavailability of pro-vitamin A compounds than white-flesh cassava. Some new yellow-flesh cassava roots in Nigeria could potentially supply about 25% of the daily vitamin A requirement if consumed in sufficient amounts [16]. Hence, processes retaining much of these micronutrients are beneficial, especially when losses are unavoidable [47]. An indirect, but efficient way to reduce nutrient losses in cassava during processing could be to significantly improve the nutrient content of the raw, unprocessed crop. Efforts to achieve this by biofortification in increasing the protein, minerals, starch and β-carotene concentrations in cassava, has been reported [4]. This has been the focus of majority of cassava breeders in research institutes in Africa and Asia. Post-harvest technique, such as solid-state fermentation has also been used to increase protein in cassava [126]. In addition, minimal processing sufficient enough to make cassava edible and safe should be encouraged. Excessive fermentation, drying, boiling, cooking, retting, and frying cause major losses in nutrients [4], and should be avoided.

7. Practical Considerations for Sustainability in Adoption of Biofortified Yellow-Flesh Cassava

To argue for the sustained cultivation of biofortified yellow cassava and the production, utilization and consumption of its products, it is important to have detailed information on properties of the biofortified variant in comparison to the conventional white-flesh cassava. The arguments focus on the advantages offered by the biofortified variants, such as substitutability, nutrition, safety for consumption, storage life and relative ease of harvest and post-harvest handling. Many of the data discussed in this work show that biofortified yellow-flesh cassava can be a suitable, and arguably better, substitute compared to the conventional white-flesh cassava. The argument can be made based on the considerations discussed below, which has evidently, and can potentially, make biofortified yellow-flesh cassava more sustainable in many aspects than white-flesh cassava.

First, biofortified yellow-flesh cassava root, flour and starch has many similar physicochemical and functional properties as found for those of white-flesh cassava, and can, therefore, serve as a substitute in any products derived from white-flesh cassava root, flour and starch. This flexibility in utilization is not possible for white-flesh cassava when pro-vitamin A nutrition is of concern, making this additional advantage of nutrition a case for sustained utilization and consumption of biofortified yellow cassava as food. This was already demonstrated by Talsma [17] in a school children feeding program in Kenya. In addition, the development of biofortified yellow-flesh cassava has evidently explored possibilities of creating value-added cassava variants without genetic modification which is still contended as ethically unacceptable. Biofortification is considered sustainable because it requires a one-time investment only, which further allows farmers flexibility in cultivation. This curbs complete dependence of farmers on manufacturers for cultivation materials supply, as is usually associated with genetically modified crops. Such flexibility is enhanced by the value addition offered by yellow-flesh cassava regarding amount of vitamin nutrient per hectare cultivated. The biofortified cassava can reach subsistent poor farmers and people in remote areas where supplementation programs, which are usually more expensive and unaffordable, scarcely reach [127].

Second, most biofortified yellow-flesh cassava variants are sweet tasting, containing mild-to-moderate toxic cyanogenic glucosides compared to the majority of white-flesh variants [7]. Lower concentrations of the toxic compound can guarantee sustainable production plans by reducing time and labor for detoxification during processing, ultimately leading to production of safer foods. In addition, lower toxicity levels could reduce toxic residues in soil and environment of rural communities and urban centers of Africa where large quantities are regularly processed. This issue is of considerable concern to proponents of environmental sustainability since the run-off from these processing centers often result in pollution and contamination of nearby water bodies and the disruption of natural ecosystem of plants and animals.

Third, post-harvest storage of biofortified yellow-flesh cassava is more sustainable, due to its robust and longer shelf-stability [6,17] compared to white-flesh cassava. This implies it can secure longer availability periods, while the farmer awaits or is engaged in a new planting cycle. Ongoing unpublished research by the authors has observed that properly waxed yellow-flesh cassava roots have much longer storage life and acceptable quality at 3–5 months in refrigeration at 3 °C, than waxed white-flesh cassava roots which developed unacceptable quality after 3–5 weeks under similar conditions.

Fourth, harvesting and post-harvest handling of some variants of biofortified yellow-flesh cassava has been reported to be less tedious than for white-flesh cassava. Farmers' response to surveys reveal they were easier to harvest and peel [127] than conventional white-flesh cassava. This advantage could reduce labor and energy costs, making manual or commercial harvesting and peeling more sustainable.

This work therefore explored comparison of the properties of both variants of cassava for the purpose of assisting farmers, processors and other interest groups in making informed choices on variants with sustainable properties. It may be surmised that, in the near future, the popularity of yellow-flesh cassava may outpace that of the conventional white-flesh cassava [128] if some measures are taken strategically. Such measures should include a more committed and robust coordination of national governments programs, research groups, and other cassava stakeholders in ensuring rapid and widespread re-orientation, adoption and dissemination of the biofortified yellow cassava and its products to the public. In addition, encouraging market-based approach in cassava value chain to attract private investors [9,128] can be helpful.

While the adoption of biofortified orange-flesh sweet potato in Sub-Saharan Africa has achieved remarkable success in countries including South Africa, Mozambique, and Uganda, such cannot be said yet of the biofortified yellow cassava. Originally, pro-vitamins-rich cassava existed in the Amazon regions of South America. They were subsequently bred [14,99] and introduced to African countries including Nigeria and Kenya through research cooperation of the International Center for Tropical Agriculture (CIAT), Colombia and the International Institute of Tropical Agriculture, Nigeria. More successes seem to have been achieved in South America regarding the yield, nutrients density [6,128] and public acceptance of biofortified yellow cassava variants than in Sub-Saharan Africa. For instance, in northeast Brazil, a survey of 760 farmers reveal 28% of farmers already preferred yellow cassava variants, with about 70% of them been familiar with the yellow varieties [127]. About 15% of the farmers cultivated the variants as early as the first year of release of the variants. Nevertheless, few successes in acceptance of biofortified yellow cassava in production of gari and fufu, for instance, has been reported in Oyo and Akwa Ibom regions of Nigeria [128]. One major gap that continues to hamper the adoption of the biofortified variants is the weakened state of the extension service in the agricultural sector of some African countries, for instance, Nigeria [9].

8. Conclusions and Recommendation

Overall, the physical properties of starch from white-flesh and yellow-flesh variants of cassava are largely similar. Morphology, thermal, crystallinity, color, and flow properties are also similar. Color differences between root and flour of white-flesh and yellow-flesh cassava are significant. However, some chemical, pasting and physicochemical properties vary significantly between the white-flesh and yellow-flesh cassava variants. Amylose content, starch content, dry matter, carotenoids, β-carotene, peak viscosity, setback viscosity and final viscosity reported in several works show variant-specific differences in these properties. Genetics play the biggest role in determining differences in characteristics between white-flesh and yellow-flesh cassava and their products. Modification of cassava flour, and its effects and benefits have not been exhaustively researched. Colored-flesh variants increasingly have the potential of matching prevalence of white-flesh variants in the future if propagation and dissemination of research results is sustained. In retrospect, the authors recommend yellow-flesh cassava for commercialized cultivation because the yellow-flesh cassava variants have largely similar properties as white-flesh cassava variants, are nutritionally valuable in some respects, and store better after harvest than white-flesh cassava. Therefore, breeding efforts to bridge the gaps in properties observed in both types of variants could hold the key to nutritional and functional uses of yellow-flesh cassava just as much as white-flesh cassava. In Sub-Saharan Africa, the challenges with adoption of the biofortified yellow-flesh cassava by farmers and the public is particularly exacerbated by conservative attitude of farmers, unwillingness to try new methods or crops, poor understanding of the advantages it offers compared to the white-flesh cassava, misinformation on the origin of biofortified variants as genetically modified, and poor commitment of the governments to consistently develop agricultural policies and extensions that will drive acceptance of the biofortified variants. Some measures suggested to resolve these challenges are serious commitment of national and regional governments in promoting the biofortified variants, nationwide dissemination of success stories of adoption, upscale of knowledge on the variant to the farmers and the public and subsidies-for-adoption programs. In addition, research bodies in Sub-Saharan Africa need to adopt workable blueprints already achieving better successes in South America and Asia with regards to development of improved biofortified variants of higher yield and nutrients density to make their cultivation more sustainable.

Author Contributions: Conceptualization, O.A., S.L., and J.M.; Resources, O.A.; Data Curation, O.A.; Writing—Original Draft Preparation, O.A.; Writing—Review and Editing, S.L., A.A. and J.M.; Visualization, J.M.; Supervision, S.L., A.A. and J.M; Project Administration, J.M., A.A.; and Funding Acquisition, S.L., A.A. and J.M.

Funding: This research was funded by the German Federal Ministry for Education and Research (BMBF), and German Federal Ministry for Economic Cooperation and Development (BMZ) with grant number 031A258A. The APC was funded by the German Federal Ministry for Education and Research (BMBF), and German Federal Ministry for Economic Cooperation and Development (BMZ).

Acknowledgments: The authors acknowledge the cooperation between the Center for Development Research (ZEF), Bonn, Germany, Institute of Agricultural Engineering, Tropics and Subtropics, University of Hohenheim, and International Institute of Tropical Agriculture (IITA), Nigeria. This work was conducted under work package 5.3 (postharvest technologies, cassava web innovations) of the BIOMASSWEB project. The authors thank Sabine Nugent, Institute of Agricultural Engineering, Tropics and Subtropics, University of Hohenheim for proofreading the manuscript. Authors are also grateful to Reinhard Kohlus, Peter Gschwind, and Theresa Anzmann of the Institute of Food Processing Engineering and Powder Technology, University of Hohenheim, for the use of the scanning electron microscope.

Conflicts of Interest: The authors of this work declare no conflict of interests in preparation and publication of this article. The article is an original work, and does not infringe the intellectual property rights of any other person or entity, hence, cannot be construed as plagiarizing any other published work.

References

1. Vimala, B.; Thushara, R.; Nambisan, B. Carotenoid Retention in Yellow-fleshed Cassava during Processing. In Proceedings of the 15th Triennial International Society for Tropical Root Crops (ISTRC) Symposium, Lima, Peru, 2–6 November 2010.

2. Bradbury, J.H.; Holloway, W.D. *Chemistry of Tropical Root Crops. Significance for Nutrition and Agriculture in the Pacific*; ACIAR: Canberra, Australia, 1988; pp. 53–77.

3. Mc Dowell, I.; Oduro, K.A. Investigation of the β-carotene Content of the Yellow Varieties of Cassava (*Manihot esculenta* Crantz). *J. Plant. Foods* **1983**, *5*, 169–171. [CrossRef]

4. Montagnac, J.A.; Davis, C.R.; Tanumihardjo, S.A. Nutritional Value of Cassava for Use as a Staple Food and Recent Advances for Improvement. *Comp. Rev. Food Sci. Food Saf.* **2009**, *8*, 181–194. [CrossRef]

5. Gegios, A.; Amthor, R.; Maziya-Dixon, B.; Egesi, C.; Mallowa, S.; Nungo, R.; Gichuki, S.; Mbanaso, A.; Mnanary, M.J. Children Consuming Cassava as a Staple Food are at Risk for Inadequate Zinc, Iron, and Vitamin A Intake. *Plant. Food Hum. Nutr.* **2010**, *65*, 64–70. [CrossRef] [PubMed]

6. Chavez, A.L.; Bedoya, J.M.; Sánchez, T.; Iglesias, C.; Ceballos, H.; Roca, W. Iron Carotene, and Ascorbic Acid in Cassava Roots and Leaves. *Food Nutr. Bull.* **2000**, *21*, 410–413. [CrossRef]

7. Gu, B.; Yao, Q.Q.; Li, K.M.; Chen, S.B. Change in Physicochemical Traits of Cassava Roots and Starches Associated with Genotypes and Environmental Factors. *Starch-Stärke* **2013**, *65*, 253–263. [CrossRef]

8. Anggraini, V.; Sudarmonowati, E.; Hartati, N.S.; Suurs, L.; Visser, R.G.F. Characterization of Cassava Starch Attributes of Different Genotypes. *Starch-Stärke* **2009**, *61*, 472–481. [CrossRef]

9. Oparinde, A.; Abdoulaye, T.; Manyong, V.; Birol, E.; Asare-Marfo, D.; Kulakow, P.; Ilona, P. A Technical Review of Modern Cassava Technology Adoption in Nigeria (1985–2013): Trends, Challenges, and Opportunities. Available online: http://ebrary.ifpri.org/utils/getfile/collection/p15738coll2/id/130268/filename/130479.pdf (accessed on 25 September 2018).

10. IITA. Nigeria Releases More Cassava with Higher Pro-Vitamin A to Fight Micronutrient Deficiency. News (2014 Archive). Available online: http://www.iita.org/2014-press-releases/-/asset_publisher/CxA7/content/nigeria-releases-more-cassava-with-higher-pro-vitamin-a-to-fight-micronutrient-deficiency?redirect=%2F2014-press-releases#.Vnwaw2tlg5s (accessed on 22 April 2016).

11. Ceballos, H.; Morante, N.; Sanchez, T.; Ortiz, D.; Aragon, I.; Chavez, A.L.; Pizarro, M.; Calle, F.; Dufour, D. Rapid Cycling Recurrent Selection for Increased Carotenoids Content in Cassava Roots. *Crop. Sci.* **2013**, *53*, 2342–2351. [CrossRef]

12. Sánchez, T.; Ortiz, D.; Morante, N.; Ceballos, H.; Pachon, H.; Calle, F.; Perez, J.C. New Approaches for Quantifying Carotenoids Content in Cassava Roots. In Proceedings of the 11th Triennial Symposium, International Society for Tropical Root Crops, African Branch, Kinshasa, Congo, 4–8 October 2010.

13. Saltzman, A.; Birol, B.; Bouis, H.E.; Boy, E.; De Moura, F.F.; Islam, Y.; Pfeiffer, W.H. Biofortification: Progress toward a More Nourishing Future. *Glob. Food Secur.* **2013**, *2*, 9–17. [CrossRef]

14. Chávez, A.L.; Sánchez, T.; Jaramillo, G.; Bedoya, M.; Echeverry, J.; Bolaños, E.A.; Ceballos, H.; Iglesia, C.A. Variation of Quality Traits in Cassava Roots Evaluated in Landraces and Improved Clones. *Euphytica* **2005**, *143*, 125–133. [CrossRef]

15. Vimala, B.; Thushara, R.; Nambisan, B.; Sreekumar, J. Effect of Processing on the Retention of Carotenoids in Yellow-Fleshed Cassava (*Manihot esculenta* Crantz) Roots. *Int. J. Food Sci. Tech.* **2011**, *46*, 166–169. [CrossRef]

16. Aniedu, C.; Omodamiro, R.M. Use of Newly Bred β-carotene Cassava in Production of Value-added Products: Implication for Food Security in Nigeria. *Glob. J. Sci. Front. Res. Agric. Vet. Sci.* **2012**, *12*, 10–16.

17. Talsma, E.F. Yellow Cassava: Efficacy of Provitamin—A rich Cassava on Improvement of Vitamin A Status in Kenyan School Children. Ph.D. Thesis, Wageningen University, Wageningen, The Netherlands, 3 July 2014.

18. UNICEF. Micronutrient Initiative. In *Vitamin and Mineral Deficiency: A Global Progress Report*; UNICEF: Ottawa, ON, Canada, 2005.

19. Vimala, B.; Nambisan, B.; Thushara, R.; Unnikrishnan, M. Variability of Carotenoids in Yellow-fleshed Cassava (*Manihot esculenta* Crantz) Clones. *Gene Conserve* **2009**, *31*, 676–685.

20. La Frano, M.R.; Woodhouse, L.R.; Burnett, D.J.; Burri, B.J. Biofortified Cassava Increases β-carotene and Vitamin A Concentrations in the TAG-Rich Plasma Layer of American Women. *Br. J. Nutr.* **2013**, *110*, 310–320. [CrossRef] [PubMed]

21. Maziya-Dixon, B.; Adebowale, A.A.; Onabanjo, O.O.; Dixon, A.G.O. Effect of Variety and Drying Methods on Physico-Chemical Properties of High Quality Cassava Flour from Yellow Cassava Roots. In *African Crop Science Conference Proceedings*; African Crop Science Society: Kampala, Uganda, 2005; pp. 635–641.

22. Ukenye, E.; Ukpabi, U.J.; Chijoke, U.; Egesi, C.; Njoku, S. Physicochemical, Nutritional and Processing Properties of Promising Newly Bred White and Yellow Fleshed Cassava Genotypes in Nigeria. *Pak. J. Nutr.* **2013**, *12*, 302–305. [CrossRef]

23. Omodamiro, R.M.; Oti, E.; Etudaiye, H.A.; Egesi, C.; Olasanmi, B.; Ukpabi, U.J. Production of Fufu from Yellow Cassava Roots Using the Odorless Flour Technique and the Traditional Method: Evaluation of Carotenoids Retention in the Fufu. *Adv. Appl. Sci. Res.* **2012**, *3*, 2566–2572.

24. Alves, A.A.C. Cassava Botany and Physiology. In *Cassava: Biology, Production and Utilization*; Hillocks, R.J., Tres, J.M., Bellotti, A.C., Eds.; CABI Publishing: Oxfordshire, UK, 2002; pp. 67–89.

25. Bokanga, M. Cassava: Post-harvest Operations. Information Network on Post-Harvest Operations. In *Post-Harvest Compendium*; FAO: Rome, Italy, 2000; pp. 1–26.

26. Moorthy, S.N. Physicochemical and Functional Properties of Tropical Tuber Starches: A Review. *Starch-Stärke* **2002**, *54*, 559–592. [CrossRef]

27. Ellis, R.P.; Cochrane, M.P.; Dale, M.F.B.; Duffus, C.M.; Lymm, A.; Morrison, I.M.; Prentice, R.D.M.; Swanston, J.S.; Tiller, S.A. Starch Production and Industrial Use. *J. Sci. Food Agric.* **1998**, *77*, 289–311. [CrossRef]

28. Defloor, I.; Leijskens, R.; Bokanga, M.; Delcour, J.A. Impact of Genotype, Crop Age and Planting Season on the Bread Making and Gelatinization Properties of Cassava (*Manihot esculenta* Crantz) Flour. *J. Sci. Food Agric.* **1995**, *68*, 167–174. [CrossRef]

29. Hoover, R. Composition, Molecular Structure, and Physicochemical Properties of Tuber and Root Starches: A Review. *Carbohydr. Polym.* **2001**, *45*, 253–267. [CrossRef]

30. Zhu, F. Composition, Structure, Physicochemical Properties, and Modifications of Cassava Starch. *Carbohydr. Polym.* **2015**, *122*, 456–480. [CrossRef] [PubMed]

31. Okigbo, B.N. Nutritional Implications of Projects Giving High Priority to the Production of Staples of Low Nutritive Quality. In The Case for Cassava (*Manihot esculenta* Crantz) in the Humid Tropics of West Africa. *Food Nutr. Bull.* **1980**, *2*, 1–10.

32. Gil, J.L.; Buitrago, A.J.A. La yuca en la alimentacion animal. In *La Yuca en el Tercer Milenio: Sistemas Modernos de Producci'on, Procesamiento, Utilizaci'on y Comercializaci'on*; Ospina, B., Ceballos, H., Eds.; Centro Internacional de Agricultura Tropical: Cali, Colombia, 2002; pp. 527–569.

33. Onitilo, M.O.; Sanni, L.O.; Daniel, I.; Maziya-Dixon, B.; Dixon, A. Physicochemical and Functional Properties of Native Starches from Cassava Varieties in Southwest Nigeria. *J. Food Agric. Environ.* **2007**, *5*, 108–114.

34. ISI. *Specifications for Tapioca Starch for Use in Cotton Textile Industries, IS 1605–1960*; Indian Standards Institution: New Delhi, India, 1970.

35. Radley, J.A. *Starch Production Technology*; Applied Science Publishers Ltd.: London, UK, 1976; pp. 189–229.

36. Radley, J.A. *Examination and Analysis of Starch and Its Derivatives*; Applied Science Publishers Ltd.: London, UK, 1976; pp. 1–213.

37. Buitrago, J.A. *La Yucca en la Alimentacion Animal*; Centro Internacional de Agricultura Tropical: Cali, Colombia, 1990.

38. Moorthy, S.N.; Rickard, J.E.; Blanshard, J.M.V. Influence of Gelatinization Characteristics of Cassava Starch and Flour on the Textural Properties of some Food Products. In Proceedings of the International Meeting on Cassava Starch and Flour, Cali, Colombia, 11–15 January 1994.

39. Moorthy, S.N.; Wenham, J.E.; Blanshard, J.M.V. Effect of Solvent Extraction on the Gelatinization Properties of Starch and Flour of Five Cassava Varieties. *J. Sci. Food Agric.* **1996**, *72*, 329–336. [CrossRef]

40. Hudson, B.J.F.; Ogunsua, A.O. Lipids of Cassava Tubers (*Manihot esculenta* Crantz). *J. Sci. Food Agric.* **1974**, *25*, 1503–1508. [CrossRef] [PubMed]

41. Rickard, J.E.; Asaoka, M.; Blanshard, J.M.V. The Physicochemical Properties of Cassava Starch. *Trop. Sci.* **1991**, *31*, 189–207.

42. Wolf, G. Is dietary β-carotene an Anti-cancer Agent? *Nutri. Rev.* **1982**, *40*, 257. [CrossRef]

43. World Health Organization, WHO. *The Global Prevalence of Vitamin A Deficiency*; Micronutrient Series Document WHO/NUT/95.3; WHO: Geneva, Switzerland, 1995.

44. Underwood, B.A. Overcoming Micronutrient Deficiencies in Developing Countries: Is There a Role for Agriculture? *Food Nutr. Bull.* **2000**, *21*, 356–360. [CrossRef]

45. Noel, W.S. Vitamin A and Carotenoids. In *Present Knowledge in Nutrition*, 8th ed.; Bowman, B.A., Russell, R.M., Eds.; ILSI Press: Washington DC, USA, 2001; pp. 127–145.

46. Maziya-Dixon, B.; Awoyale, W.; Dixon, A. Effect of Processing on the Retention of Total Carotenoid, Iron and Zinc Contents of Yellow-fleshed Cassava Roots. *J. Food Nutr. Res.* **2015**, *3*, 483–488.

47. Ceballos, H.; Luna, J.; Escobar, A.F.; Ortiz, D.; Pérez, J.C.; Sánchez, T.; Pachón, H.; Dufour, D. Spatial Distribution of Dry Matter in Yellow Fleshed Cassava Roots and Its Influence on Carotenoid Retention upon Boiling. *Food Res. Int.* **2012**, *45*, 52–59. [CrossRef]

48. Chavez, A.L.; Sanchez, T.; Ceballos, H.; Rodriguez–Amaya, D.B.; Nestel, P.; Tohme, J.; Ishitani, M. Retention of Carotenoids in Cassava Roots Submitted to Different Processing Methods. *J. Sci. Food Agric.* **2007**, *87*, 388–393. [CrossRef]

49. Oliveira, R.G.A.; de Carvalho, M.J.L.; Nutti, R.M.; de Carvalho, L.V.J.; Fukuda, W.G. Assessment and Degradation Study of Total Carotenoid and β-carotene in Bitter Yellow Cassava (*Manihot esculenta* Crantz) varieties. *Afr. J. Food Sci.* **2010**, *4*, 148–155.

50. Gregorio, G.B. Progress in Breeding for Trace Minerals in Staple Crops. *J. Nutr.* **2002**, *132*, 500–502. [CrossRef] [PubMed]

51. Ssemakula, G.; Dixon, A.G.O.; Maziya-Dixon, B. Stability of Total Carotenoid Concentration and Fresh Yield of Selected Yellow-Fleshed Cassava (*Manihot esculenta* Crantz). *J. Trop. Agric.* **2007**, *45*, 14–20.

52. Tukomane, T.; Leerapongnun, P.; Shobsngob, S.; Varavinit, S. Preparation and Characterization of Annealed-enzymatically Hydrolyzed Tapioca Starch and the Utilization in Tableting. *Starch-Stärke* **2007**, *59*, 33–45. [CrossRef]

53. Rodrıguez-Sandoval, E.; Ferna'ndez-Quintero, A.; Cuvelier, G.; Relkin, P.; Bello-Pe'rez, L. Starch Retrogradation in Cassava Flour from Cooked Parenchyma. *Starch-Stärke* **2008**, *60*, 174–180. [CrossRef]

54. Han, Y.; Xu, M.; Xingyan Liu, X.; Yan, C.; Korban, S.; Chen, X.; Gu, M. Genes Coding for Starch Branching Enzymes are Major Contributors to Starch Viscosity Characteristics in Waxy Rice (*Oryza sativa* L.). *Plant Sci.* **2004**, *166*, 357–364. [CrossRef]

55. Nabubuya, A.; Namutebi, A.; Byaruhanga, Y.; Narvhus, J.; Wicklund, T. Potential Use of Selected Sweetpotato (*Ipomea. batatas* Lam) Varieties as Defined by Chemical and Flour Pasting Characteristics. *Food Nutr. Sci.* **2012**, *3*, 889–896.

56. Ascheri, J.L.R.; Zamudio, L.H.B.; Carvalho, C.W.P.; Arevalo, A.M.; Fontoura, L.M. Extraction and Characterization of Starch Fractions of Five Phenotypes *Pachyrhizus. tuberosus* (Lam.) Spreng. *Food Nutr. Sci.* **2014**, *5*, 1875–1885.

57. Juliano, B.; Villareal, M.; Perez, M.; Villareal, P.; Loso Baños, M.; Takeda, V.; Hizukuri, S. Varietal Differences in Properties among High Amylose Rice Starches. *Starch-Starke* **1987**, *39*, 390–393. [CrossRef]

58. Ceballos, H.; Sanchez, T.; Morante, N.; Fregene, M.; Dufour, D.; Smith, A.M.; Denyer, K.; Perez, J.C.; Calle, F.; Mestres, C. Discovery of an Amylose-free Starch Mutant in Cassava (*Manihot esculenta* Crantz). *J. Agric. Food Chem.* **2007**, *55*, 7469–7476. [CrossRef] [PubMed]

59. Ikegwu, O.J.; Nwobasi, V.N.; Odoh, M.O.; Oledinma, N.U. Evaluation of the Pasting and Some Functional Properties of Starch Isolated from Some Improved Cassava Varieties in Nigeria. *Afr. J. Biotechnol.* **2009**, *8*, 2310–2315.

60. Awoyale, W.; Sanni, L.O.; Shittu, T.A.; Adegunwa, M.O. Effect of Varieties on the Functional and Pasting Properties of Biofortified Cassava Root Starches. *Food Meas.* **2015**, *9*, 225–232. [CrossRef]

61. Kossmann, J.; Lloyd, J. Understanding and Influencing Starch Biochemistry. *Crit. Rev. Biochem. Mol. Biol.* **2000**, *35*, 141–196. [CrossRef]

62. Tewe, O.O. *The Global Cassava Development Strategy. Cassava for Livestock Feed in Sub-Saharan Africa*; FAO: Rome, Italy, 2004; pp. 1–64.

63. Fukuda, W.M.G.; Guevara, C.L.; Kawuki, R.; Ferguson, M.E. *Selected Morphological and Agronomic Descriptors for the Characterization of Cassava*; International Institute of Tropical Agriculture (IITA): Ibadan, Nigeria, 2010.

64. Nassar, N.M. Wild and Indigenous Cassava, Manihot sculenta Crantz Diversity: An Untapped Genetic Resource. *Genet. Resour. Crop. Evol.* **2007**, *54*, 1523–1530. [CrossRef]

65. Moorthy, S.N.; Jos, J.S.; Nair, R.B.; Sreekumari, M.T. Variability of β-carotene Content in Cassava Germplasm. *Food Chem.* **1990**, *36*, 223–236. [CrossRef]

66. Easwari, A.C.S.; Sheela, M.N. Genetic Analyses in a Di-allele Cross of Inbred Lines of Cassava. *Madras Agric. J.* **1998**, *85*, 264–268.

67. Cach, N.T.; Lenis, J.I.; Perez, J.C.; Morante, N.; Calle, F.; Ceballos, H. Inheritance of Useful traits in Cassava Grown in Sub-humid Conditions. *Plant. Breed.* **2006**, *125*, 177–182. [CrossRef]

68. Nuwamanya, E.; Baguma, Y.; Emmambux, N.; Taylor, J.; Patrick, R. Physicochemical and Functional Characteristics of Cassava starch in Ugandan Varieties and Their Progenies. *J. Plant. Breed. Crop. Sci.* **2010**, *2*, 1–11.

69. Eke, J.; Achinewhu, S.C.; Sanni, L.; Barimalaa, I.S.; Maziya-Dixon, B.; Dixon, A. Pasting, Color, and Granular Properties of Starches from Local and Improved Cassava Varieties in High Rainfall Region of Nigeria. *Int. J. Food Prop.* **2009**, *12*, 438–449. [CrossRef]

70. Nuwamanya, E.; Baguma, Y.; Emmambux, N.; Patrick, R. Crystalline and Pasting Properties of Cassava Starch are Influenced by Its Molecular Properties. *Afr. J. Food Sci.* **2010**, *4*, 8–15.

71. Defloor, I.; Dehing, I.; Delcour, J.A. Physochemical Properties of Cassava Starch. *Starch-Stärke* **1998**, *50*, 58–64. [CrossRef]

72. Singh, J.; McCarthy, O.J.; Singh, H. Physicochemical and Morphological Characteristics of New Zealand Taewa (Maori potato) Starches. *Carbohydr. Polym.* **2006**, *64*, 569–581. [CrossRef]

73. Lindeboom, N.; Chang, P.R.; Tylera, R.T. Analytical, Biochemical and Physicochemical Aspects of Starch Granule Size, with Emphasis on Small Granule Starches: A Review. *Starch* **2004**, *56*, 89–99. [CrossRef]

74. Vamos-Vigyazo, L. Polyphenoloxidase and Peroxidase in Fruits and Vegetables. *CRC Crit. Rev. Food Sci. Nutr.* **1981**, *15*, 49–52. [CrossRef] [PubMed]

75. Nassar, N.; Vizzotto, C.S.; Da Silva, H.L.; Schwartz, C.A.; Pires-Junior, O.R. Potentiality of Cassava Cultivars as a Source of Carotenoid. *Gene Conserv.* **2005**, *15*, 267–273.

76. Ladeira, T.; Souza, H.; Pena, R. Characterization of the Roots and Starches of Three Cassava Cultivars. *Int. J. Agric. Sci. Res.* **2013**, *2*, 12–20.

77. Oduro-Yeboah, C.; Johnson, P.-N.T.; Sakyi-Dawson, E.; Budu, A. Effect of Processing Procedures on the Colorimetry and Viscoelastic Properties of Cassava Starch, Flour, and Cassava-Plantain Fufu Flour. *Int. Food Res. J.* **2010**, *17*, 699–709.

78. Falade, K.O.; Ayetigbo, O.E. Effects of Annealing, Acid Hydrolysis and Citric Acid Modifications on Physical and Functional Properties of Starches from Four Yam (*Dioscorea.* spp.) Cultivars. *Food Hydrocoll.* **2015**, *43*, 529–539. [CrossRef]

79. Agunbiade, S.O.; Ighodaro, O.M. Variation in the Physical, Chemical and Physico-functional Properties of Starches from Selected Cassava Cultivars. *N. Y. Sci. J.* **2010**, *3*, 48–53.

80. Omodamiro, R.M.; Iwe, M.O.; Ukpabi, U.J. Pasting and Functional Properties of Lafun and Starch Processed from Some Improved Cassava Genotypes in Nigeria. *Niger. Food J.* **2007**, *25*, 122–129. [CrossRef]

81. Asaoka, M.; Blanshard, J.M.V.; Rickard, J.E. Seasonal Effects on the Physicochemical Properties of Starch from Four Cultivars of Cassava. *Starch–Stärke* **1991**, *43*, 455–459. [CrossRef]

82. Asaoka, M.; Blanshard, J.M.V.; Rickard, J.E. Effects of Cultivar and Growth Season on the Gelatinisation Properties of Cassava (*Manihot esculenta*) Starch. *J. Sci. Food Agric.* **1992**, *59*, 53–58. [CrossRef]

83. Singh, N.; Kaur, L. Morphological, Thermal, Rheological and Retrogradation Properties of Potato Starch Fractions Varying in Granule Size. *J. Sci. Food Agric.* **2004**, *84*, 1241–1252. [CrossRef]

84. Singh, J.; Kaur, L. *Advances in Potato Chemistry and Technology*, 1st ed.; Academic Press: Cambridge, MA, USA, 2009; pp. 273–318.

85. Chen, Z.; Schols, H.A.; Voragen, A.G.J. Starch Granule Size Strongly Determines Starch Noodle Processing and Noodle Quality. *J. Food Sci.* **2003**, *68*, 1584–1589. [CrossRef]

86. Singh, N.; Kaur, M.; Sandhu, K.S.; Guraya, H.S. Physicochemical, Thermal, Morphological, and Pasting Properties of Starches from Some Indian Black Gram (*Phaseolus mungo* L.) Cultivars. *Starch-Stärke* **2004**, *56*, 535–544. [CrossRef]

87. Kay, D.E. *Crops and Products Digest. No. 2: Root Crops*, 2nd ed.; Tropical Product Institute: London, UK, 1987; pp. 1–308.

88. Eke, J.; Achinewhu, S.C.; Sanni, L.; Barimalaa, I.S.; Maziya-Dixon, B.; Dixon, A. Seasonal Variations in the Chemical and Functional Properties of Starches from Local and Improved Cassava Varieties in High Rainfall Region of Nigeria. *J. Food Agric. Environ.* **2007**, *5*, 36–42.

89. Leach, H.W.; Mccowen, L.D.; Schoch, T.J. Structure of the Starch Granules I. Swelling and Solubility Patterns of Various Starches. *Cereal Chem.* **1959**, *36*, 534–541.

90. Achille, T.F.; Georges, A.N.; Aphonse, K. Contribution to Light Transmittance Modeling in Starch Media. *Afr. J. Biotechnol.* **2007**, *6*, 569–575.

91. Miles, M.J.; Morris, V.J.; Ring, S.G. Gelation of Amylose. *Carb Res.* **1985**, *135*, 257. [CrossRef]

92. Emiola, L.; Delarosa, L.C. Physicochemical Characteristics of Yam Starches. *J. Food Biochem.* **1981**, *5*, 115–130. [CrossRef]

93. Numfor, F.A.; Walter, W.M.; Schwartz, S.J. Effect of Emulsifiers on the Physical Properties of Native and Fermented Cassava Starches. *J. Agric. Food. Chem.* **1996**, *44*, 2595–2599. [CrossRef]

94. Onitilo, M.O.; Sanni, L.O.; Oyewole, O.B.; Maziya-Dixon, B. Physicochemical and Functional Properties of Sour Starches from Different Cassava Varieties. *Int. J. Food Prop.* **2007**, *10*, 607–620. [CrossRef]

95. Cameron, K.; Wang, Y.; Moldenhauer, A. Comparison of Starch Physicochemical Properties from Medium-Grain Rice Cultivars Grown in California and Arkansas. *Starch-Stärke* **2007**, *59*, 600–608. [CrossRef]

96. Shanavas, S. Studies on Modified Tuber Starches Useful for Tablets and Capsules. Ph.D. Thesis, University of Kannur, Kerala, India, 2 January 2013.

97. Eliasson, A.C. *Starch in Food: Structure, Function and Applications*; Woodhead/CRC Press: New York, NY, USA, 2004; pp. 1–624.

98. Defloor, I.; Leijskens, R.; Bokanga, M.; Delcour, J.A. Impact of Genotype and Crop Age on the Breadmaking and Physicochemical Properties of Flour Produced from Cassava (*Manihot esculenta* Crantz) Planted in the Dry Season. *J. Sci. Food Agric.* **1994**, *66*, 193–202. [CrossRef]

99. Iglesias, C.; Mayer, J.; Chavez, L.; Calle, F. Genetic Potential and Stability of Carotene Content in Cassava Roots. *Euphytica* **1997**, *94*, 367–373. [CrossRef]

100. Jos, J.S.; Nair, S.G.; Moorthy, S.N.; Nair, R.B. Carotene Enhancement in Cassava. *J. Root Crops* **1990**, *16*, 5–11.

101. Bokanga, M. Distribution of Cyanogenic Potential in Cassava Germplasm. *Acta Hortic. (ISHS)* **1994**, *375*, 117–123. [CrossRef]

102. Nzigamasabo, A.; Ming, Z.H. Traditional Cassava Foods in Burundi: A Review. *Food Rev. Int.* **2006**, *22*, 1–27.

103. IITA. *Archival Report (1989–1993). Part 1. Cassava Breeding, Cytogenetics and Histology, Vol. 2, Germplasm Enhancement, Crop Improvement Division, TRIP (Tuber Root Improvement Program)*; IITA: Ibadan, Nigeria, 1993.

104. Bokanga, M.; Ekanayake, I.J.; Dixon, A.G.O.; Porto, M.C.M. Genotype-environment Interaction for Cyanogenic Potential in Cassava. *Acta Hortic. (ISHS)* **1994**, *375*, 131–140. [CrossRef]

105. Gu, B.; Li, K.; Li, Z.; Li, K. Starch Properties of Cassava root. *Chin. J. Trop. Crop.* **2009**, *30*, 1876–1882.

106. Charles, A.L.; Chang, Y.; Ko, W.; Sriroth, K.; Huang, T. Some Physical and Chemical Properties of Starch Isolates of Cassava Genotypes. *Starch-Starke* **2010**, *9*, 413–418. [CrossRef]

107. Moorthy, S. *Tuber Crop Starches*; Tech Bulletin No. 18 CTCRI; CTCRI: Trivandrum, India, 2002.

108. Ernesto, M.; Cardoso, A.P.; Nicala, D.; Mirione, E.; Massaza, F.; Cliff, J.; Haque, M.R.; Bradbury, J.H. Persistent Konzo and Cyanide Toxicity from Cassava in Northern Mozambique. *Acta Trop.* **2002**, *82*, 357–362. [CrossRef]

109. Nascimento, P.; Fernandes, N.; Kimura, M. Assessment of β-carotene Retention during Processing of Cassava. In Proceedings of the First Scientific Meeting of Global Cassava Partnership, Ghent University, Ghent, Belgium, 21–25 July 2008.

110. Oliviera, L.; Fukuda, W.; Watanabe, E.; Nutt, M.; Carvalho, J.L.; Kimura, M. Total Carotenoid Concentration and Retention in Cassava Products. In Proceedings of the First Scientific Meeting of Global Cassava Partnership, Ghent University, Ghent, Belgium, 21–25 July 2008.

111. Ortiz, D.; Sánchez, T.; Morante, N.; Ceballos, H.; Pachón, H.; Duque, M.C.; Chávez, A.L.; Escobar, A.F. Sampling Strategies for Proper Quantification of Carotenoids Content in Cassava Breeding. *J. Plant. Breed. Crop. Sci.* **2011**, *3*, 14–23.

112. Boon, C.S.; McClements, D.J.; Weiss, J.; Decker, E.A. Factors Influencing the Chemical Stability of Carotenoids in Foods. *Crit. Rev. Food Sci. Nutr.* **2010**, *50*, 515–532. [CrossRef] [PubMed]

113. Rodriguez-Amaya, D.B. *Carotenoid and Food Preparation: the Retention of Provitamin—A Carotenoid in Prepared, Processed, and Stored Foods*; John Snow Incorporated/Omni Project; Universidade Estadual de Campinas: Campinas, Brazil, 1997; pp. 1–93.

114. Smolin, L.A.; Grosvenor, M.B. *Nutrition Science and Applications*, 4th ed.; John Wiley and Sons Inc.: Hoboken, NJ, USA, 2003; pp. 1–848.

115. Clinton, S.K. Lycopene: Chemistry, Biology, and Implications for Human Health and Disease. *Nutr. Rev.* **1998**, *56*, 35–51. [CrossRef] [PubMed]

116. Dewanto, V.; Wu, X.; Adorn, K.K.; Liu, R.H. Thermal Processing Enhances the Nutritional Value of Tomatoes by Increasing Total Antioxidant Activity. *J. Agric. Food Chem.* **2002**, *50*, 3010–3014. [CrossRef] [PubMed]

117. Schierle, J.; Bretzel, W.; Buhler, I.; Faccin, N.; Hess, D.; Steiner, K.; Schuep, W. Content and Isomeric Ratio of Lycopene in Food and Human Blood Plasma. *Food Chem.* **1997**, *59*, 459–465. [CrossRef]

118. Stahl, W.; Sies, H. Uptake of Lycopene and Its Geometrical Isomers is Greater from Heat-processed than from Unprocessed Tomato Juice in Humans. *J. Nutr.* **1992**, *122*, 2161–2166. [CrossRef] [PubMed]

119. Stahl, W.; Sies, H. Lycopene: A Biologically Important Carotenoid for Humans? *Arch. Biochem. Biophys.* **1996**, *336*, 1–9. [CrossRef] [PubMed]

120. Agarwal, A.; Shen, H.; Agarwal, S.; Rao, A.V. Lycopene Content of Tomato Products: Its Stability, Bioavailability and In Vivo Antioxidant Properties. *J. Med. Food.* **2001**, *4*, 9–15. [CrossRef] [PubMed]

121. Xianquan, S.; Shi, J.; Kakuda, Y.; Yueming, J. Stability of Lycopene during Food Processing and Storage. *J. Med. Food.* **2005**, *8*, 413–422. [CrossRef] [PubMed]

122. Thakkar, S.K.; Huo, T.; Maziya-Dixon, B.; Failla, M.L. Impact of Style of Processing on Retention and Bioaccessibility of β-carotene in Cassava (*Manihot escultenta* Crantz). *J. Agric. Food Chem.* **2009**, *57*, 1344–1348. [CrossRef] [PubMed]

123. Diallo, Y.; Gueye, M.T.; Ndiaye, C.; Sakho, M.; Kane, A.; Barthelemy, J.P.; Lognay, G. A new Method for the Determination of Cyanide Ions and Their Quantification in Some Senegalese Cassava Varieties. *Am. J. Anal. Chem.* **2014**, *5*, 181–187. [CrossRef]

124. FAO/WHO. Joint FAO/WHO Food Standards Program. In *Codex Alimentarius Commission XII, Supplement 4*; FAO: Rome, Italy, 1991.

125. Black, R.E.; Allen, L.H.; Bhutta, Z.A.; Caulfield, L.E.; De Onis, M.; Ezzati, M.; Mathers, C.; Rivera, J. Maternal and Child Undernutrition: Global and Regional Exposures and Health Consequences. *Lancet* **2008**, *371*, 243–260. [CrossRef]

126. Iyayi, E.A.; Losel, D.M. Protein Enrichment of Cassava by-Products through Solid-State Fermentation by Fungi. *J. Food Technol. Afr.* **2001**, *6*, 116–118. [CrossRef]

127. Gonzalez, C.; Perez, S.; Cardoso, C.E.; Andrade, R.; Johnson, N. Analysis of Diffusion Strategies in Northeast Brazil for New Cassava Varieties with Improved Nutritional Quality. *Exp. Agric.* **2011**, *47*, 539–552. [CrossRef]
128. Ilona, P.; Bouis, H.E.; Palenberg, M.; Moursi, M.; Oparinde, A. Vitamin A Cassava in Nigeria: Crop Development and Delivery. *Afr. J. Food Agric. Nutr. Dev.* **2017**, *17*, 12000–12025. [CrossRef]

© 2018 by the authors. Licensee MDPI, Basel, Switzerland. This article is an open access article distributed under the terms and conditions of the Creative Commons Attribution (CC BY) license (http://creativecommons.org/licenses/by/4.0/).

 *sustainability*

Article

Value Addition and Productivity Differentials in the Nigerian Cassava System

Temitayo A. Adeyemo * and Victor O. Okoruwa

Department of Agricultural Economics, University of Ibadan, Ibadan, Nigeria; vokoruwa@gmail.com
* Correspondence: adeyemotemitayo@gmail.com; Tel.: +234-80-274-01925

Received: 20 September 2018; Accepted: 30 November 2018; Published: 14 December 2018

Abstract: There is an increasing need to improve value addition in order to get maximum utility from agricultural systems. Using a retrospective panel data from 482 cassava farmers covering the years 2015–2017, this study examined the effect of value addition on productivity of farmers in the cassava system in Nigeria. We analysed a non-parametric Data Envelopment Analysis to examine productivity across cassava production systems over the three year period. We also examined the impact of value addition on productivity using an endogenous switching regression to account for unobservables that determine the decision to add value and productivity of the farmers. The study found that cost and revenue outlays increased with value addition. Cassava farmers in general operated below the efficiency frontier, with total productivity declining over the 2015–2017 period. However, higher value addition farmers had better efficiency and non-reducing productivity in the periods studied. We found evidence of selection bias in the decision to add value and productivity of the farmers. The conditional and unconditional outcome estimates revealed positive gains in productivity with value addition, confirming the hypothesis that value addition increases farming households' productivity. We recommend that essential services such as extension services, agricultural training, and ease of enterprise registration that drive agricultural value addition be made available to farmers.

Keywords: cassava farmers; value addition; productivity differentials; impact; endogenous switching regression

---

1. Introduction

The practice of subsistence agriculture on marginal lands and low resource utilization is no longer feasible for sustaining farm families [1]. There has therefore arisen the need for the development of farming systems that stems from the need to integrate components and resources of farming families in order to minimize costs and maximize positive outcomes. This development of farming systems is a bid to ensure the sustainability of farmers' livelihood. Production systems have been shown to develop mainly from different ecologies of production, extent of utilization of product, extent of market access, type of cropping system, as well as extent of diversification of production [2]. Within the cassava production system in Nigeria, distinct production systems have been identified by the type of mixed cropping pattern [3,4]. However, with respect to the analysis of cassava production systems on the basis of the extent of value addition and utilization of cassava biomass, there has been a dearth of information.

The majority of agricultural produce in Nigeria is sold raw and at farm gate, leading to lower returns for the farmers [5]. It is estimated that over 50% of farm produce in the Nigerian agricultural sector is rural based and below commercial value [6]. However, potentials exist for improved returns and income from value addition in the agricultural sector [7], so that creating value has gained prominence in agriculture in recent years. This involves the development of new products and creating remunerative markets for higher value agricultural commodities [8,9]. Value addition is important for raising the livelihood of smallholders in Nigeria [10]. Important concepts and analytical frameworks such as value chains and value webs have developed from the basis of value addition processes within agriculture. Value addition within the cassava system has a great potential because of its multiproduct versatility. The cassava biomass has been shown to be able to develop multimarket, multi-industry, and trade linkages within the Nigerian economy [11]. Hence, the conceptualization of new production systems for a farming household/enterprise from the different means of value addition in the cassava system.

Productivity differences arise within and across production systems in agricultural sectors [12,13]. These differences may be a function of the extent to which farming households decide to utilize the agricultural biomass and integrate with the market. The decision process is, however, determined by access to productive resources, which in turn are determined by other variable factors [14]. The quantity and prices of production inputs and outputs differ according to the extent of value addition that defines the production system adopted by farmers. It is therefore expected that there may be productivity differentials among these production systems. This study hence examines the development of production systems based on the extent of value addition within the cassava system. The aim is to explore productivity differentials, and estimate the impact of value addition within the cassava system in Nigeria.

The benefits of value addition in agriculture have generated a number of literature. In Tanzania, the effect of farmer participation in value addition on food security [15], while [16] studied the welfare effect of wheat value addition on farmers' welfare using propensity score approach. In Kenya, the welfare impact of banana value addition on the welfare of farmers [17], while [18] explored cassava value addition using gross margin analysis in Nigeria. There is, however, a dearth of literature on productivity differentials as a result of different production systems that develop from value addition in the cassava subsector in Nigeria. This study therefore contributes to literature by first examining total factor productivity differentials across the systems that define value addition in cassava in Nigeria. Subsequently, we model the impact of value addition using an endogenous switching regression model in order to correct for selection bias.

The foregoing raises the following questions: Is there significant difference in the input and output outlay across the value-added cassava production systems in Nigeria? To what extent do productivities differ across the cassava production systems? What is the impact of value addition on the productivity of cassava farmers in Nigeria? This study therefore examines the productivity differentials across the different cassava production systems, with the aim of finding out if value addition and better utilization of cassava leads to overall increased productivity for the farmer. The aim of this paper is to create a scenario to help decision makers in investment in value addition in the cassava system. Although the main decision rests with the farming household itself, extension agents and overseeing ministries can use the results as a benchmark to guide the farmers. Knowledge of potential productivity increases may also be used to access the distance from the frontier and thus guide policy in estimating farm practices and management in terms of their productivities relative to the best practice frontier.

Hypothesis

The hypothesis to be tested is stated as:
Ho: There is no difference in productivity between value adders and non-value adders in the Nigerian cassava system

2. Theoretical and Conceptual Framework

2.1. Theoretical Framework

The basis for decision making in agricultural households was modelled by the agricultural household model of [19,20]. In the model, the agricultural household is seen as a production, consumption, and labour entity in a bid to maximize expected utility. According to [21], farming household decisions can be explained in three theoretical models, the peasant profit-maximizing model, utility maximizing theory, and the risk-averse theory. While the profit maximizing theory examines peasant farmers production choices from the point of allocative efficiency of the farming household in the 'small but efficient' hypothesis of [22]. The utility maximizing theory explores decision making of the farming household as a family and a business. In effect, it examines how farming households make production and consumption decisions subject to some constraints. The risk-averse theory, on the other hand, encompasses the risk behaviour of the farming households in decision making. The theory is related to the 'safety first' model in risk studies.

Although farming household decision could be modelled through any of these three approaches, the theory of profit maximization has been reviewed to give way to the other two theories. The basis of the profit maximization theory rests solely on allocative efficiency, where only the profit outcome is modelled without the input of the farm household decision making process. In reality, this does not work for farming households, hence the need for alternative models where the decision process of the farm family is modelled along with the expected outcome. On this basis, farming households make production decisions, such as value addition, diversification of portfolio, off-farm work, cropping pattern etc. based on either expected utility of consumption/income streams (utility maximization theory) or expected utility in the face of risk as a means of self-preservation (risk-averse theory).

The utility maximization theory was specifically inferred in this study. In the utility maximization household decision-making theory, the farming households are seen as both household and enterprise. Hence, production and consumption (welfare) decisions are subsumed in the model. The theory postulates that households seek to maximize utility subject to a set of constraints. These constraints include income constraints, production constraints, and time constraints. In this paper, we model the household decision to participate in value addition as premised on the need to realize the expected utility of welfare (income from value-added production) subject to these constraints.

The decision of a farm entrepreneur to invest or participate in an economic activity is best described by the Expected Utility Theory [23]. In this theoretical framework, farmers as Decision Making Units (DMU) choose between uncertain prospects by comparing expected utilities from each prospect. The outcomes of these choices will thereafter be seen in improved welfare, income, or productivity. Hence for the present study, the decision to add value within cassava production systems will be realized if and when the expected utility for value adding production is greater than the utility for not adding value in the production systems.

Assume that U_i and U_j are the utility of farmers in the two decision regimes; 1. value adders and 2. non-value adders. The utility is a function of the farmers is given as:

$$U = f(X, F, Z) \tag{1}$$

The linear form of the utility function for farmers in each regime is thus

$$U_i = X_i\beta + F_i\gamma + Z_i\theta + \varepsilon_j \text{ and } U_j = X_j\beta + F_j\gamma + Z_j\theta + \varepsilon_j \tag{2}$$

where U is the utility maximization function; X is a vector of socioeconomic variables, F denotes farming system characteristics, including crop characteristics; and Z are institutional and production constraints; the error terms are independently identically distributed (iid). The farmer decides to add value if he perceives that the outcome from the expected utility function $U_i > U_j$, and vice versa. The probability of the farmer's decision to add value is therefore given by:

$$P(U_i > U_j) \tag{3}$$

$$P((X_i\beta_i + F_i\gamma_i + Z_i\theta_i + \varepsilon_i) > X_j\beta_j + F_j\gamma_j + Z_j\theta_j + \varepsilon_j)) \tag{4}$$

$$P((X_i\beta_i + F_i\gamma_i + Z_i\theta_i + \varepsilon_i) - X_j\beta_j + F_j\gamma_j + Z_j\theta_j + \varepsilon_j)) > 0 \tag{5}$$

$$P\left(\left(X_i\beta_i + F_i\gamma_i + Z_i\theta_i\right) - \left(X_j\beta_j + F_j\gamma_j + Z_j\theta_j\right) + \left(\varepsilon_i - \varepsilon_j\right) > 0\right) \tag{6}$$

where β, λ, and θ are parameter estimates of independent variables. P is a probability function, which can be estimated from a variety of quantitative choice models. The outcome of the utility function in this study is the productivity of the cassava farmers, which was estimated from the non-parametric Data Envelopment Analysis (DEA).

2.2. The Nexus of Value Addition and Productivity

The conceptual framework in this study examines the relationship between the decision of cassava farmers to add value and the consequent outcome of such a decision (see Figure 1). The decision to add value within their cassava production system is determined by a number of variables. On the one hand, these variables influence the productivity of farmers; however, we are interested in how these variable determine productivity through value addition as an intervening factor. These variables include the socioeconomic characteristics of the farmer; institutional and macroeconomic environment in which the farmer is operating; crop characteristics; and the initial goal of the farming enterprise. The farmers' socioeconomic characteristics include age, gender, education, and years of experience. These factors may be inherited, or influenced by other external forces, and they dictate the responsibilities of farmers within their productive systems. Institutional and production constraints define the environment within which farmers carry out their productive activities. These include access to credit, extension contact, and macroeconomic policies which serve as catalyst to the enterprise development of the farmers. Farming system/crop characteristics determine to a large extent how much value addition can be carried out. In this study, the versatility of cassava biomass increases its value adding potential. Overall, it is expected that the decision to add value and the extent of value addition will lead to positive outcomes (such as income and productivity).

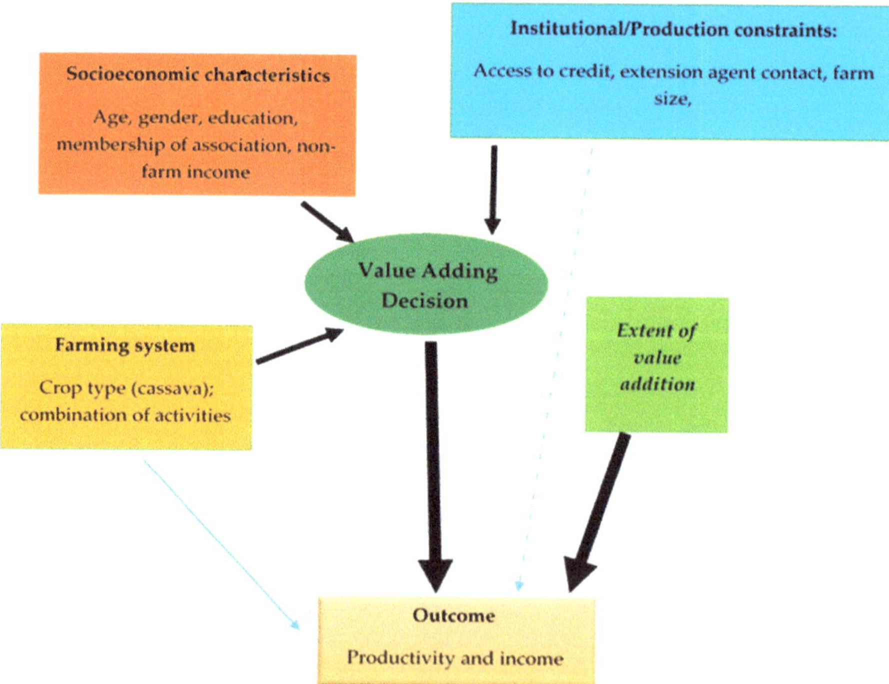

Figure 1. The Nexus of Value Addition and Productivity; Author's conceptualization.

3. Materials and Method

3.1. Study Area and Sampling Procedure

The study area is Nigeria. Cassava is produced in almost all the states in Nigeria, although the main producing zones are the forest sand guinea savannah zones [24,25]. The guinea savannah belt consists of Kogi, Kwara, Benue, Taraba, Kaduna states while the rain forest zone includes Delta, Edo, Ebonyi, Anambra, Oyo, Ogun, among others. We, however, selected Ogun and Kwara states to represent the forest and guinea savannah zones in this study (see Figure 2 for the maps of the study states).

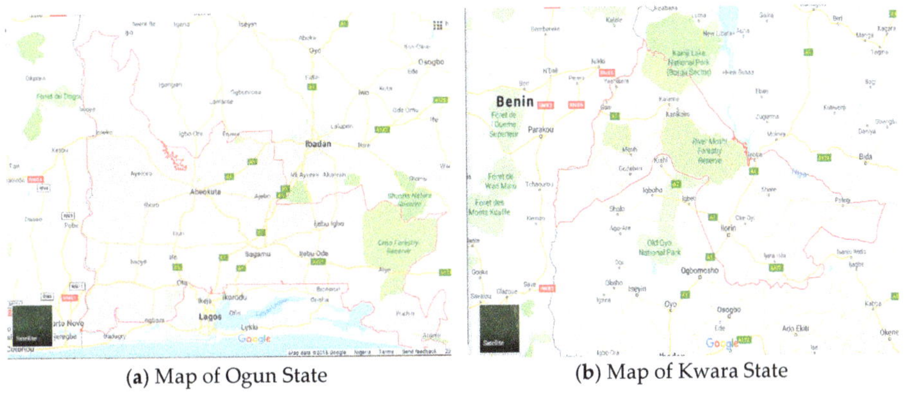

<table>
<tr><td>(a) Map of Ogun State</td><td>(b) Map of Kwara State</td></tr>
</table>

Figure 2. Maps of study states; (**a**) Ogun state and (**b**) Kwara state; Retrieved from Google Maps, 2018.

Following this, a three-stage stratified random sampling was employed to select the respondents. First, four local government areas (LGAs) corresponding to Agricultural Development Zones (ADZs) were randomly selected from each state. These LGAs are known for substantial cassava production, processing, and marketing activities. In the second stage, enumeration areas were selected from each LGA proportionate to the number of EA's in each LGA. The last stage was a random selection of cassava farmers from each enumeration area. Farmers were classified into four groups of sole producer (SP); producer/processor (PP); producer/marketer (PM); and producer/processor/marketer (PPM). As a result, the basis of analysis is the cassava farmer at different levels of value addition. Overall, a sample of 500 farming households was interviewed, but 482 (96.4%) questionnaires were useful for the analysis after data cleaning. The distribution of the respondents across the production systems were 192 (39.8%), 199 (41.29%), 42 (8.71%), and 49 (10.17%) for SP, PP, PM, and PPM respectively.

3.2. Description of Productivity Variables

We collected information using structured questionnaires on socioeconomic characteristics of the farmers such as age, marital status, gender, household size, education, and type of productive activities. Information on input and output outlay across the systems were also collected. In order to estimate the productivity changes across the systems, the information on production was collected over a period of three years (2015, 2016, and 2017). Thus, we have a retrospective panel production data, which proves useful in the absence of resources to gather information over the three year period [26,27]. The use of panel data, in this case, is important in order to capture the dynamics of productivity changes across the farmer group, and hence generate more accurate inferences [28].

Since each productive system had different types of input and output requirements, it became necessary to find a way to get a common value for the outputs in order to ensure accurate comparison across the farming systems. We used the monetary values of the output realized from the productive activities of the cassava actors. This was important in order to bring the output to the same level. The input outlay was also measured in monetary terms and discussed subsequently.

Labour cost was the cost incurred for hired labour across all the production system plus the opportunity cost of unpaid family labour, valued as the cost of hired labour. Apart from the similar labour characteristics for on-farm cassava production, additional labour costs are incurred for value addition. The variable for seed was assumed to be the raw material for production across the systems. It therefore include the price of stem cuttings for SP only, cassava tubers for the processing systems, and marketable biomass (raw tubers or finished products) for those involved in marketing.

Value addition involved mainly processing; thus we included the variable 'power', which includes every kind of source of generation of power on the farmers' resource base. Hence, we included the cost of electricity, fuel for machines, and wood and other traditional fuels for processing. Marketing functions also included the use of power, especially electricity and fuel for generating sets that would power storage functions. A minimum estimate of household electricity consumption was used for SP who were not involved in value addition. The fourth input used is transportation. We assumed that transportation cuts across all systems. SP farmers incur costs for their productive activities (such as transportation to and from farms). Also, the costs of transportation includes the costs incurred in obtaining and delivering raw materials and finished good for the value addition groups.

3.3. Data Analysis

i.   Total Factor Productivity (TFP) across Cassava Production Systems in Nigeria

The TFP index measures productivity by comparing the observed outputs in periods 1 and 2 with the maximum level of outputs that can be produced using the inputs x_1 and x_2 under a reference technology. Specifically, the study made use of the Malmquist TFP index [29], which uses a radial distance of the observed outputs and inputs in the three periods with respect to a reference technology (the year 2015). The distance measure could either be input orientated or output orientated. Input distance orientation seeks to minimize the quantity of inputs used to obtain the desired output, while the output orientation seeks to maximize output levels given a vector of inputs. The results of the input and output orientation are synonymous, however, this study made use of the input orientated Malmquist TFP index.

Input Orientated Malmquist TFP Index

The input orientated index examines how levels of inputs, x_1 and x_2, that can be used to produce the observed levels of outputs, y_1 and y_2, relative to the reference technology. Using 2015 as the reference technology, the index for each of the other two years (2016 and 2017) with respect to the reference year is given as:

$$m_i^1(y_1, x_1, y_2, x_2) = \frac{d_i^1(y_2, x_2)}{d_i^1(y_1, x_1)} \tag{7}$$

Assume that there is technical efficiency in both periods, i.e., $d_i^1(y_1 x_1) = 1$, then

$$m_i^1(y_1, x_1, y_2, x_2) = d_i^1(y_2, x_2) \tag{8}$$

This can be similarly done if the reference technology is period 2. Therefore, the input orientated Malmquist index is:

$$m_i(y_1, x_1, y_2, x_2) = \{ m_i^1(y_1, x_1, y_2, x_2) \cdot m_i^2(y_1, x_1, y_2, x_2) \}^{0.5} \tag{9}$$

The above is a measure of productivity growth when technical efficiency is assumed in the two periods. However, if there is technical inefficiency, which is the most probable case, the observed productivity change can be given as follows:

$$m_i(y_1, x_1, y_2, x_2) = \frac{d_i^2(y_2, x_2)}{d_i^1(y_1, x_1)} \left[\frac{d_i^1(y_2, x_2)}{d_i^2(y_2, x_2)} \times \frac{d_i^1(y_1, x_1)}{d_i^2(y_1, x_1)} \right]^{0.5} \tag{10}$$

Equation (4) comprises two ratios: The ratio outside the bracket is the measure of Efficiency change, while the ratio in the brackets is that of the technical change.

It is important to note that the Malmquist TFP index was composed of the following:

i. Input distance measure
ii. Constant returns to scale

Estimation of four productivity measures of technical change, price change, scale efficiency change, and total factor productivity change.

Estimation of productivity can be done either by a parametric or non-parametric approach. The parametric measure, which includes the Stochastic Frontier Analysis, involves fitting an appropriate functional form for the estimation. However, the non-parametric estimation does not require fitting a functional form. The most popular non-parametric measure is the Data Envelopment Analysis (DEA) which was used in this study.

Data envelopment analysis (DEA) has been assessed as a tool for the determinants of efficiency measures even with multi-outputs and multi-inputs scenarios. DEA is able to produce efficiency measures for each observational unit. It is also able to provide measures of productivity changes across time periods. The strengths of the non-parametric measure of DEA include the ability to estimate models without having to specify a functional form, estimate models with insufficient degrees of freedom, overcome extreme variability within the data, as well as determine productivity estimates from purely quantity data. However, it is easily affected by outliers and there is no statistical inference to be made to determine the significance of the results [30,31]. This study therefore used the DEA-Malmquist productivity measure to assess the productivity of cassava production systems over the 2015–2017 period in Nigeria. We estimated the DEA productivity measure for each production system in order to explore system-specific measures. We also estimated the productivity for the pooled data.

Since this study compared input and output outlays across different production systems, the possibilities of obtaining zero values increases. The systems had differing physical input and output outlay, which would have led to missing data when they were combined for the productivity comparison. Upholding the assumption of positivity and non-zero input/output outlay [32,33] would have meant a reduction in sample size if observations with zero values were deleted. The use of appropriate data preparation and modification of the DEA model could, however, help solve this problem. In order to successfully compare productivity across the systems, we followed the proposition of [34] in preparing the data for the DEA analysis in ensuring measurable values across groups and maintaining the positivity assumption. Therefore, the monetary values of all inputs and output outlay in the production processes were used. Specifically, the assumption of positivity was upheld across all the observations by using minimum transactional values which would have no effect on the overall efficiency outlay.

We estimated the DEA productivity measure for each of the production system, as well as for the pooled data. The DEA Malmquist measure gave the following estimates

(i) Efficiency change;
(ii) Technical change;
(iii) Pure technical efficiency change(corresponding to the VRS efficiency measure);
(iv) Scale efficiency change; and
(v) Total Factor productivity change.

The efficiency change being estimated is equivalent to the ratio of the Farell technical efficiency in period 2 to the Farell technical efficiency in period 1 [35]. This efficiency change is based on the assumption of a Constant Returns to Scale (CRS) technology of production. The technical change is the geometric mean of the shift in technology between two periods. A value greater than 1 implies a technical progress from period 1 to 2. Pure technical efficiency change is measured from the variable returns to scale technology. Scale efficiency change measures the change in productivity from a change in the scale of production and their movement towards the technologically optimum scale between two time periods. It is a measure of losses as a result of sub-optimal production size. A value greater than 1 means that the farm is nearer the optimum scale of technology in the period under consideration as opposed to the reference period.

ii.   Estimating the Impact of Value Addition on Productivity of Cassava Households

Past studies that examined productivity differentials have estimated differences in productivity by specifying production functions and comparing efficiency measures for each of the different choice regimes [36]. However, households' decision to participate in any extra investment such as value addition is an endogenous one [37,38]. This is because the decision to participate in value addition is necessarily voluntary, hence the decision to add value and the consequent level of productivity may be affected by inherent factors such as skills, experience, and innovation, which may, therefore, lead to the problem of selectivity bias. Therefore, if, for example, the cassava farmers have obtained additional skills or invested in more capital to increase their production, then their productivity is likely to increase whether or not they are involved in value addition. Hence, we cannot infer that value addition alone has led to an increase in his productivity. One of the best ways to deal with this problem is fitting a simultaneous equation model. Estimating a pooled econometric regression of both decision regimes assumes that similar explanatory variables have the same impact on the two groups, and the average effect of a value adding decision is for the whole sample [39]. This may, however, not be feasible. Moreover, an Ordinary Least Square (OLS) estimation will give consistent estimates only if the error terms are independent; if not, the estimates will be biased and it becomes important to use models that deal with the endogeneity in the data [40]. This is the basis for the use of an endogenous switching regression model. Endogenous switching regression models helps to model the counterfactual impact of choice regimes when the issue of selectivity bias arises. This is especially important with the non-random assignment of respondents in observational studies to the different choice regimes [41].

In this study, we attempted to model the expected outcomes of value addition on the productivity of cassava farmers in Nigeria. Productivity, as defined earlier, is the total revenue from the productive activities (either value adding or non-value adding) within the cassava system. First, we determine the two counterfactual regimes in our study as 0 for non-value adders and 1 for value adders. For simplicity, we merge the farmers who add value at any capacity as 'value adders' and those who do not as 'non-value adders'. For our sample, the non-value adders are the farmers in the sole producer (SP) production system. They are assumed not to add value in terms of processing or marketing to their cassava system. The value adders are those in the other three production systems (PP, PM, and PPM). The model follows that of [42–45].

Let d_i denote the latent variable that determines the value adding decision of farmers, with 1 = households in regime 1 (value adders) and 0 = households in regime 2 (non-value adders). We used the following index function to describe d_i as:

$$d_i^* = \omega Z_i + \mu_i; \tag{11}$$

$$d_i = 1 \; if \; d_i^* > 0 \tag{12}$$

$$d_i = 0 \; if \; d_i^* \leq 0 \tag{13}$$

The respondents have two possible outcomes dependent on the choice regime they belong, so that:

$$y_{1i} = \beta_1 X_{1I} + v_{1i} \quad if \; d_i = 1 \tag{14}$$

$$y_{2i} = \beta_{2i}X_{2i} + v_{2i} \quad if \; d_i = 0 \tag{15}$$

The y_{ji} are the outcome variables of the continuous equations; Z_i, X_1, and X_2 are vectors of weakly exogenous characteristics; β_1, β_2, and ω are vectors of parameters to be estimated. We also assume that the three error terms are trivariate normal, with mean zero and a covariance matrix [39].

$$\begin{bmatrix} \sigma_\mu^2 & \cdot & \cdot \\ \sigma_{21} & \sigma_1^2 & \cdot \\ \sigma_{31} & \cdot & \sigma_2^2 \end{bmatrix} \tag{16}$$

where σ_μ is the variance of error in the selection equation; while σ_1^2 and σ_2^2 are variances of error in the outcome (continuous) equation. Also, σ_{21} and σ_{31} are the covariances of μ_i and v_{1i}, and μ_i and v_{2i}, respectively. The covariance between v_{1i} and v_{1i} is not defined since in the counterfactual, we cannot observe both y_{1i} and y_{2i} at the same time.

The model above is an endogenous switching regression, in which the error term of the selection equation is correlated with the error terms in the outcome equation. The model could be estimated using a two-step approach or a maximum likelihood approach [42,44,46]. However, following [43], the model can be efficiently estimated with a single step Full Information Maximum Likelihood (FIML) estimation procedure. This study therefore used the user-written *'movestay'* command on stata 14 to estimate the FIML following studies of [37,44]. The FIML simultaneously estimates the selection and outcome equation in one step, in order to yield consistent estimates of the standard errors. The FIML model proposed by [43] and used for this study is presented as follows:

$$InL_i = \sum_{i=1} \left\{ I_i w_i \left[\ln(F(\eta_{1i})) + \ln\left(\frac{f(v_{1i}/\sigma_1)}{\sigma_1}\right) + (1 - I_i)w_i[\ln(1 - F(\eta_{2i})) + \ln(f(v_{2i}/\sigma_2/\sigma_2))] \right] \right\} \tag{17}$$

$$\text{where } \eta_{ji} = \frac{\omega Z_I + \rho_j v_{ji} / \sigma_j}{\sqrt{1 - \rho_j^2}}, \qquad for \; j = 1,2 \tag{18}$$

With rho(ρ) being the correlation coefficient between the two error terms; such that $\rho_1 = \frac{\sigma_{21}^2}{\sigma_\mu \sigma_1}$ and $\rho_2 = \frac{\sigma_{31}^2}{\sigma_\mu \sigma_2}$ are the correlation coefficients between v_1 and μ, and v_2 and μ, respectively.

The signs and significance of rho (ρ) have economic interpretations. When rho is significant, then there is evidence of endogenous switching which would then result in selection bias. Also, when ρ_1 and ρ_2 have alternate signs, the decision for each regime is based on comparative advantage [45], while it indicates hierarchical sorting if they have the same signs [37,38]. Once the parameters have been estimated, both conditional and unconditional potential outcomes can be determined. The Unconditional estimates are:

$$E(y_1|x_i) = X_{1i}\beta_{1i} \tag{19}$$

$$E(y_0|x_i) = X_{0i}\beta_{0i} \tag{20}$$

Hence, the population Average treatment effect;

$$(ATE) = E(y_{1i} - y_{0i}|X_i) \tag{21}$$

The conditional parameter estimates are:

i.   Potential outcome of farmers who are value adders and self-select into value adder groups

$$E(y_{1i}|x_i, d = 1) = X_{1i}\beta_{1i} + \sigma_1\rho_1 f(\omega Z_i)/F(\omega Z_i) \tag{22}$$

ii.   Potential outcomes of non- value adders who self-self into non-value adder groups.

$$E(y_{2i}|x_i, d = 0) = X_{2i}\beta_1 - \sigma_2\rho_2 f(\omega Z_i)/(1 - F(\omega Z_i)) \tag{23}$$

iii.   Potential outcomes of value adders if they were non-value adders

$$E(y_{1i}|x_i, d = 0) = X_{1i}\beta_1 - \sigma_1\rho_1 f(\omega Z_i)/(1 - F(\omega Z_i)) \tag{24}$$

Potential outcomes of non-value adders had they been value adders

$$E(y_{2i}|x_i, d = 1) = X_{2i}\beta_1 + \sigma_2\rho_2 f(\omega Z_i)/F(\omega Z_i) \tag{25}$$

The conditional outcomes are particularly important in this study. We are able to derive the Average Treatment Effect on the Treated (ATT) which is the impact of value addition on the outcome of value adders. The ATT is the difference in the potential outcome of value adders who self- select into value adder groups and the outcome of value adders had they been non-value adders.

$$\text{ATT} = E(y_{1i} - y_{2i} \ |d = 1) = X_i(\beta_1 - \beta_2) + (\sigma_{1\mu} - \sigma_{2\mu})\omega_1 \tag{26}$$

A similar model, but not used in this study, is the 'etregress' function in Stata for the estimation of a FIML of the Endogenous Switching Regression Model (ESRM). The 'etgress' models a linear regression model for the outcome variable and then a constrained normal distribution function in order to correct for the bias that arises from the deviation from the conditional independence assumption in the causal effect estimation procedure [47].

4. Results and Discussion

4.1. Summary Characteristics of Farmers across Cassava Production Systems

In this section, we present the results of the socioeconomic and production characteristics across the different production systems of cassava in the country. Table 1 shows that overall, there are more males (73.39%) than females (26.61%) in the cassava system. However, while SP were mostly male (91%), value addition systems (PP, PM, and PPM) had more female members among them. The overall average age of owners of enterprise was 47 years, with the oldest (about 53 years) and the youngest (about 46 years) in the PM and PP systems, respectively. Household size significantly differed across the system; averaging 7 members; except for PM households with 10 members.

There was no significant difference across the systems in terms of years of education which averaged 7 years. There were however significant differences in the number of years of experience with a mean of 17 years across the farmers. Farmers in the SP system had the highest number of years of experience at 21 years, while the PP farmers had the least at 17 years. This may be indicative of the preponderance of younger generation farmers in value addition system.

Significant differences were seen with land area holding across the systems. The average land area was 2.4 ha across the systems; however PM production system had significantly higher land area (4.43 ha), and the PP system had the least (1.85 ha). There was a general low proportion of farmers with agricultural training (24%), although up to 31% of owners in the PP systems received agricultural training and not more than 12% of the PM received agricultural training. Access to credit was generally low across the systems, however PPM system farmers had higher access to credit (about 40%) than the other groups. Registration of agricultural enterprise was extremely low across the systems (1.7%), however registration was highest (4.4%) among PPM farmers, while none of the PP farmers had registered their enterprise.

Furthermore, there was a low level of contact with extension agents (26.9%) across the production systems. However, farmers who added values at any of the levels had higher contact with extension agents, with about 37% of the PPM having extension contact. This may signify some form of endogeneity with respect to revenue generation from the systems. Although there was no significant difference in terms of membership of social group, only about 39% of the farmers had social group characteristics. Social group characteristics was however highest among the PPM (55%) and lowest among the SP (36%) farmers.

4.2. Summary Statistics of Costs and Returns across Cassava Production Systems in Nigeria

The summary of variables used to measure productivity is presented in Table 2. The pooled summary showed an increase in output revenue among the cassava farmers. However, it was noted that while labour costs reduced over the years, costs of other variable inputs (seed, power, and transport) increased. The reduction in labour cost may be as a result of increasing use of improved technologies of production across the systems. The increase in seed costs may be the result of increase production expansion and/or utilization of cassava as a result of increased awareness of its usefulness. The increase in power costs is not unconnected on the one hand to an increase in electricity tariffs and petroleum products in the course of the three periods. Transportation costs generally increased in Nigeria when prices of petrol rises; which has been the case in Nigeria in the past years since 2015.

The result in Table 2 also shows that the highest cost outlay came from the PM, followed by the PPM production system. However, the PM also had the highest revenue outlay, also followed by the PPM system. This shows the importance of access to market in agricultural systems. Access to market ensures that the farmers get remunerative prices for their produce. Although value additions in terms of transportation, storage, and communications may increase costs when compared to the other systems, the revenue is also highly significant enough. This reinforces the claims that value additions in terms of processing and marketing are important in raising the productivity of farming systems [10,18]. These significant differences in output and costs outlay already shows that there are differences in the production dynamics across the systems. This further forms an empirical basis for examining productivity differentials across the systems.

4.3. Productivity Measures across Cassava Production Systems

The Malmquist total factor productivity measure was estimated from the linear programming Data Envelopment Analysis, with the assumption of constant returns to scale and input orientation. First, we present efficiency as a measure of the productivity of the cassava farmers. Then we present the result of productivity changes across the systems for the periods under review.

The result of the mean technical efficiencies across the systems and for the pooled data is presented in Table 3. The result presents the mean technical efficiencies for the three time periods (2015, 2016, and 2017) for constant returns to scale and variable returns to scale. The results shows mean technical efficiencies of 73.3%, 71.8%, and 71.6% for the period 2015, 2016, and 2017, respectively. This shows that when compared to the reference group (a cassava farmer on the frontier), the sampled farmers have a 27.7%, 28.1%, and 28.4% chance of being on the best option frontier, that is, of increasing their output with the existing inputs for the respective years. The results also imply there was a decline in efficiency from the reference period (2015) to the other periods (2016, 2017). This follows the general decline in value added for agriculture in Nigeria which has had a growth rate of less than 1% over the years, with cassava production significantly less than the estimated potential for the country [6].

Table 1. Summary Statistics of Farmer Characteristics across Cassava Production Systems.

Variables	SP (n = 192)	PP (n = 199)	PM (n = 42)	PPM (n = 49)	POOLED (n = 482)	Chi Test
Gender of farmer (%)						
Male	91.15	54.77	92.86	62.50	73.39	77 ***
Female	8.85	45.23	7.14	37.50	26.62	
Age of farmer (mean years)	48.82 (15.38)	45.98 (11.95)	52.93 (14.50)	46.57 (10.02)	47.78 (20.14)	9.35 ***
Household size(mean)	7.08 (4.23)	6.41 (2.84)	9.62 (6.55)	8.00 (3.91)	7.11 (4.06)	15.12 ***
Years of education of farmer (mean)	7.28 (4.75)	6.59 (5.08)	5.60 (4.91)	7.47 (5.63)	6.87 (5.01)	5.57
Years of experience (mean)	21.38 (14.19)	17.50 11.58)	20.14 (12.63)	19.92 (10.19)	19.52 (12.74)	8.53 **
Land area used (ha; mean)	2.34 (4.07)	1.85 (2.26)	4.59 (8.48)	3.33 (5.28)	2.43 (4.27)	8.89 ***
Received agricultural training (%)	20.83	31.16	11.90	24.49	24.69	9.70 ***
Use Credit (%)	23.44	24.62	23.81	42.86	25.93	8.20 **
Registration of agricultural enterprise (%)	1.56	0.00	4.76	6.12	1.66	11.82 ***
Access to extension services (%)	22.40	29.65	23.81	36.73	26.97	5.39
Membership of Social group (%)	36.46	39.70	38.10	55.10	39.83	5.73

Note: Standard deviation in parentheses; Source: computed from field survey data, 2018; ** and *** are significance at 5% and 1% respectively.

Table 2. Revenue and Costs Outlays used for Productivity measures across Cassava Production Systems in Three Periods.

Variable Items	2017					2016					2015				
	SP	PP	PM	PPM	POOLED	SP	PP	PM	PPM	POOLED	SP	PP	PM	PPM	POOLED
Revenue (N)	332,414.9	570,144.2	5,496,025	794,252	984,329	297,573.1	645,213.9	4,135,661	767,969.1	823,360.7	282,737.1	707,896.9	2,915,499	546,430.5	657,614.7
Labour costs (N)	122,726.6	191,574.4	195,997.6	171,051	162,448.5	86,704.04	906,722.2	150,317.3	131,846.9	435,391.5	80,584.64	895,209	128,694.6	138,913.3	427,034.8
Seed cost (N)	19,156.25	206,835.5	227,514.3	178,809.6	131,028.1	15,020.83	182,789.7	214,835.7	142,353.8	114,642.3	14,597.66	179,402.8	191,175	115,807.9	108,315
Power (N)	9186.458	30,148.39	8566.667	39,218.37	20,839.89	7594.531	20,658.62	8463.095	27,274.71	15,064.58	6927.865	20,866.33	191,175	25,481.63	14,510.89
Transportation (N)	17,056.27	31,052.76	52,100	44,675.51	28,696.27	10,778.14	20,873.62	48,476.19	31,434.69	20,331.02	9366.276	17,705.53	42,988.1	24,053.06	17,232

SP—Sole Producer; PP—Producer Processors; PM—Producer Marketer; PPM—Producer Processor Marketer. Source: computed from field survey data, 2018.

Table 3. Mean Technical Efficiencies across Cassava Production Systems in Nigeria.

Period/Technical Efficiency	CRS					VRS				
	SP	PP	PM	PPM	POOLED	SP	PP	PM	PPM	POOLED
2015	0.836	0.825	0.770	0.894	0.732	0.865	0.881	0.840	0.937	0.806
2016	0.833	0.808	0.737	0.895	0.718	0.859	0.868	0.833	0.943	0.803
2017	0.839	0.812	0.761	0.911	0.716	0.881	0.874	0.842	0.948	0.796

Source: Computation from the DEA results output.

With respect to the different systems however, we see a decline in technical efficiency from 2015–2016 across the SP, PP, and PM systems. There was, however, an observed increase in mean technical efficiency between 2016 and 2017. For the PPM system, however, we found a steady increase in technical efficiency of 89.4%, 89.5%, and 91.1% for each of the three years respectively. This indicates that increased value addition through processing and marketing has the tendency to increase revenue and overall productivity of actors.

4.4. Productivity Growth across Cassava Production Systems

In addition to providing the technical efficiencies across the time periods, the Malmquist DEA also gave measures of changes in some productivity indices across the systems over the years. The results displayed in Table 4 gives the measures of technical change, scale efficiency change, perfect technical change, and total factor productivity change. When observed across the four production systems, we found that in 2016, only technical change increased across all the systems. This suggests that there was an improvement in technology use between 2015 and 2016 within cassava production systems in Nigeria. However, there was a general decline in scale efficiency across the system, implying that there was little in terms of scaling up among the cassava farmers between 2015 and 2016. Perfect technical efficiency measure using a Variable Returns to Scale (VRS) technology frontier improved only with SP and PPM production systems, while efficiency change improved only with PPM. Total factor productivity increased across all the systems, except for the PM system. The highest productivity change was found in the PPM system.

Table 4. Productivity Growth across Cassava Production Systems in Nigeria.

System/Growth Indices	2016					2017				
	SP	PP	PM	PPM	Pooled	SP	PP	PM	PPM	Pooled
Efficiency change	0.997	0.978	0.960	1.004	0.982	1.007	1.006	1.030	1.020	0.997
Technical change	1.005	1.027	1.037	1.019	1.020	0.978	0.974	0.950	0.933	0.980
Pure technical efficiency change	1.006	0.986	0.982	1.009	0.996	0.994	1.006	1.014	1.007	0.992
Scale efficiency change	0.991	0.992	0.977	0.995	0.986	1.013	1.000	1.016	1.013	1.005
Total factor productivity change	1.002	1.005	0.995	1.023	1.002	0.985	0.980	0.979	0.952	0.976

Source: Computed from DEA analysis of field survey data, 2018. Note; Year 2015 is the reference year.

In 2017, we observed an improvement in technical efficiency across the four systems, and the extent of improvement was highest among the PPM system. Also, scale efficiency change was positive for all production system and highest among the PPM. Perfect technical efficiency improved in all the production systems except for the SP. Total factor productivity was however found to have declined between 2015 and 2017. A plausible reason for this may be the decline in agricultural production generally as a result of conflicts and other risks within these periods in Nigeria. Moreover, inflation rose to an all-time high of 16% in 2017 as compared to the previous periods under study. This led to a general increase in costs of production while reducing the purchasing power of consumers.

From the pooled results, we observed a positive growth in total factor productivity between 2015 and 2016. However, when compared to 2015, there was a decline in total factor productivity in 2017 by 0.024. There was, however, positive scale efficiency growth among all the farmers. All the components of total factor productive also showed a decline. Therefore, we can assert that there is a decline in total factor productivity in Nigeria within the cassava system.

The above results show that there exist differences in productivity levels across the systems of cassava production. However, no causal inference can yet be made with regards to the impact of value addition within the systems. The differences shown also indicate evidence of selectivity bias in the sample. Value addition and better productivity could be the result of some intrinsic characteristics of the farmer that enhances the decision to invest in value added activities and get better returns than their counterparts.

As a result of this, we resort to the use of an endogenous switching model to estimate causal effect of value addition in the cassava system. This is expected to provide the exact extent of productivity differential with value addition without bias.

4.5. Productivity Impact of Value Addition across Cassava Production Systems

The descriptive analysis and productivity measures have shown that there are differences in productivity outcomes for the different systems (value addition/non-value addition) in the cassava system. However, to estimate a proper impact of value addition on productivity, an impact assessment needed to be carried out. As a result of assumed inherent endogeneity in the model, we used the endogenous switching regression procedure of [42] to model the outcome of value addition in cassava system on the output of cassava farmers in the study area. In the switching regression model, the variables that are contained in the covariates X_i for the outcome variable may overlap with that of the selection model, Z_i. There must, however, be some variables that identify Z_i separately from X_i [42,45]. These variables are said to determine the outcome equation only through the selection model [44]. The productivity equation is jointly estimated with the selection equation that determines value addition among the farmers.

The result of the FIML estimates is given in Table 5. First, given the significant LR test shown in the table, we are able to reject the null hypothesis of no correlation between the unobservables that determine the treatment and the outcome. Hence, a justification for the use of an endogenous switching regression model. Using an OLS regression to model the impact in this case would, therefore, have given biased estimates, and thus the need for a switching regression. The coefficient of rho for the two regimes have alternate and significant signs, implying that self-selection occurred in the value addition decision for the farmers [43]. While it is positive and significant for farmers that are value adders, it was negative and significant for the non-value adders. This suggests that the decision to add value is based on comparative advantage. This is consistent with studies of [37,45], where comparative advantage was found as the basis for adopting technology and nonfarm activities respectively.

The significance of the coefficient of rho for value adders imply that there may be unobservables that tend to lead to higher productivity rather than just adding value to cassava. Furthermore, the difference between ln sigma1 and ln sigma2 is positive suggesting that there is positive gain from value addition among farmers in the cassava system.

In the second column, we have the estimates of the joint Probit model in the joint estimation. The coefficient of gender of the farmer shows that the probability of value addition reduces for being male. This may be indicative of the socio-cultural attachment of women to agricultural value addition in general and within the cassava system in particular [48,49]. An increase in the land area available to the farmer was found to increase value addition in cassava. Land area is most times proportional to higher production in many rural systems, thus there is ample harvest that needs to be sold immediately or processed. Value addition leads to an increase in life span (processing) and better sales (marketing services) and thus may be a better option than waiting on farm-gate sales [50,51].

Table 5. Full Information Maximum Likelihood (FIML) Estimates of Impact of Value Addition of Productivity of Cassava Farmers.

	Selection Model	Productivity Equation	
	Value Adders/Non Value Adders	Value Adders	Non Value Adders
Constant	0.552 *** (0.181)	5.322 *** (0.137)	4.592 *** (0.189)
Gender of farmer (base = female)	−1.019 *** (0.094)	0.347 *** (0.076)	0.471 *** (0.117)
Age of farmer	0.005 (0.003)	0.005 (0.003)	−0.008 *** (0.003)
Land area	0.021 *** (0.010)	−0.058 *** (0.007)	−0.079 *** (0.008)
Years of education	−0.028 *** (0.008)	0.000 (0.006)	0.008 (0.007)
Agricultural training	0.187 ** (0.094)	−0.197 *** (0.074)	−0.007 (0.082)
Non-farm activities	0.075 (0.073)	0.032 (0.059)	−0.106 * (0.063)
Access to extension	0.278 *** (0.092)	0.142 ** (0.073)	−0.163 ** (0.078)
Years of experience	−0.008 *** (0.003)	−0.010 *** (0.003)	0.004 (0.003)
Access to credit	0.014 (0.083)	−0.101 (0.064)	0.045 (0.071)
Membership of social group	0.098 (0.072)	0.248 *** (0.058)	0.010 (0.063)
Marital status (base = single)	0.172 ** (0.082)		
Registration status of enterprise (base = no)	1.113 *** (0.205)		
Level of utilization (base = Low)			
Medium	0.399 *** (0.072)		
Full	0.329 *** (0.075)		
Ln Sigma1	−0.168 *** (0.042)		
Ln Sigma2	−0.188 *** (0.057)		
Rho1	0.615 *** (0.082)		
Rho2	−0.895 *** (0.034)		
LR test of independence	24.99 ***		

Note: Standard errors in parentheses; *, ** and *** are significance at 10%, 5% and 1% respectively.

Farmer's level of education was found to reduce the probability of adding value within the cassava system. This may be surprising in the light of empirical evidence on the importance of education in adopting innovation. However, [52] showed that the impact of education on agriculture may not be a direct effect on the farm family. In fact, he was able to show that with increased education and more formal skills, individuals tend to explore work option in the non-farm sector. There is alsoevidence of an inverse relationship between formal education and farmers' adoption of innovative processes [53]. The coefficient of agricultural training however showed that agricultural training increased the probability of farmers being involved in value addition. Although the descriptive shows that a smaller proportion of the farmers have agricultural training, we find that training is important in changing production perspective of the agricultural households towards a better frontier than was previously operated on [54,55]. Access to agricultural extension was found to equally be significant in increasing the likelihood of farmers to add value to their cassava products. Farmers that have access to adequate extension services have been found to be exposed to innovation that changed their perception and increased their knowledge in value addition [56,57].

Our results found that years of experience reduced the probability of adding value among the cassava farmers. This may be connected with the fact that ideas and practices that have been set in production practices over the years become harder to change with proficiency in the field and especially as the farmer grows older. Also, value addition may be considered risky and labour intensive, for which older and experienced farmers have been reported to shy away from such [58,59]. This, however, contrasts with the study of [60], where more experience farmers adopted improved technology. Married farmers were also more likely to add value than single farmers in the sample. This suggests the need for additional income for bigger household size for married individuals.

Registration status of farmers' enterprise was positively linked to value addition. While registration per se does not affect productivity, it affords the farmers access to certain market-based innovation and interventions such as insurance, credit, and subsidies that would be given only to farmers who can be tracked by their registration status [61], which could have been made possible with proper registration, the impact of which may be lower productivity.

The higher the level of utilisation of cassava within a farmer's cassava field also led to a higher probability of adding value for income generation. For example, [62] showed the utilization pathway for agricultural waste leads to increased value addition and reduction in environmental pollution.

The revenue/productivity equation is presented in the third column of Table 5. The estimates are presented separately for the value adders and non-value adders. The results show that the estimated log of output for value adders increased significantly for being a male farmer and with access to extension for the value adding farmers. The estimates of the gender of the farmer is consistent with agricultural productivity literature which shows that male-headed farming households had better productivity than their female counterparts [63,64]. This finding is synonymous with several gender related literature in productivity, where women's lack of access to productive resources meant that they had significantly lower productivities than men [65]. Access to extension was also found to improve productivity of value adders in the study. This follows productivity literature where extension services led to the dissemination of information that may be productivity-increasing for the farmers [66].

Larger land area was however found to have an inverse relationship with productivity, following the literature on agricultural productivity. For example, [67,68] found an inverse relationship between land size and productivity, which was posited as a result of diseconomies of scale for the smallholder farmers. However, [69] found a positive relationship between farm size and productivity, most likely the result of fertilizer use. Contrary to most literature, in [70,71] agricultural training was however found to negatively impact on the productivity of value adders. This may be that additional agricultural training leads to diminishing marginal returns on productivity since the farmers were already added value Years of experience also led to a loss of productivity of the value adders. This is also consistent with studies that show that older and more experience farmers are less able to exert energy and take risky decisions, with negative consequences for their productivity [59]. This is, however, different from the studies of [72], who found that years of experience improved human capital and led to increased agricultural productivity in Senegal.

Our results also reveal that social capital is important in improving the productivity of the value adders in the cassava system in Nigeria. This follows consistently with literature that examine the effect of social capital on productivity and welfare. The importance of social capital in our case is especially observed with the value adders. Hence, social capital while promoting group cohesion may help to improve adoption of innovation [73], lower transactional costs and returns higher prices [74,75], and serve as a risk management tool [76].

For the non-value adders, we found that older farmers had significantly lower productivity. Thus, apart from gaining experience as the farmer gets older, age in itself leads to loss of vigor and reduction in managerial and labour capacity of the farmer. The coefficient of land area was also negative and significant, supporting evidence of the inverse relationship between land area and agricultural productivity. Female-headed households in the non-value addition groups also had significantly lower productivity than their male counterparts. The coefficients of non-farm activities also showed an inverse relationship to productivity for non-value adders. This is supported by [77], where non-farm activities led to a loss of labour from smallholding farms and hence a reduction in productive capacity.

Moreover, the coefficient of access to extension was negative and significant in determining the productivity of the farmers. In their study, [66] found that the impact of extension on productivity came through farmers adopting the improved technology brought by extension. Hence, the negative effect of extension on productivity of the non-value adders may be suggestive of the decision not to add value in spite of possible contact with extension. However, as shown in the descriptive statistics, inadequate access to extension services may also result in low access to production information and hence low productivity [78].

4.6. Estimated Impact of Value Addition on Productivity of Cassava Farmers

We present the unconditional and conditional outcomes (Log of value of output/ha) of value addition to the cassava farmers in this section. The results are presented in Table 6. The difference between the unconditional outcomes for value adders and non-value adders gives the population Average Treatment Effect (ATE). On the other hand, the differences in the conditional outcomes give the Average Treatment Effect on a Treated (ATT) population within the study.

Table 6. Estimated Impacts of Value Addition on Productivity.

	Mean	*t*-Test
Unconditional		
$E(y_{1i}|X_i)$	5.048(0.010)	
$E(y_{0i}|X_i)$	4.395 (0.010)	
ATE	0.653 (0.014)	46.92 ***
Conditional		
$E(y_{1i}|d = 1)$	5.392 (0.011)	
$E(y_{0i}|d = 1)$	5.128 (0.013)	
ATT	0.263 (0.010)	14.91 ***

Note: Standard Errors in parentheses; *** is significance 1%.

The potential productivity measures for cassava farmers who are value adders and non-value adders, given the covariates (X_i), are 5.048 and 4.395 respectively. Therefore, the ATE of value addition in the cassava system in Nigeria is estimated at 0.653; that is, value addition within the cassava system returns an extra productivity value of 0.653 (14.85% increase) to the farmers.

In order to determine the impact of value addition on the productivity of farmers who are already value adders, the conditional outcome for the value adders ($Ey_{1i}/d = 1$) was compared to what they would have had if they were not value adders ($Ey_{0i}/d = 1$). The difference in the outcomes is the gain (or loss) for value addition. The mean log of productivity/ha for value adders is 5.392, while the outcome if they were not value adders would be 5.128. The Average Treatment Effect on the treated population of value adders (ATT) is, therefore, 0.263. This implies a positive gain of about 5% in productivity for value addition. Hence, value addition within the cassava system has the ability to enhance farmers revenue be increasing the sources and values of their production system. It is known that value addition places a premium on agricultural products, thus the extra gain from the value addition process increases overall productivity of the farmers.

5. Conclusions and Policy Recommendations

Despite the knowledge that value addition and increased biomass utilization has been shown to increase revenue base of farmers, there has been a dearth of information on the productivity differences across production systems that define these value additions. Using farm-level data from cassava farmers at different levels of value addition and cassava utilization, this study examined the productivity differentials across four identified cassava production system. The study also estimated the impact of value addition on farmers' output while solving for potential selectivity bias. Estimates of average differences in productivity and technical efficiency are not sufficient to account for the impact of the decision to add value to cassava, since they do not effectively account for other farmer characteristics that may come to play. We, therefore, modelled the impact as a selection process where an expected increase in productivity drives farmers' decision to add value, using an endogenous switching regression model. The endogenous switching regression model was able to correct for selection bias and model the impact of value addition on value adders and non-value adders in the study.

The study found that mean technical efficiencies were below the frontier for all the systems. This implies that cassava farmers still need to improve conversion of input to output more efficiently in order to be on the best practice frontier. However, technical efficiency was highest for the PPM system, which combined processing and marketing as value added activities in the cassava system. Decomposing the productivity increases, we found that while technical change was the reason for improved efficiency in 2016, scaling up was the factor that increased productivity in 2017. Therefore, we conclude that value addition in cassava production systems lead to an increase in technology use at the first and then to a scaling up effect in subsequent periods. These leads to an increase in productivity.

Estimating the effect of value addition on productivity with respect to aggregated value adding farmers versus non-value adders using the ESRM returned significant positive gains for value addition.

Our results have policy implications with respect to the decision to add value and the level of productivity of cassava farmers in Nigeria. The findings suggest that the decision to add value is dependent on the extent to which farmers have access to innovation information through extension and agricultural trainings. Hence, a renewed call for strengthening extension education and training within the Nigerian farming systems. There is a particular need to provide an avenue for farmers and research to exchange information and ideas through farmer field school, on-plot trainings, or relevant information transfer using appropriate media.

Moreover, formal registration of farmers was found to be significant in the decision to add value, hence the need to make the registration process of small and medium enterprises simple and affordable. Smallholders are largely excluded from mergers and investment opportunities since their registration status are unknown too investors. Therefore, making registration and the attendant benefits available to the farmers will enable them gain opportunities to investment. Also, the need to improve the capacity of social networks among the farmers is important. Our findings imply that social networks can help improve productivity, probably through the pathways that enable value addition. We recommend policy options that complement the efforts of these social networks, such as group training and affordable credits.

Author Contributions: Conceptualization, T.A.A. and V.O.O.; Data curation, T.A.A.; Formal analysis, T.A.A.; Methodology, T.A.A.; Project administration, V.O.O.; Supervision, V.O.O.; Visualization, T.A.A.; Writing—original draft, T.A.A.; Writing—review & editing, V.O.O.

Acknowledgments: The study is part of the Young Postdoctoral Programme of the BiomassWeb project jointly coordinated by Forum for Agricultural Research in Africa (FARA), Ghana and the Centre for Development Research (ZEF), Germany, and the University of Ibadan, Nigeria. We acknowledge the financial support from the German Federal Ministry of Education and Research (BMBF) supported with funds from the German Federal Ministry for Economic Cooperation and Development (BMZ).We also acknowledge support of all project partners.

Funding: This project was carried out under the BiomassWeb project GlobE, and was funded by the German Federal Ministry of Education and Research (BMBF) through the collaborative project "Improving food security in Africa through increased system productivity of biomass-based value webs". This project is part of the GlobE—Research for the Global Food Supply programme, Grant No (031A258H).

Conflicts of Interest: The authors declare no conflict of interest.

References and Note

1. Fan, S.; Brzeska, J.; Keyzer, M.; Halsema, A. *From Subsistence to Profit: Transforming Smallholder Farms*; International Food Policy Research Institute (IFPRI): Washington, DC, USA, July 2013.
2. Martin, G.; Martin-Clouaire, R.; Duru, M. Farming system design to feed the changing world. A review. *Agron. Sustain. Dev.* **2013**, *33*, 131–149. [CrossRef]
3. Obayelu, A.E.; Afolami, C.A.; Agbonlahor, M.U. Relative profitability of cassava-based mixed cropping systems among various production scale operators in Ogun and Oyo States Southwest Nigeria. *Afr. Dev. Rev.* **2013**, *25*, 513–525. [CrossRef]
4. Aminu, F.O.; Okeowo, T.A. economic analysis of cassava mixed farming enterprises in Epe Local Government Area, Lagos State, Nigeria. *Appl. Trop. Agric.* **2016**, *21*, 122–130.
5. Abu, B.M.; Issahaku, H.; Nkegbe, P.K. Farm-gate versus market centre sales: A multi-crop approach. *Agric. Food Econ.* **2016**, *4*, 21. [CrossRef]
6. Food and Agricultural Organisation (FAO). Nigeria at a Glance. Available online: http://www.fao.org/nigeria/fao-in-nigeria/nigeria-at-a-glance/en/ (accessed on 5 September 2018).
7. Mmasa, J.J. Value addition practices to agricultural commodities in Tanzania. *Tanzan. Country Level Knowl. Netw. Policy Brief* **2013**, *20*, 1–15.
8. Ubalua, A.O. Cassava Wastes: Treatment options and value addition alternatives. *Afr. J. Biotechnol.* **2007**, *6*, 2065–2073. [CrossRef]
9. Brees, M.; Parcell, J.; Giddens, N. Capturing vs. Creating Value. MU Agricultural Guide, University of Missouri Cooperative Extension 2010. Available online: http://extension.missouri.edu/p/G641 (accessed on 31 August 2018).
10. Lu, R.; Dudensing, R. What do we mean by value-added agriculture? *Choices* **2015**, *30*, 1–8.
11. Virchow, D.; Beuchelt, T.; Denich, M.; Loos, T.K.; Hoppe, M.; Kuhn, A. The value web approach. *Rural Focus* **2014**, *21*, 16–18.
12. Aw, B.Y.; Chen, X.; Roberts, M.J. Firm-level evidence on productivity differentials and turnover in Taiwanese manufacturing. *J. Dev. Econ.* **2001**, *66*, 51–86. [CrossRef]
13. Ghebru, H.; Holden, S.T. Technical efficiency and productivity differential effects of land right certification: A quasi-experimental evidence. *Q. J. Int. Agric.* **2015**, *54*, 1–31.
14. Besharat, A.; Amirahmadi, M. The study of factors affecting productivity in the agriculture sector of Iran. *Afr. J. Agric. Res.* **2011**, *6*, 4340–4348.
15. Kissoly, L.; Faße, A.; Grote, U. Small scale framers' integration in agricultural value chains: The role for food security in rural Tanzania. In Proceedings of the Conference on International Research on Food Security, Natural Resources Management and Rural Development, Tropentag 2015, Berlin, Germany, 16–18 September 2015.
16. Warsanga, W.B.; Evans, E.A. Welfare impact of wheat farmers' participation in the value chain in Tanzania. *Mod. Econ.* **2018**, *9*, 853. [CrossRef]
17. Obaga, B.R.; Mwaura, F.O. Impact of farmers' participation in banana value addition in household welfare in Kisii Central Sub-County. *Int. Acad. J. Soc. Sci. Educ.* **2018**, *2*, 25–46.
18. Kehinde, A.L.; Aboaba, K.O. Analysis of value addition in the processing of cassava tubers to "garri" among cottage level processors in southwestern Nigeria. In Proceedings of the International Conference of the African Association of Agricultural Economists, Addis Ababa, Ethiopia, 23–26 September 2016.
19. Singh, I.; Squire, L.; Strauss, J. (Eds.) *Agricultural Household Models: Extensions and Applications*; Johns Hopkins University Press: Baltimore, MD, USA, 1986.
20. Sadoulet, E.; de Janvry, A. *Quantitative Development Policy Analysis*; The Johns Hopkins University Press: Baltimore, MD, USA; London, UK, 1995.
21. Mendola, M. Farm Household Production Theories: A Review of "Institutional" and "Behavioral" Responses. *Asian Dev. Rev.* **2007**, *24*, 49–68. [CrossRef]
22. Schultz, T.W. *Transforming Traditional Agriculture*; University of Chicago Press: Chicago, IL, USA, 1964.

23. Meyer, J. Expected utility as a paradigm for decision making in agriculture. In *A Comprehensive Assessment of the Role of Risk in US Agriculture*; Springer: Boston, MA, USA, 2002; pp. 3–19.
24. Olomola, A.S. Competitive Commercial Agriculture in Africa Study (CCAA). World Bank Site Resource. Available online: siteresources.worldbank.org (accessed on 1 September 2018).
25. National Bureau of Statistics (NBS). Cassava production in Nigeria, 1970–2010. *NBS Stat. Bull. Federal Republic of Nigeria* **2010**.
26. Jenkins, S.P.; Siedler, T. *Using Household Panel Data to Understand the Intergenerational Transmission of Poverty*; Discussion Paper 694; German Institute for Economic Research (DIW): Berlin, Germany, 2007.
27. Assaad, R.A.; Krafft, C.; Yassin, S. *Comparing Retrospective and Panel Data Collection Methods to Assess Labor Market Dynamics*; Discussion Paper No 11052; Institute of Labour Economics (IZA): Bonn, Germany, 2017.
28. Hsiao, C. Panel data analysis—Advantages and challenges. *Test* **2007**, *16*, 1–22. [CrossRef]
29. Coelli, T.; Prasada Rao, D.S.; O'Donnell, C.; Battese, G. *An Introduction to Efficiency and Productivity Analysis*, 2nd ed.; Springer Science and Business Media: New York, NY, USA, 2005; ISBN 0387242651.
30. Stolp, C. Strengths and weaknesses of data envelopment analysis: An urban and regional perspective. *Comput. Environ. Urban Syst.* **1990**, *14*, 103–116. [CrossRef]
31. Hossain, M.K.; Kamil, A.A.; Baten, M.A.; Mustafa, A. Stochastic frontier approach and data envelopment analysis to total factor productivity and efficiency measurement of Bangladeshi rice. *PLoS ONE* **2012**, *7*, e46081. [CrossRef] [PubMed]
32. Portela, M.C.A.S.; Thanassoulis, E. Zero weights and non-zero slacks: Different solutions to the same problem. *Ann. Oper. Res.* **2006**, *145*, 129–147. [CrossRef]
33. Mazvimavi, K.; Ndlovu, P.V.; An, H.; Murendo, C. Productivity and efficiency analysis of maize under conservation agriculture in Zimbabwe. In Proceedings of the International Conference for Agricultural Economists Triennial Conference, Foz do Iguaçu, Brazil, 18–24 August 2012.
34. Sarkis, J. Preparing your data for DEA. In *Modeling Data Irregularities and Structural Complexities in Data Envelopment Analysis*; Joe, Z., Wade, C., Eds.; Springer: Boston, MA, USA, 2007; pp. 305–320, ISBN 978-0-387-71607-7.
35. Coelli, T.J.; Rao, D.S. Total Factor Productivity growth in agriculture: A Malmquist index analysis of 93 countries, 1980–2000. *Agric. Econ.* **2005**, *32*, 115–134. [CrossRef]
36. Sipiläinen, T.; Kuosmanen, T.; Kumbhakar, S.C. Measuring productivity differentials–An application to milk production in Nordic countries. In Proceedings of the 12th Congress of the European Association of Agricultural Economics–EAAE, Ghent, Belgium, 26–29 August 2008.
37. Awotide, B.A.; Abdoulaye, T.; Alene, A.; Manyong, V.M. Impact of access to credit on agricultural productivity: Evidence from smallholder cassava farmers in Nigeria. In Proceedings of the International Conference of Agricultural Economists (ICAE), Milan, Italy, 9–14 August 2015.
38. Seng, K. The Effects of nonfarm activities on farm households' food consumption in rural Cambodia. *Dev. Stud. Res.* **2015**, *2*, 77–89. [CrossRef]
39. Negash, M.; Swinnen, J.F. Biofuels and food security: Micro-evidence from Ethiopia. *Energy Policy* **2013**, *61*, 963–976. [CrossRef]
40. Hasebe, T. Copula base maximum likelihood estimation of sample selection model. *Stata J.* **2013**, *13*, 547–573. [CrossRef]
41. Alene, A.D.; Manyong, V.M. The effects of education on agricultural productivity under traditional and improved technology in northern Nigeria: An endogenous switching regression analysis. *Empir. Econ.* **2007**, *32*, 141–159. [CrossRef]
42. Maddala, G.S. Disequilibrium, self-selection, and switching models. *Handb. Econ.* **1986**, *3*, 1633–1688.
43. Lokshin, M.; Sajaia, Z. Maximum likelihood estimation of endogenous switching regression models. *Stata J.* **2004**, *4*, 282–289. [CrossRef]
44. Long, W.; Appleton, S.; Song, L. *Job Contact and Wages of Rural-Urban Migrants in China*; Discussion Paper 7577; Institute for the Study of Labour (IZA): Bonn, Germany, 2013.
45. Abdulai, A.; Huffman, W. The adoption and impact of soil and water conservation technology: An endogenous switching regression application. *Land Econ.* **2014**, *90*, 26–43. [CrossRef]
46. Lee, L.F. Some approaches to the correction of selectivity bias. *Rev. Econ. Stud.* **1982**, *49*, 355–372. [CrossRef]
47. StataCorp. *Stata: Release 13. Statistical Software*, StataCorp LP: College Station, TX, USA, 2013.

48. Wright, W.; Annes, A. Farm women and the empowerment potential in value-added agriculture. *Rural Sociol.* **2016**, *81*, 545–571. [CrossRef]

49. Masamha, B.; Thebe, V.; Uzokwe, V. Mappping cassava food value chains in Tanzania's smallholder farming sector: The implications of intra-household gender dynamics. *J. Rural Stud.* **2018**, *58*, 82–92. [CrossRef]

50. Born, H.; Bachmann, J. *Adding Value to Farm Products: An Overview*; National Center for Appropriate Technology: Butte, MT, USA, 2006.

51. Evans, E. Value Added Agriculture: Is It Right for Me. Obtenido de EDIS Document FE638, Florida Cooperative Extension Service, Institute of Food and Agricultural Sciences, University of Florida, Gainesville. 2012. Available online: http://edis.ifas.ufl.edu/pdffiles/FE/FE63800.Pdf (accessed on 18 September 2018).

52. Huffman, W. Human Capital: Education and Agriculture. Iowa State University Economic Staff Paper Series 312. 1999. Available online: http://lib.dr.iastate.edu/econ_las_staffpaper/312 (accessed on 5 September 2018).

53. Uematsu, H.; Mishra, A.K. Can education be a barrier to technology adoption? Presented at the Agricultural & Applied Economics Association 2010, AAEA CAES & WAEA Joint Annual Meeting, Denver, CO, USA, 25–27 July 2010.

54. Davis, K.; Nkonya, E.; Kato, E.; Mekonnen, D.A.; Odendo, M.; Miiro, R.; Nkuba, J. Impact of farmer field schools on agricultural productivity and poverty in East Africa. *World Dev.* **2012**, *40*, 402–413. [CrossRef]

55. Kijima, Y.; Ito, Y.; Otsuka, K. Assessing the impact of training on lowland rice productivity in an African setting: Evidence from Uganda. *World Dev.* **2012**, *40*, 1610–1618. [CrossRef]

56. Roy, R.; Shivamurthy, M.; Radhakrishna, R. Impact of value addition training on participants of farmers training institutes. *World Appl. Sci. J.* **2013**, *22*, 1401–1411. [CrossRef]

57. Ntshangase, N.L.; Muroyiwa, B.; Sibanda, M. Farmers' perceptions and factors influencing the adoption of no-till conservation agriculture by small-scale farmers in Zashuke, KwaZulu-Natal Province. *Sustainability* **2018**, *10*, 555. [CrossRef]

58. Brem, R.M.; Obare, G.A.; Owuor, G. Is value addition in honey a panacea for poverty reduction in the asal in Africa? Empirical evidence from Baringo District, Kenya. In Proceedings of the Joint 3rd African Association of Agricultural Economists (AAAE) and 48th Agricultural Economists Association of South Africa (AEASA) Conference, Cape Town, South Africa, 19–23 September 2010.

59. Danso-Abbeam, G.; Antwi Bosiako, J.; Ehiakpor, D.; Mabe, F.; Aye, G. Adoption of improved maize variety among farm households in the northern region of Ghana. *Cogent Econ. Finance* **2017**, *5*, 1416896. [CrossRef]

60. Islam, K.Z.; Sumelius, J.; Bäckman, S. Do differences in technical efficiency explain the adoption rate of HYV rice? Evidence from Bangladesh. *Agric. Econ. Rev.* **2012**, *13*, 93–104.

61. Tripp, R.; Gisselquist, D. *A Fresh Look at Agricultural Input Regulation*; Natural Resource Perspectives; Overseas Development Institute (ODI): London, UK, March 1996.

62. Wang, B.; Dong, F.; Chen, M.; Zhu, J.; Tan, J.; Fu, X.; Chen, S. Advances in recycling and utilization of agricultural wastes in China: Based on environmental risk, crucial pathways, influencing factors, policy mechanism. *Procedia Environ. Sci.* **2016**, *31*, 12–17. [CrossRef]

63. Oseni, G.; Corral, P.; Goldstein, M.; Winters, P. Explaining gender differentials in agricultural production in Nigeria. In *African Region Gender Practice Policy Brief*; The World Bank: Washington, DC, USA, 2013.

64. Kilic, T.; Palacios-Lopez, A.; Goldstein, M. Caught in a productivity trap: A distributional perspective on gender differences in Malawian agriculture. *World Dev.* **2015**, *70*, 416–463. [CrossRef]

65. Njobe, B. *Women and Agriculture: The Untapped Opportunity in the Wave of Transformation*; Background Paper; African Development Bank: Abidjan, Cote d'Ivoire, 2015.

66. Emmanuel, D.; Owusu-Sekyere, E.; Owusu, V.; Jordaan, H. Impact of agricultural extension service on adoption of chemical fertilizer: Implications for rice productivity and development in Ghana. *NJAS-Wagening. J. Life Sci.* **2016**, *79*, 41–49. [CrossRef]

67. Chand, R.; Prasanna, P.L.; Singh, A. Farm size and productivity: Understanding the strengths of smallholders and improving their livelihoods. *Econ. Political Wkly.* **2011**, *25*, 5–11.

68. Ladvenicova, J.; Miklovicova, S. The relationship between farm size and productivity in Slovakia. *Visegrad J. Bioecon. Sustain. Devel.* **2015**, *4*, 46–50. [CrossRef]

69. Singh, R.K.P.; Kumar, A.; Singh, K.M.; Chandra, N.; Bharati, R.C.; Kumar, U.; Kumar, P. Farm size and productivity relationship in smallholder farms: Some empirical evidences from Bihar, India. *J. Community Mobilization Sustain. Dev.* **2018**, *13*, 61–67.

70. Ahmad, M.; Jadoon, M.A.; Ahmad, I.; Khan, H. Impact of trainings imparted to enhance agricultural production in district Mansehra. *Sarhad J. Agric.* **2007**, *23*, 1211.
71. Ulimwengu, J.; Badiane, O. Vocational training and agricultural productivity: Evidence from rice production in Vietnam. *J. Agric. Educ. Ext.* **2010**, *16*, 399–411. [CrossRef]
72. Ndour, C.T. Effects of human capital on agricultural productivity in Senegal. *World Sci. News* **2017**, *64*, 34–43.
73. Liverpool-Tasie, L.S.; Kuku, O.; Ajibola, A. *Review of Literature on Agricultural Productivity, Social Capital and Food Security in Nigeria*; NSSP Working Paper 21; International Food Policy Research Institute (IFPRI): Washington, DC, USA, 2011. Available online: http://ebrary.ifpri.org/cdm/ref/collection/p15738coll2/id/126846 (accessed on 7 September 2018).
74. Mawejje, J.; Terje Holden, S. Does social network capital buy higher agricultural prices? A case of coffee in Masaka district, Uganda. *Int. J. Soc. Econ.* **2014**, *41*, 573–585. [CrossRef]
75. Jacques, D.C.; Marinho, E.; d'Andrimont, R.; Waldner, F.; Radoux, J.; Gaspart, F.; Defourny, P. Social capital and transaction costs in millet markets. *Heliyon* **2018**, *4*, e00505. [CrossRef] [PubMed]
76. Wossen, T.; Berger, T.; Di Falco, S. Social capital, risk preference and adoption of improved farm land management practices in Ethiopia. *Agric. Econ.* **2015**, *46*, 81–97. [CrossRef]
77. Amare, M.; Shiferaw, B. Nonfarm employment, agricultural intensification, and productivity change: Empirical findings from Uganda. *Agric. Econ.* **2017**, *48*, 59–72. [CrossRef]
78. Baloch, M.A.; Thapa, G.B. The effect of agricultural extension services: Date farmers' case in Balochistan, Pakistan. *J. Saudi Soc. Agric. Sci.* **2016**, *17*, 282–289. [CrossRef]

© 2018 by the authors. Licensee MDPI, Basel, Switzerland. This article is an open access article distributed under the terms and conditions of the Creative Commons Attribution (CC BY) license (http://creativecommons.org/licenses/by/4.0/).

Article

Determinants of Intensity of Biomass Utilization: Evidence from Cassava Smallholders in Nigeria

Temitayo Adeyemo [1,*], **Paul Amaza** [2], **Victor Okoruwa** [1], **Vincent Akinyosoye** [1], **Kabir Salman** [1] **and Adebayo Abass** [3]

[1] Department of Agricultural Economics, University of Ibadan, Ibadan 200284, Nigeria; vokoruwa@gmail.com (V.O.); voakinyosoye@gmail.com (V.A.); kksalmann@gmail.com (K.S.)
[2] Department of Agricultural Economics and Extension, University of Jos, Jos 930001, Nigeria; amazap@unijos.edu.ng
[3] International Institute of Tropical Agriculture, Tanzania Hub, Dar es Salaam 11101, Tanzania; a.abass@cgiar.org
* Correspondence: adeyemotemitayo@gmail.com; Tel.: +234-802-740-1925

Received: 27 October 2018; Accepted: 15 April 2019; Published: 30 April 2019

Abstract: The paradigm shift from value chains to value webs in the emerging bioeconomy has necessitated a review on how agricultural systems transit to value web production systems. This study examines how smallholders in the cassava system in Nigeria have been able to increase utilization of biomass in their production systems. Using a sample of 541 households, the study employed cluster analysis and ordered probit regression to examine the intensity of cassava utilization and the determinants of the intensity of utilization. The study found that over 50% of the respondents were classified as low-intensity utilization households, while ~13% were high-intensity utilization households. Land, social capital, farming experience, and asset ownership increased the probability of intensifying cassava utilization among smallholders. The study recommends strengthening assets acquisition, improving land quality and encouraging social capital development among smallholders.

Keywords: biomass utilization; intensity; cassava smallholders; Nigeria

1. Introduction

The global quest for a sustainable bioeconomy has brought to the fore the importance of engaging agricultural systems in the development of high value-added products in a sustainable way [1]. This has evolved in the need to supply both food and nonfood products from agricultural biomass, while sustaining the income and livelihood of agricultural actors and the economy as a whole [2]. The potential this has for biomass-rich developing countries is enormous [3]. In this regards, consideration for the bioeconomy is thus expected to form the core of processes in many biomass-rich economies around the world. To this end, the move is geared towards a bioeconomy such that agricultural production is no longer solely for food needs, but for many other nonfood (industrial, pharmaceutical, and energy) needs. Consequently, there have been different initiatives (projects, programs, and conceptual designs) to develop technologies and practices that ensure full utilization of agricultural biomass for food and nonfood uses. One example of this is the adaptation of the value web concept as a business model for agricultural production systems and using this concept in agriculture/biomass-based value webs [3–5].

These value webs can be viewed as the linkages between and among agricultural value chains, increased utilization of biomass and/or cascading uses of resources within an agricultural system [4,5]. Value webs suggest higher levels of value addition starting from the smallholder farmer's enclave. It implies perceiving agriculture as a business with the different income generating options linked and under the control of a single entity and basket of resources for efficiency [6]. Figure 1 is an illustration of cascading uses of cassava biomass utilization in the form of a value web production concept. Economies that currently show a tendency towards agricultural value webs include the sugarcane industry in Brazil, rice industries in China, and the oil palm waste industry in Ghana [7–9]. In these economies, full utilization of agricultural biomass predominates such that waste is minimized and maximum returns are derived from the agricultural biomass. For example, in Brazil sugarcane production is geared towards the production of sugar and molasses; while the waste from the processing as well as the farm waste during harvesting are channeled into the production of biogas [10]. This provides opportunities for employment and income generation at different value chains within the sugarcane system. On this premise, the present study thus explores how biomass-based value webs are developed in the smallholder dominated cassava system in Nigeria.

Cassava is regarded as a food security crop in many developing countries, with a potential to provide off-season calories even on low-nutrient soils [11]. The crop is native to tropical regions of South America; but is now a staple crop in many African countries [12]. With the desire to sustainably increase production to improve food security, while also providing nonfood products from cassava, many countries have begun to explore innovative processes within their agricultural systems. Countries like Ghana and Mozambique have developed innovative uses for industrial raw materials such as starch, sweeteners, beer (Mozambique and Ghana), and, in recent times, industrial bioethanol from cassava [13]. It has been reported that cassava is used to make bioethanol on a small scale and is replacing paraffin in cooking stoves in South Africa [14]. Breweries in Zambia and Mozambique currently use cassava chips at commercial scale while there is growing interest by the brewing industry in Tanzania to use cassava flour. The requirement of cassava within the emerging bioeconomy is premised on value-added production and processing stages that generate more food and nonfood products within farming systems. This in effect has led to higher levels of intensity in utilization of cassava with consequences for waste reduction, diversity of product bases, and higher income for participating farmers and overall market-led production systems. In Nigeria, cassava features prominently in the production of staple food for the teeming population and Nigeria has been reported as the largest producer of cassava in the world followed by Thailand and Indonesia [15]. Cassava makes up the bulk of dietary energy supply for many households in developing countries [16]. Its versatility in the production of food and nonfood commodities has been documented; from the production of staple foods such as fermented cassava flakes, to starch used for domestic and industrial purposes [2,11]. However, its emergence as a producer of nonfood biomass and industrial raw material is still limited in Nigeria. This is because cassava is seen by many smallholders as a reserve crop and a crop for producing some of the most consumed staples across the countries [11,15]. With the increasing population and hence demand, cassava production in the country is not sufficient to meet the food needs of its populace. This also stems from the fact that low technology adoption by the cassava smallholders leads to low productivity. In addition, the low technology of processing has been the bane of processing cassava into high valued industrial commodities [17,18].

The limitation in value addition in smallholder agriculture in Nigeria is a factor that has continually reduced the potential of the agricultural system to be a major player in the global cassava bioeconomy. The limitation is a result of many interwoven and sequential variables. Perhaps the main challenge is lack of access to market by smallholder farmers in general and cassava farmers in particular [19]. This has also been attributed to a lack of standardization of products and, hence, low competitiveness within the commodity market, which is also linked to inadequate infrastructure and technology. This has resulted in the continued classification of the Nigerian smallholder as subsistence with low production capacity and income. Hence, a cassava smallholder is mainly interested in selling his roots as fast as possible, while processing just enough for the subsistence of the farm family.

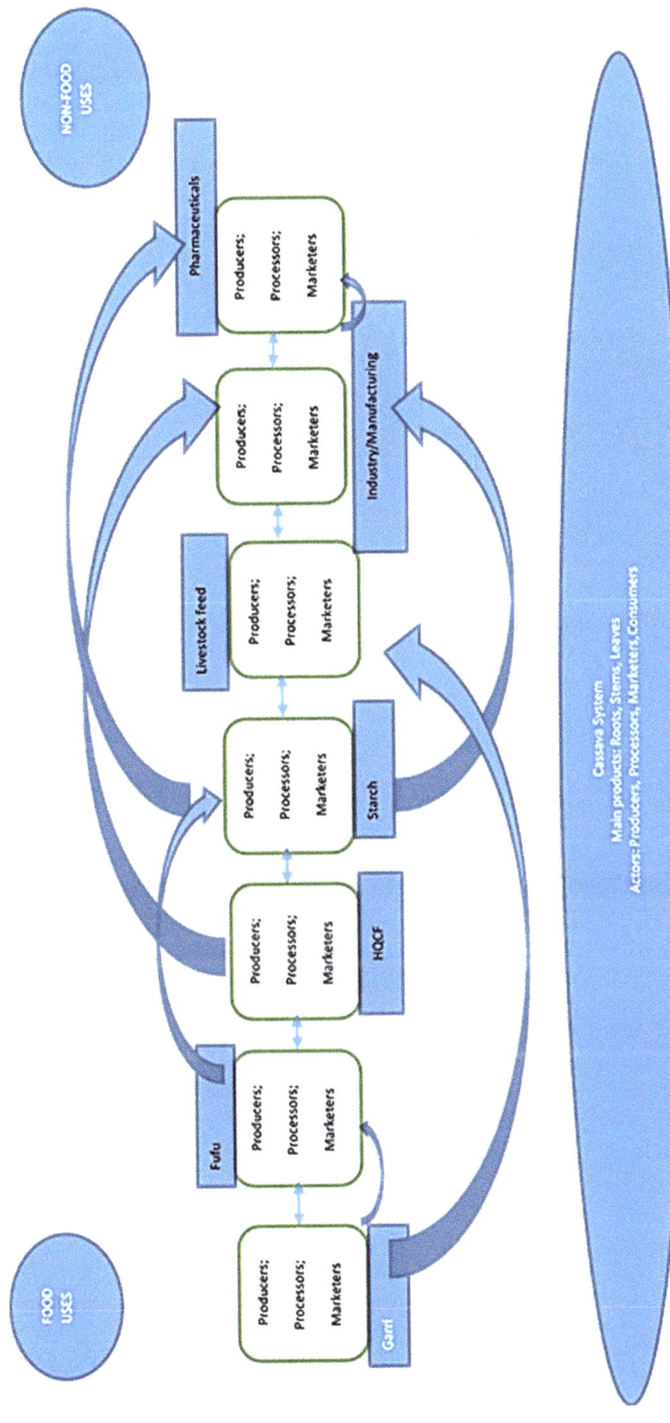

Figure 1. Illustration of cascading uses of cassava and cross cutting roles of actors in the Nigerian cassava system. Source: Author's conceptualization.

The Federal Government of Nigeria has made efforts to commercialize cassava as in the cases of cocoa and maize [20]. One such initiative was the development of the Cassava Bread Development Fund (CBDF) in 2008, which aimed to include 20% High Quality Cassava Flour (HQCF) in bread and up to 40% HQCF in other confectionery products [21]. Following the CBDF were the Agricultural Transformation Agenda (ATA) in 2011 and the present Agricultural Promotion Policy (APP) in 2015. Despite some success stories with cassava commercialization in pockets of large-scale farms, there is still a need to increase cassava utilization in view of its potential to develop the economy [22]. Since smallholders make up the majority of actors in the cassava value chain [23], there is a need to engage them in renewed efforts to increase utilization of cassava within their farming systems. We propose, in this study, that intensifying the utilization of cassava through improved production and diversified processing is a major step in value web development among cassava smallholders. On the basis of the above this study set out to establish evidence of value webs as a function of intensity of utilization in the Nigerian cassava system. The objectives are to examine the extent to which smallholders utilize cassava within their farming systems in Nigeria, classify the households into groups based on intensity of cassava utilization, and identify the determinants of the extent of cassava utilization among smallholders in Nigeria.

2. Materials and Methods

2.1. Study Area and Sampling

The study was carried out in three states representing the forest (Edo and Ogun states) and Guinea savannah (Kwara State) ecological zones in Nigeria. The Federal Republic of Nigeria is on the southern coast of West Africa, bordered by Cameroon to the East, Chad to the North East, Niger to the North, Benin to the West, and the Atlantic Ocean to the South. The ecology of Nigeria varies from the tropical rainforest in the south to the dry savanna in the North, with varying fauna and flora. The low lying coastal regions are characterized by mangroves, while the fresh water regions produce the swamp forest. Inland, the vegetation gives way to the tropical hardwood forest.

The rural population makes up over 50% of the Nigerian population [24]. In the rural areas, farming is the predominant occupation, with smallholder farming dominating. Cassava is produced in almost all states of the federation. In Northern Nigeria, cassava is produced mainly in the Guinea savannah belt—Kogi, Kwara, Benue, Taraba, and Kaduna states—while all the states in the southern Nigeria produce cassava at various capacities.

A multistage sampling procedure was used in collecting data for this study. In the first stage, three states were purposively selected from the sampling frame of cassava producing states in Nigeria. The second stage involves the random selection of two Agricultural Development Program (ADP) zones from the sampling frame of ADP zones within each selected state. Each state is divided into ADP zones and monitored by ADP officers. The third stage involves random selection of agricultural extension blocks from the zones proportionate to the size of the ADP zones earlier selected. In the fourth stage, cells were selected from each block proportionate to the size of the block. The selection of blocks and cells proportionate to the size of the ADPs and blocks, respectively, was based on the formula

$$s_i = \frac{y_i}{\sum y_i} H_i \tag{1}$$

where s_i = sample size of reference state; y_i = location population frame, and H_i = required total sample size.

In the final stage, smallholder households were randomly selected from the cell classification of the ADP. Structured questionnaires were administered to 600 smallholder cassava households; however, 541 responses were used for the analysis, while 59 responses were discarded due to inconsistencies or incompleteness.

2.2. Source and Type of Data

The main database for this study was from the household survey of cassava smallholders in the study areas. Primary data were collected for this study from a cross-section of smallholder households, since most crop farmers in Nigeria engage in mixed cropping. Determination of a cassava-based household was based on the proportion of household income that accrues to the household from cassava versus that from other crops. Data for the study was collected on household socioeconomic and demographic characteristics as well as activities carried out within the cassava holdings of the smallholder households.

2.3. Analytical Techniques

2.3.1. Intensity of Cassava Utilization among Smallholders—Cluster Analysis

A cluster analysis was used to segregate cassava smallholders into classes based on intensity of cassava utilization. The variables used in the cluster analysis were responses of the smallholder households to a list of items that correlate to utilization activities in the cassava value web system in Nigeria (see Appendix A). The study used a hierarchical cluster model to group respondents based on the activities selected so that respondents in each group were as close to each other in characterization as possible. The use of Ward's linkage method minimizes within group deviations among the respondents [25]. The Jaccard similarity measure was used because the responses to the factors used for the cluster analysis were binary in nature. With this method, the smallholder households were classified into low-level, middle-level, and high-level intensity.

2.3.2. Determinants of Intensity of Cassava Utilization—Ordered Probit Model

The ordered probit model was used to identify and compare the probabilities of smallholder households being in any of the three-ordered groups of cassava utilization intensity. The justification for using the ordered probit is that the intensity use of cassava is monotonically ordered [26,27], hence the need to examine the factors that determine inclusion in ordered outcomes. Let the ordered utilization intensity of cassava y and assume discrete values ranging from $0, \ldots, j$. The ordered probit model can be determined from a latent variable, y^*, subject to explanatory variables and specified as follows

$$y^* = \beta_i x_i' + \varepsilon_i \tag{2}$$

y^* is the unobserved discrete random variable with values ranging between 1 and 3, as shown by the categories of intensity of cassava utilization. x_i is the vector of independent variables including socioeconomic and enterprise characteristics of the smallholder households, β_i is the vector of parameters of the regression to be estimated, and ε_i is the vector of error term, which is assumed to be normally distributed and with positive probabilities.

Therefore, given the observed intensity of cassava utilization, the dependent variable takes on the following values.

$$y = \begin{cases} 1, if\ 0 < y^* \le \mu_1 \\ 2, if\ \mu_1 < y^* \le \mu_2 \\ 3, if\ \mu_2 < y^* \le \mu_3 \end{cases} \tag{3}$$

where, 1, 2, and 3 represent low, middle, and high intensity cassava utilization groups, respectively.

The explanatory variables in the model are described as

- X_1 = Age of household head
- X_2 = Age squared
- X_3 = Gender of household head (Male = 1; Female = 0)
- X_4 = Proportion of land allocated to cassava
- X_5 = Educational level of household head (Nonformal = 0; Formal = 1)

- X_6 = Household size
- X_7 = Access to agricultural credit (Yes = 1; No = 0)
- X_8 = Membership in social group (Yes = 1; No = 0)
- X_9 = Asset
- X_{10} = Nonfarm activities (Yes = 1; No = 0)
- X_{11} = number of years of farming experience
- X_{12} = Access to improved cassava variety (Yes = 1; No = 0)

The ordered probit model was analyzed by a maximum likelihood method as follows

$$\prod_{i=1}^{N} = F(X_i\beta)^{y_i}(1 - F(x_i'\beta))^{1-y_i} \tag{4}$$

The log likelihood specification for this function is thus

$$\ln L = \sum_{i=1}^{N}\left[y_i \ln F(x_i'\beta) + (1 - y_i)\ln(1 - F(x_i'\beta))\right] \tag{5}$$

Supplementary Analysis: Classification of Smallholder by Asset Ownership

The classification of farming households based on asset ownership was done through a Principal Component Analysis (PCA). The equation is given as follows:

$$C_{ij} = \sum \frac{F_i(X_{ji} - X_i)}{S_i} \tag{6}$$

where, C_{ij} = asset index value for the jth household participating in the series of 'i' activities; F_i is the weight of the ith variable from the PCA; X_{ji} is the jth household value for the ith variable; X_i and S_i are the mean and standard deviations of the values of the ith variables.). The resulting indices were thereafter used to categorise the cassava smallholders into 'Quintile' wealth categories as 'Poorest, Poor, Middle Class, Rich and Richest'.

3. Results and Discussion

This section presents results of the decision process of smallholders in utilizing cassava within the farming household decision matrix. Specifically, it shows the grouping of the smallholders based on their intensity of cassava utilization, a description of the smallholder households by their levels of participation in the cassava value web, as well as the determinants of the levels of participation in the cassava value web in the study area.

3.1. Classification of Smallholders Based on Intensity of Cassava Utilization

The result of the cluster analysis to classify the smallholders based on their intensity of cassava utilization is presented in Figures 2 and 3 and Table 1. Figure 2 is the dendrogram derived from the cluster analysis showing the classification of smallholder cassava-based households. The dendrogram shows the three distinct clusters from the observational data of households. For the purpose of initial explanation, the clusters are named A, B, and C. Cluster A is made up of 273 observations; cluster B, 195 observations; and cluster C, 73 observations.

The dendrogram does not however show which of the three clusters is higher up the hierarchy than the other in terms of the cluster groupings [28]. Hence, we do not have a clear idea of the intensity of cassava utilization among each group. The best way to make a meaningful inference from the dendrogram is to profile the groups [29]. The profiling was done by comparing values of three variables that correspond to a priori expectation of the desired groups (low, medium, and high intensity cassava utilization) across the three clusters. The results from these variables were used to score each cluster and then determine which cluster belonged to each desired group.

In this study, the profiling variables used were the number of activities carried out within the cassava system, resources (land) allocated to cassava and income from cassava-based activities. The expectation is that increased utilization of cassava suggests a higher number of activities within the cassava system. Also, increased utilization may imply more resources allocated to cassava within the smallholder's enterprise combination [30]. Finally, it is expected that, with increased cassava utilization, there would be a more diversified livelihood portfolio and hence an increase in household income [31]. The results of the profiling are shown in Table 1.

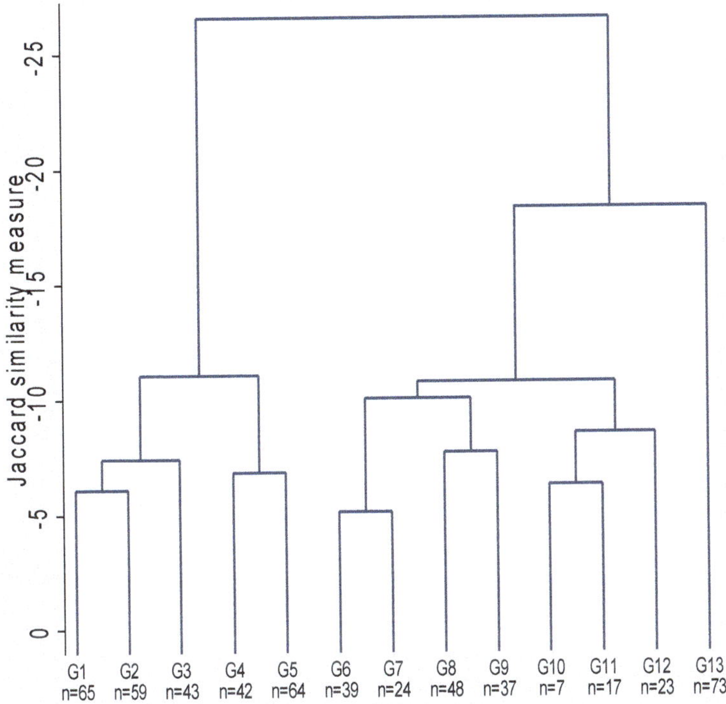

Figure 2. Dendrogram showing natural groupings of cassava-based smallholder households based on intensity of cassava utilization. Source: Author's computation using STATA statistical software.

The results presented in Table 1 show that cluster C had a rank of 1 with respect to number of activities and income but scored 2 in terms of land allocated to cassava. Cluster B, on the other hand, had a rank of 1 with respect to land allocated to cassava but a rank of 2 with respect to the other variables. However, for all three indicators, cluster A had a consistently lower rank of 3, corresponding to the lowest values for each of the indicators. We can therefore infer that by a priori expectations, cluster A corresponds to the low intensity (LI), while clusters B and C are consistent with medium intensity (MI) and high intensity (HI) cassava utilization groups, respectively. The distribution of the smallholders based on intensity of cassava utilization is shown in Figure 3. The result revealed that approximately half (50.46%) of the smallholders was LI, 36.04% was MI, and only 13.5% was HI cassava utilization groups. The results corroborate the subsistence characteristic that has been associated with the cassava industry in Nigeria [32]. This suggests that most of the policies and interventions towards commercialization of cassava as an earner of foreign revenue has not led to a significant change in the production practices of the majority of smallholders. There may be a need to justify these interventions by appraising their effects on the extent of value addition at smallholding levels.

Table 1. Distribution of selected variables for defining cluster groups.

Indicator Variables	Clusters					
	A (*n* = 273)		B (*n* = 195)		C (*n* = 73)	
	Value	Rank	Value	Rank	Value	Rank
Number of activities	4.73 (3.07)	3	6.31 (3.66)	2	8.97 (1.85)	1
Proportion of Land allocated to cassava	0.59 (0.18)	3	0.70 (0.18)	1	0.65 (0.11)	2
Income from cassava-based activities (₦)	57,364.42 (21,340.3)	3	62,021.8 (22,957.79)	2	69,368.97 (23,861.31)	1
Overall cluster rank	3		2		1	

Note: standard deviations in parentheses; Source: Computation from Field Survey, 2015.

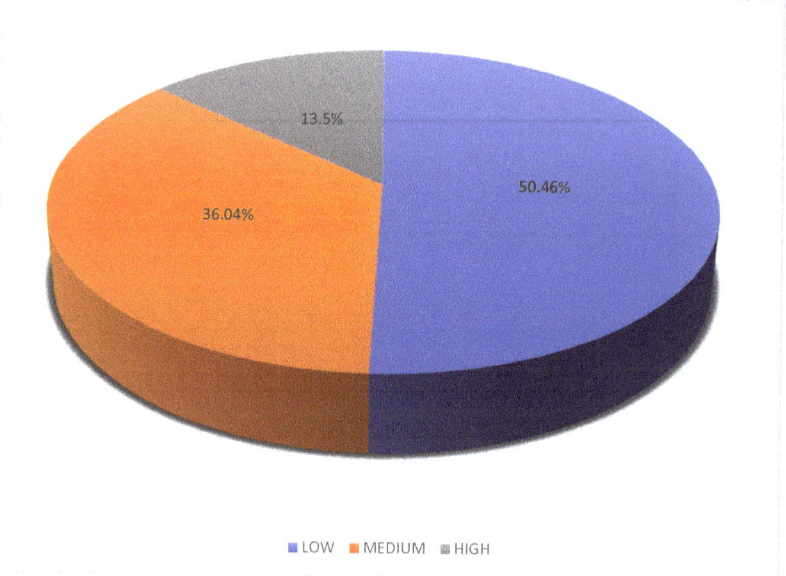

Figure 3. Percentage distribution of smallholders by intensity of cassava utilization.

3.2. Description of Farming Household Characteristics by Intensity of Cassava Biomass Utilization

The distribution of farming household socioeconomic and enterprise characteristics by their intensity of cassava utilization is presented in Table 2. The summary statistics shows no statistical difference in distribution of gender of household heads across the three groups, with overall male dominance of ~70% of the sampled household heads. The average age of the respondents was also not significantly different across the groups with mean age given as 51 years. Average household size was approximately seven across the groups, with number of dependents and income earners being four and three, respectively. A situation in which there were more dependents than income earners suggests an economic burden on the household's welfare [33]. Years of farming experience was on average estimated at 23 years, with the least being 21.64 years from the LI group and the highest being 28.12 years from the HI groups. This shows that intensity of agricultural utilization thrives not just on cassava production or educational attainment, but on the different linkages within and outside the cassava system that can be obtained from years of experience in a system [34]. The average income from cassava-based activities (₦ 60,662.98) was significantly different across the three groups; with the HI having the highest income of ₦ 69,368.97, while the MI and LI had incomes of ₦ 62,021.81 and ₦ 57,364.42, respectively.

Educational attainment showed that most household heads had primary education (31.42%), followed by secondary education (30.68%). However it is surprising that the proportion of household heads with nonformal education was highest among the HI group (42.47%), while the highest educational attainment for tertiary education was found among the MI group (11.79%). This implies that higher formal education may not be a necessary condition for smallholder households' decision to increase investment in the cassava value web [35]; rather on-farm experience and training may be more important [36]. Also, while there was a general affinity for social group membership, it was highest among the HI group (93.15%); this supports studies where social capital is hypothesized to encourage adoption of sustainable agricultural practices [37]. Access to credit was observed to be below average in the pooled observation; however, up to 67% of the HI group had access to credit for their productive activities, suggesting that access to credit encourages investment in utilizing capacity of cassava [38].

Classification of land area cultivated showed that there was a higher proportion of farming households cultivating 1.5–3 ha (57.12%); while only 15.09% cultivated <0.5 ha. This suggests an increase in land area cultivated to cassava, probably because of various governmental interventions with regards to the agricultural sector in recent years [12,14]. However, land area allocated to cassava was found to be up to 0.67 across the groups, with LI, MI, and HI allocated 0.60, 0.70, and 00.65, respectively, of their total land area to cassava. This is indicative of the renewed interest of farming households in the cassava industry and increased awareness of the potential for cassava to raise income and livelihoods of farmers and the Nigerian economy at large.

We also found that access to improved cassava variety for planting and processing was above average but increased with intensity of utilization of cassava. The results revealed that up to 77% of HI cassava households planted improved cassava varieties and 69% processed cassava varieties. This is significantly different from the LI group where ~80% planted and ~34% processed improved varieties of cassava. Approximately 65% of MI groups planted and 55% processed improved varieties of cassava in the study area. This may be suggestive of the fact that higher utilization of cassava involves some form of value addition, which entails the use of good quality seed/raw materials for maximizing utility.

Table 2. Distribution of cassava farming households by intensity of cassava utilization.

Variables	LI (*n* = 273)	MI (*n* = 195)	HI (*n* = 73)	Total (*n* = 541)	Difference Test
Sex of household head					
Male (%)	67.40	73.85	71.23	70.24	2.31
Female (%)	32.60	26.15	28.77	29.76	
Age in years (average)	51.25 (9.74)	50.32 (9.19)	51.85 (8.97)	51 (9.44)	1.74
Household size(average)	7.03 (2.73)	6.51 (2.54)	6.84 (2.00)	6.82 (2.58)	6.14 **
Income from Cassava (₦)	57,364.42 (21,340.3)	62,021.81 (22,957.79)	69,368.97 (23861.31)	60,662.98 (22,606.88)	14.60 ***
Years of farming experience (average)	21.64 (11.45)	22.84 (12.09)	28.12 (12.05)	22.94 (11.93)	17.61 ***
Education					
Nonformal education (%)	29.67	21.03	42.47	28.28	
Primary education (%)	26.74	37.95	31.51	31.42	20.38 ***
Secondary education (%)	33.70	29.23	23.29	30.68	
Tertiary education (%)	9.89	11.79	2.74	9.61	
Membership in social group (%)	75.09	70.77	93.15	75.97	14.80 ***
Access to Credit (%)	47.62	41.03	67.12	47.87	14.51 ***
Land area cultivated					
<0.5 ha (%)	15.02	18.97	10.96	15.90	5.57
0.5–1.5 ha (%)	29.67	25.13	21.92	26.99	
1.5–3 ha (%)	55.31	55.90	67.12	57.12	
Proportion of land allocated to cassava activities	0.60 (0.17)	0.70 (0.18)	0.65 (0.11)	0.67 (0.17)	10.25 ***
Nonfarm activities (%)	47.99	40.51	31.51	43.07	7.19 ***
Plant improved variety (%)	79.85	64.62	97.26	76.71	34.73 ***
Process improved variety (%)	72.89	54.87	94.52	69.32	42.57 ***

Note: LI, MI, and HI represent low intensity, medium intensity, and high intensity cassava utilization groups, respectively. Standard errors are in parentheses; **, ***: represent significance at 5%, and 1%, respectively.

3.3. Determinants of Intensity of Cassava Utilization among Smallholders

This section presents the estimates of the ordered probit regression that sought to identify the factors that determine intensity at which the smallholder households utilize cassava within their farming systems. The result is presented in Table 3. The ordered probit regression is premised on the parallel assumption [39]. Hence, we present a joint regression from the ordered probit model. However, estimates of individual determinants of intensity of utilization are obtained by predicting the marginal effect for each group.

The ordered probit model returned a likelihood ratio of −497.39, significant at 1%; implying that the overall model fit. Eight of the variables significantly explained the intensity of utilization of cassava by the smallholders. The coefficient of proportion of land allocated to cassava was positive and significant, implying that the probability of being in the high intensity cassava utilization group increased with proportion of land allocated to cassava. This is intuitively acceptable, since an expansion of land area planted to cassava may indicate an increase in the productive capacity of the smallholder and, hence, an expansion of the value adding opportunity for cassava. This is corroborated by the study of [40], where a preponderance of former smallholders had become medium-scale farmers through land expansion in Ghana and Zambia due to an increase in investment and biomass utilization. However, the effect of land on biomass utilization may be more sustainable with land intensification through investment in improved technologies of production and processing [41].

More experienced farmers were also more likely to increase utilization of cassava. The many contacts made and practices they may have learnt over the years are likely to influence the probability of taking up sustainable innovations [42]. Social capital was also found to significantly increase the probability of utilizing cassava by the smallholders. The effect of social capital is usually evident in the effect of group action in encouraging farmers to adopt sustainable practices and innovations such as proposed in our study [43].

Table 3. Determinants of level of participation in the cassava value web.

Variables	Coefficients	Marginal Effects		
		LI	MI	HI
Age of household head	−0.0733 (0.051)	0.029 (0.020)	−0.015 (0.011)	−0.014 (0.010)
Age squared	0.367 (0.315)	−0.146 (0.126)	0.075 (0.065)	0.071 (0.061)
Sex of household head (Base = Female)	0.074 (0.122)	−0.030 (0.049)	0.015 (0.025)	0.014 (0.024)
Proportion of Land allocated to cassava	0.721 ** (0.313)	−0.288 ** (0.125)	0.148 ** (0.066)	0.140 ** (0.061)
Number of years of farming experience	0.024 (0.006)	−0.009 *** (0.002)	0.005 *** (0.001)	0.005 *** (0.001)
Education (Base = Nonformal Education)	−0.072 (0.122)	0.029 (0.048)	−0.015 (0.025)	−0.014 (0.024)
Household size	−0.023 (0.023)	0.009 (0.009)	−0.005 (0.005)	−0.004 (0.004)
Access to credit (Base = No)	0.014 (0.115)	−0.006 (0.046)	0.03 (0.024)	0.003 (0.022)
Membership of social group (Base = No)	0.227 * (0.133)	−0.090 * (0.053)	0.047 * (0.028)	0.044 * (0.026)
Agroecological zone (Base = Forest)	−0.466 (0.120)	0.186 *** (0.051)	−0.096 *** (00.028)	−0.090 *** (0.025)
Income from cassava based Activities	0.340 *** (0.120)	−0.136 *** (0.048)	0.016 *** (0.007)	0.015 *** (0.007)
Asset index	0.077 ** (0.034)	−0.031 ** (0.014)	0.016 ** (0.007)	0.015 ** (0.007)
Nonfarm employment	−0.178 * (0.106)	0.071 * (0.042)	−0.005 * (0.005)	−0.004 * (0.004)
Use improved varieties	−0.264 ** (0.123)	0.105 ** (0.049)	−0.054 ** (0.026)	−0.051 ** (0.024)
Cut 1	6.280 (2.566)			
Cut 2	7.473 (2.570)			
Number of Observations	541			
Log Likelihood	−497.387			
LR chi2	69.06 ***			
Pseudo R2	0.219			

Source: Author's computation from field survey, 2015; Legend: *, **, and ***: significance at 10%, 5%, and 1%, respectively.

Our results also showed that the probability of intensity of the utilization of cassava increases with income from the smallholder's present holding. This may be due in part to the theory of 'homo economy', where farmers are assumed to be rational in decision making based on expected utility. However, it has been shown that farmers are not always rational and sociodemographic factors determine most investment decisions in the farm family [44]. Asset ownership was found to increase the probability of intensity of cassava utilization in the study area. Assets are wealth that can be converted into capital outlay, as corroborated by literature showing that asset ownership increased farmers' investment decisions in expanding production processes [45]. Engaging in nonfarm activities was, however, found to reduce the probability of utilizing cassava. This may be because nonfarm activities move labor away from the farm, thereby reducing the capacity of the farm to add value and utilize cassava. This resonates with the findings of studies where nonfarm income activities may also reduce the capacity of the farm family in managing their whole farm holdings [46]. On the other hand, income from nonfarm activities have been found to be a source of funding capital for future investment in family farms' productive activities [47].

Estimates of the marginal effects show the probability of being in any of the three cassava utilization groups for a unit change in exogenous variables. The results reveal that a unit increase in the land allocated to cassava significantly (5%) reduced the probability of households being in the LI group by up to 29%. However, a unit increase in land area allotted to cassava significantly (5%) increased the probability of participating in the MI and HI groups within the cassava biomass system by 15% and 14%, respectively. Increasing land resource allocations to cassava may imply expansion of the productive capacity of households' holdings. Smallholders have been shown to have a smaller cultivated land area than large holders; consequently, their productivities, in terms of economies of scale, are reduced. Policies that make land easily accessible in rural areas and for longer periods encourage higher investments agricultural activities. Also, this may translate to increased output, for which the smallholders are consequently able to leverage on the different value addition options inherent in the cassava industry and hence increase revenue from cassava biomass [48].

Furthermore, a unit increase in the number of years of farming experience of the household head significantly (1%) reduced the probability of the household head being in the LI by 0.1%, while it significantly increased the probability of being in the MI and HI groups by 0.5%, respectively. The more experienced the entrepreneur, the more the household can leverage on established contacts, markets, trade routes, and information to increase revenue from the cassava system level of participation within the web. This follows the results of empirical studies which reveals that elderly farmers who intended to continue faming had higher outputs than younger farmers [49]. This suggests the importance of on-field training and lessons learnt in the course of the farmer's years of experience coming in to play.

An increase in the asset index of households increased the probability of being in the HI group. The results show that a unit increase in the asset index of smallholders significantly increased the probability of being in the LI, MI, and HI groups by 18.6%, 9.6%, and 9.0%, respectively. Thus, asset ownership is probably a prerequisite for increased leverage in investing in agricultural activities with prospects of higher returns, especially with regards to accessing credit for expansion [50]. These assets are also important in the day to day activities of the smallholders, which if absent may reduce productive capacities. A unit increase in the income that accrues to the smallholders from their cassava-based activity also significantly (5%) reduced the probability of being in the LI group by 10.5%. However it increases the probability of being in MI by 1.6% and LI by 1.5%. This implies that present income may be an indicator of the extent of future income and thus economically rational actors will increase their production levels with the expectation of increased income [51].

The marginal effect for nonfarm activities revealed that employment in nonfarm activities increased the probability of being in the LI group, while it reduced the probability of being in the MI and HI groups. The implication of this is that nonfarm employment removes labor from the farm, thereby reducing the capacity of farming households to increase productive activities within the cassava system. However, low returns to agricultural activities is a factor that shifts labor away from agriculture, thereby encouraging participation in nonfarm activities to the detriment of increased participation in the agricultural system—which is evidence of structural change [52]. Social group membership was shown to reduce the probability of being in the LI by 0.9% and increase the probability of being in the MI by 0.5% and HI by 0.4%. This further shows the effect of social capital in increasing the probability of adopting best practices and innovation. This could be through availing credit facilities or supporting collective farmer decisions to invest and intensify productive activities on their holdings [53]. The coefficient of agroecological zone dummy (forest zone/Guinea savannah zone) showed that cassava smallholders in the Guinea savannah zone were more likely to be in the LI group, while those in the forest zone were more likely to be in the MI and HI groups. This may be because most processing activities in the cassava value chain in Nigeria are concentrated in the southern, forest zones of the country [54].

4. Conclusions and Policy Implications

The study examined the extent and determinants of cassava utilization as a precursor to increased value addition among smallholders in Nigeria. Findings revealed three distinct groups of cassava smallholders on the basis of utilization of cassava in low-level (LI), medium-level (MI), and high-level (HI) utilization groups. However, LI groups had the highest representation indicating a higher level of subsistence and low utilization of cassava among the farming households sampled. Households with higher levels of utilization (MI and HI) of cassava for food and nonfood products allocated more resources to cassava and obtained higher income from the cassava sector.

Overall, social capital, income, assets, land area cultivated, and farming experience were found to significantly determine the decision of the farming households for higher order cassava utilization. The probability of smallholders being in the LI increased by low land area cultivated, low level of asset ownership, nonmembership in social groups, and lack of access to improved cassava varieties as well as nonfarm income employment. The findings also revealed that variables that determined the probability of smallholders being in the MI and HI were similar in signs but different only in magnitude. Hence, the likelihood of being in the MI and HI increased with land area cultivated, farmer's experience, asset ownership, and income and membership of social groups, while it reduced with uptake of nonfarm employment.

Our findings placed asset ownership and land area cultivated as factors that positively influenced the extent of utilization of cassava in value addition processes. This brings to the fore recommendations for institutions and enabling environment that would stimulate access to productive and physical assets needed by the farmers to be involved in value additions. Enabling land rights is also important in farmers' decision to commit resources to greater utilization of cassava. Moreover, farmers that were members of social groups had a higher likelihood of higher utilization of cassava. We therefore recommend that positive group actions be encouraged among smallholder farmers through facilitating cooperatives, farmer associations, and providing interventions on a group basis, rather than creating policies that engender competitions and rural divisions.

Author Contributions: Conceptualization, T.A., P.A., A.A., V.O., and V.A.; Data Curation, T.A.; Formal Analysis, T.A.; Funding Acquisition, A.A.; Methodology, T.A. and V.A.; Project Administration, A.A. and V.O.; Resources, P.A. and A.A.; Supervision, P.A., A.A., V.O., V.A., and K.S.; Validation, P.A., A.A., V.O., V.A., and K.S.; Writing—Original Draft, T.A. and V.A.; Writing—Review & Editing, A.A., P.A., V.O., and V.A.

Funding: This research was funded by the German Federal Ministry of Education and Research (BMBF) and Ministry of Economic Cooperation and Research (BMZ), grant number FKZ 031 A258 A.

Acknowledgments: The authors acknowledge the contribution of the Federal Ministry of Education and Research (BMBF) and the German Federal Ministry for Economic Cooperation and Development (BMZ) "GlobE2-Global Food Security, Germany for providing the funding for this survey (through Contract number 8111240) to the International Institute of Tropical Agriculture (IITA) through the BiomassWeb Project. We also acknowledge the International Institute for Tropical Agriculture (IITA) for providing a research fellowship to Temitayo Adeyemo.

Conflicts of Interest: The authors declare no conflicts of interest.

Appendix A

Table A1. Activities used to define intensity of utilization of cassava in the Nigerian cassava system.

S/N	ITEM
1	Produce cassava for home consumption alone
2	Produce cassava both for home consumption and sale of cassava roots to processors
3	Process my cassava roots both for home consumption and market sales
4	Process cassava into garri
5	Process cassava into fufu
6	Process cassava into lafun
7	Process cassava into garri and fufu (or a mix of other products)
8	Process cassava into starch
9	Process cassava into high quality cassava flour
10	Sell cassava roots alone
11	Sell cassava roots and process for home consumption and market
12	Use cassava leaves and residue as manure and mulch on my farm
13	Have access to ready market for my high-quality cassava products

References

1. Devaney, L.A.; Henchion, M. If Opportunity Doesn't Knock, Build a Door: Reflecting on a Bio-economy Policy Agenda for Ireland. *Econ. Soc. Rev.* **2017**, *48*, 207–229.
2. Poku, A.; Birner, R.; Gupta, S. Is Africa Ready to Develop a Competitive Bioeconomy? The Case of the Cassava Value Web in Ghana. *J. Clean. Prod.* **2018**, *200*, 134–147. [CrossRef]
3. Virchow, D.; Beuchelt, T.; Denich, M.; Loos, T.K.; Hoppe, M.; Kuhn, A. The Value Web Approach. *Rural Focus* **2014**, *3*, 16–18.
4. Batidzirai, B.; Valk, M.; Wicke, B.; Junginger, M.; Daioglou, V.; Euler, W.; Faaij, A.P.C. Current and future technical, economic and environmental feasibility of maize and wheat residues supply for biomass energy application: Illustrated for South Africa. *Biomass Bioenergy* **2016**, *92*, 106–129. [CrossRef]
5. Virchow, D.; Beuchelt, T.D.; Kuhn, A.; Denich, M. Biomass-Based Value Webs: A Novel Perspectives for Emerging Bio-economies in Sub-Saharan Africa. In *Technological and Institutional Innovations for Marginalized Smallholders in Agricultural Development*; Gatzweiker, F.W., von Braun, J., Eds.; Springer: Berlin/Heidelberg, Germany, 2016. [CrossRef]
6. Loos, T.; Hoppe, M.; Dzomeku, B.; Scheiterle, L. The Potential of Plantain Residues for the Ghanaian Bioeconomy—Assessing the Current Fiber Value Web. *Sustainability* **2018**, *10*, 4825. [CrossRef]
7. Rípoli, T.C.C.; Molina, W.F., Jr.; Rípoli, M.L.C. Energy potential of sugar cane biomass in Brazil. *Sci. Agric.* **2000**, *57*, 677–681. [CrossRef]
8. Asibey, M.O.; Yeboah, V.; Adabor, E.K. Palm biomass waste as supplementary source of electricity generation in Ghana: Case of the Juaben Oil Mills. *Energy Environ.* **2018**, *29*, 165–183. [CrossRef]
9. Liu, Z.; Xu, A.; Long, B. Energy from combustion of rice straw: Status and challenges to China. *Energy Power Eng.* **2011**, *3*, 325–331. [CrossRef]
10. Scheiterle, L.; Ulmer, A.; Birner, R.; Pyka, A. From Value Chains to Biomass-Based Value Webs: The Case of Sugarcane in Brazil's Economy. *J. Clean. Prod.* **2018**, *172*, 3851–3863. [CrossRef]
11. Nweke, F.; Haggblade, S.; Zulu, B. *Building on Successes in African Agriculture Recent Growth in African Cassava*; International Food Policy Research Institute (IFPRI): Washington, DC, USA, 2004.
12. Allem, A.C. The origins and taxonomy of cassava. In *Cassava: Biology, Production and Utilization*; Hillocks, R.J., Thresh, J.M., Belloti, A.C., Eds.; CAB international: Wallingford, UK, 2002; Volume 1, pp. 1–16.
13. Zvinavashe, E.; Elbersen, H.W.; Slingerland, M.; Kolijn, S.; Sanders, J.P. Cassava for food and energy: Exploring potential benefits of processing of cassava into cassava flour and bioenergy at farmstead and community levels in rural Mozambique. *Biofuels Bioprod. Biorefin.* **2011**, *5*, 151–164. [CrossRef]
14. Marx, S.; Nquma, T.Y. Cassava as feedstock for ethanol production in South Africa. *Afr. J. Biotechnol.* **2013**, *12*, 4975–4983.

15. Food and Agricultural Orgnization (FAO). Cassava Outlook in Nigeria. 2018. Available online: http://www.fao.org (accessed on 4 September 2018).

16. Montagnac, J.; Davis, C.; Tanumihardjo, S. Nutritional value of cassava for use as a staple food and recent advances for improvement. *Compr. Rev. Food Sci. Food Saf.* **2009**, *8*, 181–194. [CrossRef]

17. Ehinmowo, O.; Ojo, S. Analysis of Technical Efficiency of Cassava Processing Methods among Small Scale Processors in South-West Nigeria. *Am. J. Rural Dev.* **2014**, *2*, 20–23.

18. Aminu, F.; Rosulu, H.; Toku, A.; Dinyo, O.; Akhigbe, E. Technical Efficiency in Value Addition to Cassava: A Case of Cassava-Garri Processing in Lagos State, Nigeria. *J. Agric. Sustain.* **2017**, *10*, 97–115.

19. Poole, N. *Smallholder Agriculture and Market Participation*; Practical Action Publishing: Warwickshire, UK, 2017. [CrossRef]

20. Federal Ministry of Agriculture and Rural Development (FMARD). *Agricultural Promotion Policy (2016–2020): Building on the Successes of the ATA, Closing Key Gaps*; Federal Ministry of Agriculture and Rural Development Policy and Strategy Document: Abuja, Nigeria, 2016.

21. Bank of Industry (BOI). Cassava Bread Fund. Available online: https://www.boi.ng/cassava-bread-fund/ (accessed on 24 September 2018.).

22. Rahman, S.; Awerije, B.O. Exploring the potential of cassava in promoting agricultural growth in Nigeria. *J. Agric. Rural Dev. Trop. Subtrop. (JARTS)* **2016**, *117*, 149–163.

23. Jerumeh, T.R.; Omonona, B.T. Determinants of transition in farm size among cassava-based farmers in Nigeria. *Kasetsart J. Soc. Sci.* **2018**, in press. [CrossRef]

24. World Bank 2017. Rural Population. *World* Development Indicators (WDI). Available online: http://www.worldbank.org (accessed on 30 May 2017).

25. StataCorp. *Stata: Release 13. Statistical Software*; StataCorp LP: College Station, TX, USA, 2013.

26. Jackman, S. Models for ordered outcomes. *Political Sci. C* **2000**, *200*, 1–20.

27. Greene, W.H.; Hensher, D.A. *Modeling Ordered Choices: A Primer*; Cambridge University Press: Cambridge, UK, 2010.

28. Zmuk, B. Quality of life indicators in selected European countries Hierarchical Cluster Analysis Approach. *Croat. Rev. Econ. Bus. Soc. Stat. (CREBSS)* **2015**, *1*, 42–54. [CrossRef]

29. Yim, O.; Randeem, K. Hierarchical Cluster Analysis: Comparison of three linkage measure and application to psychological data. *Quant. Methods Psychol.* **2015**, *11*, 8–21. [CrossRef]

30. Restuccia, D. Resource Allocation and Productivity in Agriculture. Plenary Panel on 'Agricultural Policy in Africa' at the Centre for the Study of African Economics (CSAE). Available online: https://www.economics.utoronto.ca/diegor/research/Restuccia_ResAlloc_Oxford.Pdf (accessed on 11 December 2017).

31. Akaakohol, M.A.; Aye, G.C. Diversification and farm household welfare in Makurdi, Benue State, Nigeria. *Dev. Stud. Res. Open Access J.* **2014**, *1*, 168–175. [CrossRef]

32. Onyenwonke, C.; Simonya, K.J. Cassava Post-harvest Processing and Storage in Nigeria: A Review. *Afr. J. Agric. Res.* **2014**, *9*, 3853–3863.

33. Ojogho, O. Determinants of Food Security among Arable Farmers in Edo State, Nigeria. *Medwell Agric. J.* **2010**, *5*, 151–156.

34. Gyau, A.; Franzel, S.; Chiatoh, M.; Nimino, G.; Owusu, K. Collective action to improve market access for smallholder producers of agroforestry products: Key lessons learned with insights from Cameroon's experience. *Cur. Opin. Environ. Sustain.* **2014**, *6*, 68–72. [CrossRef]

35. Huffman, W.E. Human capital: Education and agriculture. In *Handbook of Agricultural Economics*, 1st ed.; Bruce, L.G., Gordon, C.R., Eds.; Elsevier Science: Amsterdam, The Nertherlands, 2001.

36. Awotide, B.A.; Diagne, A.; Omonona, B.T. Impact of Improved Agricultural Technology Adoption on Sustainable Rice Productivity and Rural Farmers' Welfare in Nigeria: A Local Average Treatment Effect (LATE) Technique'. In Proceedings of the African Economic Conference, Kigali, Rwanda, 30 October–2 November 2012.

37. Liverpool-Tasie, L.S.; Kuku, O.; Ajibola, A. Review of literature on agricultural productivity, social capital and food security in Nigeria. In *NSSP Working Paper*; International Food Policy Research Institute (IFPRI): Washington, DC, USA, 2011. Available online: http://ebrary.ifpri.org/cdm/ref/collection/p15738coll2/id/126846 (accessed on 9 July 2018).

38. Iddrisu, A.; Ansah, I.G.K.; Nkegbe, P.K. Effect of input credit on smallholder farmers' output and income: Evidence from Northern Ghana. *Agric. Financ. Rev.* **2018**, *78*, 98–115. [CrossRef]

39. Williams, R. Ordinal Regression Models: Problems, Solutions and Problems with Solutions'. In Proceedings of the German Stata User Group Meeting, Notre-Dame Sociology, South Bay, IL, USA, 20 June 2008.

40. Jayne, T.S.; Chapoto, A.; Sitko, N.; Nkonde, C.; Muyanga, M.; Chamberlin, J. Is the scramble for land in Africa foreclosing a smallholder agricultural expansion strategy? *J. Int. Aff.* **2014**, *67*, 35–53.

41. Chapoto, A.; Mabiso, A.; Bonsu, A. *Agricultural Commercialization, Land Expansion, and Home-Grown Large-Scale Farmers: Insights from Ghana*; International Food Policy Reseach Institute: Washington, DC, USA, 2013.

42. Ainembabazi, J.; Mugisha, J. The Role of Farming Experience on the Adoption of Agricultural Technologies: Evidence from Smallholder Farmers in Uganda. *J. Dev. Stud.* **2014**, *50*, 666–679. [CrossRef]

43. Munasib, A.B.; Jordan, J.L. The effect of social capital on the choice to use sustainable agricultural practices. *J. Agric. Appl. Econ.* **2011**, *43*, 213–227. [CrossRef]

44. Schwarze, J.; Holst, G.S.; Mußhoff, O. Do farmers act like perfectly rational profit maximisers? Results of an extra-laboratory experiment. *Int. J. Agric. Manag.* **2014**, *4*, 11–20.

45. Quisumbing, A.R.; Rubin, D.; Manfre, C.; Waithanji, E.; van den Bold, M.; Olney, D.; Meinzen-Dick, R. Gender, assets, and market-oriented agriculture: Learning from high-value crop and livestock projects in Africa and Asia. *Agric. Hum. Values* **2015**, *32*, 705–725. [CrossRef]

46. Nasir, M.; Hundie, B. The Effect of Off Farm Employment on Agricultural Production and Productivity: Evidence from Gurage Zone of Southern Ethiopia. *J. Econ. Sust. Dev.* **2014**, *5*, 85–98.

47. Lanjouw, J.; Lanjouw, P. The Rural Non-Farm Sector: Issues and Evidence from Developing Countries. *Agric. Econ.* **2001**, *26*, 1–23. [CrossRef]

48. Hichaambwa, M.; Chamberlain, J.; Sitko, N. Determinants and Welfare Effects of Smallholder Participation in Horticultural Markets in Zambia. *Afr. J. Agric. Resour. Econ.* **2015**, *10*, 279–296.

49. Guo, G.; Wen, Q.; Zhu, J. The Impact of Aging Agricultural Labor Population on Farmland Output: From the Perspectives of Farmer Preferences. *Math. Probl. Eng.* **2015**, *2015*, 7. [CrossRef]

50. Johnson, N.L.; Kovarik, C.; Meinzen-Dick, R.; Njuki, J.; Quisumbing, A. Gender, assets, and agricultural development: Lessons from eight projects. *World Dev.* **2016**, *83*, 295–311. [CrossRef] [PubMed]

51. Samson, G.S.; Gardebroek, C.; Jongeneel, R.A. Explaining production expansion decisions of Dutch dairy farmers. *NJAS Wagening. J. Life Sci.* **2016**, *76*, 87–98. [CrossRef]

52. McCullough, E.B. Labor productivity and employment gaps in Sub-Saharan Africa. *Food Policy* **2015**, *67*, 133–152. [CrossRef]

53. Michelini, J.J. Small farmers and social capital in development projects: Lessons from failures in Argentina's rural periphery. *J. Rural Stud.* **2013**, *30*, 99–109. [CrossRef]

54. Echebiri, R.N.; Edaba, M.E.I. Production and Utilization of Cassava in Nigeria: Prospect for Food Security and Infant Nutrition. *Prod. Agric. Technol.* **2008**, *4*, 38–52.

© 2019 by the authors. Licensee MDPI, Basel, Switzerland. This article is an open access article distributed under the terms and conditions of the Creative Commons Attribution (CC BY) license (http://creativecommons.org/licenses/by/4.0/).

Article

Smallholder Agroprocessors' Willingness to Pay for Value-Added Solid-Waste Management Solutions

Olaoluwa Omilani [1], Adebayo Busura Abass [2] and Victor Olusegun Okoruwa [1,*]

[1] Department of Agricultural Economics, University of Ibadan, Ibadan 200284, Nigeria; omilani2007@gmail.com
[2] International Institute of Tropical Agriculture, Regional Hub for Eastern Africa, Dar es Salaam 11101, Tanzania; a.abass@cgiar.org
* Correspondence: vokoruwa@gmail.com; Tel.: +234-803-722-3832

Received: 4 October 2018; Accepted: 20 March 2019; Published: 23 March 2019

Abstract: The paper examined the willingness of smallholder cassava processors to pay for value-added solid wastes management solutions in Nigeria. We employed a multistage sampling procedure to obtain primary data from 403 cassava processors from the forest and Guinea savannah zones of Nigeria. Contingent valuation and logistic regression were used to determine the willingness of the processors to pay for improved waste management options and the factors influencing their decision on the type of waste management system adopted and willingness to pay for a value-added solid-waste management system option. Women constituted the largest population of smallholder cassava processors, and the processors generated a lot of solid waste (605–878 kg/processor/season). Waste was usually dumped (59.6%), given to others (58.1%), or sold in wet (27.8%) or dry (35.5%) forms. The factors influencing the processors' decision on the type of waste management system to adopt included sex of processors, membership of an association, quantity of cassava processed and ownership structure. Whereas the processors were willing to pay for new training on improved waste management technologies, they were not willing to pay more than US$3. However, US$3 may be paid for training in mushroom production. It is expected that public expenditure on training to empower processors to use solid-waste conversion technologies for generating value-added products will lead to such social benefits as lower exposure to environmental toxins from the air, rivers and underground water, among others, and additional income for the smallholder processors. The output of the study can serve as the basis for developing usable and affordable solid-waste management systems for community cassava processing units in African countries involved in cassava production.

Keywords: cassava processors; smallholders; solid waste; pollution; value-added; willingness to pay

---

1. Introduction

Annual cassava production in Africa is about 84 million tonnes, of which Nigeria produces over 30 million tonnes, as the world's largest cassava producer [1–3]. However, cassava processing is considered to contribute significantly to environmental pollution globally by depleting water resource quality through natural toxic cyanogen and the production of unpleasant odour [4]. Over 55.0% of the waste produced from cassava processing is disposed of in dumping sites, creating both environmental pollution and negative health impacts on the population in the neighbourhood of cassava processing facilities [5].

In Nigeria, cassava is mostly produced and processed by small-scale farmers at the family or village level [6]. Cassava farming provides livelihood opportunities for both men and women involved in its production and consumption cycle. Both males and females make significant labour contributions to the cassava industry, with each sex specialising in different tasks. Women are, however, the backbone of the agricultural sector, in that they account for 60% to 80% of farm labour [7].

Within the cassava industry, processing is a major contributor to environmental pollution [4]. This is so due to the massive waste generated during the processing stages. The available technology of processing cassava roots primarily involves peeling, washing, grating, fermenting, dewatering, frying, drying, and milling. Waste is often produced at each of these processing stages. The type and composition of the waste depend on the processing method and type of technology used [8].

Solid waste is unwanted solid material generated from residential, industrial or commercial activities in a given area [9]. It may be categorised according to its origin (domestic, industrial, commercial, construction, or institutional), its content (organic or inorganic), or its hazard potential (toxic, non-toxic, flammable, radioactive, infectious) [10]. Cassava waste is organic and exists in both solid and liquid forms. Peel is the first type of solid waste generated by cassava root-processing activities. Subsequently, when the peeled cassava roots are grated and dewatered, wastewater is obtained. After dewatering, the resultant semi-solid mass is sieved, and the ungrated fibre (chaff) is discarded as the final solid waste [11]. Cassava peel is the primary solid cassava waste that constitutes 20 to 25.0% of the weight of the roots [12,13]. The proportion of waste can be higher, especially during hand peeling [14].

Disposal of waste products resulting from cassava processing is often inadequate. This led to varying environmental and health hazards, causes a foul smell and an unattractive sight, and produces widespread environmental pollution at cassava processing sites [15]. Cassava waste from processing centres has contributed significantly to environmental contamination. About 75.0% of the cassava roots harvested in Africa are fermented, causing the release of cyanogenic compounds and other pollutants into farmlands, rivers, streams, and groundwater [16]. In Nigeria, cassava effluent is directed to streams, resulting in a reduction in water quality and loss of aquatic life owing to its toxic nature [16], while stagnant streams of effluent produce a strong offensive odour [17]. Dumpsites for solid waste and liquid effluent also breed insects and other organisms that lead to disease outbreak [18].

More than 60.0% of the rural population in Nigeria is engaged in cassava-based cottage industries, and millions of tonnes of cassava waste are produced [19]. Waste management has been a major challenge to processors, as wastes need to be managed and, if possible, re-utilized without causing any harm to the environment. Over 55.0% of the waste produced from cassava processing is dumped at tips near the processing sites, becomes landfill, and/or is burnt, thereby, posing a serious threat to the environment and constituting a health hazard to the processors [5]. Few cassava processors derive any financial benefit from cassava waste [5], while most of them lack either awareness or expertise regarding the conversion of waste to any form of useful resource. However, studies [13,20] have shown that agricultural waste can be profitably utilized and recycled without attendant externalities.

Against this background, we conducted a study to examine the waste management approaches used by smallholder cassava processors, their willingness to pay for value-added solid-waste management solutions, and the factors that may influence their behaviour.

The next section gives details of the theoretical framework of willingness to pay, including estimation methods. This is followed by a section on materials and methods, the study results and discussion, and finally conclusion.

2. Theoretical Framework

An essential component of every business transaction is willingness-to-pay (WTP) computation, in which buyers assess the maximum amount of financial resources they are willing to expend in exchange for the item being sold [21]. Willingness to pay is the maximum amount of money that a consumer would pay to enjoy an improvement in product quality [22]. Rodríguez et al. [23], defined it as "the difference between consumers' surplus before and after adding or improving a given product attribute" in monetary terms. In WTP surveys, a respondent is asked a series of structured questions that are designed to determine the maximum amount of money he or she is willing to pay for a product or service. Various tools are employed in measuring WTP. The three most common are the hedonic pricing method, the contingent valuation method (CVM), and the travel cost method [24,25].

Notwithstanding the debate regarding its underlying economic theory, the contingent valuation method (CVM) is considered superior and preferred to other methods because it deals with both use and non-use values [26]. It is a simple, flexible and non-market valuation method popular for cost-benefit analysis and environmental impact assessment [27,28]. The CVM involves asking sampled respondents directly whether they would be willing to pay or accept compensation for the change in preferences. The method has been adjudged to conform to the theoretical core of economics [28]. In general, we assume that person 'i' is willing to pay Y^*_i for training on improved cassava waste management, and that this payment is related to a set of the person's characteristics (X_i), so that:

$$Y^*_i = X_i\beta + \varepsilon_i \tag{1}$$

Although Y^*_i is unobserved, it is assumed to lie within the bound of such person's willingness to pay, which is (Y_{i1}, Y_{i2}). This interval is based on responses to a series of questions asked in the contingent valuation interview [29]. The likelihood of the person paying is, therefore, stated as:

$$\Pr(Y_{i1} \leq Y^*_i \leq Y_{i2}) \tag{2}$$

This equation is the basis for the use of discrete choice models for estimating the contingency valuation method.

3. Materials and Methods

3.1. Study Area and Sampling Procedure

This WTP study was carried out in three states (Edo, Kwara, and Ogun) within the humid forest and savannah zones of Nigeria (see Figure 1). A multi-stage sampling technique was used to select respondents for the research. The first stage was a purposive selection of the three states based on their high level of cassava production and processing [30]. The second stage was a random selection of two agricultural zones in each state. In the last stage, 75 respondents were randomly selected from each agricultural zone. A total of 450 copies of a questionnaire were administered to obtain information on demographic and socioeconomic characteristics of cassava processors, the quantity of waste produced, the waste management methods used, and the willingness of the processors to adopt and pay for financially-rewarding solid-waste management methods. Four hundred and three (403) copies of the questionnaire were used in the final analysis, while 47 were discarded because they contained incomplete information. Furthermore, three major technology research and extension institutions with waste management research mandate within the survey zones (humid and savannah) were purposively selected. Interviews were conducted with key informants in these institutions to identify the improved waste management technologies developed and available in the zones and Nigeria.

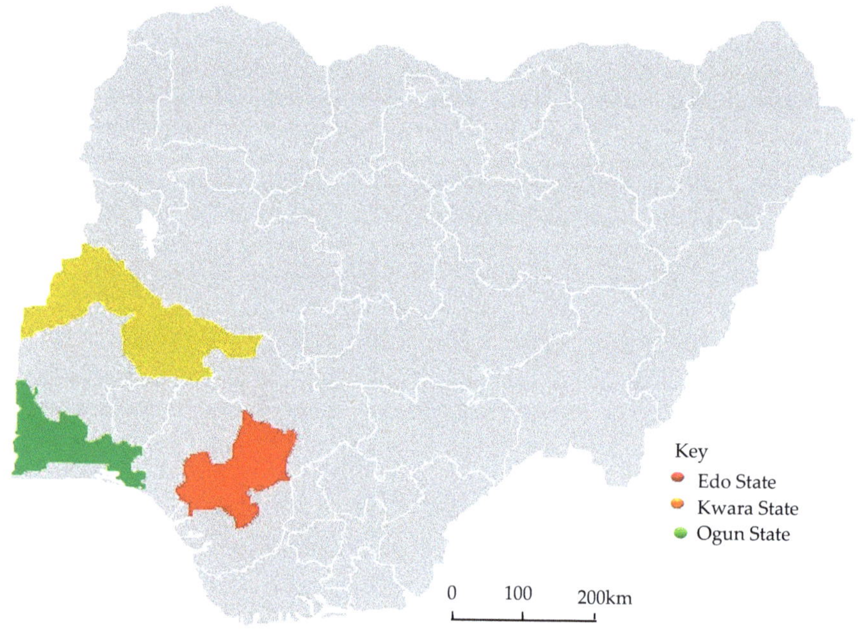

Figure 1. Map of Nigeria showing study areas.

3.2. Estimation Methods

The selected variables for this study comprised age, sex, marital status, education level, household size, years of experience, source of credit, and scale of operation. A complete description of these variables is presented in Table 1.

Table 1. Descriptive Statistics of the Respondents.

Characteristics	Percentage	Characteristics	Percentage
Sex		**Level of education (years)**	
Female	68.7	≤ 10	67.5
Male	31.3	11–20	32.5
Age (years)		**Type of education**	
≤ 30	3.2	No formal education	16.6
31–45	40.2	Arabic education	3.7
46–65	53.6	Adult education	2.2
>66	13.0	Primary education	39.7
Marital status		Secondary education	28.5
Single	3.0	Tertiary education	9.2
Married	90.6	**Source of finance**	
Divorced	2.7	Commercial bank	1.6
Widowed	3.3	Cooperative society	21.8
Household size		Money lender	7.9
1–5	36.7	Produce buyer	3.6
6–10	59.3	Friends/family	18.5
11–15	4.0	Self	57.0
Scale of operation			
<1 tonne	18.6	**Ownership structure**	
1–10 tonne	62.3	Sole	83.3
11–20 tonne	8.9	Partnership	8.7
>20 tonne	10.2	Cooperative	8.0

A logit model based on the cumulative probability function was adopted to determine the factors that influence current solid-waste management systems as well as the mean willingness to pay for a value-added solid-waste management system and the factors influencing processors' willingness to pay. This was because the model could deal with a dichotomous dependent variable. The approach for this model follows the one adopted by Yusuf et al. [31].

The logit model is specified as follows:

$$y_i = \frac{1}{1 + e^{-(\beta_0 + \beta_1 X_i)}} \tag{3}$$

where y_i represents the response of ith processor's willingness to pay for a value-added solid-waste management system, which is either 1 if yes or 0 if no.

For the model to estimate the factors influencing current use of solid-waste management system; y_i *is the 'yes or no'* response of the smallholder processor to each of the four systems (dumping, giving out free, selling wet and selling dry). B_0 is a constant; β_1 is the coefficient of the independent variables; while X_i is the price that the ith processor stated for a value-added solid-waste management system.

The mean willingness to pay of the smallholder cassava starch processors for improved solid-waste management was calculated using the formula adopted by [32] and is given by:

$$\text{Mean WTP} = \frac{1}{\beta_1} \ln(1 + \exp \beta_0) \tag{4}$$

To assess the factors influencing processors' willingness to pay for an improved solid-waste management system, the processor's responses to the WTP question—1 if yes and 0 if no—were then regressed against the socioeconomic characteristics of each processor. The regression logit model is specified as:

$$Y = \frac{1}{1 + e^{-Z_i}} \tag{5}$$

where Y represents the processor's response to the WTP question, which is either 1 if yes or 0 if no; and Z_i is equal to $\beta_0 + \beta_1 X_i$. Transformed into a linearized form, the model is expressed as:

$$Y = \beta_0 + \beta_1 X_1 + \beta_2 X_2 + \ldots + \beta_n X_n. \tag{6}$$

where Y is willingness to pay for improved waste management training, X_1 is sex (male = 0; female = 1), X_2 is household size (number), X_3 is quantity of cassava processed (kg), X_4 is processing experience in years, X_5 is years of education, X_6 is member of association (yes = 1, no = 0), and X_7 is awareness of value-added solid-waste management systems (yes = 1, no = 0).

4. Results and Discussion

4.1. Summary of Description of Cassava Processors

Table 1 presents the descriptive statistics of the respondents. The table shows that the majority (68.7%) of the interviewees were female. This result is similar to the results of [32,33], who revealed that female processors are in the majority in global cassava processing. This suggests that female processors would be predominant in the sector in the study areas. More than half of the processors fell within the age range of 46 to 65 years. The mean age of the respondents in the study was 48 years, which is similar to the mean age of cassava processors found by [34] in Ogun State. This shows that the majority of the respondents were of an active working age and may appreciate the opportunity to use value-added solid-waste management.

On average, the processors had received approximately 7 years of education. Education is expected to increase awareness of the negative impact of solid waste on humans and the environment. The mean household size of the respondents was about six persons. This is in line with the findings of [34], where the average household size for a cassava farmer in Ogun State is seven. Most (83.3%) of the respondents owned their processing centres and over half (57.0%) had used their own funds to establish their enterprises. In terms of the scale of operation, the study showed that 62.3% of the respondents' processed 1 to 10 tonnes of cassava roots monthly. In addition, liquid waste consisting of suspensions of starch, solid materials, and cyanogenic cassava compounds was produced and released into the surrounding channels, streams, and rivers, thereby, causing major environmental pollution [35–37].

4.2. Summary of Solid Waste Produced by Cassava Processors

This section presents the summary of the two main solid wastes—peel and chaff—identified in the study. The result presented in Table 2 revealed that an average of 737 ± 954 kg and 511 ± 561 kg of peel was generated per processor during the high and low processing seasons, respectively; while an average of 141 ± 131 kg and 94 ± 92 kg of chaff was produced per processor during the same respective periods. About half (50.4%) of the processors generated up to 500 kg of cassava peel per month during the peak season; while almost 45.0% of them produced 500 to 1500 kg per month during this period. During the low season, nearly 64.0% produced up to 500 kg of cassava peel per month and about 35.0% generated 500 to 1500 kg per month. Similarly, 84.0% and 69.8% of the cassava processors produced up to 150 kg of cassava chaff per month during the high and low seasons, respectively. In addition, 21.4% and 14.8% of the processors generated 300 kg and 151 kg of chaff during the high and low processing seasons, respectively. The low processing season was from October to March, while the high processing season was from April to September.

Table 2. Quantity of Solid Wastes Produced.

Quantity of Waste (kg/Processor/Month)	High Season (%)	Low Season (%)
Peel (n = 403)		
≤500	50.4	63.8
501–1000	30.8	30.9
1001–1500	14.6	4.1
>1500	4.2	1.1
Average waste in peels	737 ± 954	511 ± 561
Chaff (n = 192)		
≤150	69.8	84.0
151–300	21.4	14.8
>300	8.9	1.2
Average waste in chaff	141 ± 131	94 ± 92

4.3. Current Solid-Waste Management Systems Used by Cassava Processors

In this section, we present the summary of different cassava solid-waste management systems currently in use by the cassava processors (Table 3). We also isolate the factors that determined the use of each of the systems identified (Table 4).

4.3.1. Summary of Current Cassava Solid-Waste Management Systems Used

Waste management practices varied across the processing centres, and the processors utilised one or more waste management practices. Four management practices were identified as commonly used by the processors (see Table 3). These were dumping, gifting, selling wet waste, and selling dry waste. The processors could adopt one or more of these options, but a greater percentage of them (59.6%) dumped their waste or passed it on as a gift (58.1%).

In most community cassava-processing centres, cassava waste dumps are often a source of environmental pollution and a danger to aquatic animals, plants, and humans. A study conducted by Anikwe and Nwobodo [38] in southeastern Nigeria revealed that there were differences in soil particle size distribution between municipal dump and non-dump sites used for urban agriculture. Soil bulk density was lower, by 9 to 13.0%, while total porosity and hydraulic conductivity were higher, by 9 to 14.0% and 240 to 463.0%, respectively, at the dump sites compared with the non-dump sites. Similarly, the study by Arimoro et al. [39] on the effect of cassava effluents on macroinvertebrates downstream showed that cassava effluents caused a decrease in dissolved oxygen and pH and an increase in biochemical demand (BOD) and nitrates from the samples taken from three stations along the Orogodo River course (upstream of the cassava-impacted site, at the cassava-impacted site, and downstream of the cassava-impacted site) in the Niger Delta. In addition, another study by Izugbara and Umoh [40] on indigenous waste management practices in urban Nigeria, revealed careful waste segregation and sorting, selective burning and burying, composting and conversion as some of the common indigenous waste management practices used by the sampled respondents. In addition, a small percentage of the respondents sold the solid waste in fresh (27.8%) or dry (35.5%) form.

Table 3. Solid-Waste Management Options Used.

Waste Management Practices (n = 403)	Percentage of "Yes" Responses
Dumping	59.6
Giving out free	58.1
Selling wet waste	27.8
Selling dry waste	35.5

4.3.2. Estimates of the Determinants of Current Cassava Solid-Waste Management Systems

After identifying the current waste management systems in use by smallholder cassava processors, the study examined the factors that determined the use of these systems. The results for each management system in use by the cassava processors are presented in Table 4. The likelihood ratios were significant at 1% across the four waste management systems in use. The result also indicated that the likelihood of using dumping as a waste management system increased significantly with the quantity of cassava processed (marginal scale) and the processors who produced *fufu* (27.8%) and *garri* (36.7%). However, an increase in cooperative/partnership ownership structure rather than sole ownership and being resident in Kwara State reduced the likelihood of dumping. This implies that processors of a large quantity of cassava and those engaged in the production of *fufu* and *garri* may have found the dumping system more useful than any other waste management system for getting rid of their solid waste. However, this may not be true for processors residing in Kwara state who had a lower preference for the use of dumping as indicated by the negative sign associated with Kwara state location variable. The information obtained from the respondents in Kwara State [41] revealed that over 54.0% of the processors dispose of solid waste through the three other forms of waste management systems (giving out free, selling dry and selling wet) particular to those who rear animals, while only 13.0% used the dumping waste management system. The processors in cooperative/partnership ownership may also not have sole decision-making power in determining the use to which solid waste is put in the processing concern. This finding is in contrast with Olukanni and Olatunji [13], that found dumping as the most common waste management system used by over 90.0% of the cassava processors in Ogun State, Nigeria.

The likelihood of giving solid waste out free increased significantly with being female (1.3%), cooperative/partnership ownership structure rather than sole producer (1.1%), residency in Ogun State (27.3%), production of *garri* (50.0%) and *lafun* (15.6%). Furthermore, the likelihood of selling cassava solid waste in a wet form significantly increased with years of experience (0.6%), membership of association (10.0%), cooperative/partnership ownership structure (32.7%) and residency in Kwara State (13.8%). However, it reduced significantly for the processors resident in Ogun State (14.3%). Similar results were obtained by Ekere et al. [42] who found that gender, peer influence, land size, location of household and membership of environmental organisations explain household waste utilisation and separation behaviour. In addition, a study conducted by Sackey and Bani [5] found that three-quarters of the cassava processors in the selected districts of Ghana gave cassava peels out freely to feed ruminants, while the rest one-quarter disposed of it by dumping in open areas.

Table 4. Estimates of Determinants of the Current Cassava Solid-Waste Management System Used by the Cassava Processors.

Variables	Dumping			Giving out Freely			Selling Wet			Selling Dry		
	Coeff.	Std. Err.	Marginal Effect	Coeff.	Std. Err.	Marginal Effect	Coeff.	Std. Err.	Marginal Effect	Coeff.	Std. Err.	Marginal Effect
Sex (female = 1)	0.122	0.329	0.073	0.530 **	0.245	0.133	−0.112	0.298	−0.050	−0.378	0.259	−0.103
Household size	0.073	0.683	−0.007	−0.053	0.051	−0.014	0.038	0.060	0.002	−0.009	0.054	−0.007
Quantity of cassava processed	0.40×10^{-4} **	0.19×10^{-4}	2.26×10^{-6}	-7.06×10^{-6}	0.11×10^{-4}	-1.52×10^{-6}	-1.31×10^{-5}	1.26×10^{-5}	-4.78×10^{-6}	3.28×10^{-5} **	0.14×10^{-5}	4.45×10^{-6}
Year of experience	−0.012	0.016	−0.002	−0.012	0.013	−0.003	0.041 ***	0.015	0.006	0.006	0.013	0.001
Member of association	−0.382	0.311	−0.110	−0.091	0.235	−0.025	0.560 **	0.296	0.100	0.225	0.255	0.072
Ownership structure (coop/partnership = 1)	−0.688 *	0.396	−0.269	0.533 *	0.317	0.113	1.287 ***	0.342	0.327	0.324	0.317	0.156
State of location (Kwara = 1)	−3.910 ***	0.484	0.077	0.427	0.328	0.165	1.567	0.373	−0.138	1.512 ***	0.355	0.114
State of location (Ogun = 1)	0.337	0.468	0.044	1.340 ***	0.350	0.273	−1.305	0.456	−0.143	−0.684 *	0.388	−0.110
Producing starch	−0.778 **	0.312	−0.022	0.440	0.282	0.110	0.007	0.337	0.050	−0.476	0.301	−0.133
Producing fufu	0.874 ***	0.387	0.278	0.313	0.243	0.088	−0.245	0.282	−0.123	−0.622 **	0.253	−0.208
Producing gari	0.977 *	0.554	0.367	2.345 ***	0.467	0.500	−0.615	0.465	0.234	0.308	0.452	−0.054
Producing lafun	0.170	0.387	0.200	0.752 *	0.305	0.156	−0.458	0.350	0.078	−0.293	0.308	0.068
Constant	0.948	0.887		−2.636 ***	0.695		−1.403 *	0.768		−0.970	0.763	
	Number of obs. = 403			Number of obs. = 403			Number of obs. = 403			Number of obs. = 403		
	Likelihood = −158.04			likelihood = −239.26			Likelihood = −178.31			Likelihood = −214.43		
	Prob >chi2 = 0.0000			Prob >chi2 = 0.0000			Prob >chi2 = 0.0000			Prob >chi2 = 0.0000		
	Pseudo R^2 = 0.418			Pseudo R^2 = 0.127			Pseudo R^2 = 0.251			Pseudo R^2 = 0.182		
	LR chi2(12) = 64.33			LR chi2(12) = 69.14			LR chi2(12) = 56.52			LR chi2(11) = 44.66		
	Marginal effect Y = 0.6077			Marginal effect Y = 0.5911			Marginal effect Y = 0.2485			Marginal effect Y = 0.3374		

***, ** and * denotes 1%, 5% and 10% significance, respectively (use Palatino Linotype Font of 10 or 11).

The likelihood of selling of cassava solid waste in dry form increased with the quantity of cassava produced, albeit on a very small scale (4.45×10^{-6}) and with residency in Kwara State (11.4%). However, it reduced with being resident in Ogun State (11.0%) and producing fufu (20.8%).

4.4. Willingness to Pay for Improved Waste Management Systems

There are improved waste management systems that enable processors to make such value-added products as mushrooms, feed, ethanol, biofuel, and organic manure from cassava waste. The products generated from improved waste hold promise for environmental preservation and income generation for the cassava processors. As captured in Table 5, 61.8% of the processors indicated a willingness to pay for training in converting cassava waste into organic manure, while 56.3% indicated a willingness to pay for training in mushroom production. Similar results were obtained by Odediran and Ojebiyi [32], who found that most of the cassava processors at a technology demonstration in Ogun State were willing to adopt the technology to produce mushrooms from cassava waste.

Table 5. Responses of the Cassava Processors on Willingness to Pay for Improved Waste Management Systems.

Willingness to Pay (n = 403)	Percentage of "Yes" Responses
Conversion of cassava waste to organic manure	61.8
Production of mushrooms from cassava waste	56.3

In addition, following [31], the coefficient of the price of processors willing to pay for value-added solid-waste management (β_1) was found to be positive for both types of training, thus, suggesting that the processors were willing to pay (see Table 6). The estimated mean value of WTP was ₦504.63 ($3.00) for training on production of mushrooms from cassava waste and ₦159.15 ($1.00) for training on converting cassava waste into organic manure. Organic manure from cassava peel is not often used by the farmers in the study areas. Thus, the processors are not motivated to invest in producing manure since they believe the product will be difficult to sell. Furthermore, the use of cassava waste to produce mushrooms has limitations in the study area because mushrooms are not a staple food in Nigeria but more of a luxury food item [43].

Table 6. Mean Willingness to Pay for Training in Organic Manure and Mushroom Production from Cassava Waste.

Organic Manure Production from Cassava Waste				
Variable	Coeff.	Std. Err.	Z-Stat.	*p*-Value
Constant	−1.7472	0.2330	−7.50	0.0000
Price (β_1)	0.0011	0.0001	7.73	0.0000

Number of obs. = 403
LR chi2(1) = 211.87
Prob>chi2 = 0.0000
Pseudo R^2 = 0.3952
Log likelihood = −162.10
Mean willingness to pay = 159.15

Mushroom Production from Cassava Waste				
Variable	Coeff.	Std. Err.	Z-Stat.	*p*-Value
Constant	−1.1631	0.1835	−6.34	0.0000
Price (β_1)	0.0005	0.0001	7.48	0.0000

Number of obs. = 403
LR chi2(1) = 134.97
Prob >chi2 = 0.0000
Pseudo R^2 = 0.244
Log likelihood = −208.62
Mean willingness to pay = 540.63

Moreover, most of the processors interviewed believed that additional production activity around cassava peel would be time-consuming and distract them from their main processing activities. In many study areas, wild-mushroom hunting is common among women, youths, and children [44,45]. This decreases the importance of mushroom production from cassava waste for the processors. Therefore, it could be inferred that the considerable low WTP estimates for training on conversion of waste to economic use via manure and mushroom production can be attributed to the lack of awareness of the benefits of these two products within the study communities.

4.5. Determinants of Willingness to Pay for Improved Solid-Waste Management System

Table 7 presents the results of factors affecting WTP for the above-mentioned two types of training. Concerning training on mushroom production, the results revealed that the model was significant at 1%, while the McFadden's R squared (pseudo R^2) equalled 18.8%. The model coefficients revealed that processor's sex had a positive and significant relationship on WTP for the training, suggesting that the female gender increases the likelihood of WTP for the training by about 0.14. Similarly, the significance of household size, large scale of operation and awareness of new management system at 5% suggest that these variables are likely to increase the WTP for mushroom cultivation training by about 0.03, 0.15 and 0.32, respectively. The significance of the variable on awareness of new management systems is contrary to the conclusion of Odediran and Ojebiyi [32], who reported that there was no significant relationship between their respondents' awareness of cassava waste utilisation technologies and the WTP for them. The indication is that awareness of new management systems would likely increase the WTP for mushroom production training by a probability of about 0.32. Conversely, the quantity of cassava processed, though significant at 1%, had an opposing sign, thus, suggesting the variable's likelihood to decrease WTP by not too obvious a margin. However, the decrease in WTP associated with the quantity of cassava produced may be restricted to small-scale cassava processors. The positive significance of the variable on large-scale operation processor seems to suggest that this group of processors will be more disposed to participating in mushroom training than their small-scale operation counterparts.

Table 7. Factors Influencing the Willingness to Pay for Training in New Solid-Waste Management Options.

Variable	Organic Manure			Mushroom Production		
	Coeff.	Std. Err.	Marginal Effect	Coeff.	Std. Err.	Marginal Effect
Sex (female = 1)	0.604 **	0.258	0.137	0.556 **	0.246	0.136
Household size	0.143 ***	0.055	0.032	0.110 **	0.052	0.027
Quantity of cassava processed	-5.08×10^{-5} ***	1.15×10^{-5}	-1.15×10^{-5}	-7.83×10^{-5} ***	1.171×10^{-5}	-1.92×10^{-5}
Scale of operation (large scale = 1)	0.135	0.279	0.025	0.601 **	0.267	0.148
Ownership structure (Coop/partnership = 1)	−0.353	0.327	−0.081	−0.167	0.308	−0.041
Years of experience	−0.002	0.013	−0.001	−0.001	0.013	−0.001
Year of education	0.049 **	0.026	0.011	0.039	0.025	0.009
Member of association	−0.354	0.247	−0.079	−0.148	0.234	−0.036
Awareness of new management system	2.112 ***	0.276	0.426	1.396 ***	0.240	0.324
Constant	−1.379	0.557		−1.407	0.531	
	Number of obs. = 403			Number of obs. = 403		
	LR chi2(10) = 100.99			LR chi2 (10) = 76.97		
	Prob >chi2 = 0.0000			Prob >chi2 = 0.0000		
	Pseudo R^2 = 0.188			Pseudo R^2 = 0.1394		
	Log likelihood = −217.54			Log likelihood = −237.62		
	Marginal effect after logit Y = 0.6541			Marginal effect after logit Y = 0.5678		

***, and ** denote 1%, and 5% significant levels, respectively.

The model on processors' willingness to pay for training on organic manure was significant at 1%, while the McFadden's R squared (pseudo R^2) equalled 13.9%. Concerning the individual variables, the sex of processors had a positive relationship, with the WTP for organic manure training and was highly significant. This suggests that female respondents were more likely to be willing to pay for organic manure production training than their male counterparts. A unit increase in the household size equally increased the probability of WTP for the training by 0.03.

The significance of both awareness of new waste management systems and years of processors' education seems to indicate their importance in increasing the likelihood of WTP for this training by about 0.43 and 0.01, with the former likely to have more effect. Despite the marginal effect associated with years of education, its relevance cannot be sidelined. This is because processors with higher level/increased years of education are better exposed/accommodative to new technologies and, hence, are more likely to better appreciate the benefits to be derived from the use of organic manure, particularly on environmental and health grounds. This result is consistent with that of [45–47], who found persons with a higher level of education more willing to pay for solid waste disposal management services.

However, the relationship between the quantity of cassava processed and WTP for training in organic manure production was found to be negatively significant and with lesser effect. This indicates that any increase in the amount of cassava processed is likely to decrease the WTP for training in organic manure production. This is contrary to expectation, given the fact that those with a larger quantity of cassava to process should be the ones with more interest in finding alternative ways of disposing of solid waste, particularly if such processes will bring additional income. The observed result may arise if the processors do not find organic manure training an attractive/relevant measure for addressing their large solid waste concern. Moreover, as earlier stated, most processors believed that the additional production activities around cassava peel are time-consuming and would distract them from their main processing activities.

5. Conclusions

The study examined the willingness to pay for value-added solid-waste management systems among 403 cassava processors in the humid forest and Guinea savannah zones of Nigeria. The study showed that the solid-waste management practices currently in use by the cassava processors are dumping, gifting, and selling solid waste in wet and dry forms. The factors influencing the processor's decision on the type of waste management system were observed to include gender, membership of association, quantity of cassava processed and ownership structure. Two value-added solid-waste management options (organic manure and mushroom production) were introduced to the cassava processors, and more than half of them were willing to pay for acquiring the knowledge to use them. The factors positively influencing the WTP for expertise on conversion of solid waste to organic manure were female gender, household size, years of education, and awareness of the new options, while a possible increase in the quantity of cassava processed negatively influenced the WTP.

Similarly, while gender, household size and awareness of the new options positively influenced the willingness to pay for training in mushroom production, a possible increase in the quantity of cassava processed negatively influenced the WTP. While large quantities of waste are generated during both the high and low processing seasons, which can create significant environmental and human health hazards in the short run and in the long run, processors were willing to pay only a small amount of money to learn new methods of profitably disposing of the waste. This finding could be due to insufficient hygiene regulations and enforcement by the local authorities at cassava processing centres, which means that the processors had never been made to pay for waste disposal. Such circumstances decrease the processors' willingness to pay adequate amounts for training in new waste conversion or disposal methods.

Therefore, based on the findings of this study, it is recommended that sensitisation and subsidised training of processors in the benefits of safe waste conversion and disposal and in various new options for value-added management of solid waste should be intensified to encourage and enable cassava processors to manage agro-based waste safely. The large amount of waste generated by each processor shows the need for environmental hygiene protocols to be developed and enforced at community cassava processing centres. At the same time, the market for value-added agro-based waste products should be vigorously expanded.

Sustainability **2019**, *11*, 1759

Author Contributions: V.O.O. and A.B.A. conceived the idea, while V.O.O. and O.O. designed the study; O.O. collected and analysed the data; while the three authors wrote the paper.

Funding: This research was funded by the German Federal Ministry of Education and Research (BMBF) and Ministry of Economic Cooperation and Research (BMZ), grant number FKZ 031 A258 A

Acknowledgments: We also acknowledge the International Institute for Tropical Agriculture (IITA) for providing technical support for carrying out the field survey.

Conflicts of Interest: The authors declare no conflict of interest in this study.

References

1. Nhassico, D.; Muquingue, H.; Cliff, J.; Cumbana, A.; Bradbury, J.H. Rising African cassava production, diseases due to high cyanide intake and control measures. *J. Sci. Food Agric.* **2008**, *88*, 2043–2049. [CrossRef]
2. Manyong, V.; Dixon, A.; Makinde, K.; Bokanga, M.; Whyte, J. *The Contribution of IITA-Improved Cassava to Food Security in Sub-Saharan Africa: An Impact Study*; International Inst. of Tropical Agriculture IITA: Ibadan, Nigeria; Modern Design & Associates Ltd.: Lagos, Nigeria, 2000.
3. Ayinla, O.A. Analysis of Feeds and Fertilizers for Sustainable Aquaculture Development in Nigeria. *FAO Fish. Tech. Paper* **2007**, *497*, 453.
4. Ubalua, A.O. Cassava wastes: Treatment options and value addition alternatives. *Afr. J. Biotechnol.* **2007**, *6*, 2065–2073.
5. Sackey, I.S.; Bani, R.J. Survey of waste management practices in cassava processing to gari in selected districts of Ghana. *J. Food Agric. Environ.* **2007**, *5*, 325–328.
6. Okoya, A.A.; Oyawale, F.E.; Ofoezie, I.E.; Akinyele, A.B. Impact of industrial cassava effluent discharge on the water quality of Ogbese river, Ayede-Ogbese, Ondo state, Nigeria. *Ethiop. J. Environ. Stud. Manag.* **2016**, *9*, 339–353. [CrossRef]
7. Raney, T.; Anríquez, G.; Croppenstedt, A.; Gerosa, S.; Lowder, S.; Matuscke, I.; Skoet, J.; Doss, C. The Role of Women in Agriculture. Agricultural Development Economics Division (ESA) of the Economic and Social Development Department of the United Nations Food and Agriculture Organization (FAO). ESA Working Paper No. 11-02. 2011. Available online: http://www.fao.org/docrep/013/am307e/am307e00.pdf (accessed on 30 August 2017).
8. Osunbitan, J.A. Short term effects of Cassava processing waste water on some chemical properties of loamy sand soil in Nigeria. *J. Soil Sci. Environ. Manag.* **2012**, *3*, 164–171.
9. Aliu, I.R.; Adeyemi, O.E.; Adebayo, A. Municipal household solid waste collection strategies in an African megacity: Analysis of public private partnership performance in Lagos. *Waste Manag. Res.* **2014**, *3*, 67–78. [CrossRef] [PubMed]
10. Gilpin, A. *Dictionary of Environmental and Sustainable Development*; John Wiley and Sons: Sussex, UK, 1996; p. 201.
11. Olanbiwoninu, A.; Odunfa, S.A. Production of fermentable sugars from organosolv pretreated cassava peels. *Adv. Microbiol.* **2015**, *5*, 117–122. [CrossRef]
12. Abass, A.B.; Onabolu, A.O.; Bokanga, M. Impact of the high-quality cassava flour technology in Nigeria. In *Root Crops in the 21st Century, Proceedings of the 7th Triennial Symposium of the International Society for Tropical Root Crops-Africa Branch (ISTRC-AB), Centre International des Conferences, Cotonou, Benin, 11–17 October 1998*; Akoroda, M.O., Ngeve, J.M., Eds.; International Society of Tropical Root Crops: Cotonou, Republic of Benin, 1998.
13. Olukanni, D.O.; Olatunji, T.O. Cassava waste management and biogas generation potential in selected local government areas in Ogun State, Nigeria. *Recycling* **2018**, *3*, 58. [CrossRef]
14. Salau, M.A.; Ikponmwosa, E.E.; Olonode, K.A. Structural strength characteristics of cement-cassava peel ash blended concrete. *Civ. Environ. Res.* **2012**, *2*, 68–77.
15. Oviasogie, P.O.; Ndiokwere, C.L. Fractionation of lead and cadmium in refuse dump soil treated with cassava mill effluent. *J. Agric. Environ.* **2008**, *9*, 10–15.
16. Omotosho, O.; Amori, A. Caustic hydrogen peroxide treatment of effluent from cassava processing industry: Prospects and limitations. *Int. J. Eng. Technol. Innov.* **2015**, *5*, 121–131.
17. Onifade, T.; Adeniran, K.; Ojo, O. Assessment of cassava waste effluent effect on some water sources in Ilorin, Nigeria. *Afr. J. Eng. Res.* **2015**, *3*, 56–68.

18. Oyegbami, A.; Oboh, G.; Omueti, O. Cassava prrocessors' awareness of occupational and environmental hazards associated with cassava processing in southwestern Nigeria. *Afr. J. Food Agric. Nutr. Dev.* **2010**, *10*, 2176–2186.

19. Uhuegbu, C.C.; Onuorah, L.O.; Ayara, W.A.; Ugonabo, O. Organic kerosene production from a mixture of cassava (Manihotesculentia), maize (*Zea mays*) and Akintola leaves (*Chrollaenaodorata*). *J. Int. Acad. Res. Multidiscip.* **2015**, *3*, 54–60.

20. Sabiiti, E.N. Utilising agricultural waste to enhance food security and conserve the environment. *Afr. J. Food Agric. Nutr. Dev.* **2011**, *11*, 1–9.

21. Plassmann, H.; O'Doherty, J.; Rangel, A. Orbitofrontal cortex encodes willingness to pay in everyday economic transactions. *J. Neurosci.* **2007**, *27*, 9984–9988. [CrossRef] [PubMed]

22. Haq, M.; Mustaa, U.; Ahmad, I. Household's willingness-to-pay for safe drinking water: A case study of Abbottabad district. *Pak. Dev. Rev.* **2007**, *46*, 1137–1153. [CrossRef]

23. Rodríguez, E.; Lacaze, V.; Lupín, B. Contingent valuation of consumers' willingness-to-pay for organic food in Argentina. In Proceedings of the 12th Congress of the European Association of Agricultural Economists (EAAE), Ghent, Belgium, 1–10 August 2008.

24. Mitchell, R.C.; Carson, R.T. *Using Surveys to Value Public Goods: The Contingent Valuation Method*; Resources for the Future: Washington, DC, USA, 1989.

25. Venkatachalam, L. The contingent valuation method: A review. *Environ. Impact Assess. Rev.* **2004**, *24*, 89–124. [CrossRef]

26. Perman, R.; Ma, Y.; McGilvray, J.; Common, M. *Natural Resource and Environmental Economics*, 3rd ed.; Addison-Wesley: Boston, MA, USA, 2003.

27. Boxall, P.C.; Adamowicz, W.L.; Swait, J.; Williams, M.; Louviere, J. A comparison of stated preference methods for environmental valuation. *Ecol. Econ.* **1996**, *18*, 243–253. [CrossRef]

28. Veisten, K. Contingent valuation controversies: Philosophic debates about economic theory. *J. Socio-Econ.* **2007**, *36*, 204–232. [CrossRef]

29. Cawley, J. *Contingent Valuation Analysis of Willingness to Pay to Reduce Childhood Obesity*; Working Paper 12510; National Bureau of Economic Research (NBER): Cambridge, MA, USA, 2006; pp. 1–39.

30. Ezedinma, C.; Lemchi, J.; Okechukwu, R.; Ogbe, F.; Akoroda, M.; Sanni, L.; Okoro, E.; Okarter, C.; Dixon, A.G.O. *Status of Cassava Production in South East and South South Nigeria—A Baseline Report*; International Institute of Tropical Agriculture (IITA): Ibadan, Nigeria, 2007.

31. Yusuf, S.A.; Ojo, O.T.; Salimonu, K.K. Households' willingness to pay for improved solid waste management in Ibadan North local government area of Oyo state, Nigeria. *J. Environ. Ext.* **2007**, *6*, 57–63. [CrossRef]

32. Odediran, O.F.; Ojebiyi, W.G. Cassava processors' willingness to utilise cassava peel for mushroom production in Southwest, Nigeria. *Int. J. Agric. Policy Res.* **2017**, *5*, 86–93.

33. Davies, R.M.; Olatunji, M.O.; Burubai, W. A survey of cassava processing machinery in Oyo State. *World J. Agric. Sci.* **2008**, *4*, 337–340.

34. Obayelu, O.A.; Akintunde, O.O.; Obayelu, A.E. Determinants of on-farm cassava biodiversity in Ogun State, Nigeria. *Int. J. Biodivers. Sci. Ecosyst. Serv. Manag.* **2015**, *11*, 298–308. [CrossRef]

35. Zhang, M.; Xie, L.; Yin, Z.; Khanal, S.K.; Zhou, Q. Biorefinery approach for cassava-based industrial wastes: Current status and opportunities. *Bioresour. Technol.* **2016**, *215*, 50–62. [CrossRef] [PubMed]

36. Okunade, D.A.; Adekalu, K.O. Physico-chemical analysis of contaminated water resources due to cassava wastewater effluent disposal. *Eur. Int. J. Sci. Technol.* **2013**, *2*, 75–85.

37. Idah, P.A.; Mopah, E.J. Comparative assessment of energy values of briquettes from some agricultural by-products with different binders. *IOSR J. Eng.* **2013**, *3*, 36–42.

38. Anikwe, M.A.N.; Nwobodo, K.C.A. Long term effect of municipal waste disposal on soil properties and productivity of sites used for urban agriculture in Abakaliki, Nigeria. *Bioresour. Tech.* **2002**, *83*, 241–250. [CrossRef]

39. Arimoro, F.O.; Iwegbue, C.M.; Enemudo, B.O. Effects of cassava effluent on benthic macroinvertebrate assemblages in a tropical stream in southern Nigeria. *Acta Zool. Litu.* **2008**, *18*, 147–156. [CrossRef]

40. Izugbara, C.O.; Umoh, J.O. Indigenous waste management practices among the Ngwa of Southeastern Nigeria: some lessons and policy implications. *Environmentalist* **2004**, *24*, 87–92. [CrossRef]

41. Omilani, O. Economics of Solid Waste Management Technology among Cassava Processors in Forest and Savannah Zones of Nigeria. Unpublished Master's Dissertation, Department of Agricultural Economics, University of Ibadan, Ibadan, Nigeria, 2015; pp. 1–67.
42. Ekere, W.; Mugisha, J.; Drake, L. Factors influencing waste separation and utilization among households in the Lake Victoria crescent, Uganda. *Waste Manag.* **2009**, *29*, 3047–3051. [CrossRef] [PubMed]
43. Osarenkhoe, O.O.; John, O.A.; Theophilus, D.A. Ethnomycological conspectus of West African mushrooms: An awareness document. *Adv. Microbiol.* **2014**, *4*, 39–54. [CrossRef]
44. Beck, T.; Nesmith, C. Building on poor people's capacities: the case of common property resources in India and West Africa. *World Dev.* **2001**, *29*, 119–133. [CrossRef]
45. Chuen-khee, P.; Othman, J. Household demand for solid waste disposal options in Malaysia. *World Acad. Sci. Eng. Technol.* **2002**, *66*, 1153–1158.
46. Boateng, S.; Amoaka, P.; Poku, A.A.; Appiah, D.O.; Garsonu, E.K. Household willingness to pay for solid waste disposal services in urban Ghana: The Kumasi metropolis situation. *Ghana J. Geogr.* **2016**, *8*, 1–17.
47. Ini-mfon, V.P.; Ubokudom, E.O.; Ubong, U.S. Households' willingness to pay for improved solid waste management in Uyo metropolis, Akwa Ibom State, Nigeria. *Am. J. Environ. Prot.* **2017**, *5*, 68–72. [CrossRef]

 © 2019 by the authors. Licensee MDPI, Basel, Switzerland. This article is an open access article distributed under the terms and conditions of the Creative Commons Attribution (CC BY) license (http://creativecommons.org/licenses/by/4.0/).

sustainability

Article

Sustainability Performance of National Bio-Economies

Lisa Biber-Freudenberger [1,*], Amit Kumar Basukala [1], Martin Bruckner [2] and Jan Börner [1]

1   Center for Development Research, University of Bonn, D-53113 Bonn, Germany;
    pblkamit@gmail.com (A.K.B.); jborner@uni-bonn.de (J.B.)
2   Institute for Ecological Economics, Vienna University of Economics and Business, 1020 Vienna, Austria;
    martin.bruckner@wu.ac.at
*   Correspondence: lfreuden@uni-bonn.de; Tel.: +49-228-73-1726

Received: 28 June 2018; Accepted: 27 July 2018; Published: 1 August 2018

Abstract: An increasing number of countries develop bio-economy strategies to promote a stronger reliance on the efficient use of renewable biological resources in order to meet multiple sustainability challenges. At the global scale, however, bio-economies are diverse, with sectors such as agriculture, forestry, energy, chemicals, pharmaceuticals, as well as science and education. In this study, we developed a typology of bio-economies based on country-specific characteristics, and describe five different bio-economy types with varying degrees of importance in the primary and the high-tech sector. We also matched the bio-economy types against the foci of their bio-economy strategies and evaluated their sustainability performance. Overall, high-tech bio-economies seem to be more diversified in terms of their policy strategies while the policies of those relying on the primary sector are focused on bioenergy and high-tech industries. In terms of sustainability performance, indicators suggest that diversified high-tech economies have experienced a slight sustainability improvement, especially in terms of resource consumption. Footprints remain, however, at the highest levels compared to all other bio-economy types with large amounts of resources and raw materials being imported from other countries. These results highlight the necessity of developed high-tech bio-economies to further decrease their environmental footprint domestically and internationally, and the importance of biotechnology innovation transfer after critical and comprehensive sustainability assessments.

Keywords: bioeconomy; green economy; sustainable development; bioproductivity; high-tech bioeconomy; knowledge-based bioeconomy; primary sector; typology; cluster analysis

1. Introduction

Despite the current drop in price, many fossil fuel resources are becoming increasingly scarce [1], and their consumption is associated with climate change, and harmful effects on ecosystems and human health. At the same time, population growth and corresponding pressures on natural resources have risen beyond safe ecological limits [2]. In response to these societal challenges, countries have adopted ambitious global goals such as the 2 °C limit to global warming, the Aichi Biodiversity Targets, and the Sustainable Development Goals (SDGs). However, the unprecedented global awareness of sustainability issues has yet to be matched with appropriate action towards achieving tangible goals and targets.

An increasing number of countries look to the bio-economy as a strategy to (a) reduce reliance on fossil fuels and (b) enable sustainable development through a "biologization" of the regular economy [3–5]. As a consequence, the conceptual development of the bio-economy has been driven by the fossil fuel substitution perspective, as well as the biotechnology innovations perspective, leading to multiple definitions for the bio-economy (see [6,7]). While early publications adopted sector specific definitions or focused purely on biotechnology [8,9], the bio-economy is increasingly being understood to include all kinds of economic activity that rely on biological processes, products, and principles [10]. In this study we adopt a broad definition that includes traditionally bio-based sectors such as agriculture and forestry, but also bio-based pharmaceutics, waste treatment, energy, bio-plastics, and chemicals.

Bugge et al. [11] identified three different visions of bio-economy that have developed simultaneously: (1) the bio-technology vision focusing on research, application, and commercialization of bio-technology innovations, (2) the bio-resource vision with an emphasis on the efficient and upgraded use of raw materials, and (3) the bio-ecology vision centered around the concept of sustainability and ecosystem conservation through increased efficiency. Similarly, Hausknost et al. [12] categorized four different visions of a bio-economy along the dichotomies of sufficiency versus capitalist growth and agro-ecology versus industrial (bio-)technology. These "ideal types", however, correspond to political visions of the bio-economy and do not necessarily reflect real world conditions [13]. Behind such visions are fundamentally different views and expectations about the costs and benefits of a bio-based transformation of the economy. Proponents expect economic growth, employment opportunities, and environmental benefits [14–16], whereas critics fear the uncontrolled spread of risk technologies, unequal benefit sharing outcomes, and environmental damage [17]. A balanced view requires a better understanding and predictive tools to assess the complex cause-effect relationships involved in international trade and innovation/knowledge systems, which act as key mediators in a global bio-based transformation. Therefore, in order to strengthen social acceptance of bio-economy policies, a better understanding of sustainability implications and a consideration of this knowledge when designing bio-economy policies is necessary to avoid negative impacts from policy measures [18].

This paper aims at improving our understanding of the determinants and sustainability implications of the supply and demand of bio-based products, and our knowledge about the global bio-economy by answering the following research questions: (1) How do countries differ in terms of their bio-economy and comparative advantages? (2) Can these differences be explained by the adoption of different national policies? And (3) what are the sustainability implications of these different bio-economy pathways? First, we characterize bio-economy types based on a set of indicators that we expect to be associated with countries' strategic decisions to embark on a particular bio-based development pathway. We then show that these bio-economy types have in fact adopted considerably distinct national bio-economy strategies. However, we also find that the current and anticipated bio-economy developments bear both opportunities and risks in multiple SDG dimensions.

National bio-economies, as well as frameworks to assess their implementation, often focus on economic benefits but often ignore other aspects of sustainability such as resource and waste management along the whole value chain, and competition between biomass needs [19,20]. As in our study, a cluster approach was also applied by Philippidis et al. [21] to characterize different bio-economies of the European Union (EU) based on sector-specific backward and forward-linkages, as well as employment multipliers. Ronzon et al. [22] characterized and grouped all countries of the EU based on the productivity, turnover, and job generation of different bio-economy sectors. However, while they focused on the implications of growth in different bio-economy sectors for employment opportunities and wealth generation, we assess the sustainability of the different sectors, including an ecological dimension and link the typology to existing bio-economy policy strategies within each respective country. The enormous diversity of country contexts and bio-economy development potential around the world is likely to be reflected in the strategies adopted by governments to promote bio-based growth [3,23]. The national bio-economy policy strategy of Nigeria, for example, focuses on bioenergy, whereas India pursues high-tech, research, and innovation issues. Denmark´s strategy broadly advocates for a green economy (a sustainable economy), and Portugal concentrates on the blue economy (fisheries and other marine and coastal economic sectors, see [3]).

We expect countries' bio-economy strategies to reflect key contextual determinants, such as: (1) global comparative advantages in specific bio-economy sectors (e.g., agriculture, forestry, biotechnology, etc.), (2) current or anticipated sustainability threats, and (3) consumer, i.e., voter, preferences. Given the complexities of the cause-effect relationships involved in bringing about sustainable development, we also expect the current global mix of national contextual conditions to result in a portfolio of bio-economy strategies that requires an appropriate governance regime to minimize negative social and environmental externalities and maximize opportunities for sustainable development. Our subsequent analysis includes social development indicators, but focuses on current performance and performance trends in key environmental outcome dimensions, such as CO_2 emissions, biodiversity, and related footprint indicators.

2. Materials and Methods

We identified a set of globally available indicators supported by targeted literature e.g., [19,22,24] and data review that reflect the three key contextual dimensions introduced in the previous section (Table 1). The selection of specific indicators was mainly driven by the availability of comprehensive and up-to-date datasets, which were available in open access databases (e.g., of the Food and Agriculture Organisation, International Labour Organisation, and the World Bank). This included the overall importance of different bio-economy sectors, i.e., the primary and the high-tech sector, as well as the overall comparative advantages for these sectors, such as the availability of bioproductive land and skilled labor.

Table 1. Set of indices and indicators used for the cluster analysis to identify bio-economy types.

Indices and Indicators	Unit	Years	Source
1. Economic importance of the primary sector			
Agriculture, value added, includes agricultural as well as forest products	% of Gross domestic product (GDP)	2010 to 2016	[25]
Employment in agriculture male	% of male population	2010 to 2016	[26]
Employment in agriculture female	% of female population	2010 to 2016	[27]
Agricultural land	% of land area	2010 to 2015	[28]
Forest area	% of land area	2010 to 2015	[29]
2. Economic importance of the high-tech sector			
High-technology exports	% of manufactured exports	2010 to 2016	[30]
Patent applications, residents	Per million people	2010 to 2015	[31]
Patent applications, nonresidents	Per million people	2010 to 2015	[32]
3. Availability of bioproductive land			
Bioproductive area/Biocapacity	% of total area	2013 (2017 Edition)	[33]
Bioproductive area/Biocapacity (per capita)	ha per capita	2013 (2017 Edition)	[33]
4. Availability of skilled labor			
Gross enrolment ratio, tertiary, both sexes	% of population	2010 to 2016	[34]
Employees with an advanced education	% of population	2009 to 2016	[35]
Scientific and technical journal articles	per million people	2010 to 2013	[36]

We measured the importance of the primary and the high-tech bio-economy sectors, which was likely to reflect comparative advantages, and to some degree, social choice and voter preferences. Additionally, specific bio-economy sectors are likely to be the result of certain sustainability threats but may have their own sustainability implications that need to be considered. While we tried to capture the economic importance of the primary sector based on indicators reflecting the economic significance, employment opportunity and overall land occupied, we employed indicators reflecting economic importance in terms of international trade and patent applications for the high-tech sector. We assumed a stronger concentration of high-tech innovations in knowledge-, investment- and research-intensive sectors such as pharmacy, biotechnology, biochemical, and bioplastics. A vibrant, innovative, and economically viable bio-economy is likely to be reflected in the number of patent applications by residents, as well as non-residents and exports of high-tech products.

Comparative advantages are also reflected by the availability of resources, such as the availability of bioproductive land, which we used interchangeably with biocapacity, as well as the availability of skilled labor. According to the definition of the Global Footprint Network [37] biocapacity refers to "the capacity of a particular surface to renew what people demand. In the National Footprint Accounts, the biocapacity of an area is calculated by multiplying the actual physical area by the yield factor and the appropriate equivalence factor". Yield and equivalence factors are conversion factors used to translate hectares into global hectares. While the availability of bioproductive land is more or less determined by bioclimatic factors, the availability of skilled labor usually depends strongly on specific policies as well as on investments in education, and therefore is controlled, at least in democratic countries, by the consumer and voters' choice. Both the availability of bioproductive land and of skilled labor have certain sustainability implications in terms of land use intensification, decent work, and education.

For the availability of skilled labor, we selected indicators reflecting the share of the population enrolled in the tertiary sector, as well as employees with an advanced education, to account for immigration and emigration of trained staff; to reflect the actual output of the scientific community we applied the number of scientific publications per million people. The availability of skilled labor reflected a comparative advantage, which is likely to play an important role for the knowledge intensive high-tech sector.

Mean values were calculated, for those data with multiple years. Some indicators, which were not yet standardized by area or population size, were calculated accordingly. All values $x_{i,j}$ were normalized by calculating the z-scores $z_{i,j}$ for each country i and each indicator j (Equation (1)).

$$z_{i,j} = \frac{x_{i,j} - \min(x_j)}{\max(x_j) - \min(x_j)} \tag{1}$$

The normalized indicator values $z_{i,j}$ were combined in multivariate indices $\overline{z_i}$ calculated as mean values for each country i over all normalized indicators j (Equation (2)).

$$\overline{z_i} = \frac{1}{N} \sum_{j=1}^{N} z_{i,j} \tag{2}$$

This approach was chosen to sufficiently reflect and summarize different aspects related to the importance of different bio-economy sectors, the availability of resources including bioproductive land and skills, as well as the international role of trade in a comprehensive way. Furthermore, the combination of indicators in multivariate indices calculated as mean values across all indicators allowed us to also include indicators with incomplete datasets. In these cases, only values for available indicators were used. Countries with missing index values were excluded from the cluster analysis. All indices were combined in a new dataset and used for cluster analysis.

Initially, various clustering algorithms were applied to the dataset leading to the result-based selection of a Gaussian finite mixture model-based clustering. Different parameterized Gaussian mixture models were fitted through an Expectation-Maximization-Algorithm, setting the volume, shape, and orientation of the covariance to be either equal or variable [38]. The optimal model according to the highest BIC value (Bayesian information criterion) was selected for clustering and further analysis. The cluster-based typology was evaluated and interpreted by the calculation of the F-Statistic and t-value of the different types. The F-value indicates whether an indicator significantly contributes to the clustering of a type, by comparing the variance of the cluster to the variance of the whole dataset (the smaller the F-value, the more homogenous the cluster). The t-value indicates how a cluster is characterized by each indicator comparing the means (a positive value indicates a higher mean and a negative value a lower mean for the cluster). Additionally, boxplots as well as pairwise Wilcoxon Rank Sum Tests with Bonferroni correction were calculated and interpreted.

The clustering results were evaluated against the number and foci of existing bio-economy strategies of the different countries according to data from the German Bio-economy Council [3]. The selection of the documents by the German Bio-economy council was based on an Internet-based desk study using publicly available government documents published between 2005 and 2015. In the absence of dedicated bio-economy strategies, policy documents linking to bio-economy such as biotechnology, bioenergy, or biobased-economy/industry were included. Other papers, such as agricultural, forestry, or marine strategies were only included if they had a strong focus on bio-economy or innovative bio-based approaches.

Furthermore, sustainability performance was assessed using a selection of the SDG indicators database, as well as indicators for footprints associated with international trade (Table 2). We focused on indicators for SDGs that are frequently mentioned to be considerably affected by growth in bio-economy sectors, especially SDG 2 (no hunger), SDG 7 (affordable and clean energy), and SDG 12 (responsible consumption and production). These are among the SDGs most commonly referred to in literature [39,40] and mentioned by experts, as part of an online survey, with respect to the potential benefits from the bio-economy (e.g., increasing agricultural production, emission reduction from renewable energy production, effective waste treatment), as well as its risks, such as the competition between food and energy and biodiversity loss. However, this does not imply that there are no potential impacts on the other SDGs. Other indicators were also evaluated, but due to inconclusive or insignificant results, they were not included in this publication.

We complemented this dataset with additional data on the ecological footprint, and the ecological deficit or reserve, as well as the foreign and domestic biomass footprint that could be attributed to the consumption of food as well as non-food products. The biomass footprint for food as well as non-food purposes was calculated using the Leontief inverses [41] for the years 2000 and 2013 from the multi-region input-output (MRIO) database Eora in the harmonized 26-sector classification [42], together with the UN Environmental Program material extraction data set [43]. We classified the sectors into food and non-food categories, assuming that upstream biomass inputs of mining, non-food manufacturing as well as transport sectors are used for non-food purposes. Final products from agriculture, fishing, and food processing sectors, as well as biomass uses in the service sectors (mainly restaurants and public services) are attributed to food uses.

Sustainability performance was assessed for the current status, calculated as average values between 2010 and 2015, and for the change in performance, calculated as the change between the current status and the average values for the years 2000 to 2005. Since not all indicators were available for each year, missing years were ignored for the mean calculation.

Boxplots and pairwise Wilcoxon Rank Sum Tests with Bonferroni correction were used to compare the clusters for the different sustainability indicators. The statistical computing environment R was used for all analyses and figure preparations. The packages used were mclust 5.3 [38] for clustering, ggplot2 [44], factoextra [45], rworldmap [46], gridExtra [47], and cowPlot [48] for visualization, and Hmisc [49], matrixStats [50], and multcompView [51] for statistical evaluation.

Table 2. Indicators used for sustainability evaluation of different bio-economy types.

SDG No.	Sustainability Indicator	Unit	Years Status	Years Early Stage	Source
2.1.1	Prevalence of undernourishment	% of population	2010–2015	2000–2005	[52]
7.2.1	Renewable energy share in final consumption	% of energy	2010–2014	2000–2005	[53]
7.3.1	Energy intensity in terms of primary energy and GDP	MJ per USD const. 2011 PPP GDP	2010–2014	2000–2005	[54]
12.2.1	Material footprint	Metric tons per capita	2010	2000–2005	[55]
12.2.2	Domestic material consumption	Metric tons per capita	2010	2000–2005	[56]
na	Ecological footprint	Global ha per capita	2010–2013	2000–2005	[33]
na	Ecological deficit or reserve	Global ha per capita	2010–2013	2000–2005	[33]
na	Domestic food biomass footprint	tons per capita	2010	2000	[42,43]
na	Domestic non-food biomass footprint	tons per capita	2010	2000	[42,43]
na	Imported food biomass footprint	tons per capita	2010	2000	[42,43]
na	Imported non-food biomass footprint	tons per capita	2010	2000	[42,43]

3. Results

3.1. Bio-Economy Classification

Statistical evaluations of different Gaussian models indicated that the VVI model (diagonal distribution, varying volume and shape) with four components (clusters) received the highest BIC value and was therefore selected for further analyses. Based on the selected indicators and clustering algorithm countries were grouped into four different clusters, the Diversified Bio-economies (Diverse), the Advanced Primary Sector Bio-economies (AdvancedPrim), the High-Tech Bio-economies (HighTech), and the Basic Primary Sector Bio-economies (BasicPrim) (Figure 1). The different bio-economy clusters were considerably different from each other according to F and t-values (Table 3), and showed significant characteristics (Figure 2).

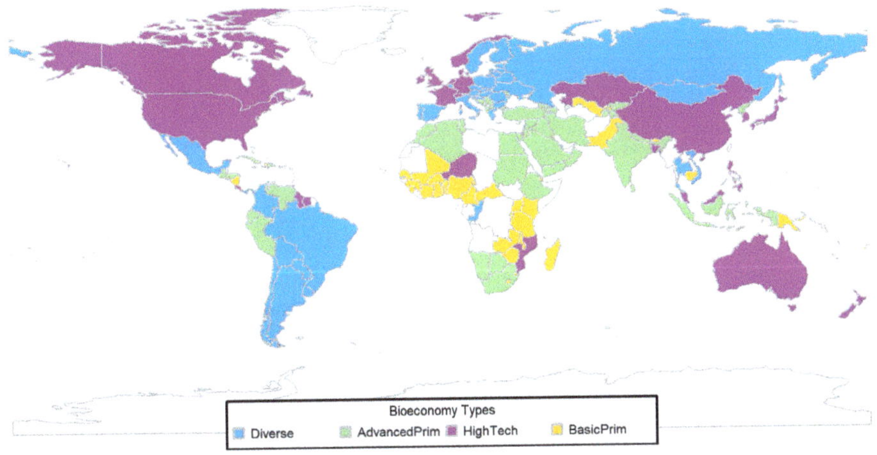

Figure 1. Map of a global bio-economy typology with countries belonging to the same cluster being displayed in the same color. Clusters: Diversified Bio-economies (Diverse), Advanced Primary Sector Bio-economies (AdvancedPrim), High-Tech Bio-economies (HighTech), and Basic Primary Sector Bio-economies (BasicPrim).

Table 3. t-values to be used for the interpretation of the clusters; cells in brackets have F-values higher than 1.2 (F-values higher than 1 indicate that the variance of a cluster is bigger than the variance of the whole dataset).

Index	Diverse	AdvancedPrim	HighTech	BasicPrim
Economic importance of the primary sector	−0.480	−0.092	−0.487	1.182
Economic importance of the high-tech sector	−0.137	−0.566	(1.513)	−0.434
Availability of bioproductive land	0.290	−0.548	(0.924)	−0.524
Availability of skilled labor	0.426	−0.246	(0.816)	−0.983

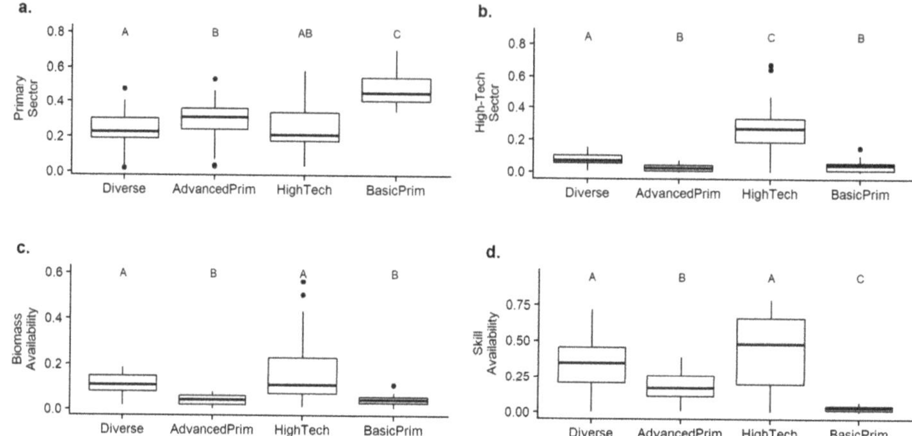

Figure 2. Boxplots comparing clustering variables for different bio-economy types (**a–d**). Letters indicate significant differences according to a pairwise Wilcox rank sum test with a significance level of 0.05. Clusters with the same letter are not significantly different.

The Diverse cluster, which included 50 countries, was mainly comprised of different countries in South and Central America (Antigua and Barbuda, Argentina, Aruba, The Bahamas, Barbados, Bolivia, Brazil, Chile, Colombia, Panama, Paraguay, St. Lucia, St. Vincent and the Grenadines, Trinidad and Tobago, Uruguay), countries in Southeast and Central Asia (Brunei Darussalam, Lebanon, Mongolia, Qatar, Samoa, Thailand, Tonga, Vietnam), eastern and southern Europe (Austria, Belarus, Belgium, Bulgaria, Croatia, Cyprus, Czech Republic, Estonia, Finland, Greece, Hungary, Italy, Latvia, Lithuania, Luxembourg, Poland, Portugal, Romania, the Russian Federation, Slovak Republic, Slovenia, Spain, Sweden, Ukraine), plus a few others (Congo, Mauritius, Mexico). Poland was identified as the most representative country of the Diverse cluster, characterized by a middling importance of both the high-tech and the primary sectors. This was matched by a mediocre availability of bioproductive land and skilled labor (the latter one not being significantly lower than the HighTech cluster).

The 45 countries belonging to the AdvancedPrim cluster were mainly found in Africa (Algeria, Sudan, Botswana, Cabo Verde, Ethiopia, Morocco, Namibia, South Africa, Tunisia, Egypt), Asia (North Korea, India, Indonesia, Iran, Iraq, Jordan, Kuwait, Kyrgyz Republic, Nepal, Oman, Saudi Arabia, Sri Lanka, Syrian Arab Republic, Tajikistan, Yemen), Central and South America (Cuba, Dominican Republic, Ecuador, El Salvador, Guatemala, Honduras, Jamaica, Peru, St. Kitts and Nevis, Venezuela) and eastern Europe (Albania, Armenia, Azerbaijan, Bosnia and Herzegovina, Georgia, Moldova, Macedonia, Turkey, Serbia). For AdvancedPrim countries, which were best represented by Botswana, the primary sector was more important for the economy than the high-tech sector, despite a relatively low availability of bioproductive land. However, the primary sector was not as important as for the BasicPrim cluster. Skill availability was at a mediocre level, which was significantly lower than for the Diverse and HighTech clusters. Corresponding to this limited availability of skilled labor, we found a rather low importance of the high-tech sector.

In the HighTech cluster, 32 countries from North America (Canada and the United States), parts of central and northern Europe (Denmark, France, Germany, Ireland, Malta, the Netherlands, Norway, Switzerland, UK), some parts of Asia (Bahrain, Bangladesh, Kazakhstan, Malaysia, Israel, Japan, Philippines, South Korea, Singapore, Timor-Leste, Vanuatu and China), South and Central America (Bermuda, Costa Rica, Grenada, Guyana, Suriname) as well as Australia, Mozambique, New Zealand, and Niger were summarized.

The high-tech sector played a very important role for the economies belonging to this cluster, best represented by countries such as Japan, South Korea, and Singapore, while the primary sector was of very little importance. This was matched by a significantly higher availability of skilled labor compared to all other clusters except the diversified cluster. The importance of the primary sector was very variable within this cluster and correspondingly, we found a high variation for the availability of bioproductive land.

Countries from Africa (Benin, Burkina Faso, Burundi, Cameroon, Central African Republic, Comoros, Côte d'Ivoire, Djibouti, the Gambia, Ghana, Guinea, Kenya, Lesotho, Madagascar, Malawi, Mali, Nigeria, Rwanda, Sao Tome and Principe, Senegal, Sierra Leone, Swaziland, Togo, Uganda, Tanzania, Zambia, and Zimbabwe), Southeast and Central Asia (Bhutan, Cambodia, Fiji, Pakistan, Papua New Guinea, Solomon Islands, Uzbekistan) and some South and Central American countries (Dominica, Haiti, Nicaragua) were included in the BasicPrim cluster (37 countries). These countries, best represented by the case of Tanzania, were characterized by a very high importance of the primary sector (significantly higher than for all other types) and a very low importance of the high-tech sector. The availability of skilled labor and bioproductive land were extremely poor (skill availability being significantly lower than for all other types). Compared to the AdvancedPrim, BasicPrim countries show a significantly higher importance of the primary sector coupled with a significantly lower skill availability.

Overall, statistical evaluation of the clusters showed that the BasicPrim cluster had the most important primary sector, followed by the AdvancedPrim cluster. This was, however, not dependent on the availability of bioproductive land, which was significantly higher for type HighTech and Diverse countries. The overall correlation between biomass availability and the importance of the primary sector was even slightly negative (Table 4, r = -0.289, $p < 0.001$). The absence of high-level skills seemed to be a much better predictor of the importance of the primary sector (r = 0.574, $p < 0.001$) than the availability of biomass. In contrast, the importance of the high-tech sector was clearly reflected by the availability of skilled labor (r = 0.401, $p < 0.001$) highest for type HighTech, followed by type Diverse.

Table 4. Spearman correlation coefficients between the different indices, ** $p < 0.001$, * $p > 0.005$.

	Primary Sector	High-Tech	Productive Land Availability
Primary Sector			
High-Tech	-0.254 *		
Productive Land Availability	-0.289 **	0.233 *	
Skill Availability	-0.574 **	0.401 **	0.294 **

3.2. Bio-Economy Strategies across Clusters

As expected, we found that existing national bio-economy strategies can be plausibly mapped to our typology of bio-economies, although we also found some apparent contradictions (Figure 3). Countries with strong high-tech sectors (type HighTech) are also the countries with the highest number of bio-economy strategies. 50% of the countries of this type had at least one political strategy that could be considered as a bio-economy strategy, and 31% had even more than one strategy. For type HighTech countries, bio-economy strategies were not only the most abundant, but also the most diversified with all types of strategies being represented, but most within the field "Bio-economy Research & Innovation" (seven countries) and "High-Tech" (seven countries). Countries that relied the most on the primary sector (especially type BasicPrim and type AdvancedPrim) showed a much lower abundance of countries with at least one bio-economy strategy (12.5% and 19%). But while for BasicPrim these strategies mainly focused on "High-Tech" (three countries, despite the low importance of the high-tech sector) and "Bioenergy" (four countries), they did not have such a specific emphasis for the AdvancedPrim countries. The focus of BasicPrim countries on "High-Tech" exemplifies how political strategies do not necessarily reflect current conditions but may rather emphasize desirable development pathways, whereas a strong focus of the same type on "Bioenergy" might be the result of an increasing or anticipated competition for land between food and energy sectors. With 30% of countries from the cluster Diverse having at least one bio-economy strategy, they were not as abundant as for type HighTech, but still more frequent than for the BasicPrim and AdvancedPrim types. Strategies of this cluster focused mainly on "High-Tech" (seven countries), which indicates the desire of these countries to perpetuate the shift from a primary sector based economy to a high-tech economy.

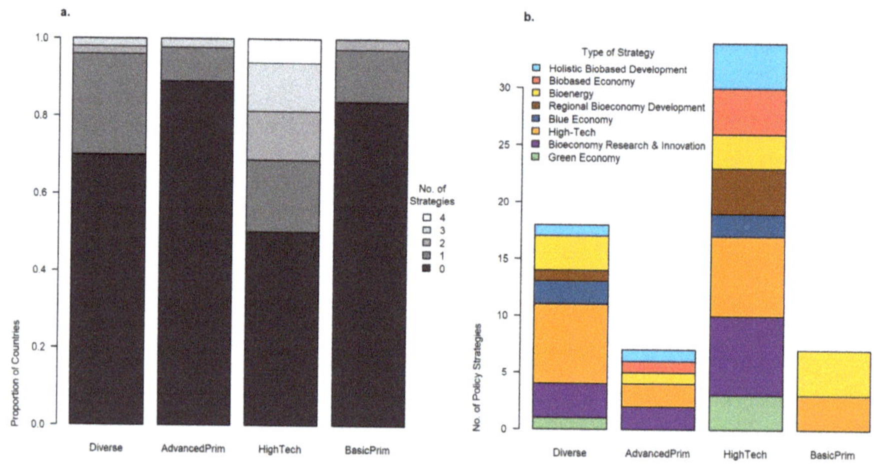

Figure 3. Matching of the bio-economy typology against (**a**) the number of different bio-economy strategies per country and (**b**) the different foci of bio-economy strategies.

3.3. Sustainability Performance of Bio-Economy Types

Results of sustainability evaluation showed partially significant differences between bio-economy types. However, all results should be interpreted with caution, as only part of the sustainability performance can be attributed to the bio-economy developments. This is not a cross-sectional study and therefore informational value about cause-effect relationships is only limited. In terms of prevalence of undernourishment, HighTech and Diverse countries had the lowest recorded rates of undernourishment (Figure 4a). BasicPrim countries showed the highest prevalence of undernourishment. For all bio-economy types, we found that undernourishment has been decreasing, indicating that hunger has been reduced globally, independently of the type of bio-economy in place (Figure 4b). However, the reduction was significantly larger in BasicPrim countries.

In terms of the renewable energy share, we found that countries of BasicPrim had the highest shares of renewable energy, significantly higher than all other types (Figure 4c). While for AdvancedPrim, and BasicPrim countries, where the share of renewable energy was decreasing, a substantial increase was found for most Diverse and HighTech countries (Figure 4d). This indicates countries especially in the global South that rely heavily on wood as an energy source are shifting away from renewable energy sources, while countries that currently rely mainly on nuclear and fossil energy sources are increasingly using renewables. Energy intensity was highest for countries of type BasicPrim (Figure 4e), and has been decreasing for all bio-economy types without any significant differences between them (Figure 4f).

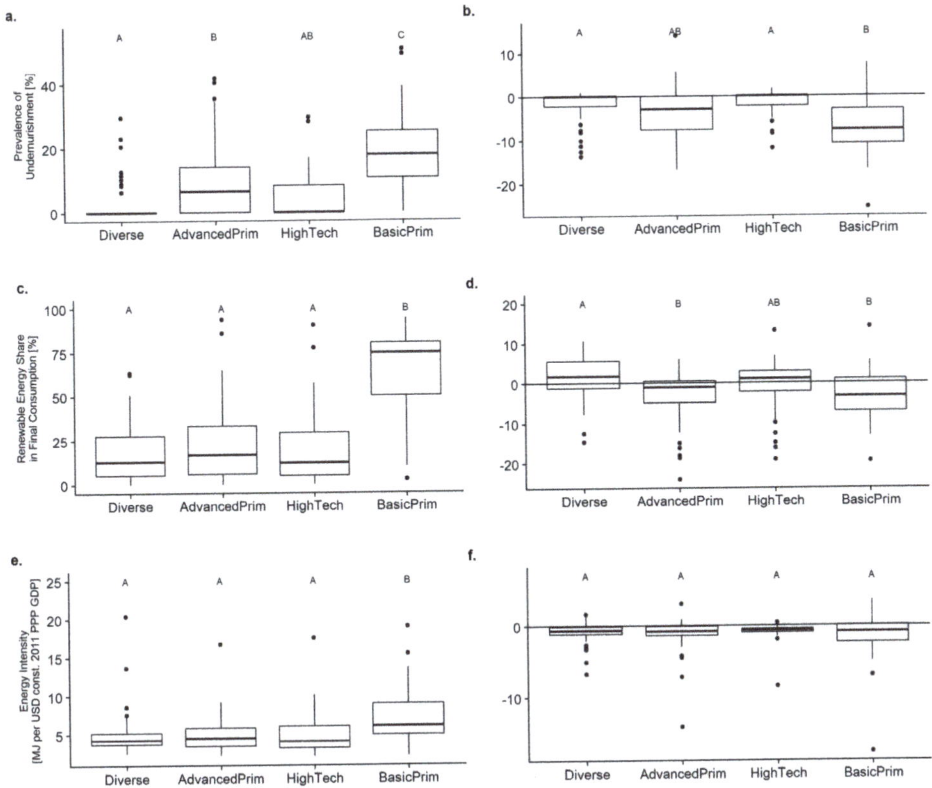

Figure 4. Comparison of the status (**a,c,e**) and trend (**b,d,f**) of different sustainability indicators related to nutrition availability and energy input for the five bio-economy types. The status was calculated as average values between 2010 and 2015 and the trend was calculated as the change between the current status and the average values for the years 2000 to 2005 as far as available. Letters indicate significant differences between types according to a pairwise Wilcoxon rank sum test ($p < 0.05$).

The material footprints were highest for countries of type HighTech and Diverse, and lowest for type BasicPrim countries (Figure 5a). Although all bio-economy types increased their material footprint (especially types Diverse and AdvancedPrim), some countries of type HighTech showed decreasing trends, which were however, not significantly different from the other types (Figure 5b). A similar pattern was found when evaluating the current domestic material consumption and the ecological footprint (Figure 5c,e). For domestic consumption, we found that besides HighTech countries, a number of countries from type Diverse also showed decreasing trends (Figure 5d,f). Overall, the analysis revealed that those countries with the highest national footprints (type HighTech and Diverse) were also the only ones showing some reductions of their footprint. Although all other bio-economy types increased their footprint, countries of type BasicPrim only increased their footprints a little during the last several years. This indicates that the overall global footprint increase did not benefit the poorest countries, but especially benefitted transitioning countries.

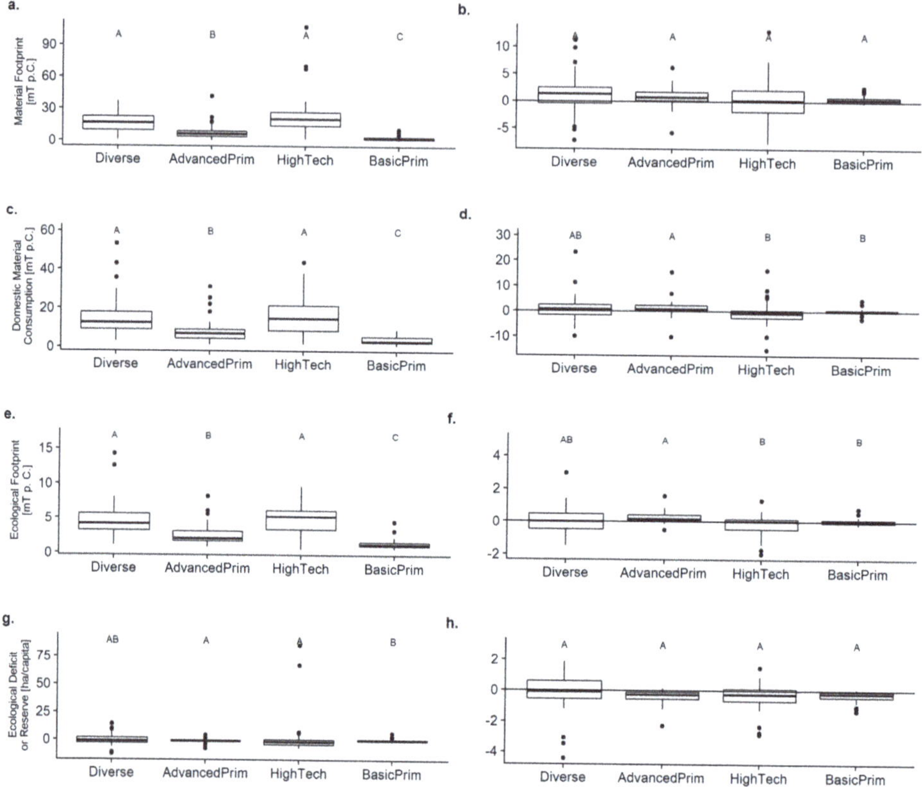

Figure 5. Comparison of the status (**a,c,e**) and trend (**b,d,f**) of different sustainability indicators related to material and ecological footprints for five bio-economy types. The status was calculated as average values between 2010 and 2015, and the trend was calculated as the change between the current status and the average values for the years 2000 to 2005 as far as available. Letters indicate significant differences according to a pairwise Wilcoxon rank sum test ($p < 0.05$).

The domestic biomass footprint for food purposes was highest for HighTech and Diverse countries, but showed no significant differences for all types (Figure 6a). However, for HighTech and Diverse types, it increased significantly more than for the other types (Figure 6b). Imported food biomass was significantly higher for HighTech and Diverse countries, and significantly lower for BasicPrim countries than for all other bio-economy types (Figure 6c). While some Diverse countries managed to decrease their domestic biomass footprints, all other types, especially BasicPrim countries, increased it (Figure 6d). The domestic biomass footprints for non-food purposes was also similar for all types (Figure 6e), and all types also decreased their footprint within a similar range (Figure 6f). The footprint due to imported biomass for non-food purposes was significantly higher for HighTech and Diverse countries than for all other types (Figure 6g), which also showed the highest increase (Figure 6h). Countries of type BasicPrim had the significantly lowest imported biomass footprint for non-food purposes (Figure 6g) with hardly any increasing trend (Figure 6h).

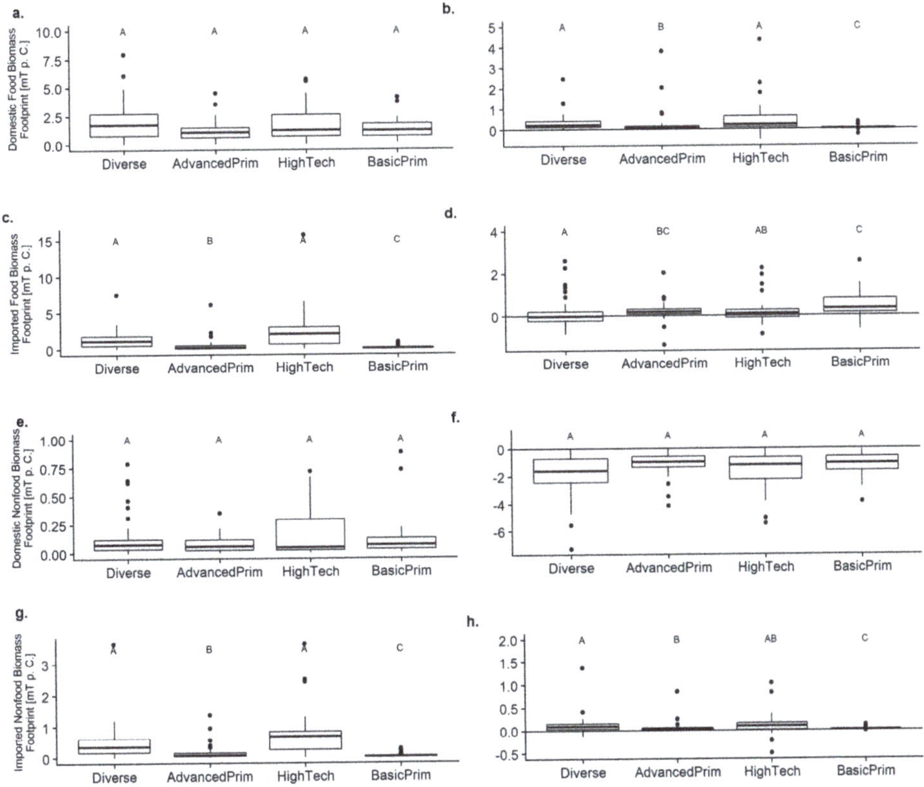

Figure 6. Comparison of the status (**a,c,e**) and trend (**b,d,f**) of different sustainability indicators related to the production and trade of biomass for the five bio-economy types. The status was calculated as average values between 2010 and 2015 and the trend was calculated as the change between the current status and the average values for the years 2000 to 2005 as far as available. Letters indicate significant differences according to the pairwise Wilcoxon rank sum test ($p < 0.05$).

Overall we found type HighTech and Diverse having similar domestic biomass footprints as all other types (Figure 6a,e, food and non-food) with increasing trends for food (Figure 6b), while domestic biomass extraction for non-food purposes remained more or less stable (Figure 6f). Imported biomass footprints were highest for HighTech and Diverse countries for food as well as non-food purposes (Figure 6c,g), but while their imported biomass footprint for food remained more or less the same (Figure 6d), they increased it for non-food purposes (Figure 6h). Countries more reliant on the primary sector (AdvancedPrim and BasicPrim) showed a very moderate or no change in their domestic as well as imported biomass footprints for non-food purposes. BasicPrim countries showed an increasing trend for imported food biomass only for food purposes. However, while their current domestic biomass footprints are on a similar level as for the other types (Figure 6a,e), their imported biomass footprints remain significantly smaller than for all other types (Figure 6c,g).

4. Discussion

In this study, we developed a global typology of bio-economies and tested if and how the resulting clusters differed from each other. We evaluated how these bio-economy types correspond to the foci of different bio-economy policy strategies and evaluated their sustainability performance using different sustainability indicators. Our results revealed four different bio-economy types which we named according to their characteristics, Diversified bio-economies (Diverse), Advanced primary sector bio-economies (AdvancedPrim), High-tech bio-economies (HighTech), and Basic primary sector bio-economies (BasicPrim).

Overall, the relative economic importance of the high-tech sector seems to be associated with the availability of skilled labor while the relative importance of the primary sector is not determined by the availability of bioproductive land. In fact, we tended to find high primary sector importance to be associated with a low availability of skilled workers. This indicates that for important high-tech bio-economies, the availability of skilled labor is a constraining factor, while the availability of bioproductive land is not of importance for a primary sector-based economy.

However, the economic importance of a sector for one country does not imply that the sector itself generates sufficient revenues or employment and education opportunities for the local population. The results simply suggest that agriculture plays an important economic role in most developing and many transitioning countries regardless of biomass availability, while the high-tech sector does not. However, based on these results it is not possible to compare the productivity or profitability of the agricultural sector between countries, nor is it possible to make any assumption about the ability of the sector to generate economic growth, or whether it can provide sufficient income or decent work to marginalized communities.

In the following, we outline potential implications of our results and discuss them in the context of other studies. Cereal production worldwide has doubled between 1960 and 2000, mainly fueled by innovations and larger inputs of water, pesticides and fertilizer [57]. But while food production in Asia and South America tripled (from one to three tons per ha) in a span of 40 years, the production across Africa remained almost constant at 1 ton per ha [58]. However, the majority of countries with the most important agricultural sector (type BasicPrim) are located in Africa, especially sub-Saharan Africa (SSA). Although these countries have managed to decrease their overall level of undernourishment during the past 17 years, they remain the countries with the highest levels of undernourishment.

Diverse countries, which are mainly located in South-America, Western Europe and South-East Asia, coincide with those countries having gone through a process of enormous agricultural intensification. In these regions, we find higher skill availability and better socio-political conditions than in type AdvancedPrim and BasicPrim countries. Substantial increases in food production, however, also had negative consequences for many countries, including land degradation, deforestation, and unsecure land use rights of local societies [57]. Based on these findings, we would argue that agricultural productivity needs to be increased and that bio-economic innovations might have the potential to do so. But at the same time, long-term sustainable bio-economic growth requires a formalized and intensive evaluation of the benefits and risks of any bio-economic innovation and policy implementation.

Increasing yields and scarcity of fossil fuels, combined with a rising awareness of the potential hazards from nuclear energy, have also resulted in the policy-supported growing use of renewable energy, especially bioenergy, in Diverse and HighTech countries. Due to the limited availability of bioproductive land, this policy had serious leakage effects all over the world, adding to the so-called food-energy-environment trilemma [59,60]. Accordingly, we found the highest and increasing imported biomass footprints for non-food purposes for these types. At the same time, we recorded an increasing trend of imported food biomass especially for countries of type BasicPrim. This trade-off and increasing pressure on food resources and the environment is also reflected by the distribution of bioenergy relevant policies, which can especially be found in countries with an important primary sector of type BasicPrim. Although these countries may not represent important global players for biofuels trade, the importance of the primary sector combined with biomass scarcity and food insecurity constitute a situation where the competition between food and energy for land becomes more pronounced than in other countries.

Increased competition between food, energy, and the environment can push innovations for more efficient use of land, biomass and other resources [61,62] but it can also increase imports of biomass, especially primary raw materials with the associated externalization effects of environmental costs [63]. This scenario is likely to take place in the more developed countries of type HighTech and Diverse, which tend to have better socio-economic conditions and higher skill availability to support these pathways. In less developed countries, especially those of type BasicPrim, increased competition is more likely to threaten food safety and natural ecosystems due to the lack of efficient policy instruments, the unequal distribution of resources including land, and the redistribution of land-use rights as a consequence of land-grabbing [61,64].

Economic sectors focusing on high-tech and high-value products are usually more knowledge-intensive but less resource-intensive, and require trained and specialized personnel. This phenomenon, is also true for the bio-economy high-tech sectors including biotechnology, biopharmaceuticals and others. It has often been argued that bio-economy offers the possibility of sustainable economic growth, especially for developing countries [14–16] and accordingly, as our results show, many countries including those of type BasicPrim, Diverse, and HighTech have issued high-tech bio-economy strategies. At the moment, however, important high-tech sectors can mainly be found in countries of type HighTech and Diverse, but not in those of type BasicPrim. Although it is not possible for us to draw conclusions about causal relationships, as this is a cross-sectoral study, in contrast to the primary sector, specific conditions seem to facilitate the growth of the high-tech sector, particularly a high availability of skilled labor. This emphasizes the need for investments in education and research to overcome the barriers for knowledge-based bio-economy growth in poorer parts of the world [65].

However, despite recent trends to improve sustainability performance in terms of domestic material consumption, the ecological footprint and the import of biomass for food purposes, countries of type HighTech and Diverse remain to have the largest footprints per capita. Reductions in material and energy consumption in developing and even in transitioning countries are very unlikely simply because resource consumption is still on a much lower level than for the more developed countries. We do not deny that some technological innovations have led to increased resource efficiency, and would argue that at least some share of the recent sustainability gain of HighTech and Diverse countries can be attributed to the development and application of high-tech and knowledge-intensive innovations. This, however, puts the countries of type HighTech in a position of high responsibility for the transfer and distribution of these innovations and new governance and regulation mechanisms might be necessary to support and facilitate this [66]. Furthermore, we highlight the parallel leakage effects that we recorded, in terms of increased biomass imports for nonfood purposes by HighTech and Diverse countries, accompanied by increased food imports by other types, especially BasicPrim.

The typology as well as the evaluation of the types' sustainability can only serve as an indication of certain sustainability trends. The limited availability and accessibility of national economic, social, and environmental data, many of which did not specifically measure particular aspects of bio-economy, resulted in a relatively rough classification of all countries of the world. Of course, not all countries within one type are equal, and values for many of the considered variables may vary considerably within types. This is also reflected by the varying degree of uncertainty when countries were assigned to one specific type (supporting Figure S1 and Table S1). We do, however, argue that the types represent crude categories of bio-economy pathways and that these types are therefore helpful to identify major trends and tendencies of bio-economy development, and to assess interdependencies while taking contextual factors into account. In terms of the sustainability evaluation, we also acknowledge the limitation that only part of the sustainability performance can be attributed to the development of the bio-economy. In some countries, where the bio-economy does not play a major role, this influence might approach even zero. However, for countries with very pronounced bio-economy sectors, we would argue that at least part of the sustainability performance is entangled with the bio-economy pathway. Sustainability evaluation approaches, particularly those focused on national policies, suffer from high uncertainties due to complex socio-ecological and socio-economic system interactions. Assessment methods for the sustainability of policies include for example the use of indicators and composite indices [67] as well as more integrated system assessments such as multi-criteria analysis, risk and vulnerability analysis or cost benefit analyses [68] with varying characteristics in terms of integration of system complexities (natural, social and economic system), temporal dimensions (ex-post vs. ex-ante) as well as spatial coverage (national vs. global) (compare with [68]). We chose a selection of indicators from different sources, including the SDG database [69] as well as different footprint indicators [33]. An exclusive focus on sustainability indicators from the SDG database would have been insufficient to account for planetary boundaries [2], while overemphasizing gains in resource efficiency instead of total consumption [69]. Interestingly, bio-economy policy strategies do not completely mirror the importance of the different sectors. While strategies of countries of type HighTech (with a high importance of the high-tech sector and high skill availability) showed the highest degree of diversification, countries of type BasicPrim (high importance of primary sector, low skill and biomass availability) show mostly a combination of strategies centering on either bioenergy or high-tech. While the focus on bioenergy is likely to be linked to actually increasing bioenergy production and growing competition for biomass between food and land, high-tech strategies might reflect preferences and strategic considerations for a transformation towards a new diversified bio-economy, but do not necessarily reflect current conditions on the ground.

The abundance of bio-economy high-tech strategies also in countries of type Diverse and the results of this studies show the importance of the availability of skilled labor to develop a knowledge-based bio-economy. This indicates, on the one hand, the need for investments for research and development, and on the other hand, supports the claim by Bauer et al. [70] that strategies on how to facilitate the engagement of small and medium-sized enterprises are currently lacking. According to the authors, recent bio-economy developments have been promoted by legislations and regulations focusing on bioenergy and biofuels, while incentives for commercial investments in non-energy biorefineries would be necessary to establish markets of these products. Furthermore, efforts in science, society, and policy to develop successful innovations and evaluate their sustainability should include distribution and transfer mechanisms to enable also poor countries to benefit from them and embark on a path of sustainability [6]. More developed countries have to acknowledge that despite their efforts to reduce the negative impacts of their economies and high consumption levels mainly through technological fixes, they continue to have the highest per-capita footprints. This level of material consumption secured by imports of agricultural, forestry, and mining commodities is leading to the externalization of environmental costs to the producing or sourcing countries [71]. An unlikely shift of developing countries away from export-oriented agricultural and forestry production and towards knowledge-intensive high-tech industries might compromise the sustainability performance of more developed countries and confront the Western Hemisphere with an increased demand on domestically available resources. Suggested sustainability indicators in the context of the SDGs are currently not sufficient to reflect the complex picture, including externalization of environmental costs and spill-over effects from national policies, and associated resource consumption trends (compare with [69]). Sustainability evaluations have to consider global environmental costs and complex interactions between bio-economy development paths, as well as international and trade dependencies.

5. Concluding Remarks

Our results highlight the potential as well as the challenges of new biotechnological developments and innovations, which are likely to trigger new material flows and processing technologies. However, a simple shift from fossil resources to renewable material and energy resources will not improve sustainability, and will increase pressures on ecosystems. New innovations from high-tech bio-economies have increased their resource efficiency, but overall, these countries continue to have the highest levels of resource consumption. The development of innovations puts them in a position of high responsibility for knowledge transfer and sustainability evaluation. True sustainable development requires a critical assessment of the risks of these innovations and developments, as well as innovation transfer and public investments, especially in developing countries.

Supplementary Materials: The following are available online at http://www.mdpi.com/2071-1050/10/8/2705/s1, Figure S1: Uncertainty of cluster-based bio-economy typology with countries belonging to the same cluster being displayed in the same color. Table S1: Probabilities for countries to belong to a specific type and uncertainty for belonging to the assigned type.

Author Contributions: Conceptualization L.B.-F. and J.B.; Methodology, L.B.-F. and M.B.; Formal Analysis, L.B.-F., A.K.B. and M.B.; Data Curation, A.K.B.; Writing—Original Draft Preparation, L.B.-F and J.B.; Writing—Review & Editing, L.B.-F., J.B., M.B. and A.K.B.

Funding: This research has been funded by the German Federal Ministry of Education and Research within the project STRIVE (Sustainable Trade and Innovation Transfer in the Bioeconomy, see www.strivebioecon.de).

Acknowledgments: We would like to thank the valuable input of our STRIVE team members and the Bioeconomy Council for the provisioning of the bio-economy policy strategy classification. We extend our gratitude to James Bruce Henderson for language editing and proof reading the final manuscript.

Conflicts of Interest: The authors declare no conflict of interest.

References

1. Shafiee, S.; Topal, E. When will fossil fuel reserves be diminished? *Energy Policy* **2009**, *37*, 181–189. [CrossRef]
2. Rockström, J.; Steffen, W.; Noone, K.; Persson, A.; Chapin, F.S.; Lambin, E.F.; Lenton, T.M.; Scheffer, M.; Folke, C.; Schellnhuber, H.J.; et al. A safe operating space for humanity. *Nature* **2009**, *461*, 472–475. [CrossRef] [PubMed]
3. Staffas, L.; McCormick, K.; Gustavsson, M. *A Global Overview of Bio-Economy Strategies and Visions*; Swedish Knowledge Centre for Renewable Transportation Fuels (f3 Centre): Göteborg, Sweden, 2013.
4. Fund, C.; El-Chichakli, B.; Patermann, C.; Dieckhoff, P. *Bioeconomy Policy (Part II): Synopsis of National Strategies around the World*; German Bioeconomy Council: Berlin, Germany, 2015.
5. Staffas, L.; Gustavsson, M.; McCormick, K. Strategies and Policies for the Bioeconomy and Bio-Based Economy: An Analysis of Official National Approaches. *Sustainability* **2013**, *5*, 2751–2769. [CrossRef]
6. Golembiewski, B.; Sick, N.; Bröring, S. The emerging research landscape on bioeconomy: What has been done so far and what is essential from a technology and innovation management perspective? *Innov. Food Sci. Emerg. Technol.* **2015**, *29*, 308–317. [CrossRef]
7. Birner, R. Bioeconomy Concepts. In *Bioeconomy: Shaping the Transition to a Sustainable, Biobased Economy*; Lewandowski, I., Ed.; Springer International Publishing: Cham, Switzerland, 2018; pp. 17–38.
8. Birch, K. The knowledge-space dynamic in the UK bioeconomy. *Area* **2009**, *41*, 273–284. [CrossRef]
9. Coleman, M.; Stanturf, J. Biomass feedstock production systems: Economic and environmental benefits. *Biomass Bioenergy* **2006**, *30*, 693–695. [CrossRef]
10. McCormick, K.; Kautto, N. The Bioeconomy in Europe: An Overview. *Sustainability* **2013**, *5*, 2589–2608. [CrossRef]
11. Bugge, M.; Hansen, T.; Klitkou, A. What Is the Bioeconomy? A Review of the Literature. *Sustainability* **2016**, *8*, 691. [CrossRef]
12. Hausknost, D.; Schriefl, E.; Lauk, C.; Kalt, G. A Transition to Which Bioeconomy?: An Exploration of Diverging Techno-Political Choices. *Sustainability* **2017**, *9*, 669. [CrossRef]
13. Goven, J.; Pavone, V. The Bioeconomy as Political Project. *Sci. Technol. Hum. Values* **2015**, *40*, 302–337. [CrossRef]
14. Komen, J. Africa Rising to a sustainable future: Economic prospects of bio-economy development in eastern Africa. In *Fostering a Bio-Economy in Eastern Africa: Insights from Bio-Innovate*; Liavoga, A., Virgin, I., Ecuru, J., Morris, J., Komen, J., Eds.; ILRI: Nairobi, Kenya, 2016; pp. 12–22.
15. Virchow, D.; Beuchelt, T.D.; Kuhn, A.; Denich, M. Biomass-Based Value Webs: A Novel Perspective for Emerging Bioeconomies in Sub-Saharan Africa. In *Technological and Institutional Innovations for Marginalized Smallholders in Agricultural Development*, 1st ed.; Gatzweiler, F.W., Ed.; Springer International Publishing: Charm, Switzerland, 2016; pp. 225–238.
16. D'Hondt, K.; Jiménez-Sánchez, G.; Philip, J. Reconciling Food and Industrial Needs for an Asian Bioeconomy: The Enabling Power of Genomics and Biotechnology. *Asian Biotechnol. Dev. Rev.* **2015**, *17*, 85–130.
17. Kröber, B.; Potthast, T. Bioeconomy and the future of food—Ethical questions. In *Know Your Food*; Dumitras, D.E., Jitea, I.M., Aerts, S., Eds.; Wageningen Academic Publishers: Wageningen, The Netherlands, 2015; pp. 366–371.

18. Meyer, R. Bioeconomy Strategies: Contexts, Visions, Guiding Implementation Principles and Resulting Debates. *Sustainability* **2017**, *9*, 1031. [CrossRef]
19. Bracco, S.; Calicioglu, O.; Gomez San Juan, M.; Flammini, A. Assessing the Contribution of Bioeconomy to the Total Economy: A Review of National Frameworks. *Sustainability* **2018**, *10*, 1698. [CrossRef]
20. Dubois, O.; Gomez San Juan, M. *How Sustainability Is Addressed in Official Bioeconomy Strategies at International, National, and Regional Levels: An Overview*; Food and Agriculture Organization of United Nations (FAO): Rome, Italy, 2016.
21. Philippidis, G.; Sanjuán, A.I.; Ferrari, E.; M'barek, R. Employing social accounting matrix multipliers to profile the bioeconomy in the EU member states: is there a structural pattern? *Span. J. Agric. Res.* **2014**, *12*, 913–926. [CrossRef]
22. Ronzon, T.; Piotrowski, S.; M'Barek, R.; Carus, M. A systematic approach to understanding and quantifying the EU's bioeconomy. *Bio-Based Appl. Econ.* **2017**, *6*, 1–17. [CrossRef]
23. Kircher, M. The transition to a bio-economy: National perspectives. *Biofuels Bioprod. Bioref.* **2012**, *6*, 240–245. [CrossRef]
24. Mungaray-Moctezuma, A.B.; Perez-Nunez, S.M.; Lopez-Leyva, S. Knowledge-Based Economy in Argentina, Costa Rica and Mexico: A Comparative Analysis from the Bio-Economy Perspective. *Manag. Dyn. Knowl. Econ.* **2015**, *3*, 213–236.
25. World Bank. World Bank National Accounts Data, and OECD National Accounts Data Files. Available online: http://data.worldbank.org/indicator (accessed on 21 September 2017).
26. ILO. Employment in Agriculture Male (% of Male Employment). Available online: http://data.worldbank. org/indicator/SL.AGR.EMPL.MA.ZS (accessed on 21 September 2017).
27. ILO. Employment in Agriculture Female (% of Female Employment). Available online: http://data. worldbank.org/indicator/SL.AGR.EMPL.FE.ZS (accessed on 21 September 2017).
28. FAO. Agricultural Land (% of Land Area). Available online: http://data.worldbank.org/indicator/AG. LND.AGRI.ZS (accessed on 21 September 2017).
29. FAO. Forest Area (% of Land Area). Available online: http://data.worldbank.org/indicator/AG.LND.FRST.ZS (accessed on 21 September 2017).
30. United Nations. High-Technology Exports (% of Manufactured Exports). Available online: http://data. worldbank.org/indicator/TX.VAL.TECH.MF.ZS (accessed on 25 September 2017).
31. WIPO. Patent Applications, Residents. Available online: http://data.worldbank.org/indicator/IP.PAT.RESD (accessed on 24 March 2017).
32. WIPO. Patent Applications, Nonresidents. Available online: http://data.worldbank.org/indicator/IP.PAT.NRES (accessed on 24 March 2017).
33. Global Footprint Network. National Footprint Accounts. Available online: http://www.footprintnetwork. org/licenses/public-data-package-free-edition-copy/ (accessed on 27 September 2017).
34. UNESCO. Gross Enrolment Ratio, Tertiary, Both Sexes (%). Available online: http://data.worldbank.org/ indicator/SE.TER.ENRR (accessed on 27 September 2017).
35. ILO. Employment by Education (Thousands). Available online: http://www.ilo.org/ilostat/faces/oracle/ webcenter/portalapp/pagehierarchy/Page3.jspx?MBI_ID=11&_adf.ctrl-state=93dm34n70_4&_afrLoop= 352234861741481#! (accessed on 27 September 2017).
36. National Science Foundation, Science and Engineering Indicators. Scientific and Technical Journal Articles. Available online: http://data.worldbank.org/indicator/IP.JRN.ARTC.SC (accessed on 24 March 2017).
37. Global Footprint Network. Glossary. Available online: https://www.footprintnetwork.org/resources/ glossary/ (accessed on 22 September 2017).
38. Scrucca, L.; Fop, M.; Murphy, T.B.; Raftery, A.E. mclust 5: Clustering, Classification and Density Estimation Using Gaussian Finite Mixture Models. *R J.* **2016**, *8*, 289–317. [PubMed]
39. El-Chichakli, B.; von Braun, J.; Lang, C.; Barben, D.; Philp, J. Policy: Five cornerstones of a globalbioeconomy. *Nat. News* **2016**, *535*, 221. [CrossRef] [PubMed]
40. European Bioeconomy Alliance (EUBA). The Crucial Role of the Bioeconomy in Achieving the UN Sustainable Development Goals. Available online: https://bioeconomyalliance.eu/crucial-role-bioeconomy-achieving-un-sustainable-development-goals (accessed on 30 July 2018).
41. Miller, R.E.; Blair, P.D. *Input-Output Analysis: Foundations and Extensions*, 6th ed.; Cambridge University Press: New York, NY, USA, 2014.

42. Lenzen, M.; Moran, D.; Kanemoto, K.; Geschke, A. Building Eora: A Global Multi-Region Input–Output Database at High Country and Sector Resolution. *Econ. Syst. Res.* **2013**, *25*, 20–49. [CrossRef]
43. UNEP. Material Extraction Dataset. Available online: https://uneplive.unep.org/ (accessed on 14 June 2017).
44. Wickham, H. *ggplot2: Elegant Graphics for Data Analysis*; Springer: New York, NY, USA, 2009.
45. Kassambara, I.; Mundt, F. factoextra: Extract and Visualize the Results of Multivariate Data Analyses. Version 1.0.5. Available online: https://CRAN.R-project.org/package=factoextra (accessed on 5 July 2018).
46. South, A. rworldmap: A New R package for Mapping Global Data. *R J.* **2011**, *3*, 35–43.
47. Auguie, B. gridExtra: Miscellaneous Functions for "Grid" Graphics. Available online: https://cran.r-project.org/web/packages/gridExtra/index.html (accessed on 5 July 2018).
48. Wilke, C.O. cowplot: Streamlined Plot Theme and Plot Annotations for 'ggplot2'. Available online: http://www.imsbio.co.jp/RGM/R_package?p=cowplot&d=R_CC (accessed on 5 July 2018).
49. Harrel, F.E., Jr.; Dupont, C. Hmisc: Harrel Miscellanous. Available online: https://CRAN.R-project.org/package=Hmisc (accessed on 5 July 2018).
50. Bengtsson, H. matrixStats: Functions that Apply to Rows and Columns of Matrices (and to Vectors). Version 0.53.1. Available online: https://CRAN.R-project.org/package=matrixStats (accessed on 5 July 2018).
51. Graves, S.; Piepho, H.-P.; Selzer, L.; Dorai-Raj, S. multcompView: Visualizations of Paired Comparisons. Version 0.1-7. Available online: https://cran.r-project.org/web/packages/multcompView/index.html (accessed on 5 July 2018).
52. FAO. Prevalence of Undernourishment. Available online: https://unstats.un.org/sdgs/indicators/database/?indicator=2.1.1 (accessed on 26 September 2017).
53. IEA; UNSD; UN Energy; SE4ALL Global Tracking Framework Consortium. Renewable Energy Share in the Total Final Energy Consumption. Available online: https://unstats.un.org/sdgs/indicators/database/?indicator=7.2.1 (accessed on 25 September 2017).
54. IEA; UNSD; UN Energy; S4All Global Tracking Framework Consortium. Energy Intensity Measured in Terms of Primary Energy and GDP. Available online: https://unstats.un.org/sdgs/indicators/database/?indicator=7.3.1 (accessed on 25 September 2017).
55. UNEP. Material Footprint, Material Footprint per Capita, and Material Footprint per GDP. Available online: https://unstats.un.org/sdgs/indicators/database/?indicator=12.2.1 (accessed on 26 September 2017).
56. UNEP. Domestic Material Consumption, Domestic Material Consumption per Capita, and Domestic Material Consumption per GDP. Available online: https://unstats.un.org/sdgs/indicators/database/?indicator=12.2.2 (accessed on 26 September 2017).
57. Tilman, D.; Cassman, K.G.; Matson, P.A.; Naylor, R.; Polasky, S. Agricultural sustainability and intensive production practices. *Nature* **2002**, *418*, 671–677. [CrossRef] [PubMed]
58. Sánchez, P.A. Tripling crop yields in tropical Africa. *Nature Geosci.* **2010**, *3*, 299–300. [CrossRef]
59. Tilman, D.; Socolow, R.; Foley, J.A.; Hill, J.; Larson, E.; Lynd, L.; Pacala, S.; Reilly, J.; Searchinger, T.; Somerville, C.; et al. Beneficial biofuels—The food, energy, and environment trilemma. *Science* **2009**, *325*, 270–271. [CrossRef] [PubMed]
60. Harvey, M.; Pilgrim, S. The new competition for land: Food, energy, and climate change. *Food Policy* **2011**, *36*, S40–S51. [CrossRef]
61. Haberl, H. Competition for land: A sociometabolic perspective. *Ecol. Econ.* **2015**, *119*, 424–431. [CrossRef]
62. Lewandowski, I. Securing a sustainable biomass supply in a growing bioeconomy. *Glob. Food Sec.* **2015**, *6*, 34–42. [CrossRef]
63. Muradian, R.; Martinez-Alier, J. Trade and the environment: From a 'Southern' perspective. *Ecol. Econ.* **2001**, *36*, 281–297. [CrossRef]
64. Havnevik, K. Grabbing of African lands for energy and food: implications for land rights, food security and smallholders. In *Biofuels, Land Grabbing and Food Security in Africa*; Matondi, P.B., Havnevik, K., Beyene, A., Eds.; Zed Books: London, UK, 2011; pp. 20–43.
65. Zilberman, D.; Kim, E.; Kirschner, S.; Kaplan, S.; Reeves, J. Technology and the future bioeconomy. *Agric. Econ.* **2013**, *44*, 95–102. [CrossRef]
66. Wield, D. Bioeconomy and the global economy: Industrial policies and bio-innovation. *Technol. Anal. Strateg. Manag.* **2013**, *25*, 1209–1221. [CrossRef]
67. Singh, R.K.; Murty, H.R.; Gupta, S.K.; Dikshit, A.K. An overview of sustainability assessment methodologies. *Ecol. Indic.* **2009**, *9*, 189–212. [CrossRef]

68. Ness, B.; Urbel-Piirsalu, E.; Anderberg, S.; Olsson, L. Categorising tools for sustainability assessment. *Ecol. Econ.* **2007**, *60*, 498–508. [CrossRef]
69. Wackernagel, M.; Hanscom, L.; Lin, D. Making the Sustainable Development Goals Consistent with Sustainability. *Front. Energy Res.* **2017**, *5*, 18. [CrossRef]
70. Bauer, F.; Coenen, L.; Hansen, T.; McCormick, K.; Palgan, Y.V. Technological innovation systems for biorefineries: A review of the literature. *Biofuels Bioprod. Bioref.* **2017**, *11*, 534–548. [CrossRef]
71. Freudenberger, L.; Schluck, M.; Hobson, P.; Sommer, H.; Cramer, W.; Barthlott, W.; Ibisch10, P.L. A View on Global Patterns and Interlinkages of Biodiversity and Human Development. In *Interdependence of Biodiversity and Human Development under Climate Change*; Ibisch, P.L., Vega, A.E., Herrmann, T.M., Eds.; Secretariat of the Convention on Biological Diversity: Montreal, QC, Canada, 2010; pp. 37–58.

© 2018 by the authors. Licensee MDPI, Basel, Switzerland. This article is an open access article distributed under the terms and conditions of the Creative Commons Attribution (CC BY) license (http://creativecommons.org/licenses/by/4.0/).

 *sustainability*

Review

Applying a Sustainable Development Lens to Global Biomass Potentials

Tina D. Beuchelt * and Michael Nassl

Center for Development Research, University of Bonn, 53113 Bonn, Germany; mnassl@uni-bonn.de
* Correspondence: beuchelt@uni-bonn.de; Tel.: +49-228-73-1847

Received: 30 September 2018; Accepted: 15 August 2019; Published: 17 September 2019

Abstract: The Sustainable Development Goals (SDGs), adopted by all UN Member States in 2015, guide societies to achieve a better and more sustainable future. Depleting fossil fuels and climate change will strongly increase the demand for biomass, as governments shift towards bioeconomies. Though research has estimated future biomass availability for bioenergetic uses, the implications for sustainable development have hardly been discussed; e.g., how far the estimates account for food security, sustainability and the satisfaction of basic human needs, and what this implies for intragenerational equity. This research addresses the gap through a systematic literature review and our own modeling. It shows that the biomass models insufficiently account for food security; e.g., by modeling future food consumption below current levels. The available biomass, if fairly distributed, can globally replace fossil fuels required for future material needs but hardly any additional energy needs. To satisfy basic human needs, the material use of biomass should, therefore, be prioritized over bioenergy. The different possibilities for biomass allocation and distribution need to be analyzed for their potential negative implications, especially for the poorer regions of the world. Research, society, business and politicians have to address those to ensure the 'leave no one behind´ commitment of the SDGs.

Keywords: biomass scenarios; global biomass; bioenergy; sustainability; food security; basic needs; intragenerational justice; equity; fairness; development

---

1. Introduction

Bioeconomies focus on the production and utilization of biological resources to generate bio-based products, including bioenergy [1]. There is a global trend to substitute biomass for fossil fuels for material or chemical use and energy, which is in part due to climate change, but is mainly driven by depleting fossil fuel stocks. Estimates for fossil fuel peaks and depletion vary [2,3], but researchers have increasingly pointed out that by 2050, hardly any oil will be available and coal reserves will be the main remaining fossil fuel [3,4]. Therefore, high-value bio-based products are receiving increasing attention, especially in the industrialized countries, and bioenergy is perceived as crucial, given its potential to combat climate change [5]. Today, almost 50 countries are pursuing bioeconomy development in their policy strategies, 15 of which have developed dedicated bioeconomy policy strategies [1]. The emerging bioeconomies are expected to lead to a strong increase in biomass demand in the next decades.

At the same time, the concept of sustainability has become very important for societal development, as recently reflected by the universal adoption of the Sustainable Development Goals (SDGs) [6]. The SDGs guide societies in their attempt to adjust their economies towards ecological and social objectives with the overall aim to "leave no one behind" [7,8]. The notions of sustainability or sustainable development comprise a societal vision of how to act within social and natural systems over the long term [9]. Many concepts and definitions of sustainability exist [6,7,10–12]. According to Baumgärtner and Quaas [10,11] (page 2057), "Sustainability aims at justice in the domain of human–nature relationships and in view of the long-term and inherently uncertain future, including (i) justice between humans of different generations ("intergenerational" justice), (ii) justice between different humans of the same generation, in particular the present generation ("intragenerational" justice), and (iii) justice between humans and nature." Both justice and equity relate to a fair balance of mutual claims and obligations within a local or global community [9]. The World Commission on Environment and Development has defined sustainable development in its report "Our Common Future," also known as the Brundtland Report, as "development that meets the needs of the present without compromising the ability of future generations to meet their own needs" [12] (page 54). From this report, Holden et al. [6] derive four sustainability dimensions: (i) The need for long-term ecological sustainability, (ii) the satisfaction of human needs, (iii) intragenerational equity; i.e., equity between humans within a country and between countries of the same generation, and (iv) intergenerational equity, which means that future generations must also be able to meet their needs.

The growing biomass demand poses new challenges to the sustainability of biomass production, the efficient use of biomass and the economies of scale in biomass mobilization [13]. Large amounts of biomass will be necessary to replace fossil fuels and to meet the future increase of food demand [13] However, the global biomass supply is limited despite biomass being renewable [14,15]. Competition exists between the alternative uses of biomass; i.e., between food, feed, fiber, bio-based materials and energy uses [16], while most planetary boundaries relevant for biomass production have already been exceeded [17]. To cover the increasing food and feed demands of a growing population in 2050, agricultural production has to increase significantly. Today, one out of nine people (820 million people) are hungry, which means they do not consume the adequate amount of dietary energy [18]. More than two billion people suffer from hidden hunger; i.e., they lack key vitamins and other micronutrients, such as iodine, iron and zinc [19]. This affects the health and well-being of the people and, as a consequence, national socioeconomic development [20]. An increasing demand for non-food biomass may impact food security with respect not only to the availability but also to the diversity, stability and access to food. This may lead to an increase in hunger given the disproportionately large energy markets compared to food markets, and the stronger economic position of those demanding more energy versus those being food-insecure [21]. Those potential effects are highly relevant for the compliance with the SDG-2, which aims to eliminate hunger by 2030. The use of biomass poses ecological, social, economic and ethical challenges regarding production, allocation and distribution.

For planning and investment purposes, as well as for governmental policies, it is important to understand how much biomass can be used by humans without putting the ecosystem at risk. In a next step, it has to be determined for what the available biomass should be used (e.g., for food, feed, fiber, materials, energy) and by whom. Several initiatives and projects estimated future biomass availability for bioenergy and came to very diverging results; i.e., they assumed that in 2050 biomass availability could range from 36 to 1458 EJ/a (Exajoule per year) [15,16,22–33]. This large variation has been explained by researchers as being due to differences in the assumptions within the applied models; e.g., regarding land area and use, cropping intensity, yield improvements or population growth [16,22,34]. Though the studies claim that their results consider the future demand for food, it is doubtful how well food and nutrition concepts were integrated in the models. Several authors have critically assessed the studies on biomass availability for bioenergy [16,22,34,35]. However, their emphasis was not on sustainable development from a socioeconomic or a holistic food security perspective.

The four dimensions of sustainable development following Holden et al. [6] can be used as an analytical perspective for examining the biomass estimates and models. The first dimension, i.e., the need for long-term ecological sustainability, is partly addressed in the models or discussed in reviews to a greater or lesser extent [16,34] The second dimension, i.e., the satisfaction of human needs, which includes food security as a human right, has not been explicitly addressed by the research on biomass potentials or the respective reviews. The fulfillment of basic non-food human needs involves housing, energy, water supply, sanitation and health care—all of which directly or indirectly depend on fossil fuels or biomass. An analysis is missing on whether the future non-food biomass supply will be able to cater for all human needs (e.g., materials, chemicals, fiber and energy), or whether certain uses will have to be prioritized. To date, research has concentrated on bioenergy and biomass availability and not on bio-based chemicals or materials [22]. However, the latter might, in future, add significantly to the biomass demand [13,22,36] and should be discussed together with the biomass potentials [13].

The question about who is to use the available biomass and for which purposes has not been part of the discussions on the biomass estimates. This intragenerational equity perspective, reflecting the third sustainability dimension, is also relevant for the SDG 10, regarding the reduction in global and national inequalities of resource use and welfare. Fair distribution and use of resources have been discussed by ethics and philosophy, but these discussions have not been linked to biomass availability. Answers to the above-mentioned knowledge gaps are urgently needed so that governments and other actors can make adequate choices about the kind of society in which we want to live and about the kind of world we want to leave to posterity [7]. Therefore, this research analyses and discusses biomass availability estimates from the perspective of the satisfaction of human needs, especially food security, and intragenerational equity. This research aims:

(i)   To understand how food security is reflected in the estimates of biomass potentials;
(ii) To identify to what extent the energetic and material use of fossil fuels can be replaced by biomass, and what this means for resource allocation, distribution and intragenerational equity.

The next section briefly describes the method, definitions and concepts used. The result section shows how far food security aspects are addressed by the estimates of biomass potentials. It identifies to what extent the energetic and material uses of fossil fuels can be covered by biomass, and what this means for the satisfaction of non-food human needs and intragenerational equity. This is followed by a discussion also presenting recommendations for future research, and conclusions with policy implications.

2. Method and Concepts

2.1. The Procedure of the Systematic Literature Review

We conducted a systematic literature review following Jesson et al. [37]. We used the PRISMA guidelines (see http://prisma-statement.org/) but adjusted them slightly for our type of review. For a flow diagram with details of the systematic review procedure see Appendix A. The data search was based on the Thomson Reuters' Web of Science (ISI Web of Knowledge) database. The Web of Science's default settings were used; i.e., time span all years (1945–2018), all indices, all languages and all document types (see Appendix A). The search terms biomass potential(s), potential(s) of biomass, and bioenergy potential(s) returned 745 findings. All titles and abstracts were screened, and the following parameters were used to exclude studies from the further review process: (i) Regionally restricted studies (ii) publications addressing only single energy plants or single biomass sources (e.g., only residues and waste), and (iii) literature focusing merely on chemical processing or economic valorization of biomass. For the remaining 35 articles, a full text analysis was done. Two further exclusion criteria were established: (iv) articles based on reviews and not on own models and estimations, and (v) all publications that did not use 2050 as a reference year for the global biomass potentials. We decided to use the year 2050 as this is when oil resources will be almost depleted and alternative uses will have to be available. Many researchers have also chosen this time frame for their models, and only few studies look at 2030 or 2100. Finally, 14 studies remained for the systematic analysis. Two studies calculated the geographical biomass potential [31,32], one study the sustainable potential [38], and the remaining eleven studies the technical potential [15,22,27,29,30,33,39–43] (see Appendix B for definition of biomass potential types). The (environmentally) sustainable potential includes more assumptions for ecological boundaries, environmental protection and long-term availability of resources.

Data Sources, Scenarios, and Models of the Studies Used in The Review

The primary data source for all reviewed studies is the database of the Food and Agriculture Organization of the United Nations (FAO): FAOSTAT. This database integrates agricultural production and land-use data (including forestry) compiled from single country surveys, satellite imaging data, projections and estimates into one global dataset. Though the quality of the dataset has been contested [44,45], it remains the only comprehensive and standardized global dataset available.

The biomass scenarios and their underlying models provide alternative narratives for how key drivers, e.g., global population, dietary changes (affecting food and feed demand), climate, economic development, crop yield improvements, and available land area, might evolve in the future, and how this might impact other dependent parameters [13]. The most prominent scenarios used in the assessments are those from the Intergovernmental Panel on Climate Change (IPCC) Special Report on Emission Scenarios. The applied modeling approach is usually based on integrating models that combine resource and demand data into a unified modeling approach (such as the Integrated Model to Assess the Global Environment (IMAGE) [31,42,46], Global Land Use and Energy Model (GLUE) [47], IIASA's Basic-Linked System model (BLS) [39]), and 'stand-alone' productivity and crop yield models, like the Lund–Potsdam–Jena model with Managed Land model (LPJmL) [30,48].

To estimate the global biomass potentials, the reviewed studies usually look, roughly summarizing, at available (agricultural) land where biomass can be produced, estimate the food requirements of the population in 2050 and the required amount of land needed for food production, and then estimate how much biomass can be produced on the remaining land. Table A2 provides an overview of the estimated biomass potentials, the type of potential and the biomass sources used in the studies. The studies have different assumptions about the availability of biomass sources; e.g., from agriculture, forestry, waste and about future land-use change. Further differences can be explained by the modeling procedure and the assumptions regarding, for example, diets, population growth and yield increases. For example, population growth projections mainly follow the 'UN medium population forecast' of 9.2 billion in 2050, but values vary from 8.7 to 11.3 billion. Cropland expansion is projected to range between 0.1 and 0.45 Gha, with the majority of the studies providing values in the lower range. Projections of global yield increases can range up to 360% [33]. Assumptions regarding cropping intensification and irrigation are hardly described in the studies but are usually included. Not all studies published details about their assumptions (including nutritional requirements) or their modeling procedure, which may skew this assessment.

2.2. Definition of Terms and Concepts Used in This Study

2.2.1. The Concept of Food Security

Being aware that the thinking and consequently definitions around food security have changed over time [49,50], this research uses the international food security definition agreed upon by all states at the World Food Summit in 1996 and again emphasized in subsequent summits and high-level UN meetings: "Food security exists when all people, at all times, have physical and economic access to sufficient safe and nutritious food that meets their dietary needs and food preferences for an active and healthy life" [51]. This definition of food security has been state-of-the-art since the turn of the millennium, and is based on four pillars; i.e., availability of food, access to food, utilization of food and stability [52,53]. The pillar of "food availability" refers to the availability of sufficient quantities of food of appropriate quality at national but also at household level. The pillar of "food access" refers to the physical and economic access of individuals to adequate resources to acquire appropriate foods for a nutritious diet. Physical and economic food access is mainly determined by the income or resource endowment of the population/household, transport and market infrastructure. The pillar of "food utilization" refers to a diet adequate in quantity, quality and diversity, fulfilling all nutritional requirements. Along with food safety, clean water, sanitation and health care, it is imperative to reach a state of nutritional well-being where all physiological needs are met. The term "nutritional requirements" refer to the amount of protein, energy, carbohydrates, fats and lipids, vitamins, minerals and trace elements (such as calcium, iron, zinc, selenium, magnesium and iodine) needed by a human being to sustain a healthy life. The pillar "food stability" refers to the access of a population, household or individual to adequate food at all times, independent of shocks or cyclical events, such as seasonal availability [52,53]. Meeting the nutritional requirements of a human being is only one aspect of food security. Many more conditions need to be met so that a person being described as 'food secure' is in line with the international food security concept which is widely used by the FAO, the UN and many civil society organizations.

2.2.2. The Concept of Allocation, Distribution and Intragenerational Equity

The term "allocation" in this paper refers to how resources (in this case biomass) are divided among different products and product uses; e.g., how much biomass is used for bioenergy or for material use, such as for construction, chemicals, plastics or fibers. This is relevant with respect to the satisfaction of human needs; i.e., one of the four sustainability dimensions [6]. The satisfaction of needs requires that no one suffers from absolute deprivation anymore; i.e., that all basic human needs are met [54]; this is also part of the SDGs. The term "distribution" refers to how goods and services are divided among people of current and future generations [55]. This is relevant for intragenerational equity, another sustainability dimension, which goes beyond the basic needs concept by targeting the relative shares of resource use and deprivation within a generation. To address intragenerational equity regarding biomass use, this research uses an egalitarian approach which entails equal resource use for each person in any society across the world independent of the natural resource base of a country; i.e., everyone gets the same share of biomass allocated for a specific use.

2.3. Estimating Biomass Availability and Uses

The data sources for the calculation of biomass availability and uses are derived from the World Energy Council [56] (see Appendix E, Table A3). The report presents two energy scenarios for 2050 with varying assumptions regarding population growth, income growth, governance and consumption behavior. The Jazz scenario is more consumer oriented and aims to achieve better energy access and affordability through economic growth. The Symphony scenario has a stronger focus on achieving environmental sustainability through coordinated policies and practices by governments. It is important to point out that neither scenario assumes that every person has access to electricity by 2050. The share of households without electricity remains high in Africa and south and central Asia, being higher in the Symphony scenario than in the Jazz scenario.

The following formulas were used to estimate how much of the energy and material requirements in 2050 could be covered by biomass.

The average minimum or maximum biomass potential *BP* across all studies is determined with

$$BP_{min/max} = \frac{1}{n} \sum_{i=1}^{n} x_i \tag{1}$$

where x_i is the biomass potential of the respective study and *i* is the reviewed study ($i=1, \dots, 14$). The minimum, respectively the maximum, value for the biomass availability potential of each study was used. If the study mentioned only one value for biomass availability, this value was used for both the minimum and maximum value.

The average biomass potential per capita BP_{pc} is determined by

$$BP_{pc} = \frac{BP}{WP} \tag{2}$$

where *WP* is the global population in 2050 as estimated by the World Energy Council [56].

For the following estimations, only those studies that were published in 2010 or beyond were used, as we assume that models improve over time with increasing experience and scientific review (e.g., [34]). The more recent studies only estimate the technical or sustainable potential, which is also more relevant for decision-makers.

The per capita energy demand covered by the biomass potentials ED_{pc} is determined by

$$ED_{pc} = \frac{BP_{pc} \times WP}{WE} \tag{3}$$

where WE is the world energy needs in 2050 as estimated by the World Energy Council [56]. This value includes energy conversion losses, etc., and is therefore higher than the value for actual "final energy consumption." Since conversion losses need to be covered also by energy supplies, for us this is the key value to use in the calculations. For the regional estimates, WE represents the respective regional energy demand in 2050, and WP the respective regional population.

The global per capita material demand covered by the biomass potentials MD_{pc} is determined by

$$MD_{pc} = \frac{BP_{pc}}{RM_{pc}} \tag{4}$$

where RM_{pc} is the regional material use of final energy consumption per capita (GJ/y). Since the World Energy Council [56] does not provide estimates for regional material needs, RM_{pc} is estimated as follows

$$RM_{pc} = \frac{FE}{WP} \times \frac{WM_{pc} \times WP}{FE} \tag{5}$$

where WM_{pc} is the per capita world material need in 2050. WM_{pc} is calculated by dividing the "final energy consumption" FE with the world population WP and then substracting the "final energy consumption per capita, excluding non-energy use." WM_{pc} is 7.6 GJ/y in the Jazz scenario and 6.1 GJ/y in the Symphony scenario, which corresponds to a global share of material uses in final energy consumption of 11% and 12%, respectively.

The estimation of RM_{pc} accounts for the different purchasing power in each region, and hence embraces more inequity in material resource use in 2050; i.e., Africa will be using much less energy for material uses than Europe. Alternatively, one could use WM_{pc} based on the normative aspect to account for the satisfaction of the basic material needs of every person across the world in the same quantity. Given the lack of better data, WM_{pc} is assumed to be a good proxy of what could be desirable as material use of energy. The motivation for this assumption stems from the second and third dimension of sustainable development; i.e., the satisfaction of basic human needs and intragenerational justice. From an intragenerational justice perspective, it is adequate to assume that per capita material demand across all regions should be the same. Despite that, the World Energy Council data estimates already embrace lower per capita energy uses in Africa and Asia than in the rest of the world, and it is not clear whether these assumed levels of energy use for Africa are actually sufficient to meet the basic needs of the whole population or not. The per capita total energy demand covered by the biomass potential after satisfying material needs based on an equal material resource distribution ($EDMD_{pc}$) is determined by

$$EDMD_{pc} = \frac{BP_{pc} - WM_{pc}}{\frac{WE}{WP} - WM_{pc}} \tag{6}$$

Taking regional differences and hence inequity in resource use into account, the per capita total energy demand covered by the biomass potential after accounting for regional difference in satisfying material needs $\left(REDMD_{pc}\right)$ is determined by

$$REDMD_{pc} = \frac{BP_{pc} - RM_{pc}}{\frac{WE}{WP} - RM_{pc}} \tag{7}$$

3. Results

3.1. How Is Food Security Accounted for in the Estimates of Future Biomass Availability?

All reviewed studies follow a kind of "food-first" approach, which means that in their scenarios, no land needed for food production is allocated to bioenergy production. Most studies estimate the amount of agricultural land needed for food production in 2050 by assuming either different "expansion-scenarios" or "non-expansion-scenarios;" i.e., increase or no increase in agriculture land compared to the current situation. These scenarios are mostly based on predictions by the FAO [57]. Some scenarios assume a cropland expansion of 9% and 19% [29,30]. Other scenarios estimate cropland expansion in hectares. Low values range between 0 and 0.1 Gha [33], medium values from 0.2 to 0.5 Gha [43], while high values are more than 0.5 Gha [40]. The projected cropland expansions are then combined in models with other parameters, such as population growth projections, yield projections, and dietary assumptions, to model the need for agricultural land for food production in 2050.

For their biomass availability models, the reviewed studies do not use the encompassing, internationally accepted concept of food security (see Section 2.2.1). However, the studies include some elements from the four pillars of food security (for a general overview, see Table A1).

3.1.1. Food Utilization: Inclusion of Nutritional Requirements in Biomass Models

Only seven out of the fourteen reviewed studies have explicitly presented their assumptions regarding food diets in 2050, and of these seven studies, four provide dietary scenarios. The main distinguishing factors in those scenarios are the total amount of kcal per capita and day and the assumed share of animal protein.

The total caloric intake per capita and day (kcal/cap/d) used as a basis by the first group of studies ranges from 2800–3170 kcal/cap/d [28–30] (Table 1). The second group uses a considerably lower range from 2410 kcal/cap/d in the vegetarian diet scenario up to 2750 kcal/cap/d in the affluent diet scenario [27,32]. The share of proteins from animal products varies considerably between the different scenario groups.

Most of the assumptions about caloric food consumption are around or below the average global food consumption levels at the turn of the millennium, when over 10% of the global population suffered from hunger [58]. The FAO [57] estimated a worldwide per capita food consumption of 2803 kcal/cap/d from 1997–1999, with an average in developing countries of 2681 kcal/cap/d and 3380 kcal/cap/d in industrialized countries. For 2050, the FAO estimates a demand for at least 3070 kcal/cap/d, with consumption of around 3000 kcal/cap/d in developing countries and 3500 kcal/cap/d in industrialized countries [21]. These estimates do not account for the amount of food needed for a food-secure global population but are just scenarios based on what people may be able to afford [21]. Most biomass scenarios include even lower levels of global food consumption than the demand prognosticated by the FAO. For industrialized countries, the models comprise consumption levels 10–30% lower than the demand expected by 2050.

Regarding the nutritional requirements, the necessary protein and energy (calories) are at least addressed by some studies (Table 1). The other nutritional requirements, such as the vitamins and minerals needed for a healthy life are not specifically included; for example, through incorporating horticultural production in the land estimates. The other relevant factors for food utilization, such as the availability of clean water, food safety or health issues, are not discussed or included in the models. All these elements, however, may be linked and negatively affected by agricultural intensification, which is typically one of the biomass model assumptions.

Table 1. Dietary scenarios: Share of animal protein and diet considerations in the reviewed studies.

Scenario	kcal/cap/d	Share of Animal Protein	Diet Considerations
Western high meat scenario; *Rich* scenario [29]	3170	44% of total protein intake; 21% of total nutrient intake.	Rapid acceleration of economic growth and consumption patterns; increases in the share of animal products, sugar and vegetable oil; requires a cropland expansion of 20%.
Current trend scenario; *TREND* scenario; *Business as usual* scenario [29,30]	2990	38% of total protein intake; 16% of total nutrient intake (on global average).	Current growth trends are maintained with strong regional differences in calories and animal product consumption; by 2050, every region is projected to attain the diet level of its country with the highest diet in year 2000; per-capita consumption of sugar and oil crops increases by 19% globally, of animal products by 7%.
Less meat scenario [29]	2990	30% of protein intake; 8% of total nutrient intake.	Total protein levels considered as nutritionally sufficient by the authors. Average protein consumption in North America and Western Europe decreases, and distribution of food categories changes: cereals, roots, pulses, vegetables and fruits categories increase above 1700 kcal/cap/d for all regions, while animal products, sugar and oil crop shares decrease, in particular in rich regions.
Fair less meat scenario; *Fair & Frugal* scenario [29,30]	2800	20% of protein intake; 8% of total nutrient intake.	Protein from animal sources reduced to 20%. Very little diet variation between world regions: richest regions reduce share of animal products, sugar, and vegetable oil. Equal food distribution.
Affluent diet scenario [27,32]	2750	Not specified. High meat and dairy products consumption	Global food requirement of dry weight grain equivalence (gr eq.): 14.4 trillion kg dry weight gr eq. needed. Of this, 73% needed for animal protein production.
Moderate diet scenario [27,32]	2410	Not specified. Some meat and high dairy products consumption.	Global food requirement 8.2 trillion kg dry weight gr eq. Of this 60% needed for animal protein production.
Vegetarian diet scenario [27,32]	2410	Not specified. Only consumption of dairy products, only small share.	Use for the global food requirement 4.5 trillion kg dry weight gr eq. needed. Of this 22% needed for animal protein production.

3.1.2. The Inclusion of Food Availability, Food Access and Food Stability in the Biomass Models

Food availability is addressed in the models through setting aside a specific area of land for food production on a global level. National and local food availability through international trade and food imports to food-deficit countries are considered by many but not by all studies. Six out of fourteen studies include an 'international food trade balance' in their estimations, which means that the gap between regional production and demand for meat and cropland products is balanced by trade; i.e., regions where the demand for primary products like cereals exceeds the regional supply are net importing regions, while regions where the biomass supply is larger than the regional demand are net exporters.

The pillar of food access is hardly addressed by the biomass models. Only two studies include estimates of the development of international food prices. These are relevant for the economic access to food by households. Other elements such as inequality, poverty, land distribution, and the transport and market infrastructure necessary for buying and trading food, are not taken into account.

The pillar of food stability is also only weakly addressed. Stable ecosystems are needed for sustained and continuous food and biomass production and to limit price fluctuations. Climate change is counteracting food security and food stability [59] but is not addressed by most studies. Only two studies use models that consider explicitly climate change and climate-change-induced yield change predictions, while three studies use IPCC scenarios to integrate climate change projections.

3.2. Biomass Availability, Allocation and Distribution

3.2.1. Global and Regional Non-Food Biomass Availability

The estimates for non-food biomass availability in 2050 range widely from 33–1548 EJ/a (Figure 1). The earlier studies tend to show higher estimates than the later studies. Most of the studies come to the conclusion that the future potential for energy from biomass is higher than the current level of around 50 EJ/a. Four studies, however, also estimate that the minimum biomass availability may be below the current usage level of biomass energy.

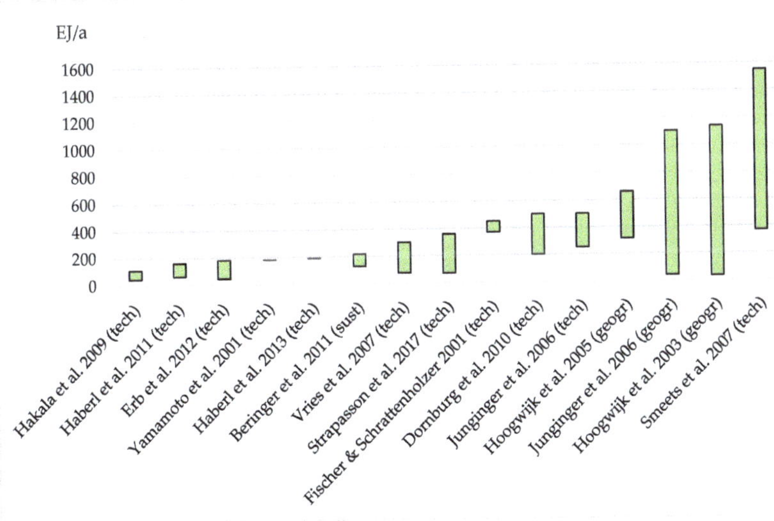

Figure 1. Range of biomass potentials (in EJ/a) and potential type (tech=technological, sust= sustainable, geogr=geographical potential) in the reviewed studies.

The estimates above 1000 EJ/a have to be viewed very critically, as other studies that addressed the planet's maximum capability to produce new biomass concluded that biomass availability cannot be sustained for human use over time. The maximum range for the global energy extraction from biomass is estimated to be between 1080 EJ/a and 1368 EJ/a [14–17]. This is based on the net primary production; i.e., the maximum available biomass for human use, which would include the use of all resources, such as forests, savannah regions and protected areas.

The average minimum availability of biomass across all reviewed studies is 151 EJ/a; highest availability is estimated to be 500 EJ/a. In studies published before 2010, this value is 18% to 35% higher, respectively, and in studies published after 2010, between 23% and 50% lower (Table 2). The assumptions with respect to the minimum and maximum potentials usually vary, regarding, for example, diets, yield improvements, and land and forestry use. The technical or sustainable potentials reveal consistently lower average non-food biomass availability than when the geographic potentials are included.

Table 2. Minimum, maximum and per capita availability of non-food biomass in 2050.

Biomass Potentials	Mean Value (Published Before 2010)	Mean Value (Only Tech or Sust Potential) [1]	Mean Value (Published 2010 or Later)
Minimum biomass potential (EJ/a)	178	148	116
Maximum biomass potential (EJ/a)	674	434	268
Per capita minimum biomass potential (GJ/cap/a)—Jazz scenario	20.4	17.0	13.3
Per capita maximum biomass potential (GJ/cap/a)—Jazz scenario	77.4	49.9	30.8
Per capita minimum biomass potential (GJ/cap/a)—Symphony scenario	19.0	15.8	12.4
Per capita maximum biomass potential (GJ/cap/a)—Symphony scenario	71.9	46.3	28.6

[1] Tech = technical potential, sust = sustainable potential.

Given the estimated global energy needs in 2050 of 879 EJ/a (Jazz scenario) and 696 EJ/a (Symphony scenario), the projected biomass availability shows that energy needs can be covered by anything between 13% and 97% depending on the assumptions about developments in society, agriculture and ecological sustainability and the year of publication (Table 3). The studies published after 2010 show a much lower share, between 13% and 39%, which is more likely to be a realistic one. The range is still considerably large and requires a more in-depth look at the assumptions, as implications for future food security and agricultural systems, as well as impacts on the environment and the society, can be very great.

Table 3. Share of global energy demand which can be covered by biomass in 2050 grouped according to publication year or estimated potential.

	Published Before 2010		Only Tech or Sust Potential [1]		Published 2010 or Later	
	Jazz	Symphony	Jazz	Symphony	Jazz	Symphony
Global energy demand covered by the min. biomass (%)	20.2	25.5	16.8	21.2	13.2	16.7
Global energy demand covered by the max. biomass (%)	76.7	96.8	49.4	62.4	30.5	38.5

[1] Tech = technical potential, sust = sustainable potential.

The projected regional distribution of future biomass availability also shows a wide range of values (Table 4), though most studies do not present regional estimates. Poor data availability at the regional level may also limit the reliability of the data. As an example, the estimated non-food biomass potential for Africa ranges from 25–369 EJ/a, while other studies reveal even lower values, such as 2.5–9 EJ/a [24,60].

Table 4. Overview of regional biomass availability for 2050 (EJ/a).

Source	Min/Max [1]	Africa	Europe	Eastern Eur./CIS [2]	Asia	Oceania/ Pacific	North Amer.	Latin Amer.	Sum All Regions
Fischer & Schrattenholzer 2001	min	100	22	31	58	20	40	83	354
	max	124	27	38	77	26	50	103	445
Haberl et al. 2011	n.a.	24.6	3.59	14.25	20.9	1.89	15.55	23.99	105
Smeets et al. 2007	min	44	15	50	46	42	34	59	290
	max	369	64	205	193	109	193	235	1368
Mean value	min	56.2	13.5	31.8	41.6	21.3	29.9	55.3	249.7
Mean value	max	172.5	31.5	85.8	97.0	45.6	86.2	120.7	639.3

[1] Min = lowest estimate, max = highest estimate. [2] CIS = Commonwealth of Independent States.

3.2.2. Allocating and Distributing Non-Food Biomass to Cover Energetic and Material Demands

Regarding the biomass availability per capita, there is some variation depending on the future societal and energy scenarios (Jazz or Symphony scenarios, see Section 2 for details), but the more recent studies indicate much lower biomass availability for energy and material purposes, ranging from 12–29 GJ/cap/a compared to 17–57 GJ/cap/a using all studies (Table 2). Average global per capita energy needs in 2050 are estimated to be between 74 GJ/a (Symphony Scenario) and 101 GJ/a (Jazz Scenario). The lower estimates for biomass availability would then cover around 13–17% of the global per capita energy demand, and the higher estimates would cover 31–39% (Table 4).

However, the satisfaction of material needs is considered as a top priority of other energy uses, as it is currently hardly possible to replace fossil fuels for material use without using biomass, while technologies to replace fossil fuel energy for industry, electricity or heating with solar, wind or water power are very much advanced. Based on data by the World Energy Council, the global material use of final energy consumption is estimated to be 7.8 GJ/cap/a or 11% in the Jazz scenario for 2050, and 6.1 GJ/cap/a or 12% in the Symphony scenario. The final energy consumption does not take conversion and other losses into account, which amount to 30%. Therefore, a much higher supply of energy in the first place is needed and the presented estimates show the upper bound.

This material demand can be easily covered at the global level by the estimated biomass availability, as there is a surplus of biomass in both the Jazz and the Symphony scenario (Table 5). At the regional level, most regions have no difficulty in fulfilling their material energy requirements if a globally equal biomass distribution exists. An exception is North America, which would not be able to meet all regional material energy needs through biomass, should only the minimum biomass availability be feasible. Europe would just be able to manage in the Jazz scenario, but this would basically leave no leverage for any other biomass use.

The question is now, how much biomass will be available for non-material energy needs when all human material needs are accounted for in the same way; i.e., the amount of material energy distributed to each human being is the same and the unequal regional material consumption levels are not taken into account. At the global level, this is 6–9% for the minimum estimates and 23–30% for the maximum estimates. At the regional level, Sub-Saharan Africa shows the highest values, ranging from 19% to 96%, and North America with the lowest values from 3% to 13% at best, followed closely by Europe (Table 5).

Table 5. Share of Energetic and Material Demand for Fossil Fuels which can be Replaced with Biomass [1].

	Global		Sub-Saharan Africa		Middle East & North Africa		Latin America & Caribbean		North America		Europe		Asia (Incl. Pacific)	
	Jazz	Sym	Jazz	Sym	Jazz	Sym	Jazz	Sym	Jazz	Sym	Jazz	Sym	Jazz	Sym
Per capita energy demand covered by the min. biomass potential in 2050 (%) [a]	13.2	16.7	43.9	52.8	9.3	11.1	11.5	14.6	6.1	7.3	8.1	9.3	14.4	18.7
Per capita energy demand covered by the max. biomass potential in 2050 (%) [a]	30.5	38.5	102	122	21.5	25.6	26.5	33.8	14.1	16.9	18.7	21.4	33.2	43.1
Per capita energy demand for material uses covered by the minimum biomass potential (%) [b]	170	204	548	635	119	131	158	179	81.2	90.7	107	122	182	226
Per capita energy demand for material uses covered by the maximum biomass potential (%) [b]	393	472	1267	1468	275	303	365	414	188	210	248	281	419	522
Per capita total energy demand covered by the min. biomass potential with equal satisfaction of material needs (%) [c]	5.4	8.5	18.1	26.9	3.8	5.7	4.7	7.5	2.5	3.7	3.3	4.7	5.9	9.5
Per capita total energy demand covered by the max. biomass potential with equal satisfaction of material needs (%) [c]	22.7	30.3	75.7	96.1	16.0	20.2	19.8	26.6	10.5	13.3	13.9	16.9	24.8	34.0
Per capita total energy demand covered by the min. biomass potential with unequal satisfaction of material needs [2] (%) [d]	n.a.	n.a.	35.9	44.5	1.5	2.6	4.2	6.5	−1.4	−0.7	0.5	1.6	6.5	10.4
Per capita total energy demand covered by the max. biomass potential with unequal satisfaction of material needs [2] (%) [d]	n.a.	n.a.	93.5	113.6	13.7	17.2	19.2	25.6	6.6	8.8	11.1	13.8	25.3	34.9

[1] using only studies published in 2010 or later. [2] Using estimated regional material needs, which are based on poverty and income developments, population growth, etc., in 2050. [a] corresponds to EDpc, [b] corresponds to MDpc, [c] corresponds to EDMDpc, [d] corresponds to REDMDpc (see Section 2.3).

This picture would look even worse if one were to ask how much of the globally available biomass per person would be available at the regional level for non-material energy uses once all regional material demands predicted in 2050 have been fulfilled. The rich regions would then continue to consume per capita much more than other regions; e.g., North America would consume 6.7 times more material energy than Sub-Saharan Africa, and Europe 5.1 times more. This is especially critical, as it is very likely that poverty and inadequate fulfillment of material needs in Africa and other poor regions would persist given their low income-levels. Based on a projected unequal material resource consumption, Sub-Saharan Africa would be able to cover 36–45% of its non-material energy needs with the minimum biomass share available. North America would not be able to cover any of its non-material energy needs at all, and Europe only around 1–2%. Using the estimates for the maximum biomass availability, Sub-Saharan Africa would be able to fully cover its energetic needs, while North America and Europe could cover around 10% with biomass, and would need other energy sources for the remaining 90% energy requirements.

4. Discussion

The emerging bioeconomies in Europe, North America and elsewhere will require large amounts of biomass in the future. Key questions are, therefore, how much biomass will be available in future for food and non-food uses? For what shall the non-food biomass be used? According to whose needs or national or regional consumption levels? There is increasing research available to answer the first question about availability, but the question of future food security has not been sufficiently addressed in the models of future non-food biomass availability. There is much less research even, on the other questions.

4.1. Limitations Regarding Food Requirements and Food Utilization in the Biomass Scenarios

In the reviewed biomass potential estimates, food security is reduced to mainly the production of calories and proteins through future yield and land-use assumptions. This approach resembles the food security concept used in the 1970s; e.g., by the World Food Conference in 1974, which is now outdated [61]. Given the importance of global food security especially highlighted in the SDGs, and the clear prioritization of food security when moving towards bioeconomies, the limited consideration of the current internationally accepted and standardized food security concept, such as that adopted by the World Food Summit in 1996, is surprising. Even other studies hardly discuss the concept of food security in relation to biomass availability and use; an exception is the report of the German Advisory Council on Global Change (WGBU) with the input study by Faaij [24,46].

The dietary and land-use scenarios used in the biomass models have to be reconsidered with respect to desirability and feasibility from a food security perspective. These studies use diets of 2450–3150 kcal/capita/d, a quantity which is lower than the current food consumption levels in industrialized countries. Therefore, industrialized countries have to reconsider the data they rely on for their bioeconomy strategies when aiming to maintain current food consumption levels. A change in food habits in the OECD countries, as implicit of basically all biomass potentials, is unlikely to materialize. The scenario of a global vegetarian diet is of course creating higher values for non-food biomass availability, but is definitely not an option for policy makers and neither globally nor nationally implementable until 2050. The suggestion of a weekly "meat free day" by the Green Party led to a public outcry in Germany, and significantly decreased the popularity of the party at that time. So how realistic is the implementation of a dietary shift with reducing food calories by 10% to 30% as proposed by the researchers? Which governmental party will explain to its voters that they have to eat less meat to be able to continue their material consumption and drive their cars as they are used to? Biomass availability scenarios should be built on implementable assumptions. It is not at all likely that the above scenario can ever be implemented or is desirable in a society. In a democracy, the state cannot prescribe what and how much to eat, and if food consumption has to be reduced for producing bioenergy or bioplastic, this becomes highly questionable from a moral and equity perspective, as this will be at the expense of the poor.

From the perspective of low-income countries, the question emerges whether the lower range of calories, e.g., 2450 kcal/capita/d, is sufficient, as most employment opportunities involve hard physical labor, especially in the agricultural sector, which is one of the mainstays for most economies. Strauss [62] points out the correlation between caloric input and labor productivity in agriculture, and shows that in developing countries a caloric consumption of over 4500 kcal/capita/d still leads to an increase in labor productivity. A sugar cane harvester in Latin America needs around 3900 kcal/capita/d of food (personal communication, sugar cane plantation manager, 2018). It is very likely that a vegetarian or low meat diet in low-income countries will lead to nutritional deficiencies, as the choice of food is limited there [63] and alternative protein and iron sources are hardly available, extremely costly, often not of good quality, and not a typical part of a diet. For example, around one billion people are anaemic due to iron deficiency; in some countries in Africa over 60% of the population is [64]. The diversity of food sources is also important for a balanced nutrition, so adequate and diverse horticultural production should be integrated in the models to fulfil the requirements for vitamins, minerals and micronutrients.

While some biomass availability scenarios assume the same food consumption level for each individual globally (e.g., some scenarios with less meat, the fair and frugal scenario), other models assume that existing inequalities in consumption will remain in 2050 (e.g., business as usual scenarios), while others do not specify their assumptions (e.g., those that only set a certain amount of land aside for food production). It is likely that the latter are built on maintaining an unequal global food consumption (including undernourishment and malnourishment), as the FAO projections typically estimate future demand for food and not the amount needed to adequately satisfy all food needs of the global population [21]. As the FAO [57] highlights, the higher the inequality in food consumption is in a country, the more calories per person need to be incorporated in future demand estimations if the objective is to reduce or eliminate undernourishment. In other words, for a country with high inequality, even an average per capita caloric consumption of 3100 kcal can still mean that 5% of the population are undernourished.

These estimates of future non-food biomass availability thus imply that in wealthier regions people eat much more than they may need, while in other regions food calories can be below the physical needs of the population. This is a realistic assumption but implies severe conflicts in the future about non-food biomass and global food security. This will be counteracting the efforts to eradicate hunger as globally agreed upon in the SDGs, and is inacceptable from an intragenerational equity perspective. Relying on these studies for future global non-food biomass availability means accepting undernourishment, while at the same time, biomass is used in the rich countries for bioenergy or other uses. If the objective is to achieve global food security before any other biomass use, food caloric availability needs to be much higher than the modeled values, and the FAO data cannot be used, as it assumes that by 2050 poverty and inequality will continue to exist [21].

4.2. Limitations Regarding Food Access, Availability and Stability in the Biomass Scenarios

The pillar of food access would need to better addressed in future models. The development of international food prices has been neglected. Food and biomass prices will be influenced in 2050 by changes in the oil and fossil fuel prices, and there are studies that already claim this relation [65,66]. The latter prices are likely to increase significantly as resources deplete further and alternative energy sources are not developed fast enough to fill the gap [2]. Increasing crop prices are predicted through an expansion of biofuel production along with a net decrease in availability and access to food, especially in Africa [67]. The price developments and effects need to be included in future scenarios, along with poverty levels and inequality in income, land tenure and other resources that affect the physical and economic access to food.

Regarding food availability, the question of international trade of food and non-food biomass and its effects on national availability should be considered in future scenarios. Modeling the future development of the global food and biomass trade is uncertain, as the global agricultural market is influenced not only by subventions and trade barriers [25] but also by the purchasing power of nations. Adding scenarios that depict different market developments and account for purchasing power is needed to better understand the effects of future biomass use on food security by a nation or whole region, especially on low-income, in-food-deficit countries.

The stability of food supplies is at risk due to climate change, environmental degradation, and disease or pest outbreaks [68]. Climate change entails risks and uncertainties for future food security in all its dimensions, as agriculture is sensitive to climate variability and change. Especially in some regions, climate change may slow down the progress towards food security [59]. Therefore, biomass availability models need to integrate climate change effects.

There are also doubts about the effects of agricultural intensification as part of most models. Agricultural intensification is one of the key drivers of biodiversity loss on an unprecedented scale, due to habitat loss and pollution caused by synthetic pesticides and fertilizers, which may also affect ecosystem services, such as food production [69,70]. Whether the agricultural intensification assumptions with their potential effects would be within the planetary boundaries is not certain, but it is rather unlikely, since phosphorus and nitrogen, the key fertilizer elements in agriculture, are far beyond the safe operating space [71]. By definition, the technical potential estimated in nearly all studies does not entail an environmentally sustainable perspective. It is the decision of each researcher how far environmental aspects are considered. It is somewhat surprising that even the more recent research concentrates on the technical and not on the sustainable potential, since ecological sustainability and the need to maintain ecosystem services, especially of rainforests, have already been discussed for decades at the national and international level. Topics also discussed include problems associated with agriculture on peatlands or in biodiversity hotspots, planetary boundaries, land degradation or environmental problems associated with agricultural intensification. Those studies that included an ecologically sustainable potential came to much lower biomass availability levels [24,38]. Ecological sustainability is important not only for food stability but also for the sustainability objective of intergenerational equity to ensure that future generations also have continued and stable access to biomass to fulfill their food and non-food needs. Therefore, the inclusion of ecological sustainable biomass scenarios in addition to technically possible scenarios should become the standard in future studies.

4.3. Large Range of Biomass Availability Estimates

As also indicated by other recent studies, the maximal possible non-food biomass availability will be somewhere around 250–270 EJ/a in 2050 due to biospheric constraints [15,16,35,72]. Searle and Malins [34] derived 60–120 EJ/a as the limit to long-term biomass availability, which is in line with the minimum values of biomass availability of the more recent studies. The range for the technical potential is still large, and implications for future food security, agricultural systems, environment and society can vary greatly. In addition, the reviewed studies insufficiently addressed nutritional requirements and food security, so future available biomass will be lower than the currently technical non-food biomass estimates.

The very high sensitivity of the results with respect to assumptions and modeling techniques implies uncertainties. The assumptions on yields, land use or rehabilitation of degraded land indicate different opinions about what is technically possible, practically feasible or ecologically desirable. They therefore relate to a very different normative perspective of the world; i.e., from a more technologically- and economically-oriented perspective to a more ecological perspective. Any biomass use beyond this range would either mean an expansion of biomass use at the cost of food security, or the conversion of precious, conservation-worthy landscapes like rainforests; or a significant increase in production through irrigation and agricultural intensification far beyond what is estimated to be sustainable, ecologically recommended and practically feasible [24]. The practical feasibility of significant agricultural intensification in low- and lower-middle income countries in the next thirty years needs to be questioned, as these regions lack the necessary socio-economic pre-conditions, such as functional institutions, available infrastructure and financing. They have been affected for decades by problems, such as lack of knowledge among farmers, a dysfunctional extension system, volatile markets, and a lack of roads, marketing infrastructure, and of access to inputs, credits and insurance. Ongoing land degradation [73] and climate change leading to further decreasing yields [56,57] are two additional factors which require a cautious, if not critical look at the estimates of the future potential biomass availability.

4.4. Biomass Availability Is Insufficient to Cover Energy Needs after Satisfying Material Needs

Switching to a (global) bioeconomy "will entail high demand for biomass not only for bioenergy, but also for bio-materials such as plastics that are presently derived from fossil sources" [13].

With a projected depletion of oil stocks by 2050, it is surprising that little attention is paid to the question of where biomass shall come from to substitute fossil fuels. Most research looks at a specific sector, e.g., bioenergy, and every sector claims that there is enough biomass available while ignoring other sectors' needs. For that reason, we think that material/chemical uses of biomass have to be prioritized over energetic uses for electricity or fuel. Our calculations are rough and simplistic but add a new perspective given the strong focus on bioenergy. The data used and the assumptions, especially those using the Jazz and Symphony scenario, will need refining in future research and when better data on material, chemical and other biomass uses is available. While Dornburg et al. [22] suggest that in future, biomass may meet up to 30% of the projected global energy demand, our findings also show that this only may occur if the maximum biomass estimates are used. It could be only 13% based on the minimum values, which is to be expected. This share would be even lower if food security were to be adequately taken into account. There is probably enough non-food biomass available at the global level to fulfill the material energy needs. This is good news, since for material uses, fossil fuels can still hardly be substituted without the use of biomass.

4.5. Intragenerational Equity and Biomass Availability

The industrialized countries will have roughly enough biomass for the predicted material use in 2050 and, depending on the data, equity assumptions and scenarios used, some biomass will remain for additional bioenergy. If resource consumption in Europe and North America continues at the current high levels, basically all available biomass in the future will be needed to fulfill material demands, at least in a world where intragenerational equity is the norm. Hence, any bioenergy consumption higher than the values presented in Table 5 will either require a reduction in consumption levels in North America or Europe, or inevitably lead to an unfair global resource consumption at the cost of biomass availability for poorer nations, and possibly affect global food security via rising prices.

A significant reduction in material consumption in industrialized countries until 2050 is unlikely, since the political agenda is still based on a growth paradigm. Therefore, when prioritizing the fulfillment of material energy needs, there is likely to be no, or only a very small amount of biomass available to fulfill non-material energy needs. Future investments in the bioenergy sector should be kept at a minimum, and the focus for decision-makers, politicians and investors alike would need to be on material uses of biomass.

There is also the threat that biomass use will be increasingly less fair; i.e., that rich countries will consume biomass at the cost of food security or basic material uses of poor countries, when a significant reduction in resource consumption is not taking place at the same time. It is questionable whether market prices will transmit the necessary signals of resource stock depletion in time, as external costs are commonly excluded. Politics needs to address ways to reduce energy usage, be it in production, distribution or consumption [2].

The presented estimations assume either a globally equal share of the material energy needed or a regionally different share with poorer countries consuming less. Both assumptions may be flawed, as in many poorer countries, basic human needs and rights such as adequate housing, food, health, schooling, etc., have not yet been fully addressed. Higher material consumption levels than a globally equal share might be needed to develop a poor region like Sub-Saharan Africa to an acceptable state. Unfortunately, there is no data on how much energy for materials, industry, etc., would be needed to reach an acceptable state.

Several authors conclude that the rise in the use of biomass (biofuels) requires international cooperation, regulations, certification mechanisms and sustainability criteria regarding the use of land, sustainable production and the mitigation of environmental impacts caused by biomass production [16,22,34]. There is a need to go beyond this. Not only does food security need to be ensured in all agricultural production areas and along international value chains, the question about biomass allocation and distribution also needs to be internationally agreed upon.

4.6. Future Research Recommendations

To address food security in biomass availability estimates, future models should integrate several aspects. First, there is the need to include a "zero hunger" target as stated by the SDGs. Second, the advice of global nutrition experts on balanced diets should be integrated taking into account the needs of the population in developing countries that include physically hard-working people, diseases, and health and sanitation problems which may inhibit nutritional uptake. Future research should address the question of how much food will be needed in 2050 if all people are to have stable access to sufficient, nutritious and safe food. The assumption of unequal food consumption levels should always be accompanied by a scenario that assumes a sufficient intake of food. Third, the question of purchasing power, price developments and international trade of food and non-food biomass; and the effects on a global scale, but ideally also at regional and national levels, should be included in future scenarios, especially depicting the effects on low-income, food-deficient countries. To address food stability, climate change effects should always be considered, and more research on the ecological sustainable biomass potential is urgently needed. The key question is really that if all this is considered, how much biomass will be available if food security is appropriately accounted for?

The research presented here could be further strengthened, as it is built on available data and scenarios by the World Energy Council, which constructed plausible future energy scenarios but did not estimate energy scenarios that contain intragenerational equity or the satisfaction of basic human needs. It also may have missed some model assumptions, as not all were explicitly mentioned in the reviewed studies, especially regarding the estimated food requirements, which would have required contacting all authors individually. Also, no solutions to the identified bottlenecks of the models could be offered, as this would have gone beyond the objective of this research. The purpose of the research was to assess the biomass potentials with respect to their inclusion of food security, to show key tendencies and to derive options for future biomass allocation and distribution based on available data, especially from a sustainability perspective. The results are considered to be 'food for thought' about the fair and best uses of biomass for researchers, civil society and among decision-makers.

More research is needed to identify the necessary material energy requirements to satisfy all basic human needs at the global level, and which material energy requirements are required for acceptable outcomes based on human rights. This would involve estimates about global energy needs if all people were to have, for example, access to electricity, water and health services or appropriate housing. These estimates would then be a social boundary that defines the thresholds below which human well-being is endangered. Future research needs hence to identify how much biomass will be available for extra uses (such as bioenergy) if at least all global basic human needs are covered, and global poverty and hunger have been eradicated.

5. Conclusions

A conversion of the current fossil fuel-based economies into bio-based economies is constrained by the overall limited availability of biomass. This systematic review adds a sustainable development perspective on the estimates and models of future biomass availability. The focus is on the satisfaction of human needs through biomass, especially of food security and intragenerational equity, which includes options regarding the allocation and distribution of biomass. Though food security is key for most bioeconomy strategies and international agreements, an internationally accepted concept of food security based on the pillars' availability, access, utilization and stability was not included in the models for estimating non-food biomass potentials. Dietary assumptions were partly based on food consumption levels below the current food consumption in the OECD countries, and partly included globally unequal consumption levels. This has various implications. If the OECD countries were to adjust their economies based on the assumption of high biomass availability, they would have to reduce food consumption—or people in other countries would have to. The dietary assumptions imply that physically hard-working people in low-income countries would not receive sufficient calories for their activities. Unequal food consumption assumptions are very likely to entail the continued persistence of undernourishment until 2050, while biomass is used for material or energetic uses. This is neither acceptable from a sustainable development and intragenerational justice perspective, nor does it meet the SDG targets. It is not certain whether the technical potential can be materialized within the planetary boundaries. Estimates for an ecological sustainable biomass potential are rare and indicate a very low availability of biomass. However, they would be a guiding value to ensure food stability, and thereafter the availability of sufficient, safe and nutritious food over time. In conclusion, if biomass availability estimates had accounted for food security in its four dimensions, the availability of biomass for any use would be even lower than the range 116–268EJ/a of the more recently published biomass potentials.

Material uses of fossil fuel can be replaced by biomass but so far not by other renewable energy sources. Though there is research in this direction, e.g., solar biofuels derived by synthetic biological processes, it is unlikely that energy-efficient technologies that, for example, convert CO_2 from the air into plastic, are ready for the market by 2050. Given the relatively high energy demand for material uses compared to the limited availability of non-food biomass, it does make sense to prioritize the material uses of biomass over bioenergy to fuel cars or generate electricity. However, if this takes place, not much biomass will be left for bioenergy. With increasing development in low- and low-middle income countries, it is likely that material consumption will increase even more than is assumed here. First- or second-generation biofuels may be still an option for land-abundant, fertile countries, but are also questionable from a global distributional perspective. Cascading uses of biomass, and especially the use of waste, for energy generation is much more appropriate. These should be the key element of any bioenergy strategy, and the target for industrial and financial investments.

Increasing energy efficiency is another important element, but despite many calls in the past, progress has been limited. Here, governments need to provide incentives for industry and prevent undesirable behavior; e.g., though taxation. Moreover, new concepts to reduce the total consumption of energy and materials in industrialized countries need to become part of the agenda of politicians, businesses, and civil society, but also of researchers. Policies need to provide many more guiding frameworks for the economic system, including the bioeconomy more than is currently the case, so that economies develop in a sustainable direction agreed upon in the SDGs. If a fairer distribution of biomass use were to become a global norm, the bioeconomy, energy, and economic, climatic and social policies of the OECD countries would need to change significantly to account for at least a limited biomass availability. This implies reconsidering an economic system which, so far, is built on growth, unlimited global availability of inputs, and excludes external costs.

Since the Brundtland report in the late 1980s, it has been reiterated that economy, ecology and society need to be considered and addressed together. Adding a sustainable development perspective to future biomass availability enriches the discussion about what is technically feasible with what may be socially needed. Further research needs to determine how much biomass for bioenergy would really be left if food security and a decent minimum standard of global human well-being were to be incorporated in the estimates. Furthermore, how much biomass would be available if we were to additionally incorporate a stronger environmental sustainability perspective, as envisioned by the SDGs. Other questions are more practically oriented: Which policies and changes are needed at national and multilateral levels to ensure global food security before any other biomass use, given the tremendous income, power, and hence, energetic use differences between nations? Which economic policies and state regulations at the global and national level are needed to ensure the satisfaction of material needs of all humans while powerful economic sectors favor bioenergy? This entails that distribution questions, especially regarding resource use, are raised to the multilateral level and receive more global attention to build a peaceful, sustainable and equitable world.

Author Contributions: Conceptualization and methodology was developed by T.D.B. The formal analysis was shared by T.D.B. and M.N.; the writing and editing were done mainly by T.D.B. with some contributions by M.N.

Funding: This research was funded by the project "Improving food security in Africa through increased system productivity of biomass-based value webs" (BiomassWeb), which is funded by the German Federal Ministry of Education and Research (BMBF) based on the decision of the Parliament of the Federal Republic of Germany (grant number FKZ 031A258A) and by the project "Implementation of food security criteria within biomass sustainability standards" which is funded by the German Federal Ministry of Food and Agriculture through the Fachagentur für Nachwachsende Rohstoffe (FNR) (grant number FKZ 22025616).

Conflicts of Interest: The authors declare no conflict of interest. The founding sponsors had no role in the design of the study, in the collection, analyses, or interpretation of data, in the writing of the manuscript, or in the decision to publish the results.

Appendix A. Flow Diagram, Search Parameters and Details of the Systematic Literature Search

Search parameters:

- Searched in all databases being part of the Web Of Science database.
- Search terms: TOPIC = (biomass potential), OR TOPIC = (biomass potentials) OR TOPIC = (potential of biomass), OR TOPIC = (potentials of biomass) OR TOPIC = (bioenergy potentials), OR TOPIC = (bioenergy potential).
- Time span: All years (1945–2018).
- Search language = Auto.
- Date last searched: 5 September 2018.

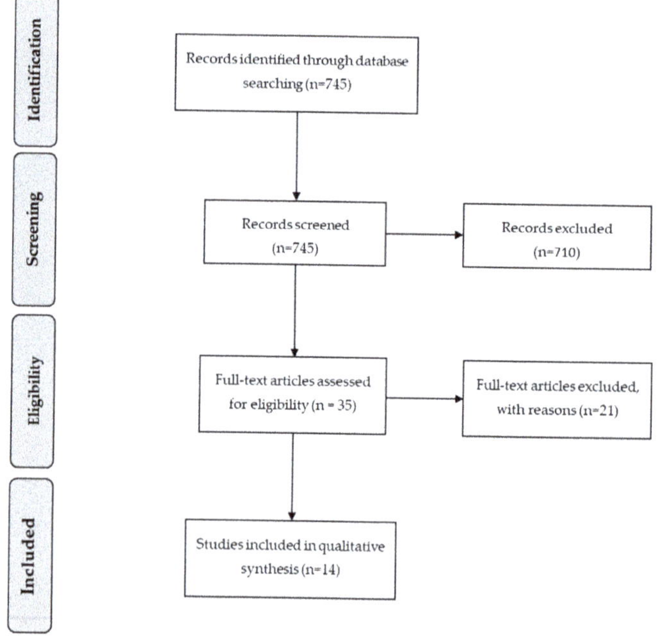

Figure A1. Flow Diagram of the Systematic Literature Search Procedure.

Appendix B. Definitions of Biomass Potential Types

Types of biomass potentials are generally discussed in terms of a hierarchical sequence of the upper limits of energy availability; i.e., theoretical, geographical, technical, economic, implementation/realistic and sustainable potentials. It is to be noted that these terms may be interpreted in different ways by different studies, and this definitional fuzziness hampers transferability and increases the risk of misunderstanding. The most basic potential type is the "theoretical potential". It is limited only by the fundamental physical and biological barriers of the net primary productivity of biomass produced on the Earth's total surface by the process of photosynthesis [24,31,33].The "geographical potential" is the fraction of the theoretical potential limited to the energy stored in terrestrial biomass (i.e., excluding oceans, rivers, etc.) [24,33]. Most studies, but not all, also include 'availability,' 'accessibility' and/or 'suitability' for bioenergy production of terrestrial biomass products as a limiting factor in their definitions. The "technical potential" describes the fraction of the geographical potential that is left after losses from the conversion of the primary energy to secondary energy sources have been subtracted [24,33]. While the terms theoretical and geographical potential are used relatively consistently, the term "technical potential" lacks a universally used definition. In a broad understanding, the "technical potential is the geographical potential reduced by the losses of the conversion of the primary energy to secondary energy sources" [31], which means it is only reduced by conversion efficiency as a result of the level of advancement of agricultural and industrial-energy technology. Other authors include a range of further limiting factors, such as the demand for land for food production, housing and infrastructure, and the conservation of forests (e.g., [33]) The "sustainable potential" is the fraction of the technical potential that remains after considering ecological limitations [24].

Table A1. Use of the food security concept and nutritional requirements in the reviewed studies.

	Food Security Definition	Caloric (Energy) Requirements	Protein Requirements	Vitamin and Micronutrient Requirements	Development of International Food Prices	Role of Governance	International Food Trade Balance Included
Beringer et al. 2011	-	-	-	-	-	-	-
Dornburg et al. 2010	-	-	-	-	yes	-	-
Erb et al. 2012	-	yes	yes	-	-	yes	yes
Fischer & Schrattenholzer 2001	-	-	-	-	-	-	-
Haberl et al. 2011	-	yes	-	-	-	-	yes
Haberl et al. 2013	-	-	-	-	-	-	-
Hakala et al. 2009	-	yes	-	-	-	-	-
Hoogwijk et al. 2003	-	yes	-	-	-	-	-
Hoogwijk et al. 2005	-	yes	-	-	-	-	yes
Junginger et al. 2006	-	yes	-	-	-	-	-
Smeets et al. 2007	-	yes	yes	-	yes	-	yes
Strapasson et al. 2017	-	yes	-	-	-	-	-
Vries et al. 2007	-	-	-	-	-	-	yes
Yamamoto, Fujino et al. 2001	-	-	-	-	-	-	yes

Appendix D

Table A2. Overview of the reviewed studies: Estimated biomass potential type and range, biomass sources and description of assumptions.

	Estimated Biomass Potential Type [1]	Range of Biomass Potential Estimates (Minimum and Maximum) (EJ/a)	Biomass Sources [2]	Assumed Cropland Expansion	Well Described Assumptions for Land Use
Beringer et al. 2011	sust	126–216	EC	10–30%, 142–454 Mha	-
Dornburg et al. 2010	tech	200–500	EC, FR, AR	-	-
Erb et al. 2012	tech	46–181 [3]	EC, AR	9%, 19%	yes
Fischer & Schrattenholzer 2001	tech	370–450	EC, F, AR, W	280 Mha	yes
Haberl et al. 2011	tech	64–161	EC, AR	9.2–19.1%	yes
Haberl et al. 2013	tech	190	-	-	-
Hakala et al. 2009	tech	44–110	EC, AR	-	-
Hoogwijk et al. 2003	geogr	33–1135	EC, FR, AR, W	-	yes
Hoogwijk et al. 2005	geogr	311–657 [3]	EC, AR, FR	-	yes
Junginger et al. 2006	tech	40–1100	EC, FR, AR, W	0–400 Mha	-
Smeets et al. 2007	tech	367–1458	EC, F, AR	100–200 Mha	yes
Strapasson et al. 2017	tech	70–360	EC, F, AR, FW, W, R	12 Mha/a	yes
Vries et al. 2007	tech	75–300	EC	-	-
Yamamoto, Fujino et al. 2001	tech	182	EC, R (+W)	439 Mha	yes

[1] Potential types: sust=sustainable; tech=technical; geogr= geograhical. For a general definition of each potential type see Appendix B. [2] Biomass sources: EC = energy crops; F = forestry; AR = agricultural residues, FR = forestry residues; W = waste; R = general residues (e.g., MSW, etc.); FP = fishery products [3] Total sum is calculated from subtotals provided in the publication.

Appendix E

Table A3. Data sources for our own estimations. As described in Section 2.3, the data is derived from the World Energy Council [58] based on its two different scenarios for the world in 2050; i.e., the Jazz and the Symphony scenarios (details are in the report). The data here stems from the "Regional Summary" (page 249) and the table "The world in 2050" (page 252).

	Scenario	
	Jazz	**Symphony**
World population in 2050 (million)		
Global	**8703**	**9374**
Sub-Saharan Africa	1648	1961
Middle East & North Africa	551	601
Latin America & Caribbean	577	603
North America	594	619
Europe	819	853
Asia (incl. Pacific)	4513	4738
World energy needs (EJ)		
Global	**879**	**696**
Sub-Saharan Africa	50	46
Middle East & North Africa	79	67
Latin America & Caribbean	67	51
North America	130	105
Europe	135	114
Asia (incl. Pacific)	418	314
Final energy consumption (EJ/a)		
Global	**629**	**491**
Sub-Saharan Africa	37	33
Middle East & North Africa	57	49
Latin America & Caribbean	45	36
North America	90	73
Europe	94	75
Asia (incl. Pacific)	306	224
Final energy consumption per capita (GJ/a) excluding non-energy uses		
Global	64.5	46.3

References

1. German Bioeconomy Council. *Bioeconomy Policy (Part III) Update Report of National Strategies Around the World*; Office of the Bioeconomy Council: Berlin, Germany, 2018.
2. Chapman, I. The end of Peak Oil? Why this topic is still relevant despite recent denials. *Energy Policy* **2014**, *64*, 93–101. [CrossRef]
3. Owen, N.A.; Inderwildi, O.R.; King, D.A. The status of conventional world oil reserves—Hype or cause for concern? *Energy Policy* **2010**, *38*, 4743–4749. [CrossRef]
4. Shafiee, S.; Topal, E. When will fossil fuel reserves be diminished? *Energy Policy* **2009**, *37*, 181–189. [CrossRef]

5. Souza, G.M.; Ballester, M.V.R.; Cruz, B.C.H.; Chum, H.; Dale, B.; Dale, V.H.; Fernandes, E.C.M.; Foust, T.; Karp, A.; Lynd, L.; et al. The role of bioenergy in a climate-changing world. *Environ. Dev.* **2017**, *23*, 57–64. [CrossRef]

6. Holden, E.; Linnerud, K.; Banister, D. Sustainable development: Our Common Future revisited. *Glob. Environ. Chang.* **2014**, *26*, 130–139. [CrossRef]

7. Meadowcroft, J. Who is in Charge here? Governance for Sustainable Development in a Complex World. *J. Environ. Policy Plan.* **2007**, *9*, 299–314. [CrossRef]

8. Barbier, E.B.; Burgess, J.C. The Sustainable Development Goals and the systems approach to sustainability. *Econ. E J.* **2017**. [CrossRef]

9. Stumpf, K.; Baumgärtner, S.; Becker, C.; Sievers-Glotzbach, S. The Justice Dimension of Sustainability: A Systematic and General Conceptual Framework. *Sustainability* **2015**, *7*, 7438–7472. [CrossRef]

10. Baumgärtner, S.; Quaas, M. Sustainability economics General versus specific, and conceptual versus practical. *Ecol. Econ.* **2010**, *69*, 2056–2059. [CrossRef]

11. Baumgärtner, S.; Quaas, M. What is sustainability economics? *Ecol. Econ.* **2010**, *69*, 445–450. [CrossRef]

12. WCED. *Report of the World Commission on Environment and Development: Our Common Future. A/42/427*; UN General Assembly: New York, NY, USA, 1987.

13. Scarlat, N.; Dallemand, J.F.; Monforti-Ferrario, F.; Nita, V. The role of biomass and bioenergy in a future bioeconomy: Policies and facts. *Environ. Dev.* **2015**, *15*, 3–34. [CrossRef]

14. Running, S.W. A measurable planetary boundary for the biosphere. *Science* **2012**, *337*, 1458–1459. [CrossRef]

15. Haberl, H.; Erb, K.H.; Krausmann, F.; Running, S.; Searchinger, T.D.; Kolby Smith, W. Bioenergy: How much can we expect for 2050? *Environ. Res. Lett.* **2013**, *8*, 31004. [CrossRef]

16. Popp, J.; Lakner, Z.; Harangi-Rákos, M.; Fári, M. The effect of bioenergy expansion: Food, energy, and environment. *Renew. Sustain. Energy Rev.* **2014**, *32*, 559–578. [CrossRef]

17. Rockström, J.; Steffen, W.; Noone, K.; Persson, Å.; Chapin, F.S., III; Lambin, E.F.; Lenton, T.M.; Scheffer, M.; Folke, C.; Schellnhuber, H.J.; et al. A safe operating space for humanity. *Nature* **2009**, *461*, 472–475. [CrossRef]

18. FAO. *The State of Food Security and Nutrition in the World: Building Climate Resilience for Food Security and Nutrition*; FAO: Rome, Italy, 2018.

19. WHO; FAO. *Guidelines on Food Fortification with Micronutrients*; World Health Organization and Food and Agriculture Organization of the United Nations: Geneva, Switzerland, 2006.

20. WHO. *Diet, Nutrition and the Prevention of Chronic Diseases: Report of A Joint Who/Fao Expert Consultation*; WHO Technical Report Series No 916; WHO: Geneva, Switzerland, 2003.

21. Alexandratos, N.; Bruinsma, J. *World Agriculture Towards 2030/2050: The 2012 Revision*; ESA Working paper No. 12–03; FAO: Rome, Italy, 2012.

22. Dornburg, V.; van Vuuren, D.; van de Ven, G.; Langeveld, H.; Meeusen, M.; Banse, M.; van Oorschot, M.; Ros, J.; Jan van den Born, G.; Aiking, H.; et al. Bioenergy revisited: Key factors in global potentials of bioenergy. *Energy Environ. Sci.* **2010**, *3*, 258. [CrossRef]

23. Chum, H.; Faaij, a.; Moreira, J.; Berndes, G.; Dhamija, P.; Dong, H.; Gabrielle, B.; Goss Eng, A.; Lucht, W.; Mapako, M.; et al. Bioenergy. In *Renewable Energy Sources and Climate Change Mitigation.: Special Report of the Intergovernmental Panel on Climate Change*; Edenhofer, O., Pichs-Madruga, R., Sokona, Y., Seyboth, K., Matschoss, P., Kadner, S., Zwickel, T., Eickemeier, P., Hansen, G., Schlφmer, S., Eds.; Cambridge University Press: New York, NY, USA, 2011; pp. 209–332.

24. WGBU. *Welt im Wandel: Zukunftsfähige Bioenergie und nachhaltige Landnutzung*; Wissenschaftlicher Beirat der Bundesregierung Globale Umweltveränderungen (WBGU): Berlin, Germany, 2008.

25. Zeddies, J.; Bahrs, E.; Schönleber, N.; Gamer, W. *Globale Analyse und Abschätzung des Biomasse–Flächennutzungspotentials*; Institut für Landwirtschaftliche Betriebslehre. Universität Hohenheim: Stuttgart, Germany, 2012.

26. *BMVBS Globale Und Regionale Räumliche Verteilung von Biomassepotenzialen*; BMVBS-Online-Publikation, 27/2010; BMVBS: Berlin, DE, Germany, 2010.

27. Hakala, K.; Kontturi, M.; Pahkala, K. Field biomass as global energy source. *Agric. Food Sci.* **2009**, *18*, 347–365. [CrossRef]

28. Erb, K.H.; Haberl, H.; Krausmann, F.; Lauk, C.; Plutzar, C.; Steinberger, J.K.; Müller, C.; Bondeau, A.; Waha, K.; Pollack, G. *Eating the Planet: Feeding and Fuelling the World Sustainably, Fairly and Humanely: A Scoping Study*; Institute of Social Ecology: Vienna, Austria, 2009.

29. Erb, K.H.; Haberl, H.; Plutzar, C. Dependency of global primary bioenergy crop potentials in 2050 on food systems, yields, biodiversity conservation and political stability. *Energy Policy* **2012**, *47*, 260–269. [CrossRef]

30. Haberl, H.; Erb, K.H.; Krausmann, F.; Bondeau, A.; Lauk, C.; Müller, C.; Plutzar, C.; Steinberger, J.K. Global bioenergy potentials from agricultural land in 2050: Sensitivity to climate change, diets and yields. *Biomass Bioenergy* **2011**, *35*, 4753–4769. [CrossRef]

31. Hoogwijk, M.; Faaij, a.; Eickhout, B.; de Vries, B.; Turkenburg, W. Potential of biomass energy out to 2100, for four IPCCSRES land-use scenarios. *Biomass Bioenergy* **2005**, *29*, 225–257. [CrossRef]

32. Hoogwijk, M.; Faaij, A.; van Den Broek, R.; Berndes, G.; Gielen, D.; Turkenburg, W. Exploration of the ranges of the global potential of biomass for energy. *Biomass Bioenergy* **2003**, *25*, 119–133. [CrossRef]

33. Smeets, E.M.W.; Faaij, A.; Lewandowski, I.; Turkenburg, W. A bottom-up assessment and review of global bio-energy potentials to 2050. *Prog. Energy Combust. Sci.* **2007**, *33*, 56–106. [CrossRef]

34. Searle, S.; Malins, C. A reassessment of global bioenergy potential in 2050. *Gcb Bioenergy* **2015**, *7*, 328–336. [CrossRef]

35. Haberl, H.; Beringer, T.; Bhattacharya, S.C.; Erb, K.H.; Hoogwijk, M. The global technical potential of bio-energy in 2050 considering sustainability constraints. *Curr. Opin. Environ. Sustain.* **2010**, *2*, 394–403. [CrossRef]

36. World Energy Council. *World Energy Resources. 2013 Survey;* World Energy Council (WEC): London, UK, 2013.

37. Jesson, J.; Matheson, L.; Lacey, F.M. *Doing Your Literature Review;* SAGE Publications: London, UK, 2011.

38. Beringer, T.I.M.; Lucht, W.; Schapphoff, S. Bioenergy production potential of global biomass plantations under environmental and agricultural constraints. *Gcb Bioenergy* **2011**, *3*, 299–312. [CrossRef]

39. Fischer, G.; Schrattenholzer, L. Global bioenergy potentials through 2050. *Biomass Bioenergy* **2001**, *20*, 151–159. [CrossRef]

40. Junginger, M.; Faaij, A.; Rosillo-Calle, F.; Wood, J. The growing role of biofuels opportunities, challenges and pitfalls. *Int. Sugar J.* **2006**, *108*, 618.

41. Strapasson, A.; Woods, J.; Chum, H.; Kalas, N.; Shah, N.; Rosillo-Calle, F. On the global limits of bioenergy and land use for climate change mitigation. *Gcb Bioenergy* **2017**, *9*, 1721–1735. [CrossRef]

42. De Vries, B.J.M.; van Vuuren, D.P.; Hoogwijk, M.M. Renewable energy sources: Their global potential for the first-half of the 21st century at a global level: An integrated approach. *Energy Policy* **2007**, *35*, 2590–2610. [CrossRef]

43. Yamamoto, H.; Fujino, J.; Yamaji, K. Evaluation of bioenergy potential with a multi-regional global-land-use-and-energy model. *Biomass Bioenergy* **2001**, *21*, 185–203. [CrossRef]

44. Krausmann, F.; Erb, K.-H.; Gingrich, S.; Lauk, C.; Haberl, H. Global patterns of socioeconomic biomass flows in the year 2000: A comprehensive assessment of supply, consumption and constraints. *Ecol. Econ.* **2008**, *65*, 471–487. [CrossRef]

45. Slade, R.; Saunders, R.; Gross, R.; Bauen, A. *Energy from Biomass: The Size of the Global Resource;* Imperial College Centre for Energy Policy and Technology and UK Energy Research Centre: London, UK, 2011.

46. Faaij, A. *Bioenergy and Global Food Security: Externe Expertise Für Das WBGU–Hauptgutachten Welt Im Wandel: Zukunftsfähige Bioenergie und Landnutzung;* WBGU: Berlin, DE, Germany, 2008.

47. Yamamoto, H.; Yamaji, K.; Fujino, J. Evaluation of bioenergy resources with a global land use and energy model formulated with SD technique. *Appl. Energy* **1999**, *63*, 101–113. [CrossRef]

48. Ackerberg, D.; Lanier Benkard, C.; Berry, S.; Pakes, A.; James, J.H.; Edward, E.L. Chapter 63 Econometric Tools for Analyzing Market Outcomes. In *Handbook of Econometrics;* Elsevier: Amsterdam, The Nederland, 2007; pp. 4171–4276.

49. Maxwell, D.G. Measuring Food Security: The Frequency and Severity of "Coping Strategies. *Food Policy* **1996**, *21*, 291–303. [CrossRef]

50. Pinstrup-Andersen, P. Food security: Definition and measurement. *Food Secur.* **2009**, *1*, 5–7. [CrossRef]

51. World Food Summit. *Rome Declaration on World Food Security;* World Food Summit: Rome, Italy, 13–17 November 1996. Available online: http://www.fao.org/docrep/003/w3613e/w3613e00.htm (accessed on 25 September 2018).

52. FAO. *An Introduction to the Basic Concepts of Food Security;* EC FAO Food Security Programme, FAO: Rome, Italy, 2008.

53. FAO. *Food Security;* Policy Brief, Issue 2; FAO: Rome, Italy, 2006.

54. Ip, K.K.W. *Global Distributive Justice*; Oxford Research Encyclopedia of International Studies; Oxford University Press: Oxford, UK, 2017.

55. Daly, H.E. Allocation, distribution, and scale: Towards an economics that is efficient, just, and sustainable. *Ecol. Econ.* **1992**, 185–193. [CrossRef]

56. World Energy Council. *World Energy Scenarios: Composing Energy Futures to 2050*; World Energy Council: London, UK, 2013.

57. FAO. *World agriculture: towards 2015/2030. An FAO perspective*; FAO: Rome, Italy, 2013.

58. FAO. *The State of Food Insecurity in the World*; FAO: Rome, Italy, 1999.

59. Wheeler, T.; Braun, J. Climate Change Impacts on Global Food Security. *Science* **2013**, *341*, 508–513. [CrossRef]

60. Zeddies, J.; Schönleber, N. *Literaturstudie Biomasse Flächen und Energiepotenziale*; Institut für Landwirtschaftliche Betriebslehre. Universität Hohenheim: Stuttgart, DE, Germany, 2014.

61. Maxwell, D. Measuring food insecurity: The frequency and severity of "coping strategies". *Food Policy* **1996**, *21*, 291–303. [CrossRef]

62. Strauss, J. Does Better Nutrition Raise Farm Productivity? *J. Political Econ.* **1986**, *94*, 297–320. [CrossRef]

63. Sanders, T.A.B. The nutritional adequacy of plant-based diets. *Proc. Nutr. Soc.* **1999**, *58*, 265–269. [CrossRef]

64. WHO; UNICEF. Focusing on Anaemia, Towards an Integrated Approach for Effective Anaemia Control. 2004. Available online: http://www.who.int/nutrition/publications/micronutrients/WHOandUNICEF_statement_anaemia/en/ (accessed on 6 July 2019).

65. De Gorter, H.; Drabik, D.; Just, D.R. How biofuels policies affect the level of grains and oilseed prices: Theory, models and evidence. *Glob. Food Secur.* **2013**, *2*, 82–88. [CrossRef]

66. Malins, C. *Thought for food A Review of the Interaction Between Biofuel Consumption and food Markets*; Cerulogy: London, UK, 2017.

67. Rosegrant, M.W.; Msangi, S.; Ringler, C.; Sulser, T.B.; Zhu, T.; Cline, S. *International Model for Policy Analysis of Agricultural Commodities and Trade (IMPACT): Model Description*; International Food Policy Research Institute: Washington, DC, USA, 2008.

68. Sundström, J.F.; Albihn, A.; Boqvist, S.; Ljungvall, K.; Marstorp, H.; Martiin, C.; Nyberg, K.; Vågsholm, I.; Yuen, J.; Magnusson, U. Future threats to agricultural food production posed by environmental degradation, climate change, and animal and plant diseases a risk analysis in three economic and climate settings. *Food Sec.* **2014**, *6*, 201–215. [CrossRef]

69. Sánchez-Bayo, F.; Wyckhuys, K.A.G. Worldwide decline of the entomofauna: A review of its drivers. *Biol. Conserv.* **2019**, *232*, 8–27. [CrossRef]

70. Firbank, L.; Bradbury, R.B.; McCracken, D.I.; Stoate, C. Delivering multiple ecosystem services from Enclosed Farmland in the UK. *Agric. Ecosyst. Environ.* **2013**, *166*, 65–75. [CrossRef]

71. Steffen, W.; Richardson, K.; Rockström, J.; Cornell, S.E.; Fetzer, I.; Bennett, E.M.; Biggs, R.; Carpenter, S.R.; de Vries, W.; Wit, C.A.d.; et al. Sustainability. Planetary boundaries: Guiding human development on a changing planet. *Sci. (New York N.Y.)* **2015**, *347*, 1259855. [CrossRef]

72. Smith, W.K.; Zhao, M.; Running, S.W. Global Bioenergy Capacity as Constrained by Observed Biospheric Productivity Rates. *BioScience* **2012**, *62*, 911–922. [CrossRef]

73. Behrend, H. Land Degradation and Its Impact on Security. In *Land Restoration: Reclaiming Landscapes for a Sustainable Future*; Chabay, I., Frick, M., Helgeson, J., Eds.; Elsevier AP: Amsterdam, The Nederland, 2015; pp. 13–26.

© 2019 by the authors. Licensee MDPI, Basel, Switzerland. This article is an open access article distributed under the terms and conditions of the Creative Commons Attribution (CC BY) license (http://creativecommons.org/licenses/by/4.0/).

Article

Governance of the Bioeconomy: A Global Comparative Study of National Bioeconomy Strategies

Thomas Dietz [1,*], Jan Börner [2,3], Jan Janosch Förster [2] and Joachim von Braun [2]

[1] Institute of Political Science, Westfälische Wilhelms-Universität Münster, 48151 Münster, Germany
[2] Center for Development Research (ZEF), Rheinische Friedrich-Wilhelms-Universität Bonn, 53113 Bonn, Germany; jborner@uni-bonn.de (J.B.); jforster@uni-bonn.de (J.J.F.); jvonbraun@uni-bonn.de (J.v.B.)
[3] Institute for Food and Resource Economics (ILR), Rheinische Friedrich-Wilhelms-Universität Bonn, 53115 Bonn, Germany
* Correspondence: thomas.dietz@uni-muesnter.de

Received: 25 July 2018; Accepted: 31 August 2018; Published: 6 September 2018

Abstract: More than forty states worldwide currently pursue explicit political strategies to expand and promote their bioeconomies. This paper assesses these strategies in the context of the global Sustainable Development Goals (SDGs). Our theoretical framework differentiates between four pathways of bioeconomic developments. The extent to which bioeconomic developments along these pathways lead to increased sustainability depends on the creation of effective governance mechanisms. We distinguish between enabling governance and constraining governance as the two fundamental political challenges in setting up an effective governance framework for a sustainable bioeconomy. Further, we lay out a taxonomy of political support measures (enabling governance) and regulatory tools (constraining governance) that states can use to confront these two political challenges. Guided by this theoretical framework, we conduct a qualitative content analysis of 41 national bioeconomy strategies to provide systematic answers to the question of how well designed the individual national bioeconomy strategies are to ensure the rise of a sustainable bioeconomy.

Keywords: bioeconomy; governance; development policy; innovation; technology; bio-based

1. Introduction

The bioeconomy is based on the idea of applying biological principles and processes in all sectors of the economy and to increasingly replace fossil-based raw materials in the economy with bio-based resources and principles. An innovative and sustainable use of bio-based resources in different sectors of the economy (i.e., a bio-based transformation) provides opportunities for achieving a number of different Sustainable Development Goals (SDGs), which have been designed to improve social, economic, and ecological living conditions. Particularly, this applies to sustainable solutions to current climate change risks [1]. However, recent studies emphasize the dependence of a sustainable bioeconomy on technical, economic, and social prerequisites that the bioeconomy itself cannot create [2]. Experts, therefore, increasingly demand the development of a comprehensive governance framework for the bioeconomy to ensure the emergence of sustainable bio-based transformations [3,4].

Previous research on this topic is mostly organized around case studies, which focus on the governance of selected segments of the bioeconomy in individual countries or in small samples of countries [5,6]. The detailed contribution by Pannicke et al. [7] on the governance of the German wood industry may serve as an example. However, a broader perspective that provides a comparative global overview about national bioeconomy politics is still missing.

Overall, we found 41 states worldwide that currently pursue explicit political strategies to expand and promote their bioeconomies. In this paper, we provide a systematic overview of 41 of these national bioeconomy strategies in existence at the time of this research. What types of bioeconomies are individual states striving for? Why does the development of a sustainable bioeconomy require an effective governance framework? Which political means are available to states to promote transformations towards sustainable bioeconomies, and how do individual states design their national bioeconomy strategies in order to meet this demand for a sustainable governance framework? In the following sections, we will address these research questions. In doing so, we aim to not only develop an overview of national bioeconomy policies, but also to develop an information tool that enables national and international policy makers to learn from other countries' bioeconomic strategies.

Our considerations rest on a comprehensive understanding of the bioeconomy. We distinguish between four bio-based transformation paths: (1) substitution of fossil fuels with bio-based raw materials; (2) productivity increase in bio-based primary sectors; (3) increasing efficiency in biomass utilization; and (4) value creation and addition through the application of biological principles and processes separate from large-scale biomass production.

Whether or not the bioeconomic development along these four pathways will have a positive impact on the achievement of SDGs is uncertain. One key challenge is that bio-based transformations may involve high conversion costs [8]. Path dependencies and economic incentive systems that stem from the fossil fuel era and pre-biotechnological production processes might hamper investments in a progressive bioeconomy. The question of how politics can support the rise of the bioeconomy through appropriate political means (enabling governance), therefore, presents the first key challenge for the development of a sustainable bioeconomy. In principle, states have a wide range of different mechanisms at their disposal to promote their bioeconomies. These mechanisms may include a bio-based research and development strategy, enhancing the competitiveness of bio-based products through subsidies, or implementing awareness-raising campaigns to increase societal participation in bio-based transformation including more responsible and sustainable consumption.

However, technical progress rarely offers only positive opportunities, but usually also leads to new risks. This is also the case for the bioeconomy. Scholars interested in studying the bioeconomy point to goal conflicts between SDGs that can result from bio-based transformations. Today, the discussion about conflicting goals goes far beyond the original "food versus fuel" debate in the field of bioenergy development and includes issues such as global equity concerns, water scarcity, land degradation and land use change. The identification and effective political management of conflicting goals, therefore, represents the second major challenge for the development of a sustainable governance framework for the bioeconomy. To address this, a number of different public and private governance tools exist that states can use to minimize tradeoffs and promote synergies in bio-based transformation processes (constraining governance).

However, how do individual states really react to these two fundamental governance challenges, and which means do they concretely employ to make their bioeconomies sustainable? Our results suggest the following: today a great number of states have set the goal of developing and expanding their bioeconomies. Further, states are willing to provide comprehensive political support to their bioeconomies to achieve this goal. Currently, states are highly active in addressing the first abovementioned governance challenge (enabling governance). On the other hand, our results show that the political management of conflicting goals has not yet reached the same level of attention. Only a minority of national bioeconomy strategies even mention the potentially negative consequences of bio-based transformations for sustainable development, and those states that are pursuing a more sustainable strategy mostly opt for soft political approaches to manage these conflicts. Overall, states address the second fundamental challenge of developing a sustainable bioeconomy (constraining governance) to a considerably lesser degree than the first challenge (enabling governance).

The paper consists of two sections: the first section lays out the conceptual foundations for our empirical study. We begin with a brief note on the concept of governance. Subsequently, we characterize the four different transformation paths along which bio-based transformations are likely to proceed. We then discuss the two key governance challenges for a sustainable bio-based transformation and present a set of key governance mechanisms that governments can use to support the development of a sustainable bioeconomy. Based on this theoretical framework, the second section presents our empirical analysis of a total of 41 national bioeconomy strategies. Here, we show which bio-based transformation path (or which combination of transformation paths) the states follow strategically, which of the governance mechanisms specified in the first section the states apply to promote their bioeconomies, which goal conflicts they identify, and how they attempt to regulate them. Finally, we summarize the results of the study and present perspectives for further research.

2. Concepts

2.1. A Short Note on the Concept of Governance

Governance can be understood as the process by which societies adapt their rules to new challenges [9]. Governance has a substantial dimension (what are the rules?), a procedural dimension (how are the rules developed?) and, finally, a structural dimension (the procedural rules and institutions that determine rule-making, how the rules are implemented and enforced, and how conflicts over rules are resolved). Societal adaption of rules to new challenges can be spontaneous and informal at the level of social relationships and networks. However, modern societies also delegate governance functions to specialized institutions, which set and enforce the rules in formally organized procedures. Such institutions first and foremost include the state at local, regional, and national level, but may also include inter- and supranational organizations, as well as private standard setters, which together build an interacting and overlapping governance system of plural authorities. In this sense, the UN Commission has defined the term governance as "[…] the sum of the many ways individuals and institutions, public and private, manage their common affairs. It is a continuing process through which conflicting or diverse interests may be accommodated and cooperative action may be taken. It includes formal institutions and regimes empowered to enforce compliance, as well as informal arrangements that people and institutions either have agreed to or perceive to be in their interest..." (p. 1) [10].

2.2. The Concept of Four Bio-Based Transformation Paths

The course and effects of bioeconomic transformation processes depend, among other aspects, on the development level, resources and political system of a given state (see Figure 1).

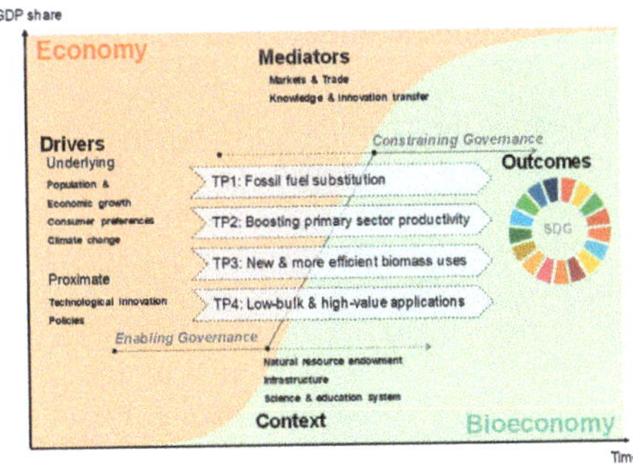

Figure 1. Conceptual diagram of transformative pathways in the bioeconomy (developed by the authors).

Transformation processes can be triggered by the interaction of driving forces, such as population growth and technological innovation, or by political or social action. Depending on the country context and its interaction with other economies, for example in the form of trade and knowledge transfer, bioeconomic transformation can proceed along one or more of the four paths depicted in Figure 1 with different possible effects.

Transformation Path 1 (TP1): In the past, this rather intensely researched TP has often been triggered by temporarily increased oil prices, subsidies, and environmental policies. For example, biofuel policies in the EU and US have led to increased demand for bioenergy, with direct and indirect effects on land use worldwide depending on land availability and the effectiveness of environmental and economic governance systems [11–13].

Transformation Path 2 (TP2): If technological innovation increases productivity in agriculture, forestry, or even fishing, it can release transformative forces that open up new production methods or locations. In the past, and globally, according to the so-called Borlaug hypothesis, this has repeatedly led to an easing in food markets despite increasing population growth [14]. However, regional and local boosts in agricultural productivity have also been shown to increase demand for land in ecological sensitive biomes, leading to losses in globally valued ecosystem services [11,15].

Transformation Path 3 (TP3): Innovation in downstream sectors often aims to increase the efficiency of biomass use and waste stream recycling. Such innovation can be associated with "rebound effects", i.e., increased demand due to improved provision. In the long term, however, the impact depends on supply dynamics, consumer behavior and the regulatory environment [16,17].

Transformation Path 4 (TP4): Biological principles and processes can be used largely independently of biomass streams' industrial applications, such as in the case of enzymatic synthesis and "biomimicry". Many countries with bioeconomic ambitions have high expectations for this knowledge and technology-intensive TP (see Section 2). Corresponding transformative processes result, inter alia, from providing cheaper and more environmentally friendly production methods or completely new products.

The above-mentioned transformation pathways can be driven by both production (supply) and consumption (demand) dynamics. We focus primarily on supply side dynamics in this paper. However, it is noteworthy that promoting sustainable consumption through regulations and incentive systems is one among many of the governance challenges of the sustainable bioeconomy.

3. Governing the Bioeconomy: Theoretical Framework

3.1. Governance to Promote Sustainable Bioeconomic Dynamics

The four paths of bio-based transformation presented in the last section offer opportunities as well as risks for a sustainable transformation of our existing economic and social systems. As shown above, one of the major opportunities of a comprehensive bio-based transformation is the possibility of promoting sustainable growth across economic sectors. However, a sustainable bio-based transformation cannot be taken for granted.

Current literature on bioeconomy repeatedly emphasizes the great potential of the bioeconomy for sustainable developments towards SDG achievement, but simultaneously points out that the realization of these potentials is facing considerable hurdles. Some researchers argue that the path dependence of economic and political development is the root cause of the problem [18]. This means that previous decisions in politics, economics, and society—taken before the bio-based transformation paradigm emerged—have shaped the economic system in a way that today hampers the development of a bio-based economy even though it may bring about significant sustainability gains.

First, problems of path dependencies may arise from a lack of adaptation of existing institutional frameworks to the specific needs of the bioeconomy. Indeed, the political and legal institutions (such as intellectual property rights, consumer protection, environmental rights), which govern our current economic systems, have developed over long periods, during which the technological possibilities of the current bioeconomy were unknown. Given this, the chances are high that existing institutions are poorly aligned to the institutional demands of a rapidly developing and innovative bioeconomy. Institutional path dependencies might thus lead to a situation in which the bioeconomy faces high regulatory and transaction costs, which, in turn, may prevent the transformative dynamics of the bioeconomy from unfolding.

Further, problems of path dependency occur at the level of industrial organization and production. Many existing value chains are specialized in an efficient use of fossil-based resources and pre-biotechnological production processes. The same applies to existing infrastructure (transport systems), on which these economic activities are based. Naturally, this leads to lock-in effects [19,20]. Even if bio-based transformations promise long-term sustainability gains for both individual companies and society as a whole, companies currently avoid incurring the costs of changing their organizational structures and methods of production towards bio-based processes, since under the given conditions such changes would still compromise their competiveness. To conclude, it seems that current economic systems that have been shaped through the utilization of fossil-based resources and pre-bioeconomy production techniques are not yet able to provide the necessary incentives to leverage comprehensive bio-based transformations.

Both points have in common that they conceptualize path dependency problems as problems of economic incentives that ill-inform individual economic decisions. From these rational choice-based approaches, a structural approach can be distinguished. From a sociological perspective, both our identity and knowledge about the world is defined by culture, social norms, and ideology and, ultimately, these social structures also determine our economic conduct [21].

Obviously, normative and cognitive structures that incrementally manifest in a given society are even harder to change than economic incentives. At the level of social structures, path-dependency problems limiting bioeconomic dynamics may, therefore, be even stronger than at the level of economic institutions, organizations and production techniques. Misinformation, including limited knowledge, about the properties of bio-based products or a conceptual reduction of the bioeconomy to risk technologies can undermine consumer confidence (a phenomenon well known from the debate around genetically modified organisms). The bioeconomy has an influence on almost all areas of social life. It changes what we eat, how we live, how we move, how we dress, and much more. Consumption patterns in all these areas are deeply rooted in the cultural habits of societies and, therefore, extremely difficult to change [8].

In conclusion, it can be said that not only the economic institutions, organizations, and production techniques that evolved in the era of fossil resource utilization but also the societal structures that developed during this period may hamper the emergence of a dynamic bioeconomy. Against this background, it is not surprising that scholars interested in bioeconomy research currently regard the creation of an appropriate governance framework that is capable of overcoming the various path-dependency problems as one of the most pressing political challenges in the development of a sustainable bioeconomy.

However, which specific governance mechanisms can governments use to address this challenge? One governance tool, often discussed in this context, presents the implementation of a comprehensive research and development strategy to promote investments in technological innovations whose costs and risks private actors are not willing to incur under the given conditions. [5] Further, political support measures can aim at increasing the competitiveness of bio-based products through subsidies, thereby creating markets for the bio-economy that do not independently develop in the economy [22]. Industrial location policies may have similar effects [23]. Political support measures such as the creation of favorable legal frameworks, state-supported training of the labor force or the promotion of industry clusters are all intended to make it more attractive for companies to invest in the bioeconomy. This form of political support for the bioeconomy also includes measures for strategic international research collaborations and foreign direct investment. Finally, states can promote bio-based transformation at a societal level through deliberate political campaigns to increase the legitimacy and acceptance of the bioeconomy [8].

Table 1 provides an overview of such governance mechanisms that states can use to promote bio-based transformative processes. In the following empirical section of this paper, this serves as a typology for the policy instruments that states actually intend to use to promote their respective bioeconomies.

Table 1. Overview of the means for enabling governance.

(I)   Promoting research and development for a bio-based transformation

- Funding of research projects
- Establishment of specific research facilities
- Promotion of research networks and strategic partnerships
- Promotion of knowledge and technology transfer (science-praxis-nexus)

(II)   Improving the competitiveness of the bioeconomy through subsidies

- Quotas for the bioeconomy
- Promotion of bio-based public procurement
- Promotion of sustainable consumption behavior
- Tax benefits
- Specific credit programs

(III)   Industrial location policies for bio-based industries

- Promotion of industry clusters in the field of bioeconomy
- Promotion of knowledge and technology transfer between research and industry
- Promotion of labor education in the field
- Creation of appropriate intellectual property rights
- Promotion of foreign direct investment (FDI) in the field

(IV)   Political support for bio-based social change

- Promote public dialogues to increase understanding of the functioning of the bioeconomy
- Promote public dialogues on technological risks in the field of bio-economics

3.2. Governance of Risks and Goal Conflicts

The creation of a favorable political framework within which the bioeconomy can thrive presents one major governance challenge. However, political support measures alone will not suffice to ensure the development of a sustainable bioeconomy. The problem is that, as much as the bioeconomy can contribute to the achievement of a range of different SDGs, it can also undermine the achievement of SDGs [24,25]. An effective political regulation of these conflicting objectives presents the second major challenge for a sustainable governance of the bioeconomy.

The concept of bioeconomy rests on the idea of applying biological principles and processes in all sectors of the economy and to increasingly replace fossil-based raw materials in the economy with biogenic resources. However, the question whether or not bioeconomic transformations will either lead to more sustainability or produce new sustainability risks remains debated. The following table (Table 2) provides an overview of some common aspects of this debate.

Table 2. Possible opportunities and risks of bioeconomic transformation.

Sustainability Dimension (SDG)	Opportunities	Risks
Food security (SDG 2)	Increase via higher yields and new production methods	Reduction due to food price increases
Poverty/inequality (SDG 1, 10)	Reduce via transfer of technology and leapfrogging	Increase via exclusion from technical progress
Natural resources (SDG 7, 14, 15)	Conserve by improving production methods	Degrade/loss through inefficient production and overuse
Health (SDG 3)	Improve through new and refined forms of therapy	Risk/damage through improper use of risky technologies
Climate Change (SDG 13)	Mitigate through emissions reductions	Exacerbate through direct and indirect land use change

Sources: [26–28].

Both the above-mentioned optimistic and critical views on the impact of bioeconomic transformation on SDGs achievements (Table 2) depend strongly on assumptions about how and in which contexts new bio-based technologies and principles will be used. We illustrate this point in the following examples.

Example 1: The EU promotes biofuels with the aim of reducing emissions (SDG 13). This can lead to a global loss of tropical forests through direct and indirect land use change, but also to the spread of environmentally hazardous and health-threatening production methods (which conflicts with SDGs 3, 14, 15). Both technological innovation (e.g., improving production of biomass at marginal sites with higher yields) and governance mechanisms (e.g., implementing existing legislation to prevent illegal deforestation or misuse of agrochemicals or incentive systems for sustainable production) can help alleviate this conflict.

Example 2: Developed countries promote bio-based applications in chemical or pharmaceutical sectors (SDG 3). Due to restrictive patent rights and often lengthy and costly licensing procedures, the associated benefits accrue only to the affluent segment of the world's population. This might create a conflict with SDG 10. This conflict could be mitigated by innovation transfer, more efficient administrative structures and a more inclusive patent system.

These two examples show how narratives of the bioeconomy that highlight the potentially associated risks often assume that regulations constraining the bioeconomy are ineffective, or that existing technologies and processes that might be able to increase the efficiency of the bioeconomy remain inaccessible. On the other hand, perspectives that highlight the opportunities inherent in bioeconomic developments assume that efficient biotechnologies will evolve and diffuse and that appropriate governance frameworks can be set up to regulate the remaining potentially negative effects of the bioeconomy.

The political support measures that enable the evolution and diffusion of efficient biotechnologies have been discussed above (enabling governance). In the following, we focus on the question of what states can do to constrain economic activities related to the bioeconomy where necessary (constraining governance). Looking into this issue of regulating the bioeconomy, it strikes us that various governments and non-government actors have already developed a variety of rules to govern bioeconomic activities in different areas of the bioeconomy. For example, multi-stakeholder initiatives such as the Global Bioenergy Partnership or the United Nations Voluntary Guidelines on the Responsible Governance of Tenure, Land, Fisheries, and Forests in the Context of National Food Security both aim to ensure the priority of the right to food in the bioeconomy to prevent land grabbing. Other examples include the International Draft Standard DIN EN ISO 14046: 2015-11, which sets out guidelines for determining the water footprint of products based on a life cycle assessment, or the United Nations Convention on Biological Diversity, which aims to connect the bioeconomy to conservation initiatives.

Given this relatively well-developed normative basis, the central challenges in developing an effective regulatory framework for the bioeconomy clearly emerge in the later stages of the governance cycle, i.e., in the implementation and enforcement of the existing rules [29]. The adoption of regulations into state legislation is one possibility, but it presupposes the existence of functioning state enforcement mechanisms, which do not exist in many emerging and developing countries. In addition, state regulations operate only within the territory of a state, but they have no reach to regulate cross-border economic processes, and they have less influence again on global economic dynamics, both of which are becoming increasingly important in the global bioeconomy. An expansion of international law might provide a solution, but is itself subject to major compliance problems due to the absence of an authority beyond the individual states that could enforce compliance with international law [30]. Of course, states can refrain from a pure legal enforcement logic and create positive incentives to regulate a global bioeconomy (e.g., payment for ecosystem services [31]), and support softer instruments, such as private standards and certification systems along global value chains [32].

Ultimately, an effective form of regulation for the bioeconomy can only be created through the use of a combination of different public and private mechanisms. We summarize the individual regulatory approaches that states may support to achieve this goal in Table 3 below.

Table 3. Overview of regulatory mechanisms.

(I)	State regulation of the bioeconomy
(II)	Governmental development of positive incentives (e.g., payments for environmental services)
(III)	Government support for private standards and certifications
(IV)	International cooperation (through international organizations and regimes)

4. Methods

We conducted a qualitative document analysis [33] of national bioeconomy strategy documents using ATLAS.TI software. We provide an overview of the countries and documents analyzed in Appendix A at the end of this article. The tables above (Tables 1–3, and Figure 1) served as a codebook guiding the systematic coding of the strategy documents. We have used Table 1 as providing the themes to analyze the enabling governance means for achieving national development goals as well as contributing to addressing selected global sustainability goals contained in Table 2. Table 3 serves as a heuristic conceptual overview of possible regulatory mechanisms grouped into four (I–IV) dimensions. The methods used draw mainly upon techniques of qualitative content analysis [34]. The analytic procedure entailed selecting and appraising passages contained within the policy documents with regard to the themes of the codebook and connecting them to other lines, quotations about political means chosen to address a certain issue. This, for example, is related to the finding of anticipated negative impacts of implementing the bioeconomy policy on land and water resources and the governance means chosen to address them. Such document analysis yielded data in the form of excerpts, quotations, or entire passages chosen according to the major themes and categories from the codebook [35].

5. Results and Discussion

Having laid out our preferred indicators to distinguish and classify national strategies, in this section we now discuss our findings from the empirical analysis of national bioeconomy strategies. Specifically, our empirical analysis of 41 different national bioeconomy strategies aims to contribute to answering the following three questions:

1. Type of bioeconomy: Which of the four bio-based transformation pathways or combinations of transformation paths are individual countries pursuing in their strategies?

2. Enabling governance: Which means of governance do countries employ in their political strategies to overcome problems of path dependencies in the development of a sustainable bioeconomy?

3.   Constraining governance: Which goal conflicts in the development of a sustainable bioeconomy have the individual countries identified in their strategies, and which political means have the individual strategies used to regulate these goal conflicts and reduce resulting risks?

5.1. Types of Bioeconomy

Practically all countries with explicit bioeconomy strategies aim to foster transformation processes along at least two of the pathways outlined in Figure 2. In countries that explicitly envision only two transformation pathways, particular emphasis is often placed on the efficient provision of biomass for TP1, both domestically and for trading partners, as in the case of Brazil.

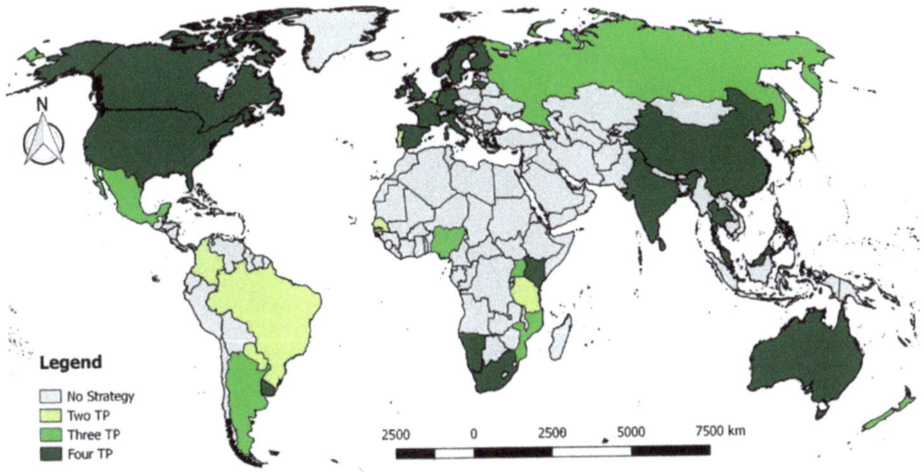

Figure 2. Transformative pathways by country.

By contrast, the majority of industrial nations, as well as some emerging economies, envisage or currently implement more diversified strategies along all four TPs. In the majority of cases, the selection of and focus on individual TPs in the examined strategies reflects three aspects: the respective resource availability of the countries (e.g., availability or scarcity of agricultural area); historically developed pioneering roles in special technology and research areas (e.g., biotechnology); or country-specific development deficits to be overcome. For example, the German bioeconomy strategy specifically focuses on applications in the field of recycling waste streams and the more efficient or cascading use of biomass (TP2). In turn, China's bioeconomy strategy relies strongly on bio-based substitution of fuels and materials (TP1).

5.2. Strategies to Enable the Bioeconomy

How do the individual states intend to promote their bioeconomies politically, and what concrete political means do they use to do so? In this context, Figure 3 below shows the intentions of the individual states to provide political support to their bioeconomies. In Table 2 of our conceptual framework, we distinguished between four political support measures that states can draw upon in promoting their bioeconomies. Our analysis of these national strategies is based on those categories, and reveals that the individual states are indeed intensively using all these means to strategically promote the development of their bioeconomies.

It becomes clear that almost all states with an explicit bioeconomy strategy rely on at least three of the political support measures identified, and the majority of states even deploy all four measures mentioned above. In other words, they pursue a targeted research and development strategy for bio-based transformation and want to improve the competitiveness of their bioeconomy through subsidies. In addition, many countries pursue active industry location policies aimed at improving the overall conditions for bio-based industries, and plan to improve the acceptance of the bioeconomy through education and other capacity-building and awareness-raising campaigns. Thus far, we can state that many countries with bioeconomic ambitions declare comprehensive bioeconomies as a strategic political goal (see Figure 2) and are prepared to intensively promote this development politically (see Figure 3). Overall, this suggests that the bio-based transformation may gain momentum in the coming years.

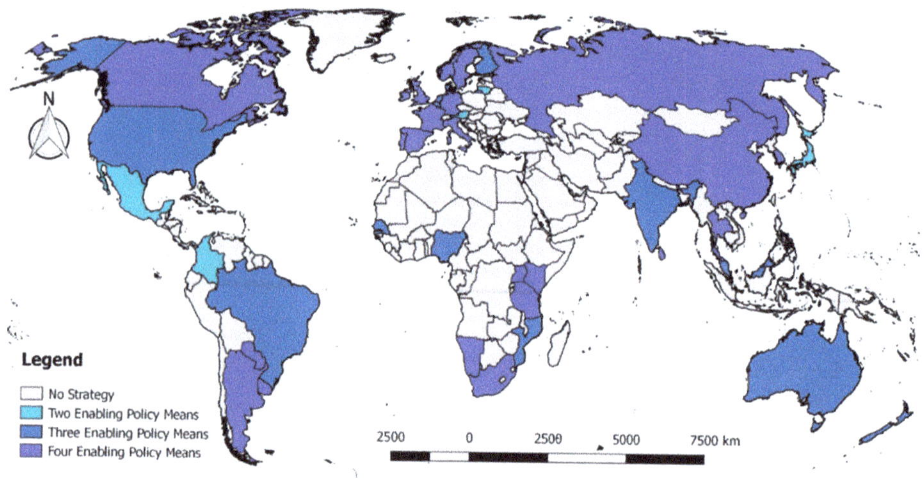

Figure 3. Enabling policy means in national bioeconomy strategies.

5.3. How Do States Regulate Their Bioeconomies?

The complex task of creating expedient regulatory measures for managing conflicting interests throughout the development of a bioeconomy is the second governance challenge. Figure 4 shows the extent to which national bioeconomy strategies give political answers to the risks and potentially related goal conflicts mentioned in Table 2 above.

Most national strategies pay little or no attention to risks and goal conflicts (26 out of 41 states). This includes countries with potentially large bioeconomies, such as the USA, Russia, Brazil, and Argentina. In contrast, China and a few African states explicitly recognize the need to manage risks as a crucial political challenge in shaping a sustainable bioeconomy. Overall, European states show the highest political sensitivity to potential risks and goal conflicts.

Table 4 compares the identification of conflicting goals in national strategies. It shows that states are particularly concerned with negative impacts of the bioeconomy on land and water resources, as well as on global food security. This reflects the discourses about the sustainability risks associated with the first generation of biofuels. Other negative effects potentially associated with the bioeconomy, such as inequality and poverty, climate, or health risks, have only played a minor role in national strategies so far.

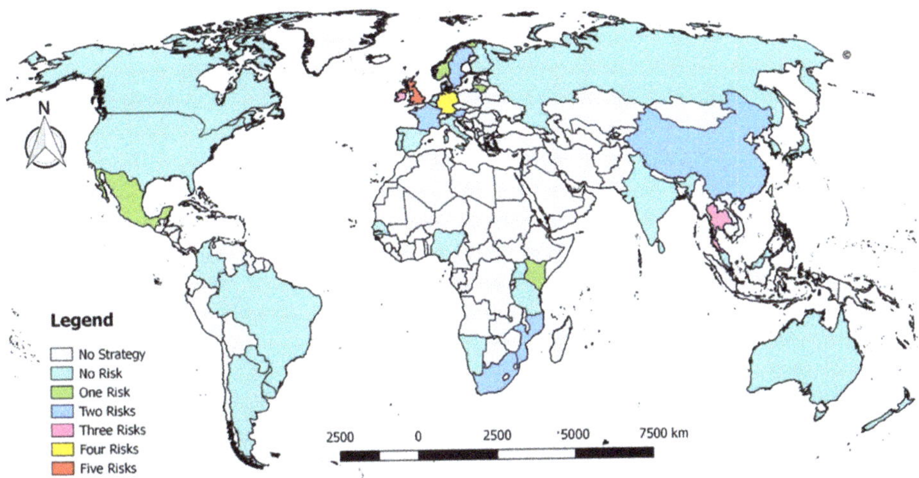

Figure 4. Anticipated risks in national strategy documents of 41 countries.

Table 4. Overview of conflicting goals and associated risks identified in national bioeconomy strategies.

Country	Nutrition	Poverty/Inequality	Nat. Res. (Air)	Nat. Res. (Forests)	Nat. Res. (Land)	Nat. Res. (Water)	Health	Climate
Austria	●				●			
Denmark	●				●			
France	●			●	●			
Germany	●	●		●	●			●
Ireland	●				●		●	
Kenya				●	●	●		
Lithuania					●			
Mexico				●	●	●		
Mozambique	●				●			
Norway					●	●		
South Africa	●		●		●	●		
Sweden	●				●			
Thailand	●			●	●			●
United Kingdom	●	●	●	●	●	●	●	●
China	●				●	●		
Total	12	2	2	6	15	7	2	3

Our content analysis also shows (see Table 5) that states rely heavily on soft regulatory means, such as self-regulation of global value chains through private standards and certification regimes, to manage bioeconomy-related risks. In addition, most states advocating more comprehensive regulation to avoid conflicting goals (as in the case of Germany) aim to intensify international cooperation in this field. Despite this, the need to react to bioeconomic conflicts of interest by means of concrete legislative amendments was not a central focus of the national bioeconomy strategies examined. Our analysis also does not reveal a broad willingness of countries with bioeconomy strategies to safeguard the protection of natural resources through the development of positive incentives, such as the widely discussed instrument of payments for ecosystem services [34].

Table 5. Overview of regulatory mechanisms by country.

Country	State Regulation	Creation of Positive Incentives by Governments	Private Standards and Certifications	International Cooperation	Total
Austria			●		1
Denmark			●		1
European Union			●		1
France	●	●	●	●	4
Germany	●	●	●	●	4
Ireland	●	●	●	●	4
Kenya			●		1
Lithuania	●		●	●	3
Mexico					
Mozambique			●	●	2
Norway			●		1
South Africa	●	●		●	3
Sweden			●	●	2
Thailand			●	●	2
United Kingdom	●	●	●	●	4
China	●	●	●	●	4
Total	8	6	14	10	

5.4. Regional Developments

The last sections have provided a global overview of national bioeconomy strategies. In the following, we complement this view by a short regional assessment. In doing so, it becomes clear from the various figures and maps presented above, that European states have developed the most advanced sustainable bioeconomy strategies, notably the UK and Germany. These results reflect the role of the European Union as an active partner in promoting bioeconomic transformations. It strikes us that most Eastern European Countries are, so far, absent from these developments. Despite the fact that compared to other regions European countries have developed the most advanced bioeconomy strategies, in Europe a substantial governance gap still exists between promoting and regulating the bioeconomy.

The Western Hemisphere presents a further world region in which most individual states are currently advancing comprehensive bioeconomy strategies. Different from the European bioeconomy strategies, which have at least partly integrated some measures to regulate the bioeconomy, regulatory aspects that deal with potential sustainability risks associated with the rise of the bioeconomy are almost completely absent in the strategies drafted by countries located in the Western Hemisphere. The gap between promoting and regulating the bioeconomy is, therefore, even greater here than in Europe. Overall, our results make clear that both North and South American countries are currently undertaking significant efforts to enhance their bioeconomic sectors.

Again, a different picture emerges in Asia and Australia. In this region, we find many states—especially major states such as China, India, Russia, and Australia—that have adopted advanced bioeconomy strategies. However, we also find a significant number of states without explicit bioeconomy strategies. Different from the states located in the Western Hemisphere, among the Asian states at least two states (China and Thailand) pay some attention to the sustainability risks associated with a rise of the bioeconomy.

In Africa, we find the smallest share of countries with bioeconomy strategies. Nevertheless, the countries located in the southern parts of Africa show with their strategies that they see very large potential in the bioeconomy to foster their economic developments in a sustainable way. Among these countries, South Africa and Mozambique stand out in having developed the most advanced bioeconomy strategies. They also include some regulatory aspects. Overall, there is still very large potential for African states to develop more explicit bioeconomy strategies.

6. Concluding Remarks

Summarizing the results of our analysis, it is evident that many countries seek to develop and expand their bioeconomies. In order to achieve this, states are willing to support their bioeconomies through comprehensive political means. It is also clear that countries around the world have embraced the first major governance challenge of enabling bio-based transformation. However, the second challenge of deploying political means to address the potential risks and goal conflicts of bio-based transformation does not appear to be wholeheartedly addressed. Only a minority of states even mentioned the potentially negative implications of bio-based transformation for sustainable development. Those states pursuing comprehensive strategies rely largely on soft political means of risk mitigation and conflict management.

The notion of governance includes the process of how societies adapt their rules to new challenges [9]. In this article, we explored the question of how nation-states globally aim to adapt their rule systems to the governance challenges associated with an emerging bioeconomy. This raises further questions: why are the respective national strategies different? How effectively do individual states implement their strategies? What are the real impacts on SDG achievement that follow when states implement their bioeconomy strategies? In conclusion, it can be said that national governments widely regard the development of a modern bioeconomy as a central strategy to promote their economies and to ensure sustainable development worldwide. However, to achieve these goals, national bioeconomies need an effective and globally coordinated governance framework. Future research should contribute to identifying key ingredients of such a framework and support their effective implementation, for example by documenting implementation processes and outcomes in all relevant sustainability dimensions.

A prerequisite for creating effective governance arrangements is the development of comprehensive approaches for measuring and assessing the bioeconomy [36]. Inadequate monitoring and a lack of impact assessment could otherwise lead to over- or under-regulation of the bioeconomy. The risks associated with the business-as-usual scenario of a fossil-fuel-based future global economy must be confronted with the bioeconomy-specific risks in order to comprehensively assess risks and conflicting goals [35]. This exceeds the scope of this chapter, but we strongly emphasize the need to investigate these issues in future research.

Author Contributions: All authors substantially contributed to the conception, writing, and revisions of the paper. Thomas Dietz is the lead author.

Funding: We would like to thank the German Ministry of Education and Research (BMBF) and the Ministry for Culture and Science of the State of North Rhine-Westphalia for their funding and support. We acknowledge support by Open Access Publication Fund of University of Muenster.

Acknowledgments: We are grateful to Hugo Rosa and Maximilian Meyer for their research assistance to the STRIVE project at the Centre for Development Research (ZEF).

Conflicts of Interest: The authors declare no conflict of interest.

Appendix A

Table A1. Overview of national bioeconomy strategies.

Country	Title	Author
Austria	FTI-strategy for a bio-based industry in Austria	Federal Ministry for Traffic, Innovations and Technology
	Bioeconomy—Position Paper	Austrian Association for Agriculture, Life and Environmental Sciences with BIOS Science Austria
Belgium	Bioeconomy in Flanders—The vision and strategy of the Government of Flanders for a sustainable and competitive bioeconomy in 2030	Flemish government
France	The new face of industry in France	Ministry for Economic Regeneration
	Les usages non alimentaires de la biomasse	Interministerial
	A Bioeconomy Strategy for France—Goals, Issues and Forward Vision	French Republic
Germany	National Policy Strategy on Bioeconomy	Federal Ministry of Food and Agriculture
	Bioeconomy—Baden Württenberg's path towards a sustainable future	Federal state of Baden-Württenberg, with Federal Association BIOPRO
	National research strategy bioeconomy 2030	Federal Ministry of Education and Research
Ireland	Harnessing Our Ocean Wealth	Ministry for Agriculture, Food and the Marine
	Delivering our Green Potential—Government Policy Statement on Growth and Employment in the Green Economy	Government of Ireland
	Towards 2030—Teagasc's Role in Transforming Ireland's Agri-Food Sector and the Wider Bioeconomy	Teagasc—The Agriculture and Food Development Authority (Intersectoral)
Italy	BIT—Bioeconomy in Italy: A Unique Opportunity to Reconnect the Economy, Society and the Environment	Government of Italy
Lithuania	National Renewable Energy Action Plan	Lithuanian Government
Netherlands	Green Deals Overview	Ministry of Economic Affairs
	2012 Bioenergy Status Document	Ministry of Economic Affairs
Portugal	Estrategía Nacional para o Mar (2013–2020)	Government of Portugal

Table A1. *Cont.*

Country	Title	Author
Russia	State Coordination Program for the Development of Biotechnology in the Russian Federation until 2020 "BIO 2020" (Summary)	Government of the Russian Federation
Spain	The Spanish Bioeconomy Strategy—2030 Horizon	Ministry of Economy and Competitiveness
Denmark	Growth Plan for Water, Bio and Environmental Solutions	The Danish Government
Denmark	The Copenhagen Declaration for a Bioeconomy in Action March 2012	The Danish Council for Strategic Research
Finland	The Finnish Bioeconomy Strategy	Interministerial document
Norway	Research Programme on Sustainable Innovation in Food and Bio-based Industries	The Research Council of Norway
Norway	National strategy for biotechnology	Ministry of Education and Research
Norway	Marine Bioprospecting—a source of new and sustainable wealth growth	Interministerial document
Norway	Familiar resources—undreamt of possibilities—The Government's Bioeconomy Strategy (English Summary)	Interministerial document
Sweden	Swedish Research and Innovation Strategy for a Bio-based Economy	The Swedish Research Council for Environment, Agricultural Sciences and Spatial Planning (commissioned by the Swedish Government)
Great Britain	A UK Strategy for Agricultural Technologies	Interministerial document
Great Britain	UK Bioenergy Strategy	Interministerial document
Great Britain	UK Cross-Government Food Research and Innovation Strategy	Interministerial document
Kenya	A National Biotechnology Development Policy	Republic of Kenya
Kenya	Strategy for developing the Bio-Diesel Industry in Kenya (2008–2012)	Ministry of Energy (Renewable Energy Dept.)
Mozambique	Politica e Estrategia de Biocombustiveis	Council of Ministers
Namibia	National Programme on Research, Science, Technology and Innovation	National Commission on Research, Science and Technology (government)
Nigeria	Official Gazette of the Nigerian Bio-fuel Policy and Incentives	Federal Republic of Nigeria

Table A1. *Cont.*

Country	Title	Author
Senegal	Lettre de Politique de Développement du Secteur de L'Energie	Interministerial document
	Biofuels in Senegal—The Jathropha program	Enda Energy, Environment, Development Programme (NGO) (sourced from Ministry of Agriculture)
South Africa	The Bio-Economy Strategy	Department of Science and Technology
	A National Biotechnology Strategy for South Africa	Unspecified
	Public Perceptions of Biotechnology in South Africa	HSRC, Human Sciences Research Council (TIA, Technology Innovation Agency)
Tanzania	National Biotechnology Policy	Ministry of Communication, Science and Technology
Uganda	Biomass Energy Strategy (BEST) Uganda	Ministry of Energy and Mineral Development (support UNDP)
	National Biotechnology and Biosafety Policy	Ministry of Finance, Planning and Economic Development
	The Renewable Energy Policy For Uganda	Ministry of Energy and Mineral Development
Canada	Growing Forward 2 In Newfoundland and Labrador	Government of Newfoundland and Labrador
	British Columbia Bio-Economy	Minister of Jobs, Tourism and Innovation
Mexico	Estrategia Intersecretarial de los Bioenergéticos	Interministerial document
USA	Farm Bill	Congressional Research Service
	Strategic Plan for a Thriving And Sustainable Bioeconomy	Bioenergy Technologies Office—U.S. Department of Energy
	National Bioeconomy Blueprint	The White House
Argentina	Biotecnología argentina al año 2030: Llave estratégica para un modelo de desarrollo tecno-productivo	Ministry of Science, Technology and Productive Innovation
Brazil	Plano Decenal de Expansão de Energia 2023	Ministry of Mines and Energy
	Política de Proteção de Desenvolvimento da Tecnologia	Brazilian Government
Colombia	Politica para el Desarrollo Commercial de la Biotecnología a partir del Uso Sostenible de la Biodiversidad	Council for Economic and Social Policy (Interministerial)
Paraguay	Politica y Programa Nacional de Biotecnología Agroprecuaria y Forestal del Parauay	Agriculture Ministry
Uruguay	Plan Sectorial de Biotecnología 2011–2020	Interministerial document

Table A1. *Cont.*

Country	Title	Author
China	12th Five-year Plan (2011–2015) on Agricultural Science and Technology Development	Ministry of Agriculture
	National Modern Agriculture Development Plan	Ministry of Agriculture
	13th Five-Year Plan for Environmental Protection	State Council of the People's Republic of China
	13th Five-Year Plan For economic and social development of the People's Republic of China (2016–2020)	Central Committee of the Communist Party of China
	13th Five-Year Plan for the Environmental Health Work of National Environmental Protection	Ministry of Environmental Protection
	The National Medium- and Long-Term Program for Science and Technology Development (2006–2020)	National Development and Reform Commission
	13th Five-Year Plan for Energy Saving and Emission Reduction	General Office of the State Council
	13th Five-Year Plan for Bioindustry Development.	State Council of the People's Republic of China
	Policies to Promote Quick Development of Biological Industry. 2009	State Council of the People's Republic of China
	13th Five-year Plan for National Strategic Emerging Industries	State Council of the People's Republic of China
	13th Five Year Plan of Renewable Energy Development	State Council of the People's Republic of China
India	National Biotechnology Development Strategy 2015–2020	Ministry of Science & Technology
	The Bioenergy Roadmap (2012)	Ministry of Science & Technology
Japan	The 3rd Fundamental Plan for Establishing a Sound Material-Cycle Society 2013	Ministry of the Environment
Malaysia	National Biomass Strategy 2020: New wealth creation for Malaysia's biomass industry Version 2.0	National Innovation Agency of Malaysia
	Bioeconomy Transformation Programme	Ministry of Science, Technology and Innovation (Commissioner)
	Biotechnology for Wealth Creation and Social Wellbeing	Ministry of Science, Technology and Innovation
South Korea	Biotechnology in Korea (2013)	Ministry of Science, ICT and Future Planning (Commissioner)
	Status of Biotechnology in Korea	Biotech Policy Research Center
	Vision 2015: Korea's Long-term Plan for S&T Development	Ministry of Science and Technology
	Biovision 2016—For Building a Healthy Life and a Prosperous Bioeconomy	Ministry of Science and Technology

Table A1. *Cont.*

Country	Title	Author
Sri Lanka	National Biotechnology Policy	Ministry of Science and Technology
Thailand	Thailand's National Biotechnology Policy Framework (2012–2021)	Ministry of Science and Technology
	Alternative Energies Development Plan 2012–2021	Ministry of Energy
	National Roadmap for the Development of Bioplastics Industry (2008–2012)	Ministry of Science and Technology
Australia	National Collaborative Research Infrastructure Strategy	Department of Industry, Innovation, Climate Change, Science, Research and Tertiary Education
	Opportunities for Primary Industries in the Bioenergy Sector—National Research, Development and Extension Strategy	Rural Industries Research and Development Corporation (Semi-Government agency)
	2011 Strategic Roadmap for Australian Research Infrastructure	Department of Industry, Innovation, Climate Change, Science, Research and Tertiary Education
New Zealand	2014 Sector Investment Plan—Biological Industries Research Fund	Ministry of Business, Innovation and Employment
	The Business Growth Agenda	Ministry of Business, Innovation and Employment

References

1. De Besi, M.; McCormick, K. Towards a Bioeconomy in Europe: National, Regional and Industrial Strategies. *Sustainability* **2015**, *7*, 10461–10478. [CrossRef]
2. Pfau, S.; Hagens, J.; Dankbaar, B.; Smits, A. Visions of Sustainability in Bioeconomy Research. *Sustainability* **2014**, *6*, 1222–1249. [CrossRef]
3. Von Braun, J.; Birner, R. Designing Global Governance for Agricultural Development and Food and Nutrition Security. *Rev. Dev. Econ.* **2016**, *21*. [CrossRef]
4. El-Chickakli, B.; von Braun, J.; Barben, D.; Philp, J. Policy: Five cornerstones of a global bioeconomy. *Nature* **2017**, *535*, 221–223. [CrossRef] [PubMed]
5. Bosman, R.; Rotmans, J. Transition Governance towards a Bioeconomy: A Comparison of Finland and The Netherlands. *Sustainability* **2016**, *8*, 1017. [CrossRef]
6. Purkus, A.; Röder, M.; Gawel, E.; Thrän, D.; Thornley, P. Handling Uncertainty in Bioenergy Policy Design—A Case Study Analysis of UK and German Bioelectricity Policy Instruments. *Biomass Bioenergy* **2015**, *79*, 54–79. [CrossRef]
7. Pannicke, N.; Gawel, E.; Hagemann, N.; Purkus, A.; Strunz, S. The political economy of fostering a wood-based bioeconomy in Germany. *Ger. J. Agric. Econ.* **2015**, *64*, 224–243.
8. Bröring, S.; Baum, C.M.; Butkowski, O.; Kircher, M. Kriterien für den Erfolg der Bioökonomie. In *Bioökonomie für Einsteiger*; Pietszch, J., Ed.; Springer Spektrum: Wiesbaden, Germany, 2017; pp. 161–177. ISBN 9783662537626.
9. Stone-Sweet, A. Judicialization and the construction of governance. *Comp. Polit. Stud.* **1999**, *32*, 147–184. [CrossRef]

10. Commission on Global Governance. A New World. In *Our Global Neighborhood*; Oxford University Press: Oxford, UK, 1995; Chapter 1. Available online: http://www.gdrc.org/u-gov/global-neighbourhood/chap1.htm (accessed on 13 June 2018).

11. Ceddia, M.G.; Bardsley, N.O.; Gomez-y-Paloma, S.; Sedlacek, S. Governance, agricultural intensification, and land sparing in tropical South America. *Proc. Natl. Acad. Sci. USA* **2014**, *110*, 7242–7247. [CrossRef] [PubMed]

12. Ceddia, M.G.; Sedlacek, S.; Bardsley, N.O.; Gomez-y-Paloma, S. Sustainable agricultural intensification or Jevons paradox? The role of public governance in tropical South America. *Glob. Environ. Chang.* **2013**, *23*, 1052–1063. [CrossRef]

13. Searchinger, T.; Edwards, R.; Mulligan, D.; Heimlich, R.; Plevin, R. Do biofuel policies seek to cut emissions by cutting food? *Science* **2015**, *347*, 1420–1422. [CrossRef] [PubMed]

14. Lobell, D.B.; Baldos, U.L.C.; Hertel, T.W. Climate adaptation as mitigation. The case of agricultural investments. *Environ. Res. Lett.* **2013**, *8*, 1–12. [CrossRef]

15. Angelsen, A.; Kaimowitz, D. *Agricultural Technologies and Tropical Deforestation*; CABI Publishing in Association with Center for International Forestry Research (CIFOR): New York, NY, USA, 2001; ISBN 0851994512.

16. Herring, H.; Roy, R. Technological innovation, energy efficient design and the rebound effect. *Technovation* **2007**, *27*, 194–203. [CrossRef]

17. Smeets, E.; Tabeau, A.; van Berkum, S.; Moorad, J.; van Meijl, H.; Woltjer, G. The impact of the rebound effect of the use of first generation biofuels in the EU on greenhouse gas emissions: A critical review. *Renew. Sustain. Energy Rev.* **2013**, *38*, 393–403. [CrossRef]

18. Gawel, E.; Purkus, A.; Pannicke, N.; Hagemann, N. Die Governance der Bioökonomie–Herausforderungen einer Nachhaltigkeitstransformation am Beispiel der holzbasierten Bioökonomie in Deutschland. Available online: http://nbn-resolving.de/urn:nbn:de:0168-ssoar-47319-9 (accessed on 13 June 2018).

19. Unruh, G.C. Escaping Carbon Lock-in. *Energy Policy* **2002**, *30*, 317–325. [CrossRef]

20. Unruh, G.C. Understanding Carbon Lock-in. *Energy Policy* **2000**, *28*, 817–830. [CrossRef]

21. Finnemore, M. *National Interests in International Society*; Cornell University Press: Ithaca, NY, USA, 1996.

22. Dabbert, S.; Lewandowski, I.; Weiss, J.; Pyka, A. *Knowledge-Driven Developments in the Bioeconomy: Technological and Economic Perspectives*; Springer: Berlin, Germany, 2017. [CrossRef]

23. Cooke, P. *Growth Cultures: The Global Bioeconomy and Its Bioregions*; Routledge: London, UK, 2007; ISBN 978-0-415-39223-5.

24. Kleinschmit, D.; Arts, B.; Giurca, A.; Mustalahti, I.; Sergent, A.; Pülzl, H. Environmental concerns in political bioeconomy discourses. *Int. For. Rev.* **2017**, *19*, 41–55. [CrossRef]

25. Fritsche, U.; Rösch, C. Die Bedingungen einer nachhaltigen Bioökonomie. In *Bioökonomie für Einsteiger*; Pietszch, J., Ed.; Springer Spektrum: Wiesbaden, Germany, 2017; pp. 177–203. ISBN 366253763X.

26. Von Braun, J. Bioeconomy–Science and Technology Policy to Harmonize Biologization of Economies with Food Security. In *The Fight against Hunger and Malnutrition*; Sahn, D., Ed.; Oxford University Press: London, UK, 2015; pp. 240–262. [CrossRef]

27. Von Braun, J. "Land Grabbing". Ursachen und Konsequenzen internationaler Landakquirierung in Entwicklungsländern. *Z. Außen Sicherh.* **2010**, *3*, 299–307. [CrossRef]

28. Swinnen, J.; Riera, O. The global bio-economy. *Agric. Econ.* **2013**, *44*, 1–5. [CrossRef]

29. Förster, J.J.; Downsborough, L.; Chomba, M.J. When policy hits reality: Structure, agency and power in South African water governance. *Soc. Nat. Resour.* **2017**, *30*, 521–536. [CrossRef]

30. Dietz, T. *Global Order Beyond Law–How Information and Communication Technologies Facilitate Relational Contracting in International Trade*; Hart Publishing: Oxford, UK, 2014; ISBN 9781849465403.

31. Börner, J.; Baylis, K.; Corbera, E.; Ezzine-de-Blas, D.; Honey-Rosés, J.; Persson, U.M.; Wunder, S. The effectiveness of payments for environmental services. *World Dev.* **2017**, *96*, 359–374. [CrossRef]

32. Auld, G.; Balboa, C.; Bernstein, S.; Cashore, B. *The Emergence of Non-State Market Driven (NSMD) Global Environmental Governance: A Cross Sectoral Assessment*; Cambridge University Press: Cambridge, UK, 2009. [CrossRef]

33. Mayring, P. Qualitative Inhaltsanalyse. In *Handbuch Qualitative Forschung: Grundlagen, Konzepte, Methoden und Anwendungen*; Flick, U., von Kardoff, E., von Rosenstiel, L., Wolff, S., Eds.; Beltz–Psychologie Verl. Union: München, Germany, 1991; pp. 209–2013. ISBN 9783621280747.

34. Labuschagne, A. Qualitative Research: Airy Fairy or Fundamental? The Qualitative Report. 2003. Available online: https://nsuworks.nova.edu/tqr/vol8/iss1/7/ (accessed on 3 August 2018).
35. Wild, P.J.; McMahon, C.; Darlington, M.; Liu, S.; Culley, S. A diary study of information needs and document usage in the engineering domain. *Des. Stud.* **2009**. [CrossRef]
36. Wesseler, J.; von Braun, J. Measuring the Bioeconomy: Economics and Policies. *Annu. Rev. Resour. Econ.* **2017**, *9*, 275–298. [CrossRef]

© 2018 by the authors. Licensee MDPI, Basel, Switzerland. This article is an open access article distributed under the terms and conditions of the Creative Commons Attribution (CC BY) license (http://creativecommons.org/licenses/by/4.0/).

MDPI

St. Alban-Anlage 66

4052 Basel

Switzerland

Tel. +41 61 683 77 34

Fax +41 61 302 89 18

www.mdpi.com

Sustainability Editorial Office

E-mail: sustainability@mdpi.com

www.mdpi.com/journal/sustainability

www.ingramcontent.com/pod-product-compliance
Lightning Source LLC
LaVergne TN
LVHW071619170726
843515LV00009B/2437